Geographica Historica

Begründet von ERNST KIRSTEN
Fortgeführt von ECKART OLSHAUSEN und VERA SAUER

Band 43

Ethnic Constructs, Royal Dynasties and Historical Geography around the Black Sea Littoral

---

Edited by
Altay Coşkun
With the assistance of Joanna Porucznik
and Germain Payen

Franz Steiner Verlag

Bibliografische Information der Deutschen Nationalbibliothek:
Die Deutsche Nationalbibliothek verzeichnet diese Publikation in der Deutschen
Nationalbibliografie; detaillierte bibliografische Daten sind im Internet über
<http://dnb.d-nb.de> abrufbar.

© Franz Steiner Verlag, Stuttgart 2021
Druck: Memminger MedienCentrum, Memmingen
Gedruckt auf säurefreiem, alterungsbeständigem Papier.
Printed in Germany.
ISBN 978-3-515-12941-1 (Print)
ISBN 978-3-515-12944-2 (E-Book)

Dedicated
to the lasting scholarship
and singular humanity of

Heinz Heinen
(14 September 1941 – 21 June 2013)

ZUM GELEIT

Der vorliegende Band kam auf die Initiative von Altay Coşkun zustande. Er enthält vornehmlich Studien, die im Zuge mehrerer Tagungen in den Jahren 2015 bis 2019 entstanden sind und der Pflege und Weiterführung des wissenschaftlichen Vermächtnisses von Heinz Heinen dienten. Thematisch kreisen sie um einen der Forschungsschwerpunkte Heinens, den nördlichen Schwarzmeerraum mit seinen ethnischen, kulturellen und politischen, insbesondere dynastischen Verflechtungen. Ein zentrales Anliegen der Autor*innen ist es, überkommene Vorstellungen von Ethnizität und Kultur und deren Einfluss auf politische Verhältnisse zu hinterfragen.

Vielfältig sind die historisch-geographischen Bezüge der Beiträge, deren Veröffentlichung in den *Geographica Historica* mehr als gerechtfertigt ist, zumal sie aus dem Blickwinkel ihrer Thematik der Forschung einer Region neue Aspekte hinzufügen, der bereits frühere Bände der Reihe gewidmet sind.

Darüber hinaus ist diese Publikation eine willkommene Möglichkeit, unserer persönlichen Verbundenheit mit Heinz Heinen Ausdruck zu verleihen.

Eckart Olshausen und Vera Sauer

CONTENTS

A. Studies in Group Identity and Ethnicity Constructs around the Ancient Black Sea Coast

B. Studies in the Royal Dynasties of Pontos and the Bosporan Kingdom

PREFACE

As is the case with most books, there are some threads of its pre-history that authors or editors are keen to talk about. The longer story would take us back to the early 1960s, when my former supervisor and mentor Heinz Heinen discovered his interest in Russian politics, language and culture in the hottest phase of the Cold War. As the first Chair of Ancient History at the University of Trier (1971–2006), he developed his department into an international hub for ancient Black Sea studies. It felt intuitively right to me that the Greeks called the Black Sea *Pontos Euxeinos* – 'Hospitable Sea'. Heinen's intellectual skills and love for the ancient world were paired with a sense of humour, generous hospitality and, perhaps most of all, a deep respect for the student, colleague or simply the human who was engaging in a discussion with him. He thus fostered open discourse between scholars across the ideological boundaries that deeply divided the West and the Soviet-dominated East (cf. Heinen 1996; Cojocaru et al. 2014), an effort that resulted in many reflections on the roots of the divide, especially the scholarly work and biography of Michail Rostovtzeff and the effect of Marxism or Leninism on the course of Russian Classical studies (e.g., Heinen 1980; 2006a; 2006b; 2008). Besides, many lasting friendships and multiple research cooperation arose along the way. However, I have told this story elsewhere (Coşkun 2014).

I gained my first insights into this kind of dialogue as a 2nd-year undergraduate student (1992), but shifted my own research towards the Black Sea only much later, when I began collaborating with Heinen on Roman diplomacy and the dynasties of the Graeco-Roman world, briefly in 1999/2000 as his assistant and again as his research associate from 2002 to 2008. Initially, I concentrated on Anatolia and the theory of Roman *amicitia* (Coşkun 2008 and *APR*), while Heinen's focus was on the Mithradatid house that connected Pontos on the southern littoral of the Black Sea with the Kimmerian Bosporus on the opposite side (e.g., Heinen 1994; Coşkun & Heinen 2004; Heinen 2006a). In 1997, the first chapter drafts of an envisioned monograph on the dynastic history of the Bosporus from 63 BC to AD 68 materialized (cf. Coşkun 2016; 2020c; in preparation). His several commitments to his students and administrative duties, besides his dedication to the study of ancient slavery since the later 1990s, prevented him from following through on this plan, especially when his life was cut short by an aggressive cancer.

A year after his death (21 June 2013), his widow Marie-Louise Heinen entrusted me with his unfinished book chapters on the Bosporus. The best I could think of was to leverage them into a broad international cooperation, in order to acquire the support and expertise to one day publish them in a setting that would at least come close to the original book design, without falling short of the required expertise. To this end, I began building a network of advisors and collaborators, both from among his former friends and, as Heinen would have liked it, also including many young colleagues with their fresh ideas on ancient Black Sea studies. I could draw on the previous contributors to my *Amici Populi Romani* database (*APR*) and further on the *Interconnectivity* workshop that I co-organized with Victor Cojocaru in Iaşi (8–12 July 2013). We had designed it to honour Heinen's achievements, but eventually held it with sorrowful hearts to commemorate him a few days after his funeral (cf. Cojocaru et al. 2014).

A series of workshops and conference panels followed to discuss old traditions and new trends in ancient Black Sea studies, with a special emphasis on, but not limited to, Heinen's main ideas, the reflection of ideological implications, and his demand for a sober and diverse methodological approach, paired with strong encouragement for intra- and interdisciplinary cooperation.

4–5/7/2015 (with Andrea Binsfeld): *Colloquium in Memory of Prof. Dr. Heinz Heinen*, St. Vith, Belgium

6–11/7/2015 (with Victor Cojocaru & Alexander Rubel): *Mobility in Research on the Black Sea Region*, Archaeological Institute of the Romanian Academy, Iaşi Branch, Romania

5–8/4/2017: *Recent Research in Ancient Black Sea Studies*, Panel at the 113th Annual Meeting of CAMWS, University of Waterloo, ON.

16–18/7/2017 (with Victor Cojocaru): *Advances in Ancient Black Sea Studies: Methodological Innovation, Interdisciplinary Perspectives and International Cooperation*, Archaeological Institute of the Romanian Academy, Iaşi Branch, Romania.

23/7/2018 (with Joanna Porucznik and Krzysztof Nawotka): *Power, Status and Symbols in the Black Sea Area in Antiquity*, Institute of History, University of Wrocław, Poland.

12/11/2018 (with Germain Payen): *Recent Research in Ancient Black Sea Studies in Canada and Beyond. Colloquium Ponticum Canadiense*, University of Waterloo, ON.

2/8/2019 (with Nick Sekunda): *Black Sea Study Day: The Northern Black Sea Coast on the Fringes of the Roman Empire*, Sopot near Gdańsk, Poland.

Many of the papers given on those occasions have been published elsewhere (such as in Cojocaru & Rubel 2016), while others are still being developed for a volume dedicated specifically to the Bosporan kingdom (Coşkun in preparation). The present collection assembles 14 original studies on the history, archaeology and geography of the ancient Black Sea region, many of which were first discussed at one of the abovementioned gatherings. When combined, they cover the Euxine coastlines of all four hemispheres, while addressing problems from the archaic to the Byzantine period with a panoply of methodological approaches.

(A) The first five papers (I / Mordvintseva, II / Porucznik, III / Harland, IV / Oller Guzmán and V / Podossinov) try to overcome essentialist views on cultures and ethnicities, demonstrating how much more can be learned about the past and the present, if we regard such notions not as stable and closed entities, but as highly fluid and permeable concepts. In fact, they are best understood as social constructs that one way or another work within ideological frameworks, ancient or modern, and sometimes tell us more about those who speak of them than about what they are supposed to describe. The Orientalism debate, the Postcolonial turn and many other constructivist approaches have gradually allowed such wisdom to penetrate the Humanities and Social Sciences for some time now, but their reception is heavily delayed in ancient Black Sea studies: European nationalisms and Marxist materialism appear to have cast longer shadows on this part of the ancient world than elsewhere. While this is particularly true for Russian, Soviet and post-Soviet scholarship, such perspectives are by no means limited to eastern Europeans (cf. Coşkun 2020c).

The fundamental role that the Russian scholar Michail Rostovtzeff has been playing for more than a century is disproportionate to the limited accessibility of his publications, especially in the West. A clear description of his world-historical analysis, placed within its historical and cultural context, thus opens this book (I / Mordvintseva). And I recommend the study of the Olbian *chora* (II / Porucznik) as a second introduction to this volume, thanks to its lucid survey of scholarship on intercultural encounter and (in)considerate use of physical evidence.

(B) Four further chapters are the result of my colleagues' and my interest in dynastic history along the shores of the Pontos Euxeinos. At the same time, most of these studies illustrate the potential of questioning pre-conceived ideas of ethnicity and their assumed or effective influence on politics (VI / Dana, VIII / Ballesteros Pastor, IX / Coşkun & Stern). The investigation of Pharnakes I (VII / Payen) traces the Mithradatid dynasty's pre-history on its way to becoming the leading player in the Black Sea. As such, it could as well have been grouped with the next part.

(C) Feeling the need for short-termed adjustments to the overall book plan (see below), I have contributed three chapters on the historical geography of Pontos and Kolchis. These exemplify how quickly research in political or cultural history leads to controversial questions on toponymy, settlement history or political geography, while also illustrating how many details of our ancient literary accounts have remained underexplored. Too often modern scholars have quarried them, looking for the information they were expecting, while missing subtle points that ancient authors were making. Even worse than this traditional 'positivism' is a bequest of Marxist materialism, a strong tendency to downplay or even discard literary evidence as unreliable or ideologically distorted – as if documents, such as coins, inscriptions and artefacts, were not subject to similarly purposeful distortions. I would hence like to show how reading ancient authors in context provides at least glimpses of the world 'through their eyes'. Many problems disappear, while new ones may emerge. In other words, accounting for every source individually rather than selecting or rejecting according to our preconceived ideas is a path that still promises to yield many new insights. Similar emphasis on the subjective perspective of ancient authors are also prevalent in the earlier chapters, such as the one that deconstructs ethnic hierarchies in ancient civilizations (III / Harland), explores paradoxical descriptions of barbarians (V / Podossinov) or discovers clusters of confrontation between Greek settlers and indigenous people (IV / Oller Guzmán).

Many chapters compare material evidence with the literary or documentary tradition, e.g., in an effort to illustrate assumed ethnic markers (I / Mordvintseva), to anchor the sense of threat as reflected in historiographical accounts also in architectural remains (IV / Oller Guzmán), to shed light on the Achaimenid agenda of Pharnakes II (VIII / Ballesteros Pastor) or in the context of Roman imperial propaganda as displayed in the friezes of the *Ara Pacis Augustae* (IX / Coşkun & Stern).

(D) Three contributions primarily focus on archaeological data, also showing the vibrancy and methodological diversity of archaeological fieldwork along the Black Sea coasts – by far the most intensive area of research in terms of manpower and financial resources. The first of these studies is on cult rituals in the *chora* of Olbia (II / Porucznik) and has been grouped with part A. Another chapter soberly challenges the perceived view that Christianization reshaped the urban structure in the 4th to 5th centuries AD – an unbiased reassessment of the evidence appears to tell a different story (XIII / Ruscu). Third comes the final chapter of this volume (XIV / Elton), which offers a long-term perspective from the Classical to the Byzantine age and thus briefly revisits many of the historical periods addressed throughout the book, while investigating the crops that farmers cultivated in Pontos. It is innovative for its combination of biology, geology and cultural history.

Science and technology have left their traces also in other studies: osteology contributed to the scrutiny of Olbian rituals (II / Porucznik) and satellite images hugely benefitted my own research on Kolchian geography (XII / Coşkun; cf. Coşkun 2020a and 2020b), just as the maps that my student Stone Chen has skilfully drawn for this volume (printed at the end of this volume), beginning with the summary map 'Key Settlements on the Black Sea Littoral'. The investigation of farming in Pontos (XIV / Elton) yields the well-documented result that periods of climate change, which was a reality in the past as it is in the presence, ultimately affected the choice of crops to a much lesser degree than major political reversals and the new fiscal and economic conditions that these entailed. As such, our volume closes with an example of the fresh insights that historical research may expect from new technologies in the future, while, at the same time, implicitly endorsing the relevance of the most traditional concern of historical studies: political power – its protagonists, the structures within which these operated and the effect it wielded on historical societies.

This preface provides me with the opportunity to thank those who have contributed in so many different ways to produce this book, to develop its much broader research agenda or to rekindle the passion for collaborative research on the ancient world of which the Black Sea region formed an integral part.

I start with Heinz Heinen for the immeasurable support, guidance and inspiration he gave me ever since we first met in 1991. Close by his side, I mention Marie-Louise Heinen for her ceaseless moral support and heart-warming affection.

Next, I would like to thank all the co-organizers and participants of the workshops mentioned before as well as the authors of the studies presented here. To many of them, I am indebted for more than entrusting me their research papers; many gave me advice, offered hospitality or shared literature. I refrain from repeating all their names and refer the readers instead to the short CVs assembled at the end of this volume.

This restraint notwithstanding, I wish to mark out Valentina Mordvintseva, a model of dedication (in her roles as daughter, mother and grandmother no less than as colleague and professor), bestowed with a mysterious source of energy. I mention Luis Ballesteros Pastor for the friendship we have been enjoying since our first encounter in Trier in 2007, which goes beyond discovering ever new facets of Mithradatic history. I first made friends with Alexandr Podossinov during his visits to Trier in the 1990s, lost touch but happily reconnected with him in Moscow in 2017; he did not hesitate to offer a contribution. Our shared interest in Dynamis and the *Ara Pacis* allowed me to learn much from Gaius Stern, to benefit from his generous editorial support and to be inspired by his devotion to exploring the ancient and modern worlds and sharing new insights.

While working intensively on this book, devastating news reached me twice, first of the passing of my friend Mackenzie Lewis (7 March 2020). As a scholar deeply invested into ancient colonial history and archaeology, he actively contributed to my research workshops at Waterloo and gave me encouraging feedback (not only) on my Leukothea piece (XII), which I would like to dedicate to him. Not much after this loss, I was saddened by the likewise premature death of Federicomaria Muccioli. Our friendship goes back to my undergraduate years in Trier; he last hosted me at Bologna in 2018. One of my next publications on Hellenistic history (a passion we shared), will be dedicated to his memory.

We lost four paper commitments towards the conclusion of the present volume, at least in part due to the corona pandemic, which continues imposing unusual restrictions on all of us. Three of these would have strengthened and diversified part C on historical geography (to which I originally planned to contribute only one paper). One of them was meant to explore the geography of the Bosporan kingdom and would thereby have addressed the most sensational discovery in recent years, the Kuban Bosporus, a second straight connecting the Maiotis (Sea of Azov) with the Pontos Euxeinos. Its two straights thus carved out Phanagoreia as an island (Zhuravlev & Scholtzhauer 2016; Schlotzhauer et al. 2017; cf. Dan 2016, 270f.; Papuci-Władyka 2018, 312 and see Map 1 at the end of this volume). While hoping to include contributions by this team in one of my subsequent Black Sea volumes, I do not want to fail to thank Udo Schlotzhauer and Anca Dan for kindly receiving me in Berlin (2017) and Paris (2019) respectively, and for the many valuable books they gifted. The latter trip to France is memorable also for other reasons, the generous hospitality of Madalina & Dan Dana in Paris and of Suzan and Alexandru Avram in Le Mans, besides the opportunity of visiting Notre Dame a few days before it went up in flames.

Germain Payen and I connected some eight years ago due to our shared interest in Asia Minor studies. I am glad I enlisted his support for my Black Sea studies agenda, which brought him to Waterloo as a postdoctoral fellow (September 2017 to December 2018). Much of the support for this book he has provided as a postdoc at Cologne University. I look forward to continuing our cooperation and friendship in whichever format in the future. Joanna Porucznik was a postdoctoral research fellow at Wrocław University when I met her first at a Humboldt Conference hosted by the Russian Academy in 2017. I immediately benefitted from her many talents, including not getting lost in Moscow. She was quickly appointed assistant professor at Opole University, and I foresee that many other institutions will want to have her. Germain and Joanna both joined me repeatedly at workshops or even co-organized them with me in 2018. They gave feedback on some of the papers and helped me with formatting others. Germain prepared the index of names. Joanna took it on herself to unify the transcription of Russian titles in all bibliographies and translated into Russian all abstracts from English, with the support of Olga Olszewska (Wrocław), whom I include in my expression of gratitude. Cordial thanks further go to Stone Chen for his diligence and aesthetic ambitions in drafting five maps for this volume.

I got in touch with the series editors of *Geographica Historica* Eckart Olshausen and Vera Sauer in Spring 2019, and received more than kind encouragement, useful advice and mature guidance. I very much appreciate the efficient and diligent review process as well as the bibliographical support they provided during the pandemic library closures of 2020. My cooperation with the Franz-Steiner Verlag was as pleasant as previously, thanks to the dedication of Katharina Stüdemann and Sarah-Vanessa Schäfer.

Much of the research that I have been conducting on the ancient Black Sea would not have been possible without the institutional support of the University of Waterloo, my academic home since 2009, as well as the financial support that my project 'Ethnic Identities and Diplomatic Affiliations in the Bosporan Kingdom' is receiving from the Social Sciences and Humanities Council of Canada (2017–2022).

My prefaces usually close by acknowledging one of the two women who have mattered most in my life, my mother Brunhilde and my wife Dorothea. This time, both of them are to be named: less so for the typos they picked in some of the chapters than for patiently and lovingly allowing me to be away, whether absorbed in books or off to a conference: my mother regularly took generous care of the logistics of my European travel base in Herzogenrath, while my wife never fails in giving me peace of mind by keeping our children safe and happy.

Altay Coşkun
Waterloo, August 2020

REFERENCES

Cojocaru, V., Coşkun, A. & Dana, M. 2014: 'Preface – Building Bridges', in *iidem* (eds.), *Interconnectivity in the Mediterranean and Pontic World during the Hellenistic and Roman Periods. The Proceedings of the International Symposium (Constanţa, July 8–12, 2013)*, Cluj-Napoca, 9–15.

Cojocaru, V. & Rubel, A. (eds.) 2016: *Mobility in Research on the Black Sea. (Iaşi, July 5–10, 2015)*, Cluj-Napoca.

Coşkun, A. & Heinen, H. 2004: 'Amici Populi Romani. Das Trierer Projekt Roms auswärtige Freunde stellt sich vor', AncSoc 34, 45–75.

Coşkun, A. (eds.) APR: see 'Notes on Abbreviations' below.

Coşkun, A. (eds.) 2008: Freundschaft und Gefolgschaft in den auswärtigen Beziehungen der Römer (2. Jh. v.Chr.–1. Jh. n.Chr.), Frankfurt.

Coşkun, A. 2014: 'Interconnectivity – In honorem & in memoriam Heinz Heinen (1941–2013). With a Complete Bibliography of His Scholarly Publications', in Cojocaru, Coşkun & Dana 2014, 25–71.

Coşkun, A. 2016: 'Heinz Heinen und die Bosporanischen Könige – Eine Projektbeschreibung', in Cojocaru & Rubel 2016, 51–71.

Coşkun, A. 2020a: '(Re-) Locating Greek & Roman Cities along the Northern Coast of Kolchis. Part I: Identifying Dioskourias in the Recess of the Black Sea', VDI 80.2, 354–376.

Coşkun, A. 2020b: '(Re-) Locating Greek & Roman Cities along the Northern Coast of Kolchis. Part II: Following Arrian's Periplous from Phasis to Sebastopolis', VDI 80.3, 654–674.

Coşkun, A. 2020c: 'The Bosporan Kings in-between the Mithridatic Tradition and Friendship with Rome: the Usurpation of Asandros Revisited', forthcoming in D. Braund, A. Chaniotis & E. Petropoulos (eds.), Roman Pontos, Athens 2020.

Coşkun, A. (ed.), in preparation: Studien zur Herrschaft über das Bosporanischen Reich. Mit bisher unveröffentlichten Kapiteln von H. Heinen. (preliminary title)

Dan, A. 2016: 'The Rivers Called Phasis', AWE 15, 245–277.

Heinen, H. 1980: Die Geschichte des Altertums im Spiegel der sowjetischen Forschung, Darmstadt.

Heinen, H. 1994: 'Mithradates von Pergamon und Caesars bosporanische Pläne. Zur Interpretation von Bellum Alexandrinum 78', in R. Günther & S. Rebenich (eds.), E fontibus haurire. Beiträge zur römischen Geschichte und zu ihren Hilfswissenschaften (=FS H. Chantraine), Paderborn, 63–79.

Heinen, H. 1996: 'Russen im römischen Trier', in K. Eimermacher & A. Hartmann (ed.), Deutschrussische Hochschulkooperation: Erfahrungsberichte, Bochum, 134–139.

Heinen, H. 2006a: Antike am Rande der Steppe. Der nördliche Schwarzmeerraum als Forschungsaufgabe, Stuttgart.

Heinen, H. 2006b: ‚Michael Ivanovich Rostovtzeff (1870–1952)‘, in L. Raphael (ed.), Klassiker der Geschichtswissenschaft, vol. 1: Von Edward Gibbon bis Marc Bloch, Munich, 172–189.

Heinen, H. 2008: ‚La tradition mithridatique des rois du Bosphore, de Rostovtzeff à l'historiographie soviétique‘, in J. Andreau & W. Berelowitch (eds.), Michel Ivanovitch Rostovtzeff, Bari, 137–152.

Papuci-Władyka, E. 2018: 'An Essay on Recent Archaeological Research on the Northern Black Sea Coast', in M. Manoledakis, G.R. Tsetskhladze & I. Xydopoulos (eds.), Essays on the Archaeology and Ancient History of the Black Sea Littoral, Leuven, 273–332.

Schlotzhauer, U., Žuravlev, D., Dan, A., Gehrke, H.-J., Kelterbaum, D. & Mommsen, H. 2017: 'Interdisziplinäre Methoden in der Landschaftsarchäologie und andere archäometrische Untersuchungen am Besipiel der griechischen Kolonisation im Nordpontus', in V.I. Molodin & S. Hansen (eds.), Multidisciplinary Approach to Archaeology: Recent Achievements and Prospects. Proceedings of the International Symposium (June 22–26, 2015, Novosibirsk), Novosibirsk, 392–404.

Zhuravlev, D. & Schlotzhauer, U. (eds.) 2016: Asian Bosporus and Kuban Region in Pre-Roman Time. Materials of the International Round Table, June 7–8, 2016, Moscow.

ABBREVIATIONS AND OTHER EDITORIAL NOTES

Abbreviations of ancient authors, epigraphic corpora and reference works normally follow the *Oxford Classical Dictionary* (for similar lists, see the introduction to *DNP* vol. 3 or to *BNP* vol. 1), except that spelling has been adjusted to the modernized style adopted for this volume: e.g., we write Ail(ianos) instead of Ael(ianus) and Thuk(ydides) instead of Thuc(ydides). Journal titles normally follow *l'Année Philologique* (https://about.brepolis.net/aph-abbreviations), although abbreviations may have been extended or titles have been written in full in cases of rarer works. Abbreviations specific to only a single article are listed at the top of the according chapter bibliography. Departing from these principles or in addition to the items mentioned, we have used the following abbreviations:

ANRW	*Aufstieg und Niedergang der Römischen Welt.*
APR	A. Coşkun (ed.), *Amici Populi Romani. Prosopographie der auswärtigen Freunde Roms = Prosopography of the Foreign Friends of Rome*, Version 01–09, Trier 2007–2008 & Waterloo, ON 2010–2019. URL: http://www.altaycoskun.com/apr.
Arr. *PPE*	Arrian, *Periplus Ponti Euxini* (rather than *Peripl. M. Eux.*).
BA	R.J.A. Talbert (ed.), *Barrington Atlas of the Greek and Roman World*, Princeton 2000.
BA Directory	R.J.A. Talbert (ed.), *Map by Map Directory to Accompany Barrington Atlas of the Greek and Roman World*, Princeton 2000.
BE	*Bulletin épigraphique.*
BNJ	*Brill's New Jacoby* (online).
BNP	*Brill's New Pauly* (print or online).
CAH²	*Cambridge Ancient History*, 2nd ed.
CIRB	V. Struve et al., *Corpus inscriptionum regni Bosporani* (*Korpus Bosporskikh nadpiseï*), Moscow 1965 (in Russian).
DNP	*Der Neue Pauly* (print or online).
EA	*Epigraphica Anatolica*
EAH	R.A. Bagnall et al. (eds.), *Encyclopedia of Ancient History*, 10 vols., Hoboken, NJ 2013. Also online: 1st ed. 2012, 2nd ed. 2014–2020.
EncIr	*Encyclopaedia Iranica*. URL: http://www.iranicaonline.org.
FGrH	Jacoby, F. 1923–1954: *Die Fragmente der griechischen Historiker*, I–III, Berlin & Leiden. (Cf. Brill online edition with English translation, besides, *BNJ*)
IGDOlbia	Dubois, L. 1996: *Inscriptions grecques dialectales d'Olbia du Pont*, Genève.
IOSPE	Latyschev, B. 1885–1901: *Inscriptiones Antiquae Orae Septentrionalis Ponti Euxini Graecae et Latinae*, I–II, IV, Saint Petersburg. (rather than *IPE*)
IOSPE I²	Latyschev, B. 1916: *Inscriptiones Antiquae Orae Septentrionalis Ponti Euxini Graecae et Latinae*, vol. I, 2nd ed., Saint Petersburg (repr. Hildesheim 1965).
LGPN I–V.C	P.M. Fraser et al. (eds.), *Lexicon of Greek Personal Names,* vols. I–V.C, Oxford 1987–2018. (Cf. https://www.lgpn.ox.ac.uk/)
LSJ	Liddell, H.-G. & Scott, R.: *A Greek-English Lexicon.* Revised and Augmented Throughout by Sir H.S. Jones, with the Assistance of R. McKenzie. With a Revised Supplement, Oxford 1996 (multiple reprints).

PIR² *Prosopographia Imperii Romani. Saec. I. II. III*, Berlin, vols. I–VIII.2, 2ⁿᵈ ed. 1933–2015.

Plin. *NH* Pliny the Elder, *Naturalis Historia* (rather than *HN*).

Pomp. Trog. *Prol.* Pompeius Trogus, *Prologi Historiarum Philippicarum.*

Röm. Mitt. *Mitteilungen des Deutschen Archäologischen Instituts.*

Strab. *Geogr.* Strabo, *Geography.*

TAVO *Tübinger Atlas des Vorderen Orients.*

VDI *Vestnik Drevneï Istorii*, Moscow. URL: http://vdi.igh.ru/?locale=en.

All Internet links have been checked again between August and October 2020.

LIST OF FIGURES

Coloured Maps at the end of the volume

A. Studies in Group Identity and Ethnicity Constructs around the Ancient Black Sea Coast

‚IRANER‘ UND ‚SARMATEN‘
IN DER WELTSICHT MICHAEL ROSTOVTZEFFS

Valentina Mordvintseva

Abstract: ‛Iranians’ and ‛Sarmatians’ in the World View of Michael Rostovtzeff: Michael Rostovtzeff designed a model of cultural history of the North Pontic region that span from the Archaic period to the early Middle Ages, covered much of the Eurasian territory and tried to integrate the literary, epigraphic, numismatic and archaeological sources available in the early-20th century. In essence, he reduces the cultural development of the 1st millennium BC with an antagonism between a dominant Greek and a recessive Iranian culture, the latter represented by the ‘Scythians’, who were receptive of Hellenizing influences. In contrast, the period of Roman political domination is defined by a ‘Sarmatization’ in ethnic and cultural terms, which Rostovtzeff also calls ‘new Iranization’ and ‘Barbarization’ nearly without distinction. He extends this second phase into Late Antiquity, when ‘waves’ of Iranian migrations are seen as thoroughly impacting Western European civilization. This synthesis commanded so much respect that it continues influencing cultural history in Russian scholarship (and beyond), especially in Classical and Scythian-Sarmatian archaeology. Rostovtzeff's ideas root in concepts prevalent in his days. First, the etiological myth of the Russian Empire had already been well-established by the end of the 19th century, where the steppe corridor of Eurasia was considered a ‘world axis’, regularly inviting mass migration from East to West, substantially affecting the ethnic composition of the European population. At the same time, the steppes of Eastern Europe were seen as a ‘buffer zone’ that slowed down or mitigated the impact from the Far East. Second, the predominant role that Rostovtzeff ascribes to the common people in the context of cultural change ultimately goes back to Marxist theory. Third comes the direct association of certain elements of the archaeological material with specific ethnic groups, whence a change of material culture is regularly explained with the migration of peoples. Since such paradigms tend to be internalized at the early stages of a scholar's socialization, they are slow to change, even when current international scholarship draws on more nuanced socio-cultural concepts and research methodology, largely incompatible with Rostovtzeff's premises and conclusions.

Абстракт: «Иранцы» и «Сарматы» в образе мира Михаила Ростовцева: Созданная М. И. Ростовцевым модель культурно-исторического развития Северного Причерноморья была комплексной и на тот момент непротиворечивой. Описанные им культурно-исторические процессы охватывали несколько исторических эпох (от греческой архаики до раннего средневековья) и значительную территорию, включавшую практически весь Старый Свет. Им был обобщен весь доступный к тому времени материал, сведены вместе разные группы источников – письменные, археологические, эпиграфические, нумизматические, для интерпретации которых привлечен широкий круг аналогий. Масштаб работ Ростовцева был настолько велик, стиль изложения настолько уверенным, а культурно-историческая модель настолько убедительна, что они определили пути развития нескольких направлений в отечественной науке, в частности антиковедения и скифо-сарматской археологии вплоть до настоящего времени. При этом на сложение представлений М. И. Ростовцева о роли иранцев и сарматов в формировании культур Южной России, видимо, повлияли несколько современных ему концептов. Это, во-первых, уже вполне сложившийся к концу 19 в. этиологический миф Российской империи, где степной коридор Евразии

рассматривался как своеобразная «мировая ось», по которой постоянно происходило
движение народов с востока на запад, что приводило к частичному изменению
этнического состава населения. Степи Восточной Европы выполняли при этом
функцию «буферной зоны», своеобразного котла, в котором задерживались и
творчески перерабатывались культурные импульсы с Востока. Во-вторых, это
марксистское в своей основе представление о преобладающей роли народных масс в
культурогенезе. В-третьих, это прямая ассоциация отдельных элементов
археологического материала с конкретным этносом. Развитие материальной культуры,
а именно появление в ней новых категорий предметов или предметов нового облика,
происходило, по мнению исследователя, в результате миграций, когда трансфер
инноваций в культуру-реципиент происходил физически вместе с потоками мигрантов.
Соответственно в основе трансформации материальной культуры лежали
исключительно этнические изменения. Поскольку все эти концепты во многом
являются парадигмами, критика которых затруднена вследствие их формирования у
исследователей уже на ранних этапах их социализации, они до сих пор превалируют в
современной археологической литературе. Это препятствует получению нового
позитивного знания при анализе археологических источников.

I. EINFÜHRUNG

Das heute immer noch maßgebliche Modell einer kulturhistorischen Entwicklung
des nördlichen Schwarzmeerraums in der Antike stammt von Michael Rostovtzeff
(1870–1952), dem nach wie vor bekanntesten aller russischen Althistoriker. Insbesondere als Begründer der sozioökonomischen Ausrichtung der Klassischen
Archäologie hat er sich in der westlichen Altertumswissenschaft einen bleibenden
Namen gemacht. In den ersten zwei Jahrzehnten des 20. Jahrhunderts widmete er
sich aber auch intensiv der griechisch-römischen Geschichte Südrusslands, bevor
ihn die Oktoberrevolution zur Flucht in den Westen zwang.[1] Mithin basieren auch
diejenigen seiner nordpontischen Arbeiten, die er erst später in Berlin oder Princeton veröffentlichte, auf dem Material, das er noch in den vorrevolutionären Jahren
gesammelt und bearbeitet hatte.[2] Die im Ansatz schon von seinen Vorgängern
erkannten kulturhistorischen Prozesse in Südrussland versah Rostovtzeff erstmals
mit den Schlagwörtern *Barbarisierung*, *Iranisierung* und *Sarmatisierung*, die in
der Folgezeit Schlüsselrollen spielen sollten.

In der jüngeren anthropologischen Literatur des mittleren und zum Teil auch
noch späteren 20. Jahrhunderts wurden solche Prozesse überwiegend unter dem
allgemeineren Begriff der ‚Akkulturation‘ gefasst.[3] Er impliziert den Transfer
bestimmter kultureller Inhalte durch *aktive Teilnehmer* an den Prozessen durch
Transferwerkzeuge (Gegenstände und Interaktionen) an *passive Teilnehmer*. Das
Konzept von den ‚aktiven‘ Gebern und ‚passiven‘ Empfängern ist freilich recht
begrenzt und im Laufe der letzten Jahrzehnte starker Kritik ausgesetzt gewesen,
weil es die vielfältigen Wechselwirkungen unsachgemäß ausblendet.[4] Kulturtrans-

1   Zuev 1991, 167; Heinen 2008, 139. Zu seinen späteren Arbeiten s. Shaw 1992.
2   Mordvintseva 2013, 205f.
3   Dobesch 2004, 54.
4   Ulf 2009, 82.

fer kann sowohl bei direktem Kontakt unter ‚Kulturgruppen' als auch indirekt stattfinden, in der Regel durch Netzwerkkommunikation. Sowohl Objekte als auch Ideen können übertragen werden, wobei diese ihren ehemaligen Bedeutungszusammenhang weitgehend beibehalten oder auch ganz verlieren bzw. gegen einen neuen eintauschen können. Gleichzeitig mag die Verbreitung neuer kultureller Elemente unter äußerem Druck oder freiwillig, bewusst oder unbewusst sowie mit unterschiedlichem Intensitätsgrad erfolgen. Als Ergebnisse solcher Prozesse lassen sich teils eklektische Mischungen oder Hybridisierungen, teils eine weitreichende, ja sogar vollständige Assimilation an eine ehemals fremde Kultur beobachten.[5]

Für die Skizzierung der bis heute nachwirkenden kulturgeschichtlichen Analyse Rostovtzeffs ist es indes angemessen, am einfacheren Akkulturationsmodell festzuhalten, da es seinem eigenen Verständnis recht nahekommt. Ziel des vorliegenden Aufsatzes ist nicht, die Theorie des Gelehrten einer grundlegenden Kritik zu unterziehen (dies habe ich an anderer Stelle versucht), sondern vielmehr die – überwiegend auf Russisch publizierte – Lehre prägnant in ihrem Argumentationszusammenhang darzustellen. Auf diese Weise hoffe ich, die von Heinz Heinen entwickelten biographischen und ideologiekritischen Ansätze sinnvoll zu ergänzen.[6]

II. HINTERGRUND

Das öffentliche Interesse an der Geschichte der Völker, die in der Antike das Territorium Eurasiens bewohnten, erwachte im Russischen Reich im späten 17. Jahrhundert während seiner Integration in das Europäische Welt-System. Zu dieser Zeit diente die ‚alte und glorreiche Geschichte' der geopolitischen Stärkung der europäischen Staaten und der Begründung ihrer territorialen Ambitionen. Schon in der zweiten Hälfte des 18. Jahrhunderts wurde die ‚Russische Geschichte' von Wasilij Tatiščev, einem jüngeren Zeitgenossen von Peter I., veröffentlicht.[7] Diese Arbeit fasste erstmals die Informationen antiker Autoren, mittelalterlicher Chroniken und späterer Quellen über diejenigen südlichen Gebiete zusammen, die vor kurzem dem Russischen Reich hinzugefügt worden waren. Tatiščev behandelte Fragen zu den Unterschieden der Völker, die in der Antike ‚die Länder des gegenwärtigen Russlands bewohnten', sowie zum ständigen Wechsel der Völker und zu den Gründen für solcherlei Veränderungen.[8] Die Stämme der Skythen, Sarmaten und Slawen betrachtete er als ‚die drei wichtigsten für das öffentliche

---

5   Woolf, Greg 1994; 1998; Dobesch 2004; Bekker-Nielsen 2006; Guldager Bilde & Stolba 2006; Guldager Bilde & Petersen 2008.
6   Mordvintseva 2013; Heinen 1999; 2006; 2008; cf. Coşkun & Heinen 2004, 62–69; Coşkun 2020; Halamus 2017, 162, Anm. 14.
7   Tatiščev 1768.
8   Tatiščev 1768, 88.

Lehren', womit das gesellschaftliche Verstehen der eigenen Geschichte gemeint war.[9]

In einer anderen zusammenfassenden Arbeit beschrieb Nikolaj Karamzin die Steppe des russischen Südens als ein Gebiet, in dem sich die aus dem tiefen Asien kommenden Völker ständig gegenseitig ersetzten.[10] Sergej Solov'jov, der Rektor der Moskauer Universität in den 1870er Jahren, sah in den Besonderheiten der Landschaft des russischen Südens die Gründe für die Eigenartigkeit seiner historischen Entwicklung: ‚die Massen der Nomaden gingen seit undenklichen Zeiten durch das Breite Tor zwischen dem Uralgebirge und dem Kaspischen Meer und besetzten die Länder im unteren Teil der Wolga, Don und Dnjepr; die Alte Geschichte sieht sie hier ständig vorherrschen'.[11] Aus seiner Sicht war ja der historische ‚Zweck' der sesshaften Bevölkerung des nördlichen Schwarzmeergebietes, diese unzähligen ‚Horden' davon abzuhalten, weiter nach Westen, nach Europa vorzudringen.

Während des 19. Jahrhunderts gab es eine intensive Vermehrung der Quellen zur Geschichte des nördlichen Schwarzmeergebietes. Infolge der Ausgrabungen antiker Städte und vielfältiger Gräberfelder wurden eine Fülle materieller Überreste gefunden: Architekturkomplexe, Skulpturen und toreutische Arbeiten sowie eine große Zahl von Münzen und Inschriften, welche mehr oder weniger direkt Zeugnis von der Antike gaben. Diese neuen Materialien bereicherten die Informationen der schriftlichen Quellen erheblich, teils bestätigend und teils korrigierend. In der zweiten Hälfte des 19. Jahrhunderts wurden die ersten Sammlungen der epigraphischen,[12] numismatischen[13] und archäologischen[14] Zeugnisse herausgegeben. Die Veränderung des Umfangs und der Struktur der Quellenbasis erforderte eine neue Synthese sowie die Entwicklung eines entsprechenden konsistenten Modells der kulturhistorischen Prozesse im nördlichen Schwarzmeergebiet.

Der erste groß angelegte Versuch war die von Graf Ivan Tolstoj und Nikodim Kondakov vorbereitete sechsbändige Publikation ‚Russische Antike in Kunstdenkmälern' (1889–1899), die bald von Salomon Reinach französisch übersetzt und unter dem Titel *Antiquités de la Russie méridionale* im Westen veröffentlicht wurde.[15] Das Werk enthält umfangreiche Zusammenfassungen der Texte antiker Autoren, die abwechselnd mit Informationen zu griechischen und lateinischen Inschriften sowie mit Beschreibungen und Auswertungen numismatischer und archäologischer Zeugnisse, vor allem ‚Kunstdenkmäler', zusammengestellt wurden. Für die Rekonstruktion der Geschichte des nördlichen Schwarzmeergebietes verwendeten die Autoren zum ersten Mal auch die Nachrichten der chinesischen

9   Tatiščev 1768, 72.
10   Karamzin 1818.
11   Solov'jov 1851, 3.
12   Pomjalovskij 1881; Latyšev 1887; *IOSPE* I 2.
13   Koene 1848; 1857; Buračkov 1884; Orešnikov 1888.
14   *DBK* 1854; *DGS* 1866–1872; *CR St. Petersburg* 1862–1918.
15   Kondakov u.a. 1891.

Quellen über die Völker, die in der Antike in Zentralasien lebten, die kurz zuvor von Iakinf Bičurin veröffentlicht wurden.[16] Tolstoj und Kondakov sahen ihre Aufgabe in der Darstellung der ‚gemeinsamen unaufhörlichen Kontinuität' der Altertümer Russlands, der ‚historischen Bildung und Entwicklung der altrussischen Kunst', zu dem ‚der hellenisierte Skythe, der Meister aus Korsun, der genuesische Händler und der Deutsche aus Moskau ihren Beitrag geleistet hätten.[17] Ausführlich charakterisierten Tolstoj und Kondakov die Prozesse der Wechselwirkung zwischen den ‚bekanntesten barbarischen Stämmen' (Skythen, Sarmaten), die von ihnen als ‚Volksmassen' wahrgenommen wurden. Als spezifischer ‚ethnischer' Marker der nomadischen Kultur Eurasiens benannten sie den Tierstil – die ‚eigenartige Kunst der Nomaden'. Ihrer Meinung nach waren die Gegenstände des Tierstils ‚zusammen mit den Völkern' aus Sibirien durch Zentralasien nach Südrussland eingedrungen.[18]

Die Arbeit von Tolstoj und Kondakov verfolgte die Hauptaufgabe, die Ursprünge und Entwicklung der ‚altrussischen Kunst' zu erfassen, ging aber weit darüber hinaus. Sie stellte das erste kulturhistorische Modell für die nördliche Schwarzmeerregion auf der Grundlage einer umfassenden Quellensynthese dar, der weitere folgen sollten. Erstmals wurde dabei auch der Versuch unternommen, die einzelnen Elemente der antiken materiellen Kultur Südrusslands als Kennzeichen für bestimmte Ethnien zu werten, die in der klassischen Literatur Erwähnung finden. Die Forschungen von Tolstoj und Kondakov spiegeln so den allgemeinen Zustand der wissenschaftlichen Methodik am Ende des 19. Jahrhunderts gleichwie die damals einflussreiche Weltanschauung der Slawophilie oder des Panslawismus. In ihrem Werk wurde der Mythos von der Eroberung Skythiens durch die Sarmaten entworfen und der Bevölkerung der südrussischen Steppen eine historische Rolle für die Bildung der Kulturen des mittelalterlichen und letztlich auch des modernen Europa zugewiesen.

III. DAS KULTURHISTORISCHE MODELL
VON MICHAEL ROSTOVTZEFF

1. Die Hellenisierung der Barbaren unter griechischer Vorherrschaft

Michael Rostovtzeffs Forschungen konnten unter anderem auf den Quellenkatalogen Tolstojs und Kondakovs aufbauen sowie an ihre ethnisierenden Deutungen anknüpfen. Die wichtigsten Subjekte der antiken kulturhistorischen Prozesse im Süden Osteuropas waren (nicht nur) in Rostovtzeffs Sicht auf der einen Seite die im Laufe der griechischen Kolonisierung der nördlichen Schwarzmeerregion entstandenen Staaten, vor allem die mächtigen Poleis Olbia und Chersonesos Taurika sowie das Bosporanische Reich, das die griechischen Städte beiderseits des Kim-

---

16 Bičurin 1851.
17 Tolstoj & Kondakov 1889, i–iii.
18 Tolstoj & Kondakov 1890, 132, 147. Exemplarische Illustrationen folgen weiter unten, Fig. 4.

merischen Bosporus vereinte;[19] auf der anderen Seite standen die ‚barbarischen'
Gesellschaften in der Nachbarschaft der griechischen Städte, welche in der Literatur vor allem Skythen, Sarmaten, Maioten oder Taurer genannt werden. Das Zu-
sammenwirken dieser ‚lokalen' kulturpolitischen Gruppen wurde oftmals vor dem
Hintergrund der Bildung, Entwicklung und Krise dualer Welt-Systeme gedeutet,
zuerst einer griechisch-persischen Dichotomie, dann einer römisch-parthischen
bzw. -sassanidischen, welche schließlich auch auf den nördlichen Schwarzmeerraum als ein Grenzgebiet übertragen wurde.

Gemäß Rostovtzeff beeinflussten die nordpontischen Griechen aktiv die um-
liegenden ‚Barbaren' im Wesentlichen durch den Handel.[20] Eine Folge dieses Zu-
sammenwirkens war die ‚Hellenisierung des Geschmacks der Skythen'.[21] Diese
Entwicklung manifestiere sich insbesondere in der Verbreitung von Bildern
skythischer Götter, ein Phänomen, das der ‚iranischen Natur ihrer Religion' eigentlich zuwiderlaufe.[22] In einigen Fällen, wie etwa bei den Sinden oder Krim-
Skythen, zeigt sich die ‚Hellenisierung' der lokalen Bevölkerung nach Rostovtzeff
im Import der Staatlichkeit ‚unter der Leitung von halb-griechischen Dynasten',
der auf diplomatische Beziehungen und dynastische Eheverbindungen beruhe.[23]
Am ‚tiefgreifendsten' habe sich der Prozess der ‚Hellenisierung' in der Bevölkerung der Taman-Halbinsel und des unteren Kubangebietes vollzogen, d.h. bei den
Stämmen, die in der Nähe der griechischen Zentren lebten.[24] Unter den archäologischen Belegen für derartige ‚Hellenisierungsprozesse' führt Rostovtzeff die im-
portierten Objekte aus den ‚barbarischen' Siedlungen und Grabanlagen an, welche
sich zum Teil in großer Entfernung von den griechischen Zentren befanden. Darüber hinaus benennt er die Herstellung von ‚Hybridobjekten', d.h. von Gegenständen, welche ‚barbarische' Formen und ‚griechische' Herstellungstechniken
oder Dekoration verbanden,[25] sowie ferner die Anwendung griechischer Bautech-
nik in barbarischen Siedlungen.

2. Romanisierung, ‚Barbarisierung' und ‚Sarmatisierung'

Die Rolle Roms in den kulturhistorischen Prozessen im nördlichen Schwarzmeergebiet wird von Rostovtzeff ebenfalls in den Blick genommen, vor allem hinsichtlich der Städte des Bosporanischen Reiches,[26] das er ‚Vorposten der römischen
Welt und der römischen Militärmacht' sowie ‚Puffer-Staat' und ‚Agent-Staat'

---

19   Rostowzew 1993, 30.
20   Rostowzew 1993, 86.
21   Rostovtzeff 1929, 34.
22   Rostovtzeff 1918b, 55.
23   Rostovtzeff 1918b, 76, 90. Vgl. auch den Beitrag von Madalina Dana, Kapitel 6 in diesem
     Band.
24   Rostovtzeff 1918b, 90f.
25   Rostowzew 1993, 86f.
26   Rostovtzeff 1918b, 76, 79, 109.

nennt.[27] Dem russischen Forscher zufolge resultierte die Errichtung des römischen ‚Protektorats‘ aus der Schwächung des ‚kulturellen Widerstandes der Griechen‘, welche wiederum zu ihrer ‚Barbarisierung‘ geführt habe.[28] Die ‚Romanisierung‘ des Bosporanischen Reiches äußere sich am deutlichsten in der Art der Münzprägungen.[29] Zum Einfluss Roms auf die ‚barbarischen‘ Gesellschaften des nördlichen Schwarzmeergebietes nahm Rostovtzeff nur indirekt Stellung, indem er etwa bemerkte, dass ‚auch Sarmaten zu Nachbarn Roms wurden‘.[30] Sein Hauptaugenmerk gilt den Auswirkungen, welche die ‚Barbaren‘ auf die griechische Zivilisation hatten, insbesondere die Vertreter der iranischen Welt, in welcher den Stämmen der Sarmaten die zentrale Rolle zugeschrieben wird. Diese setzt er sowohl mit den ‚Volksmassen‘ sowie einzelnen Stämmen als auch mit der ‚vorherrschenden Gruppe der Bevölkerung‘ gleich, welche auch die politische Elite der Gesellschaft gebildet habe.[31]

‚Barbaren‘ ist also bei Rostovtzeff ein Sammelbegriff für die lokale (vor- und nicht-griechische) Bevölkerung im nördlichen Schwarzmeergebiet, welche er als ‚überwiegend iranisch‘ ansieht.[32] Die ‚Iraner‘ Rostovtzeffs waren demnach Vertreter der iranischen Sprachgruppe, ‚einer riesigen iranischen Kulturwelt‘.[33] Dazu gehörten Skythen und Sarmaten, d.h. ‚nördliche Iraner‘,[34] sowie ‚tatsächliche Iraner‘, also Perser, ‚Vertreter der alarodischen Rasse‘ (womit in etwa die vorderasiatischen Stämme der Armenier und Syrer gemeint sind) sowie speziell auch Repräsentanten der mithradatischen Dynastie samt ihren Untertanen.[35]

Unter dem ‚gemeinsamen Namen der Sarmaten‘ erscheinen bei Rostovtzeff die aktiven Teilnehmer im Prozess der ‚Sarmatisierung‘, die auch ‚neue Iraner‘,[36] ‚Fremdlinge aus dem Osten‘, ‚nördliche‘ und ‚östliche Nachbarn der Skythen‘ genannt werden. Die Iraner-Sarmaten hätten sich von den Iraner-Skythen durch ihre ‚größere Grobheit und Wildheit‘ unterschieden. Als ‚Sarmaten‘ werden somit verschiedene nomadische Stämme zusammengefasst, die im nördlichen Schwarzmeergebiet (wie die Jazygen, Roxolanen, Siraken, Aorsen, Alanen usw.) oder auch in weiter abgelegenen östlichen Gebieten lebten (s. Karte = Fig. 1). Im Gegensatz zu den meisten modernen Wissenschaftlern und entgegen der antiken narrativen Tradition unterscheidet Rostovtzeff Sarmaten von Sauromaten, welche er als die Bewohner des Territoriums östlich des Tanaïs betrachtet.[37]

---

27 Rostovtzeff 1918b, 116.
28 Rostovtzeff 1918b, 80.
29 Rostovtzeff 1918b, 123.
30 Rostovtzeff 1918b, 104.
31 ‚Volksmassen‘ und Stämme: Rostovtzeff 1918b, 14f.; Rostowzew 1993, 36. Eliten: Rostovtzeff 1918a, 81; 1929, 67.
32 Rostovtzeff 1925, 170.
33 Rostovtzeff 1918b, 7, 12.
34 Rostovtzeff 1918b, 15, 33, 41, 94, 116; Rostowzew 1993, 39. Vgl. auch Heinen 1993, 54 Anm. 14.
35 Rostovtzeff 1918b, 15, 42, 78; 1925: 276, 302, 611, 616.
36 Rostovtzeff 1918b, 15, 41, 93.
37 Rostovtzeff 1918b, 89, 93; 1925, 111, 113; Rostowzew 1993, 39. Vgl. auch Heinen 1993, 62 Anm. 34.

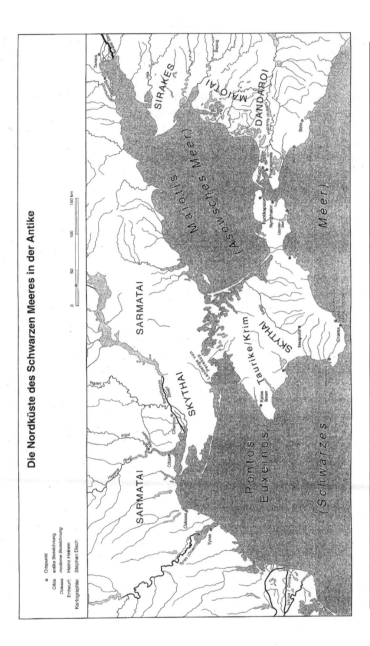

Fig. 1: Die Nordküste des Schwarzen Meeres in der Antike. Autor: Heinz Heinen. Kartograph: Stephan Disch, Trier ca. 1998.

Als spezifisches Unterscheidungsmerkmal führt er das Fehlen von Indizien eines Matriarchats bei den Sarmaten an, welches antike Autoren indes den Sauromaten[38] und Maioten[39] zugeschrieben hätten.[40] Rostovtzeffs Begriffe ‚Barbaren‘, ‚Iraner‘ und ‚Sarmaten‘ lassen sich in der folgenden taxonomischen Reihe darstellen: ‚Sarmaten‘ sind Teil der ‚Iraner‘, die wiederum zu den ‚Barbaren‘ gehören. Es ist nicht immer klar, welche Bedeutung diese Begriffe in jedem Einzelfall haben. Diese Unsicherheit färbt auch auf die unterstellten Prozesse der ‚Barbarisierung‘, ‚Iranisierung‘ und ‚Sarmatisierung‘ ab, als deren aktive Teilnehmer jeweils ‚Barbaren‘, ‚Iraner‘ und ‚Sarmaten‘ unterstellt werden. Als passiv bei diesen Vorgängen erscheinen in den Schriften Rostovtzeffs die Vertreter der antiken Zivilisation, vor allem die Griechen, aber auch andere ‚Barbaren‘ wie die ‚Skythen‘ oder die ‚lokale Bevölkerung‘ des Kuban- und unteren Dnjeprgebiets sowie schließlich manche Einwohner Westeuropas seit der Völkerwanderungszeit. Als archäologische Kennzeichen einer solchen ‚Barbarisierung‘ oder ‚Sarmatisierung‘ werden das Auftreten der Beerdigung von Pferden mit Geschirr, viele nicht-griechische Grabbeigaben, einschließlich Waffen, goldene Flitter und Armbänder angeführt.[41]

3. ‚Barbarisierung‘ und ‚Sarmatisierung‘ des Bosporanischen Reiches

Den Prozess der ‚Barbarisierung‘ bezog Rostovtzeff aber hauptsächlich auf die Bevölkerung des Bosporanischen Reiches, dem er eine genauere Untersuchung widmete. Den iranischen Einfluss bemerkte er in den Namen und der Kleidung der auf den Grabstelen genannten bzw. dargestellten verstorbenen Bewohner von Pantikapaion.[42] Zudem verweist er auf die Ikonographie und Namen der Könige auf den späten bosporanischen Münzen,[43] ferner auf die toreutischen Arbeiten und Schmuckobjekte,[44] weiterhin Grabriten, einschließlich der männlichen Waffenbestattungen und Pferdebestattungen mit Geschirr,[45] sowie überhaupt die verschiedenen Waffenarten.[46] Den Anfang des Prozesses der ‚Iranisierung‘ datiert Rostovtzeff ins 3. Jahrhundert v.Chr.[47] Die iranischen Kulturelemente seien vor allem durch physisches Eindringen von Iranern in das Bosporische Reich gelangt, was zu einer ‚Iranisierung der Bevölkerung durch Blut und Ansichten‘ geführt habe.[48]

---

38 Hdt. 4.110–117; cf. Corcella 2007, 658f. with further references (ancient and modern).
39 Strab. *Geogr.* 11.5.1f. (503f.C) (with Roller 2018, 648f.); Polyain. 8.55.
40 Rostovtzeff 1925, 113.
41 Rostovtzeff 1925, 243f. S. die Abbildungen im Folgenden.
42 Rostovtzeff 1918b, 125; 1925, 163; Rostowzew 1993, 115. Zu den onomastischen Daten vgl. Halamus 2017, 162.
43 Rostovtzeff 1918b, 115. S. jetzt auch den Beitrag von Madalina Dana, Kapitel 6 in diesem Band.
44 Rostovtzeff 1925, 172.
45 Rostovtzeff 1918b, 128f; 1925, 235.
46 Rostovtzeff 1918b, 124.
47 Rostovtzeff 1925, 579.
48 Rostovtzeff 1918b, 111, 122, 125.

Im Laufe der Zeit sei die ‚Iranisierung‘ der griechischen Bevölkerung jenes Reiches verstärkt worden. Aus in spätantiken Kammergräbern entdeckten Gemälden hat er den Schluss gezogen, dass die Bewohner Pantikapaions einen halbnomadischen Lebensstil gepflegt hätten: „Jeden Frühling bewegte sich eine ganze Reihe von Wagen mit Zelten, Vieh, landwirtschaftlichen Geräten, Frauen und Kindern aus Pantikapaion in die Steppe.‘[49]

Ähnliche Merkmale werden auch bei der Beschreibung des Prozesses der ‚Sarmatisierung‘ des Bosporanischen Reichs aufgeführt, das zuletzt ‚keine griechische oder griechisch-thrakische Siedlung war wie zuvor, sondern griechisch-sarmatisch wurde‘.[50] Von den Arbeiten Rostovtzeffs geht die Vorstellung aus, dass die ‚Iranisierung‘ jenes Reiches ausschließlich durch das Eindringen der Sarmaten erfolgt sei. Allerdings äußert der Forscher in seinem Hauptwerk *Skythien und der Bosporus* den Gedanken, dass auch andere Ursachen zu dieser Entwicklung geführt haben könnten. So weist er insbesondere auf die Bedeutung der von der pontischen Südküste herkommenden ‚iranischen‘ Dynastie des Mithradates VI. Eupator hin.[51] Vor allem mit dem Wirken dieses Königs verbindet Rostovtzeff die Beobachtung, dass in den vorangehenden 300 Jahren des ‚Kampfes und Zusammenlebens zwischen Griechen und Skythen‘ keine so starke ‚Iranisierung‘ des Bosporanischen Reiches wie in der späteren sarmatischen Ära stattgefunden habe.

Mithin werden die Prozesse der ‚Barbarisierung‘, ‚Iranisierung‘ und ‚Sarmatisierung‘ der Bevölkerung der griechischen Städte des nördlichen Schwarzmeergebietes von Rostovtzeff durch die gleichen Merkmale beschrieben, ja sie verschmelzen in seiner Darstellung zu ein und demselben Phänomen. Territorial beschränkt er diesen Prozess auf die Nordküste des Schwarzen Meeres, so dass er geradezu einen lokalen Charakter erhält. Demgegenüber ist der Zeitrahmen breit angelegt und reicht vom 3. Jahrhundert v.Chr. bis zum 3. Jahrhundert n.Chr. Bei anderen *passiven Teilnehmern* bzw. anderen ‚Barbaren‘ verwendet Rostovtzeff die Begriffe ‚Iranisierung‘ und ‚Sarmatisierung‘ indes nicht direkt. Allerdings scheint er doch auf die gleichen Prozesse anzuspielen, wenn er über das Auftreten der neuen iranischen bzw. sarmatischen Ethnien in der barbarischen Welt des nördlichen Schwarzmeergebietes oder über ihre ‚Mischung‘ mit der lokalen ‚vorsarmatischen‘ Bevölkerung schreibt. In diesem Sinne ähneln die beschriebenen Vorgänge der ‚Iranisierung‘ und ‚Sarmatisierung‘ den Ereignissen in den griechischen Staaten. Die wichtigsten *passiven Teilnehmer* in der ‚barbarischen‘ Welt waren in dieser Rekonstruktion die Skythen bzw. fast die gesamte nicht-griechische Bevölkerung des nördlichen Schwarzmeergebietes.

---

49   Rostovtzeff 1918b, 129. Siehe auch Rostowzew 1993, 121f.
50   Rostovtzeff 1925, 245, 613, 616.
51   Rostovtzeff 1925, 616.

4. Die ‚Sarmatisierung' der Skythen nach Rostovtzeff

Im Gegensatz zur ‚Sarmatisierung' der recht komplexen Gesellschaft des Bosporanischen Reiches beschreibt Rostovtzeff den Akkulturationsprozess der Skythen einfacher, da er diese als ethnisch verwandt betrachtet. Diese Vorannahme erfordert freilich eine ausführlichere Argumentation. Diese basiert auf der Idee, dass sich regelmäßige ‚Wellen' der neuen iranischsprachigen Stämme von Osten nach Westen bewegt hätten.[52] Diese Wellen ‚erschienen allmählich in den Steppen Südrusslands, setzten sich, bewegten sich wieder, einige eroberten andere, vermischten sich mit den vorherigen Bewohnern und Schritt für Schritt verdrängten sie die alten Herren – Skythen, die sich ohne Zweifel mit ihnen vermischten'.[53]

Fig. 2: Phaleren von Pferdegeschirren im nördlichen Schwarzmeergebiet, 3.–1. Jh. v.Chr.
(nach Mordvinceva 2001, Kat. 2–4, 6, 7, 54, 69).
Nr. 1–3, 6: Fedulov Schatz. Nr. 4: Starobelsk Schatz. Nr. 5: Uspenskaja. Nr. 7: Jančokrak Schatz.

52   Rostovtzeff 1918b, 15, 38, 93; 1925, 612, 615; Rostowzew 1931, 43, 604.
53   Rostovtzeff 1925, 612.

Als Hauptargument für das Konzept der Eroberung Skythiens durch die Sarmaten dient die Annahme einer dramatischen Veränderung der materiellen Kultur im nördlichen Schwarzmeergebiet während des späten 3. Jahrhunderts v.Chr., also ‚gerade zu der Zeit, in welcher die schriftlichen Quellen erstmals Sarmaten erwähnen'.[54] Dies spiegele sich vor allem im Auftauchen von ‚neuen Gegenständen' im Grabinventar wider, also hauptsächlich in den Prestigeobjekten. Unter den gemeinsamen Merkmalen der materiellen Kultur von Sarmaten listet Rostovtzeff bestimmte Waffenformen auf, darunter ein kurzes Schwert mit Ringknauf und ein mit Schuppenpanzer kombiniertes Kettenhemd sowie persönliche Gegenstände wie Gürtelsatz mit Kunstschnallen oder kleine geometrische goldene Flitter.

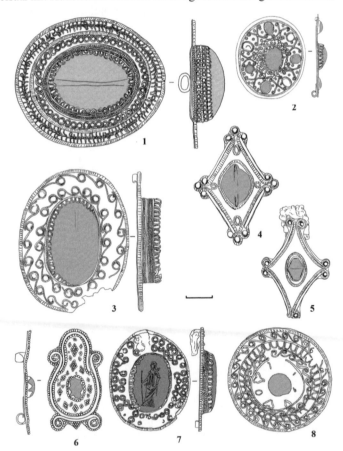

Fig. 3: Polychrome Broschen aus dem Kubangebiet, 3.–1. Jh. v.Chr. (nach Mordvintseva u.a. 2010, Kat. 44, 57, 227, 294, 103, 464, 471, 604). – Nr. 1: Mezmaj. Nr. 2: Dinskaja. Nr. 3: Rassvet. Nr. 4: Razdolnaja. Nr. 5: Verkhnij. Nr. 6: Lenina. Nr. 7: Brjukhovetskaja. Nr. 8: Elitnyj.

54   Rostovtzeff 1925, 309; 1929, 41.

Des Weiteren verweist er auf die ‚schriftlichen Zeichen' (Tamgas oder Tamgen), die vor allem auf den ‚asiatischen Kesseln' zu beobachten sind.[55] Als Indikatoren der ersten Welle der Ankömmlinge aus dem Osten betrachtet Rostovtzeff die Phaleren des Pferdegeschirrs (Fig. 2), für die er eine griechisch-skythische Herstellung vermutete,[56] ferner Gürtelhaken[57] und polychrome Broschen (Fig. 3).[58] Charakteristisch für die zweite ‚sarmatische Welle' seien Gegenstände des ‚neuen' Tierstils (Fig. 4), die sich vom skythischen Tierstil vor allem durch Polychromie und Sujetsatz, besonders die Vorliebe für Tierkampfszenen, unterscheide.[59]

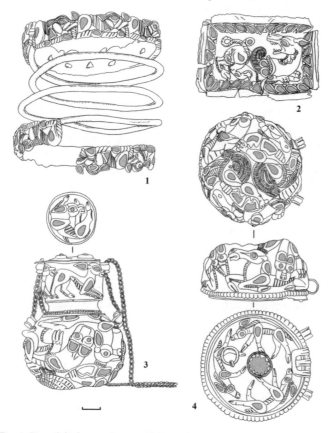

Fig. 4: Tierstilobjekte aus dem nördlichen Schwarzmeergebiet, 1.–2. Jh. n.Chr. (nach Mordvintseva 2003, Kat. 61, 64, 65, 84). – Nr. 1, 3, 4: Chochlač. Nr. 2: Kubangebiet.

55 Rostovtzeff 1925, 615.
56 Rostovtzeff 1929, 42, 44; Rostowzew 1931, 606. Zum Vergleich siehe Mordvinceva 2001, 36f.
57 Rostovtzeff 1929, 42.
58 Rostovtzeff 1922, 132; 1929, 42f., 45. S. auch Treister 2002; Mordvintseva 2010.
59 Rostovtzeff 1929, 55. S. auch Mordvintseva 2003; Zasetskaya 2019.

Diese jüngere Gruppe von Objekten des Tierstils nennt er ‚sarmatisch' und später auch ‚yuezhisch',[60] wobei er die enge Verbindung dieser Kunst mit dem Partherreich betont.[61] Daneben reflektiere sich die Nähe zu den Parthern auch in den Besonderheiten der sarmatischen Religion (Feueranbetung), der patriarchalischen Lebensweise und der militärischen Gesellschaftsorganisation. Neue Gegenstände erschienen erst mit den Ankömmlingen als Folge ihrer ‚Infiltration' in das Milieu der skythischen Aristokratie, so dass die ‚kulturellen Veränderungen noch nicht besonders zu beobachten sind'.[62]

Die ‚frühe sarmatische Welle' hat Rostovtzeff mit verschiedenen archäologischen Befundgruppen in Zusammenhang gebracht (s. Karte = Fig. 5). An erster Stelle ist hier die Orenburg-Gruppe vom Prochorowka Hügel im Verwaltungsgebiet Orenburg (nordöstlich des Kaspischen Meeres, außerhalb des Kartenausschnitts) zu nennen.[63] Im ersten Kurgan wurden eine militärische Rüstung, Waffen, polychromer Schmuck, Tierstilgegenstände und achämenidische Phialen gefunden. Die darunter befindlichen direkten iranischen Importe haben zu der Bestimmung dieses Bestattungskomplexes als sarmatisch geführt. Im Prochorowka Kurgan 1 sah Rostovtzeff eine kulturelle Verwandschaft mit anderen osteuropäischen Bestattungsgruppen. Hierzu zählt er die Hügel neben dem Dorf Elizavetovskaja (Jelisawetowskaja) am Don. Eine weitere Gruppe liegt im Kubangebiet und schließt den Stawropol Schatz, die Buerowa Mogila (bei Phanagoria), die Gräber bei den Dörfern Beslenejewskaja und Kurdzhipskaja (Kurdschips) mit ein.[64] Seiner Meinung nach war die ganze Steppe östlich des Flusses Don schon am Ende des 4. Jahrhundert v.Chr. in den Händen von Sarmaten.[65] Die zweite ‚sarmatische Welle' ist durch Gruppen von Kurganen im unteren Dongebiet (‚Hochlač' Hügel, Nowotscherkassk) und im Kubangebiet besonders am sogenannten ‚Goldenen Friedhof' (Kurganenkette von Ust-Labinskaja bis zu Kasanskaja Dorf) vertreten.

Die Merkmale der ‚Sarmatisierung' und ‚Iranisierung' der lokalen Bevölkerung im nördlichen Schwarzmeergebiet bilden in den Werken Rostovtzeffs eine Einheit. Die ‚Sarmatisierung' der Skythen versteht er als eine ‚neue Iranisierung'. Dieser Prozess, der sich gleichzeitig mit der ‚Sarmatisierung' der griechischen Stadtbevölkerung jener Gegend vollzogen habe, hatte in seinen Augen einen kontinentalen Maßstab und umfasste ein sehr umfangreiches Gebiet von den nördlichen Grenzen Chinas und den zentralasiatischen Staaten bis zum Ungarischen Plateau und den westeuropäischen Territorien. Nach Rostovtzeff ebbte die ‚sarmatische Welle' nicht innerhalb des nördlichen Schwarzmeergebietes in ihrer Bewegung nach Westen ab, sondern ‚bewegte sich weiter: ihre Spuren sehen wir in der ganzen Steppe Südrusslands nordwestlich des Kuban und weiter westlich der Do-

---

60   Rostovtzeff 1929, 106.
61   Rostovtzeff 1929, 124.
62   Rostovtzeff 1925, 458, 550; 1929, 42.
63   Rostovtzeff 1925, 593.
64   Rostovtzeff 1922, 125, 128f.
65   Rostovtzeff 1929, 21.

Fig. 5: Karte Südrusslands aus Rostowzew 1931, 143. Vgl. Rostowzew 1993,
Karte 1; Heinen 2006, 9.

nau, bis tief in Westeuropa'.[66] Ein Beweis für diese Reichweite ist die Entstehung von Tierstilobjekten in Europa, die im Bosporanischen Reich, so Rostovtzeff, hergestellt wurden und vermittels Sarmaten und Goten den ornamentalen Stil Europas in Spätantike und Frühmittelalter beeinflussten. Das Aufkommen von Sarmaten in Europa habe zur Bildung des romanischen und auch gotischen Stils beigetragen und ,leistete seinen Beitrag zur Geschichte der westeuropäischen Kunst'.[67]

Es ist anzumerken, dass Rostovtzeff auf den Wellen der ,Sarmatisierung' bestand, ohne freilich Elemente des Latène-Kreises im nördlichen Schwarzmeergebiet systematisch zu berücksichtigen, geschweige denn zu erklären. Dennoch hat er sie wiederholt im Kontext der skythischen materiellen Kultur erwähnt.[68] Potentieller ,westlicher Einfluss' auf das nördliche Schwarzmeergebiet findet also keinen Platz in seinem Blickfeld, und dies auch nicht, wenn er die gewaltigen Invasionen der östlichen Nomaden nach Westen in der Spätantike beschreibt. Trotz der relevanten Daten aus narrativen, epigraphischen und archäologischen Quellen blieb der Prozess der ,Latènisierung' in Rostovtzeffs vermeintlich universalem historisch-kulturellen Modell unberücksichtigt.[69]

---

66 Rostovtzeff 1925, 457f., 616f.
67 Rostovtzeff 1918b, 98, 132.
68 Rostovtzeff 1918b, 16, 41, 56, 63; 1925, 138, 458, 526, 613.
69 Zur ,Latènisierung des nördlichen Schwarzmeergebiets' s. Treister 1993; Falileyev 2013; Mordvintseva 2017.

IV. ZUSAMMENFASSUNG UND AUSBLICK

Die von Michael Rostovtzeff entworfene Kulturgeschichte des nördlichen Schwarzmeerraums kann für ihre Entstehungszeit als komplexes und weitgehend konsistentes Modell gelten. Die von ihm beschriebenen kulturhistorischen Prozesse umfassen die Epochen von der griechischen Archaik bis zum frühen Mittelalter und zudem ein gewaltiges Territorium, das fast die gesamte Alte Welt einschließt. Zu diesem Zweck führte er das damals verfügbare literarische, epigraphische, numismatische und archäologische Quellenmaterial einer umfassenden Synthese zu, welche zudem aus einer breiten Palette von Analogien schöpfte. Die scheinbare Evidenz dieses großen Entwurfs, der durch die schiere Belegfülle, Stringenz der Argumentation und Ausstrahlungskraft des einzigartigen russischen Altertumswissenschaftlers zu bestechen vermochte, ebnete ihm den Weg auf mehrere Felder der nationalen und internationalen Forschung, insbesondere der klassischen und skythisch-sarmatischen Archäologie, wo das Werk Rostovtzeffs bis heute einflussreich geblieben ist.[70]

Seine Vorstellungen von der Rolle der Iraner und Sarmaten in der Kulturgeschichte Südrusslands wurden offenbar von zeitgenössischen Konzepten beeinflusst. Dies war erstens der schon bis zum Ende des 19. Jahrhunderts voll ausgebildete ätiologische Mythos des Russischen Reiches, in dem die Steppe Eurasiens als eine Art ‚Weltachse‘ angesehen wurde: Durch sie hätten sich ständig Völker von Osten nach Westen bewegt und so die ethnische Zusammensetzung der jeweiligen Bevölkerung verändert. Die Steppen Osteuropas erfüllten gleichzeitig die Funktion der ‚Pufferzone‘, einer Art Kessel, in dem kulturelle Impulse aus dem Osten verzögert und kreativ bearbeitet worden seien. Zweitens war die Vorstellung eines dualen Mediterran-Orientalischen Weltsystems nach der Teilung der antiken Oikumene zwischen Rom und Parthien durch Augustus und Phraates V. am Euphrat (20/19 v.Chr.) prägend.[71] Für Rostovtzeff hatte dieser Prozess schon unter der Herrschaft des Mithradates VI. Eupator begonnen und sollte auch die spätere Unterordnung des Bosporanischen Reiches unter Rom überdauern. Drittens war es eine damals weit verbreitete, wohl ursprünglich marxistische Idee, den ‚Volksmassen‘ eine entscheidende Rolle in der Kulturgenese zuzusprechen. Viertens ist Rostovtzeffs Methode durch eine direkte Assoziation materieller Elemente mit einem spezifischen Ethnos geprägt. Dabei sei die Entwicklung der materiellen Kultur, nämlich die Entstehung neuer Kategorien von Objekten oder Objekten neuer Form, als Folge von Migrationen zu erklären; der Transfer bzw. die Innovationen habe sich durch das Einströmen von Migranten ergeben.[72]

Derartige Konzepte sind vielfach Paradigmen, die schon früh in der individuellen Sozialisierung individueller Forscher grundgelegt werden. Nur selten öffnen sich Forscherpersönlichkeiten später in ihrem Leben für grundlegende Kritik, egal wie einschlägig die neue Evidenz sein mag. Im konkreten Fall hat dies dazu ge-

---

70 S. z.B. Klejn 2016; Skripkin 2017; Zasetskaya 2019.
71 Cf. Overtoom 2016, 141.
72 Rostovtzeff 1929, 42, 66; vgl. 1925, 612.

führt, dass die kaum mehr zeitgemäßen Prämissen Rostovtzeffs weiterhin die moderne archäologische Literatur bestimmen, obwohl mittlerweile vor allem in den westlichen Sozial- und Kulturwissenschaften viel differenziertere Modelle für Kulturtransfer entwickelt worden sind.[73] Ganz eklatant ist etwa, dass praktisch alle von Rostovtzeff postulierten Zeichen der ‚Iranisierung' und ‚Sarmatisierung' lediglich die Kultur – und hier insbesondere die Bestattungspraktiken – der gesellschaftlichen Elite betreffen, was seiner Idee von Volksmassen widerspricht. Heute ist uns viel bewusster, dass lokale Eliten oft in einem multiethnischen oder hybriden Rahmen interagierten, der stark von wirtschaftlichen und politischen Faktoren beeinflusst wird.

Die Erhöhung des ‚iranischen Einflusses' auf die Kulturen der nördlichen Schwarzmeerregion in der sarmatischen Ära, insbesondere in den ersten Jahrhunderten n.Chr., sollte also aus gegenwärtiger Sicht der Dinge nicht vorschnell durch ‚regelmäßige Wellen der Invasionen von neuen Stämmen aus dem Osten' erklärt werden, sondern eher durch die Verstärkung der Netzwerkverbindungen lokaler Eliten mit den parthischen Eliten im Iran und seiner Satelliten.

Abkürzungen

CR St.Petersburg 1862–1918: *Comptes-Rendu de la Commission Impériale Archéologique*, St. Petersburg.

DBK 1854: *Drevnosti Bospora Kimmerijskogo, khranjaščiesja v Imperatorskom Muzee Ėrmitaža* (Antiquities of the Cimmerian Bosporus Stored in the Imperial Museum of the Hermitage), St. Petersburg.

DGS 1866–1872: *Drevnosti Gerodotovoj Skifii. Sbornik opisanij arkheologičeskikh raskopok i nakhodok v černomorskikh stepjakh* (Antiquities of Herodotus Scythia. Collection of Descriptions of Archaeological Excavations and Finds in the Black Sea Steppes), St. Petersburg.

Bibliographie

Bekker-Nielsen, T. (ed.) 2006: *Rome and the Black Sea Region: Domination, Romanisation, Resistance*, Aarhus.

Bičurin, I. 1851: *Sobranie svedenij o narodakh, obitavšikh v Srednej Azii v drevnie vremena* (The Collection of Information about the Peoples Who Lived in Central Asia in Ancient Times), St. Petersburg.

Burachkov, P.O. 1884: *Obščij katalog monet, prinadležaščikh ėllinskim kolonijam, suščestvovavšim v drevnosti na severnom beregu Čjornogo morja, v predelakh nynešnej Južnoj Rossii* (The General Catalogue of Coins Belonging to the Hellenic Colonies that Existed in Ancient Times on the Northern Shore of the Black Sea, within the Current Southern Russia). Part 1, Odessa. 3

Corcella, A. 2007: 'Book IV', in D. Asheri, A. Lloyd & A. Corcella, *A Commentary on Herodotus Books I–IV*, ed. by O. Murray & A. Moreno, Oxford, 543–721.

---

73 S. oben mit Anm. 3–5.

Coşkun, A. 2020: ‚The Bosporan Kings in-between the Mithridatic Tradition and Friendship with Rome: the Usurpation of Asandros Revisited‘, demnächst in D. Braund, A. Chaniotis & E. Petropoulos (eds.), *Roman Pontos*, Athens.

Coşkun, A. & Heinen, H. 2004: ‚*Amici Populi Romani*. Das Trierer Projekt Roms auswärtige Freunde stellt sich vor‘, *Ancient Society* 34, 45–75.

Dobesch, G. 2004: ‚Zentrum, Peripherie und „Barbaren“ in der Urgeschichte und der Alten Geschichte‘, in H. Friesinger & A. Stuppner (eds.): *Zentrum und Peripherie – Gesellschaftliche Phänomene in der Frühgeschichte. Materialien des 13. Internationalen Symposiums „Grundprobleme der frühgeschichtlichen Entwicklung im mittleren Donauraum“*, Wien, 11–93.

Falileyev. A. 2013: ‚Going Further East: New Data, New Analysis‘, in J.L. García Alonso (ed.), *Continental Celtic Word Formation. The Onomastic Data*, Salamanca, 85–98.

Guldager Bilde, P. & Stolba, V.F. (eds.) 2006: *Surveying the Greek Chora: The Black Sea Region in a Comparative Perspective*, Aarhus.

Guldager Bilde, P. & Petersen, J.H. (eds.) 2008: *Meetings of Cultures in the Black Sea Region: Between Conflict and Coexistence*, Aarhus.

Halamus, M. 2017: ‚Barbarization of the State? The Sarmatian Influence in the Bosporan Kingdom‘, *VDI* 77, 160–167.

Heinen 1993: s. Rostowzew 1993.

Heinen, H. 1999: ‚Rostovtzeff et la Russie méridionale‘, in A. Marcone (ed.), *Rostovtzeff e l'Italia*, Neapel, 45–61 = H. Heinen, *Ausgewählte Schriften*, Stuttgart 2006, 305–319.

Heinen, H. 2006: ‚Michael Ivanovich Rostovtzeff (1870–1952)‘, in L. Raphael (ed.), *Klassiker der Geschichtswissenschaft*, Band 1: *Von Edward Gibbon bis Marc Bloch*, München, 172–189.

Heinen, H. 2008: ‚La tradition mithridatique des rois du Bosphore, de Rostovtzeff à l'historiographie soviétique‘, in J. Andreau & W. Berelowitch (eds.), *Michel Ivanovitch Rostovtzeff*, Bari, 137–152.

Karamzin, N.M. 1818: *Istorija gosudarstva Rossijskogo* (History of the Russian State), Vol. 1, St. Petersburg.

Klejn, L. 2016: *Pervyj vek: sokrovišča sarmatskikh kurganov* (The First Century: Treasures of Sarmatian Barrows), St. Petersburg.

Koene, B.V. 1848: *Issledovanija ob istorii i drevnostjakh goroda Khersonesa Tavričeskogo* (Research on the History and Antiquities of the City of Chersonesos Taurica), St. Petersburg.

Koene, B.V. 1857: *Opisanie muzeja knjazja V. V. Kočubeja i issledovanie ob istorii i numizmatike grečeskikh poselenij v Rossii, ravno kak i tsarstv Pontijskogo i Bospora Kimmerijskogo* (Description of the Museum of Prince V.V. Kočubej and a Study of the History and Numismatics of the Greek Settlements in Russia, as well as the Kingdoms of Pontos and of the Kimmerian Bosporus), St.-Petersburg.

Kondakov, N., Reinach, S. & Tolstoj, I. 1891: *Antiquités de la Russie méridionale*, Paris.

Latyšev, V.V. 1887: *Issledovanie ob istorii i gosudarstvennom stroe Ol'vii* (Research on the History and the Political System of Olbia), St. Petersburg.

Mordvinceva, V. 2001: *Sarmatische Phaleren*, Rahden, Westf.

Mordvintseva, V. 2003: *Polikhromnyj zverinyj stil* (Polychrome Animal Style), Simferopol.

Mordvintseva, V.I. 2010: ‚Prikubanskie broši (po materialam Krasnodarskogo gosudarstvennogo istoriko-arkheologičeskogo muzeja-zapovednika)‘ (Kuban Brooches (Based on the Materials of the Krasnodar State Historical and Archaeological Museum-Reserve)), *Bosporskie issledovanija (Bosporan Studies)* 23, 302–320.

Mordvintseva, V. 2013: ‚The Sarmatians: The Creation of Archaeological Evidence‘, *Oxford Journal of Archaeology* 32.2, 203–219.

Mordvintseva, V. 2017: ‚Kulturno-istoričeskie protsessy v „varvarskikh“ sotsiumakh Kryma 3 v. do n.ė. – 3 v.n.ė.‘ (Cultural-Historical Processes in „Barbarian“ Societies of the Crimea 3rd c. BC – 3rd c. AD), in A. Ivantchik & V. Mordvintseva (eds.), *Krymskaja Skifija v sisteme kul'turnykh svjazej meždu Vostokom i Zapadom (3 v. do n.ė. – 7 v.n.ė.)* (Crimean Scythia in a

System of Cultural Relations between East and West (3 c. BC – 7 c. AD)), Simferopol, 183–224, 290–299.

Orešnikov, V.A. 1892: ‚Obozrenie monet, najdennykh pri Khersonesskikh raskopkakh v 1888 i 1889 godakh‘ (A Review of the Coins Found during the Excavations in Chersonesos in 1888 and 1889), *Drevnosti Juzhnoj Rossii. Materialy po arkheologii Rossii (Antiquities of Southern Russia. Materials on the Archaeology of Russia)* 7, St. Petersburg, 36–46.

Overtoom, N.L. 2016: ‚The Rivalry of Rome and Parthia in the Sources from the Augustan Age to Late Antiquity‘, *Anabasis* 7, 137–174.

Pomjalovskij, I.V. 1881: *Sbornik grečeskikh i latinskikh nadpisej Kavkaza* (Collection of Greek and Latin Inscriptions of the Caucasus), St. Petersburg.

Roller, D.W. 2018: *A Historical and Topographical Guide to the Geography of Strabo*, Cambridge.

Rostovtzeff, M.I. 1918a: *Kurgannye nakhodki Orenburgskoj oblasti ėpokhi rannego i pozdnego ėllinizma* (Early and Late Hellenistic Barrow finds from Orenbourg region). Materialy po arkheologii Rossii (Materials on the Archaeology of Russia) 37, St. Petersburg.

Rostovtzeff, M.I. 1918b: *Ėllinstvo i iranstvo na juge Rossii* (Hellenism and Iranism in South Russia), Petrograd.

Rostovtzeff, M.I. 1925: *Skifija i Bospor. Kritičeskij obzor pamjatnikov literaturnykh i arkheologičeskikh* (Scythia and Bosporus. A Critical Review of the Literary and Archaeological Monuments), Leningrad.

Rostovtzeff, M.I. 1929: *The Animal Style in South Russia and China*, Princeton.

Rostowzew, M. 1931: *Skythien und der Bosporus.* Band 1: Kritische Übersicht der schriftlichen und archäologischen Quellen, Berlin.

Rostowzew, M. 1993: *Skythien und der Bosporus*, Band 2: *Wiederentdeckte Kapitel und Verwandtes.* Auf der Grundlage der russischen Edition von V. Ju. Zuev mit Kommentaren und Beiträgen von G.W. Bowersock übersetzt und herausgegeben von H. Heinen in Verbindung mit G.M. Bongard-Levin und Ju. G. Vinogradov, Stuttgart.

Shaw, B. 1992: ‚Under Russian Eyes‘, *JRS* 82, 216–228.

Skripkin, A.S. 2017: *Sarmaty* (Sarmatians), Volgograd.

Solov'jov, S.M. 1851: *Istorija Rossii s drevnejšikh vremjon* (History of Russia from Ancient Times), St. Petersburg.

Tatiščev, V.N. 1768: *Istorija Rossii s samykh drevnejšikh vremjon* (The History of Russia from the Most Ancient Times), Band 1.1, Moskau.

Tolstoj, I. & Kondakov, N. 1889: *Klassičeskie drevnosti južnoj Rossii. Russkie drevnosti v pamjatnikakh iskusstva* 1 (Classical Antiquities of South Russia. Russian Antiquities in Monuments of Art 1), St. Petersburg.

Tolstoj, I. & Kondakov, N. 1890: *Drevnosti vremjon pereselenija narodov. Russkie drevnosti v pamjatnikakh iskusstva* 3 (Antiquities of the Great Migration Period. Classical Antiquities of South Russia. Russian Antiquities in Monuments of Art 3), St. Petersburg.

Treister, M. 1993: ‚The Celts in the North Pontic Area: a Reassessment‘, *Antiquity* 67 (257), 789–804.

Treister, M. 2002: ‚Late Hellenistic Bosporan Polychrome Style and Its Relation to the Jewellery of Roman Syria (Kuban Brooches and Related Forms)‘, *Silk Road Art and Archaeology* 8, 29–72.

Ulf, Ch. 2009: ‚Rethinking Cultural Contacts‘, *AWE* 8, 81–132.

Woolf, G. 1994: ‚Becoming Roman, Staying Greek: Culture, Identity and the Civilizing Process in the Roman East‘, *Proceedings of the Cambridge Philological Society* 40, 116–143.

Woolf, G. 1998: Becoming Roman. The Origins of Provincial Civilization in Gaul, Cambridge.

Zasetskaya, I.P. 2019: *Iskusstvo zverinogo stilja sarmatskoj ėpokhi (2 v. do n.ė. – načalo 2 v.n.ė.)* (The Art of Animal Style of the Sarmatian Epoch (2nd c. BC – Early 2nd c. AD)), Simferopol.

Zuev, V.Ju. 1991: ‚Tvorčeskij put' M.I. Rostovtseva‘ (The Creative Path of M.I. Rostovtzeff), *VDI* 1, 166–175.

CULT AND FUNERARY PRACTICES
IN OLBIA AND ITS *CHORA*

A Methodological Approach to the Study of Cultural Identity
in Urban and Rural Communities

Joanna Porucznik

Abstract: This chapter discusses the cultural identity of urban and rural communities inhabiting Olbia Pontike and its rural territory in the northern Black Sea region. Cult and funerary practices are analysed in order to detect the expression of a group's self-definition during different phases of Greek settlement in that area. The traditional approach according to which clear ethnic markers can be found in archaeological material is rejected in favour of a more balanced approach that takes into consideration the flexibility of culture and the creation of collective identities. As demonstrated in this study, the differences between urban and rural zones are possible to discern through the popularity of Orphic-Dionysiac cults, the spread of mortuary practices in both the town and countryside and the attachment to old traditions that were visible, especially during the Roman period. These discernible trends are examined in relation to political, economic and demographic changes in the region over an extended period of time.

Абстракт: Культ и погребальные практики в Ольвии и ее хоре: методологический подход к изучению культурной идентичности в городских и сельских общинах: В этой главе обсуждается культурная идентичность городских и сельских общин, населяющих Ольбию Понтийскую и ее сельскую территорию в северном Причерноморье. Культовые и погребальные практики анализируются с целью исследования того, как данная группа жителей выражала самоопределение на разных этапах греческого поселения в этом районе. Традиционный подход, согласно которому в археологическом материале можно найти четкие этнические признаки, отвергается в пользу более сбалансированного подхода, учитывающего «гибкость» культуры и создание коллективных идентичностей. Как показано в этом исследовании, разницу между городскими и сельскими зонами можно найти благодаря популярности орфико-дионисийских культов, распространению погребальных обычаев в городе и сельской местности и привязанности к старым традициям, особенно в римский период. Эти заметные тенденции рассматриваются в связи с политическими, экономическими и демографическими изменениями в регионе в течение продолжительного периода времени.

I. INTRODUCTION

The archaeological, epigraphic and literary material from the city of Olbia and its necropolis provides a rich source of information concerning the religious life of the inhabitants. It was closely intertwined with cultural, political and social changes that can be observed in this region over a long period of time (see Fig. 1 for a map).[1] A comprehensive analysis of cult and funerary practices may help us describe the expression of a group's self-definition during different phases of Greek settlement in the area. Of special importance is the material gathered from the Olbian rural territory, since it permits the possibility of discerning differences between urban and rural zones through such phenomena as the popularity of certain cults, the spread of mortuary practices in the town and countryside, and the attachment to old traditions visible in certain periods both in the city and the *chora*.

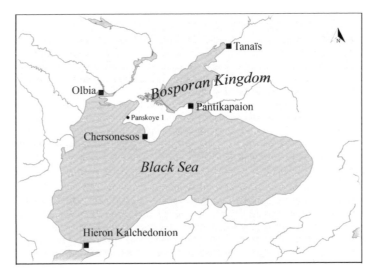

Fig. 1: Map of the Black Sea Region. © Joanna Porucznik, Opole 2020.

A comprehensive study of the archaeological material excavated at rural settlements has been made available through a series of monographs published several decades ago in Kiev.[2] Even though recent research has re-assessed previous ideas as to the chronology, development and extent of the Olbian *chora*,[3] these works remain both an important and the most complete study of Olbian rural settlements. Rural necropoleis have recently attracted scholarly attention providing important

---

1   For the Olbian necropolis, see Parovich-Peshikan 1974; Kozub 1974; Skundova 1988; Papanova 2006; for Olbian religion, see Rusyaeva 1992; for the Olbian *temenos*, see Rusyaeva, Kryzhitskiĭ & Krapivina 2006; Buĭskikh 2015.
2   Krÿzhitskiĭ 1987; Krÿzhitskiĭ et al. 1989; 1990.
3   See Kryzhitskii 2000; Kryžickij 2006; Bylkova 2005; see also Buĭskikh & Buĭskikh 2010; Bujskich & Bujskich 2013 for a revised chronology of the archaic rural settlements.

observations and comparative analyses of the archaeological material.[4] The relative neglect of such studies in previous research was most probably due to the fact that the location of rural cemeteries as well as their association with a concrete settlement was often uncertain.[5]

The archaeological material from both rural and urban Olbian necropoleis, sanctuaries and other cultic places such as the so-called zol'niks (open cult areas),[6] which will be discussed later in this paper, provides a valuable source of information about a group's self-awareness and its attachment to a given cultural framework. The introduction of certain cults, such as the cult of Apollo, which was brought to Olbia by the settlers from Miletos, may demonstrate the relationship between the inhabitants of an *apoikia* and its mother city. Furthermore, bonds with a new territory could be developed and strengthened through specific cults, some of which may have been introduced intentionally, to create a sense of collective identity. An example of this is the cult of Achilles: he was very prominent in the rural territory and worshipped in extra-urban sanctuaries that demarcated the borders of the city.[7] Also, certain funerary rites and objects may have been open to a cross-cultural understanding, such as kurgans, zol'niks, niche graves, and mirrors found in a funerary context.[8]

II. METHODOLOGY

Those looking at the cultic and mortuary practices both in the city and *chora* have frequently placed the main focus of analysis on the ethnic attribution of the archaeological material, instead of trying to understand the way in which local urban and rural communities were formed, and how they functioned and interacted with each other.[9] It has been proven that the search for clear ethnic markers in the 'colonial' environment of the North Pontic region cannot be successful and should be avoided.[10] The insistence on a sharp dichotomy between the Greeks and non-

---

4   Snýtko 2013, 91f.; the author identifies 14 rural necropoleis dated to the late archaic and early classical periods (the 6th–5th centuries BC) and 12 necropoleis dated to the Hellenistic period (the 4th–3rd century BC); however, the necropolis at the settlement of Koshary should also be included in that list; see Papuci-Władyka & Redina 2011, 289–295.
5   See Krÿzhitskiï et al. 1989, 217; from this period, the most comprehensive study of funerary rites and religion in the Olbian *chora* can be found in Kryzhitskii 1987.
6   For the origin of the term 'zol'nik', see Rusyaeva 2006a, 95 with n. 1.
7   See Buïs'kikh 2004.
8   See the discussion by Kryžickij 2006, 109f.
9   See the overview by Gerasim 2013; for the analysis of hand-made pottery, see Marchenko 1988; for the discussion on the ethnic affiliation of kurgans from the rural territory of Olbia dated between the end of the 6th and the middle of the 5th century BC, see Grechko 2010, who argues for their Scythian origin, and Snitko 2011, who points to their Greek character. For the discussion on treating semi-dugouts and dugouts as ethnic indicators, see Krÿzhitskiï 1982, 148; Kuznetsov 1999.
10   See Guldager Bilde & Petersen 2008; Petersen 2010.

Greeks that is still promoted in Black Sea archaeology is undoubtedly the remains
of a culture-history approach that for many decades enjoyed popularity, especially
among Eastern European scholars.[11]

Unlike Western scholarship, which concentrated on the Greco-Roman Mediterranean world for a long time, a particular interest in the Greek-Barbarian relationship among Soviet scholars is visible in several publications and conferences
held during the Soviet period.[12] The impact of local indigenous peoples on the
formation of Greek *apoikiai* was often either exaggerated or, conversely, strongly
denied.[13] Most importantly, the Eastern European approach to the study of ethnicity was little affected by the perspectives and methods that were developed in
Anthropology in Western Europe and North America. The discipline's interest in
phenomena such as acculturation and multiculturalism was aroused as a consequence of the effects of migration, decolonisation and globalisation from the
1960s onwards.[14] Accordingly, the post-colonial discourse that emerged in the
late-20th century in Western universities, with such influential works as Homi
Bhabha's *The Location of Culture*,[15] had little impact on the way in which cultural
encounters were studied among Eastern European historians, archaeologists and
classicists.[16]

Throughout the Soviet period, the idea of an ethnos was deeply rooted in
Marxist theory. This notion was mainly based on the historical principle that identified a certain ethnic group through its long history. Ethnic groups were perceived as social units defined by common language, territory, history, ethnonym,
culture and self-awareness. Consequently, ethnic identities were surmised to be
stable over a long time, although they, too, are situational and flexible by nature,
just as all other kinds of collective identities.[17] Ethnic groups were understood as
being continuous from generation to generation and connected through specific

11 Shennan 1991; see also Antonaccio 2010, 33f. who points out how difficult it is to escape
    from the old culture-history method in the study of ethnicity.
12 Lordkipanidze 1979; 1981; for the overview of post-war Soviet research on the Greek-
    Barbarian relationship in western Crimea, see Novikova 2012.
13 See Novikova 2012; an example of this is the lengthy debate about the ethnic and cultural
    affiliation of contracted burials discovered in the northern cemetery of Chersonesos; see the
    discussion by Stolba 2011, 332f.
14 Jones 1997, 8; the so-called 'ethnic revival' among indigenous post-colonial societies since
    the 1950s has also influenced attitudes towards the study of cultural and ethnic identity; see
    Hutchinson & Smith 1996, v.
15 Bhabha 1994. According to his concept, *hybridity* is perceived as a *third space* for communi-
    cation and negotiation between the colonised and the coloniser, a physical and psychological
    place in which creativity and innovation occur; this resembles the idea of the *middle ground*
    that is discussed by White with regard to a Native American community and European colo-
    nisers in the 18th and 19th centuries (White 1991). Both concepts have gained popularity in
    classical studies and have started to develop independently (see Antonaccio 2003 for the con-
    cept of hybridity, and Antonaccio 2013 for the middle ground theory in classical studies).
16 See Gosden 2004; Hurst & Owen 2005.
17 Dragadze 1980.

forms of social organisation. It is visible that the development of Eastern anthropological thought was closely related to the culture-history approach that was applied in archaeology, for example, by Gustaf Kossinna, whose works significantly influenced the ideology of national socialism, and by Gordon Childe, who actively promoted the Marxist approach outside of Soviet scholarship.[18]

According to the culture-history approach, ethnic identity is directly reflected in material culture, which in turn resulted in the idea of so-called 'static indigenism', namely the belief that groups of people lived in one place for thousands of years. As a consequence, every change in material culture was interpreted as an effect of diffusion or local and cultural evolution.[19] Such a 'stagnant' approach to ethnicity was rejected by the structural-functionalists in the West, who paid more attention to flexibility and the situational nature of ethnic groups and their cultural boundaries, rejecting at the same time any connotations of ethnicity related to genetics.[20] In his seminal work *Ethnic Groups and Boundaries,* the social anthropologist Fredrik Barth developed the idea of ethnicity as a self-defining system, strengthened through active maintenance of cultural boundaries that are perceived as a sense of a culture's distinctiveness, which creates the self-conceptualisation of a society.[21] Keeping Barth's approach in mind, flexibility, fluidity and the situational nature of a group's identity can explain certain practices that developed among Black Sea communities more satisfactorily. This is especially achievable when cultural trends are observed from a broader chronological and geographic perspective, in which constant movement of people and ideas as well as changes of the political, economic and environmental structures are taken into account.

Another important matter discussed by anthropologists and archaeologists is a distinction between ethnic and cultural identity, a distinction that should be drawn when studying past societies, since these two aspects of a person's self-definition may manifest themselves on different levels of social interaction. It has to be taken into account that, within the same cultural milieu, not only ethnicity but also other kinds of identity related to such criteria as status, gender and age may have been manifested by the inhabitants.[22] In certain instances, the expression of these criteria may have been more important than ethnic identity. The necropoleis of Pantikapaion are a good example of this. The analysis of their archaeological ma-

---

18 Trigger 1989, 344–352; Dolukhanov 1995, 332.
19 Shennan 1991, 30.
20 Shennan 1991; for an interesting discussion of Eastern and Western anthropologists about methodology in the 1970s, see Geller et al. 1975. For Marx's legacy to archaeologists and social scientists, see Patterson 2003. Notably, a similar methodological problem is apparent in the debate on La Tène art objects, perceived as an indicator of Celtic identity; for an overview, see Harding 2007, 1–6.
21 Barth 1969. This kind of approach was also prompted by Cohen, who treated ethnicity as an extremely instrumental construct based on economic and political rather than psychological reasons; Cohen 1996, 83f.
22 As argued by Antonaccio 2010, 33, cultural identity, unlike ethnic identity, transcends social criteria such as class, gender and age.

terial has shown that the grave assemblages were arranged according to the gender and social status of the deceased, rather than their ethnicity.[23] Similar conclusions have been drawn by Jane Hjarl Petersen in her analysis of the early-4th-century-BC kurgan complex in Kerkinitis. The expression of ethnicity by maintaining traditional burial customs appears less important than the expression of social status through various prestige markers, such as kurgan construction, derived from different cultural spheres.[24] When analysing burial assemblages, Petersen also points out the importance of family traditions and personal preferences in burial practices that were not ethnic-specific and should rather be interpreted cross-culturally.[25]

In this paper, cultural identity is understood as a feeling of belonging to a particular group of people (for example, an ethnic group) who have their own distinctive culture. As William Sewell argues, culture should be understood as a concrete and bounded system of beliefs and practices which correspond with a given society. This system, however, does not have to be stable and fully coherent, since culture itself is often contradictory and loosely integrated, which, in the case of the Greek cultural framework, was enhanced by the high mobility of people, who travelled, traded and founded *apoikiai* overseas.[26] What is needed for a culture to function is at least a thin coherence of shared practices which creates reciprocal comprehension among fellow group members rather than homogeneity.[27]

Therefore, it seems realistic to assume that ethnic and other social groups can function within the same cultural milieu and try to express their own identity as long as it does not disturb the sense of minimal cultural coherence in a given community. This process frequently includes such strategies as cooperation and negotiation, as well as power relations and resistance to other oppositional groups, which are especially evident when the balance between coherence and difference is disturbed. As a result, cultural identity should be perceived as an ongoing process of formation and re-formation which greatly depends on changeable social, economic and political circumstances.[28]

III. THE CITY AND THE CHORA

It has to be noted that neither the city nor the *chora* had territories that were fixed and clearly defined geographically during the whole period of their existence. Both Olbia and its *chora* experienced periods of prosperity and decline, which

---

23   Fless & Lorenz 2005a; 2005b.
24   Petersen 2010, 141.
25   Petersen 2010, 110–114.
26   Sewell 1999, 44–47; Antonaccio 2003, 58; Hall 2003; as argued by Antonaccio 2010, 33, such cultural practices as the use of a particular type of pottery or speaking a particular dialect are related to cultural identity and cannot constitute strict criteria for ethnicity.
27   Sewell 1999; Antonaccio 2003, 57–59.
28   See Hall 1996; Sewell 1999, 57; Ober 2003.

reflected the political and economic situation in the North Pontic region, climate change, the relationship between the city and its *chora*, and their relations with the steppe world. Demographic changes were also important factors that shaped the situation in that region. Causing factors were, to start with, the settlement of a new *apoikia* by Milesian settlers, the continuous movement of people between the city and the *chora*, and further waves of migration of non-Greek peoples.

The final process of the formation of the Olbian *chora* in the Lower Bug region took place on the verge from the 6th to the 5th centuries BC. There are over a hundred rural settlements known from that period spreading along the Dnieper and Bug estuaries with Olbia situated roughly in the middle of its rural territory.[29] Around the middle of the 5th century BC, a drastic reduction of the *chora* is visible in the archaeological material. This is likely to result from the development of the Olbian *asty* and its significance as a political, administrative and economic centre; such a situation often results in the migration of people from rural territories to the city due to socio-economic factors, causing the development of the sub-urban zone.[30] Olbian rural settlement started to expand again at the end of the 5th century BC. That was also the time when Scythian settlements were established in the neighbouring territory of the Lower Dnieper region, among which the Kamenskoe hillfort appears to be one of the best-studied sites.[31] It is most probable that the Greek settlement of Glubokaya Pristan' (Sofievka 2) located on the right side of the Dnieper Estuary served as a transit point for trade between Olbia and the steppe world (Fig. 2, no. 13). Defensive constructions have also been recorded at the site, which is exceptional for the rural territory of that period.[32]

Written sources recorded the campaign of Zopyrion, a general of Alexander, and the siege of Olbia in 331 BC. The Scythians attacked his army, which certainly contributed to the failure of the Pontic campaign and the total defeat of Zopyrion.[33] The ensuing period up to the middle of the 3rd century BC is considered the peak of the *chora*'s extension in geographic, demographic and economic terms.[34] The settlement can be divided into two zones – the coastal zone depending on fishing and agriculture, and the inland/steppe zone dedicated to cattle breeding.[35] During the second half of the 3rd century BC, the northern Black Sea region suffered an economic and political crisis, which is visible in the gradual decline of

---

29 Kryzhytskyy & Krapivina 2003, 513; Krÿzhitskiï et al. 1990, 120; Busjkich & Bujskich 2013, 2 with fig. 1.
30 Krÿzhitskiï et al. 1990, 42f.; Krÿzhitskiï et al. 1989, 95; as pointed out by the authors, there is no evidence to connect the reduction of the rural territory with Scythian migration and political conflicts as suggested by some scholars (see Vinogradov's theory of a Scythian protectorate over Olbia, Vinogradov 1989, 90–109, and a critique by Cojocaru 2008).
31 Alekseev 2003, 214; 285.
32 Krÿzhitskiï et al. 1990, 70–72; according to Bylkova, the settlement dates to ca. 400–275 BC (Bylkova 2005, 219f.).
33 Krÿzhitskiï et al. 1989, 97; Macrobius, *Saturnalia* 1.11.33; Justin 12.2.
34 Krÿzhitskiï et al. 1990, 21; Kryzhytskyy & Krapivina 2003, 517.
35 Kryzhytskyy & Krapivina 2003, 518.

the city and the considerable reduction of the amount of Olbian as well as Scythian settlements. Factors that caused the crisis were complex and included climatic and economic changes as well as pressure from the peoples of both the West (as attested in the Protogenes inscription: *IOSPE* I^2 32) and the East (which can be connected with the movement of Sarmatian tribes)[36].

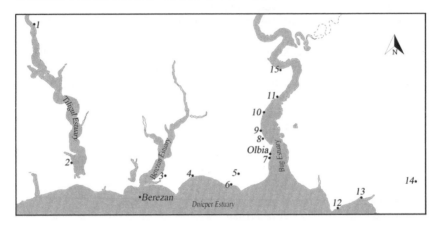

Fig. 2: Geographic Locations of Sites Discussed in Chapter III. © Joanna Porucznik, Opole 2020.
1. Sofievka; 2. Koshary; 3. Bolshaya Chernomorka; 4. Kutsurub; 5. Adzhigol'skaya Balka; 6.
Petukhovka; 7. Shirokaya Balka; 8. Chertovatoe; 9. Katelino; 10. Kozÿrka; 11. Staraya Bogdanovka; 12. Zolotoï Mÿs; 13. Glubokaya Pristan'; 14. Belozerka; 15. Didova Khata.

The political and demographic situation in the city and its *chora* changed considerably during the Roman period. The attack of the Getai and Dacians under Burebista in the middle of the 1st century BC destroyed the city, which was then abandoned for several decades.[37] Gradually, a new community emerged in the city, which may have included a Sarmatian group connected with the political activity of the Scythian king Pharzoios and his heir Inismeus (Inensimeus),[38] besides the surviving rural population that took part in the restoration of the city.[39]

The revival of the *chora* can be observed at the turn from the 1st century BC to the 1st century AD, when new fortified settlements appeared. Defensive structures and the strategic location of settlements in relation to the urban centre reflect

---

36 For the critique of the idea of the Sarmatian invasion being the only determining factor behind the crisis, see Mordvintseva 2013; see also Krÿzhitskiï et al. 1989, 100.

37 Krapivina 2007a, 161.

38 Tokhtas'ev 2013.

39 Krapivina 2007a, 161f. assumes that all the new inhabitants who took part in the rebuilding of Olbia (including the non-Greek population from the rural territories) received full citizenship.

the unstable political situation in the region during that time.[40] From then on, the *chora* functioned up to the Gothic invasion in the middle of the 3rd century AD, whereas the *asty* survived as a small settlement up to the 3rd or 4th centuries AD.[41]

It is problematic to define the borders of the Olbian *chora* precisely.[42] Such criteria as material culture, funerary rites, architecture, and the use of Greek coinage have frequently been used to identify the Greek nature of rural settlements and to define the borders of the Olbian *chora*. In certain situations, however, these criteria are not as clear as one may expect. An example of this is the site of Koshary that is located on the right bank of the Tiligul Estuary (ca. 40 km east of modern Odessa; Fig. 2, no. 2). Based on a field survey, the settlement was initially classified as barbarian.[43] However, the systematic excavations carried out between 1998 and 2008 by the Polish-Ukrainian mission at the settlement and its necropolis have proven that Koshary was a Greek settlement that functioned within the Olbian administrative system from the late-5th or early-4th up to the middle of the 3rd century BC.[44] The great prevalence of Olbian coins discovered at the settlement also proves economic links with Olbia.[45] An Olbian monetary decree from the second quarter of the 4th century BC *(IGDOlbia* 14 = *IOSPE* I^2 24), found at Hieron Kalchedonion (modern Anadolukavağı in Turkey; Fig. 1), sheds some further light on the far reach of Olbia's monetary policies. The document regulated the coinage system that had to be used for trade within the Olbian territory.[46]

Similarly, the settlement of Belozerka located on the right bank of the Dnieper River in the eastern part of the distant Olbian *chora* was initially identified as a Scythian settlement (Fig. 2, no. 14).[47] However, archaeological investigation of the site has revealed that Belozerka resembles a typical Olbian rural settlement with regard to its material culture (including pottery, graffiti and terracotta figurines), architecture (a combination of stone and mud brick constructions), cult practice, and the way of life of the inhabitants, which was based on farming, cattle-breeding, fishing and trade.[48] Also, a considerable amount of Olbian coins along with a 3rd-century BC hoard of Olbian coins found at the settlement points to close economic ties with the Greek *polis*.[49] As Bylkova rightly points out, the attribution of the settlement to the Olbian *chora* does not automatically determine

---

40 Krÿzhitskiï et al. 1990, 96f.; for a brief overview of the history of the Olbian *chora* in a Western language, see also Bujskich 2006.
41 Krÿzhitskiï et al. 1990, 97.
42 Krÿzhitskiï et al. 1989, 8f.; 143; Kryzhitskii 2000. See also the discussion by Bylkova 2005.
43 Buïskikh 1986, 22.
44 Papuci-Władyka et al. 2009; 2010; Papuci-Władyka & Redina 2011.
45 There is one coin of Tyras recorded at Koshary which is exceptional and may point to the importance of Koshary as a local trading point (Bodzek 2008, 17f.; 2011, 58).
46 Krÿzhitskiï et al. 1989, 143.
47 Bylkova 1996, 101; 2011. The settlement functioned from the beginning of the 4th until the first quarter of the 3rd century BC.
48 Bylkova 1996, 117.
49 Bylkova 1996, 101.

the ethnic and social structure of the population, because the archaeological material does not provide enough evidence to do so.[50] Since there are no known contemporary settlements east of Belozerka, it appears that the inhabitants of Belozerka were a frontier community mediating their activities between the sedentary and nomad zones, whereas the settlement may have served as a local centre of trade and production.[51]

Notably, the archaeological material from Olbian rural settlements demonstrates a lack of homogeneity among the inhabitants in terms of their status and economic situation.[52] The comparison of the archaeological material and architecture of the two late archaic settlements of Staraya Bogdanovka 2 (14.5 km north of Olbia; Fig. 2, no. 11) and Kutsurub 1 (20 km south-west of Olbia; Fig. 2, no. 4) has demonstrated that a close economic relationship with the *polis* seems to have been reflected in the condition of a settlement. The material from Staraya Bogdanovka 2 includes such objects as imported Greek pottery, 23 Olbian coins and seven building complexes with above-ground constructions,[53] whereas Kutsurub 1 displays a simplistic architecture as well as a homogeneous and modest material assemblage including two Olbian coins and a considerable amount of hand-made pottery.[54] Notably, dug-out structures were in use at Kutsurub 1 up to the 3rd century BC, which indicates the archaic and conservative character of this settlement.[55] The difference in the development of these two rural settlements may

---

50   Bylkova 1996, 117; 2011, 54. For the same reason, the conclusion drawn by Kryzhytskyy & Krapivina 2003, 543 that the Olbian *chora* was inhabited by the Greeks in an environment of barbarians does not seem convincing.

51   Bylkova 2011, 47.

52   Cf. Kryžickij 2006, 100, who argues that the social and legal status of the first groups of settlers was equal.

53   Marchenko 2013, 22–57.

54   Marchenko 2013, 58–69. Marchenko 1988 analyses Olbian hand-made pottery and understands it as an ethnic marker. Hand-made pottery is a common phenomenon at northern Black Sea sites. It has to be noted, though, that a quantitative study of hand-made pottery is difficult from a methodological point of view, since, as a rule, it tends to break into more sherds compared to wheel-made pottery. Therefore, an analysis based on an estimated number of complete vessels would be more accurate and would undoubtedly demonstrate a lower percentage of hand-made pottery at Black Sea Greek settlements; see the discussion by Stolba 2012, 340; Petersen 2010, 84f. with n. 62.

55   Marchenko 2013, 77. The use of semi-dugouts and dugouts (especially during the first stage of a Greek settlement) was a common phenomenon not only in the northern Black Sea area, but also in other regions such as Kolchis, the Bulgarian coast, and Italy. As far as the Olbian *chora* is concerned, they were recorded for the first time at Shirokaya Balka 1 (Fig. 2, no. 7) and were attributed to a local non-Greek population called 'the Kallipidian culture' named after the Herodotean Kallipidai (who were described as 'Greek Scythians': Hdt. 4.17.1). However, further investigation revealed Greek material at the settlement (KrÿzhitskiÏ et al. 1990, 25f.). As a rule, they were gradually replaced by stone buildings; nevertheless, the use of semi- and dugout structures in the 5th century BC is also recorded. For the discussion on treating semi-dugouts and dugouts as ethnic indicators, see KrÿzhitskiÏ 1982, 148; Kuznetsov 1999; Tsetskhladze 2004.

suggest that Staraya Bogdanovka 2 established a more prosperous relationship with the city, resulting in trade contacts and the exchange of goods and ideas, as well as allowing for the constant movement of people between the village and the city. However, the difference in the economic development of the settlements was not necessarily directly reflected in the ethnic composition of their inhabitants, as suggested by Marchenko (who argues that Kutsurub 1 was inhabited by a considerable group of non-Greeks), but rather in their lifestyle.[56]

An onomastic analysis of names recorded in the countryside may also provide important information regarding the self-definition of the inhabitants. The appearance of the Iranian, most probably Scythian name Κοκονακος on a 4th-century BC judicial *defixio* from the rural settlement of Kozÿrka 12 (*SEG* LII 742) may indicate specific preferences in name-giving among the inhabitants.[57] Other examples of Iranian names recorded in the Olbian *chora* include: Χιμυνακες (5th century BC), Κοκονακος (2nd half of the 4th century BC) and the partly preserved Φανισαλ- (1st quarter of the 3rd century BC).[58] There are also Iranian names preserved on a lead *defixio* dated to the early-Hellenistic period, found on the south hill of the Olbian necropolis: Θατόρακος, Καφακης and perhaps Ἀτάκης.[59] Notably, the name Συχοτος, which is inscribed on a bowl found at the Belozerka settlement, is also non-Greek and does not have any analogies in the northern Black Sea region.[60] Also, a 4th-century graffito on a sherd from the settlement of Kozÿrka 2 may point to a local peculiarity of the language spoken in the countryside (*SEG* LI 980). The graffito displays: ΠΡΑΣΙΛΑΤΟΣΕΙΜΙ ('I belong to Prasilas'), which might be an example of an irregular, non-Ionian form of the genitive.[61] This seems to suggest that, unlike in the city where the Ionian dialect was preserved until the Roman period, the Greek language spoken among rural communities may have had its local idiosyncrasies.

IV. DEVELOPING LOCAL IDENTITY

Joint migration and the foundation of a new *apoikia* led to the formation of new cohesive community with a sense of common identity. It is important to point out that the settlers were rarely a homogeneous group of people who migrated from their common mother city. It is more likely that they were groups of Greeks and presumably also a certain amount of non-Greeks of different origins led by the

---

56 Marchenko 2013, 94.
57 Tokhtas'ev 2002; Marchenko 2013, 111.
58 Tokhtas'ev 2013, 576.
59 Vinogradov 1994; For Iranian names in the prosopography of Olbia, see Vinogradov 1981 with corrections made by Stolba 1996; Cojocaru 2004; Tokhtas'ev 2013.
60 The graffito reads ΣΥΧΟΤΟΟΥ; however, the reading ΟΙΧΟΤΟΟΥ is also possible; see Bylkova 2011, 50.
61 Tokhtas'ev 2000 points out the influence of Koine Greek and a possible consonant shift. See *LGPN* vol. 4 for *Prasilas*.

founder.[62] Consequently, the settlers became members of a new community through their common experience of the participation in the foundation process. The 'colonial' society was a new entity unified under the umbrella of an inclusive cult. The absorption of the settlers into the social, religious and political order was reflected in the city's *nomima,* which were customary institutions applied by a newly established community.[63] This process involved a range of strategies such as the invention and manipulation of tradition, shaping the past and establishing local collective memory.[64]

However, Greek *apoikiai* also tended to integrate into a broader Panhellenic identity. This was expressed through ritual (such as participation in the Pythian Games at Delphi), participation and dedications in the Panhellenic sanctuaries in Olympia and Delphi, and the use of the Panhellenic mythical framework in the expression of 'colonial' origin.[65] Similarly, rural settlements may have been founded by a leading group of settlers who, for instance, had the same geographical origin and thus had social links already established among them. It is also possible that in certain situations such links were created spontaneously as a reaction to the foundation process of a new city.[66]

The development of a community's identity through the creation of local history is visible at the cemetery of Koshary. The analysis of the spatial development of the necropolis demonstrates that the sacred space was intentionally arranged in connection with older, Bronze-Age kurgans and the route leading to the steppe. One of the largest kurgans seems to have demarcated the northern border of the cemetery.[67] Of special importance is barrow 55, contemporary to the cemetery (4th–3rd centuries BC). It was a central point of the sacred area, around which other burials were located over time. The barrow was originally demarcated by a surrounding circular trench of ca. eight metres circumference, which was an element typical of kurgans found in steppe and forest-steppe regions of that period.[68] The circle had two entrances situated opposite each other on its east-west axis. The entrances may have served as a passage to the sacred area of the barrow. There were two niche graves located in the centre of the barrow, one of a young man and the other of a child. The area that spreads 3–4 meters away from the trench was

62 Malkin 2009, 378. Moreover, the complex ethnic composition of the Ionians expressed by Herodotos (1.146f.) also demonstrates that the settlers may not have been as ethnically coherent as one would expect according to modern criteria; the Herodotos passage has also been commented on by Krÿzhitskiï et al. 1989, 51.
63 Malkin 2009, 386–390; 2003, 67.
64 For the idea of 'intentional history', which is the projection in time of the elements of self-categorisation that creates a group's distinctive identity, see Foxhall et al. 2010; for local history and the construction of the past in cities of the northern and western Black Sea region, see Dana & Dana 2007.
65 See Malkin 2003; 2009.
66 Cf. Krÿzhitskiï et al. 1989, 40.
67 Redina 2007, 95f.
68 Chochorowski 2008, 26 with n. 4.

left empty whereas, further on, a circle of graves centred around barrow 55 has been recorded. This suggests a special socio-religious status of the deceased young man, who may have been perceived by the local community as a local hero or a mythical forefather.[69]

It has also been suggested that the arrangement of graves in relation to the barrow within the boundaries of the necropolis seems to reflect the social system and the ethno-cultural composition of the local society. Analysis of the age, gender and status of the deceased buried around barrow 55 has revealed that they roughly reflect a natural biological structure of an average population linked by social ties, which demonstrates that the deceased represented a local community.[70] The estimated timeframe for these graves, based on radiocarbon dating, is 121 years.[71] Funerary rites recorded in them indicate that both Greek and steppe traditions were maintained among the users of the necropolis.[72] This may suggest that participation in local social structures was not limited to those who followed traditional Greek customs. It is important to observe that the archaeological material from the settlement demonstrates that Koshary was strongly influenced by an Olbian cultural model which may have had a unifying effect on the inhabitants. The settlement held close economic ties with the *polis*, which is confirmed by the use of Olbian coins (as mentioned above) and trade goods imported from the city.[73]

V. OLBIAN KURGANS

The appearance of kurgans in the territory of the Olbian *chora* at the end of the 6th century through to the middle of the 5th century BC has been discussed on several occasions with regard to their ethnic affiliation. The funerary assemblages found in these kurgans were variously interpreted as belonging either to the Greek or barbarian cultural milieus.[74] However, an ethnic-oriented interpretation of these kurgans does not contribute much to the discussion about the importance of these funerary structures in the local landscape and their symbolic meaning for the inhabitants of the *chora*. As in the case of Koshary, the conceptual understanding of the use of kurgans and barrows goes far beyond simple categorisation as 'Greek'

---

69   Chochorowski 2008, 28.
70   Chochorowski 2008, 31.
71   Chochorowski 2008, 28.
72   Steppe tradition includes the orientation of the deceased to the west; the deposition of a Scythian quiver in a grave of a young woman, which may perhaps be interpreted as a sign of prestige (grave no. 111); organic bedding in graves; arrows placed at the left thigh of the deceased; binding of the ankles recorded at a grave of a young warrior (grave no. 107) and a female grave (no. 108); Chochorowski 2008, 31–33.
73   Papuci-Władyka & Redina 2011.
74   Grechko 2010 argues for their Scythian origin, whereas Snitko 2011 points to their Greek character.

or 'non-Greek'. Kurgans were well known both in the Greek and steppe traditions and it seems more reasonable to perceive their use cross-culturally.[75]

In the urban necropolis, kurgans appeared in large quantities in the early Hellenistic period. This was contemporaneous with the economic and cultural acme of the city that took place after the siege of Zopyrion in 331 BC.[76] The largest kurgans were situated at the perimeter of the necropolis and near the main roads leading to the steppe; this suggests that they may have also served as local landmarks and watchtowers, which may have been connected with the apotropaic function of kurgans.[77] Their arrangement along the boundaries with the steppe territory was most probably planned by the city's officials or religious associations rather than by individual citizens.[78] Similar locations of kurgans can also be observed at Koshary, where older Bronze-Age burial mounds were used to demarcate the sacred space of the cemetery. The re-use of old kurgans may have related to the economic condition of the inhabitants, who could not afford burial constructions as elaborate as in the city.[79] Such a practice also demonstrates how these kurgans, which were elements of the local landscape, were consciously incorporated into a new cultural sphere and acquired new symbolic significance.

The introduction of kurgans in Olbia was contemporaneous with the appearance of a series of decrees honouring prominent citizens, such as Kallinikos, son of Euxenos around 325 BC (*SEG* XXXII 794). It is possible to assume that such citizens were honoured by the city with a burial under a kurgan. It should not be excluded that after the siege of Zopyrion, the most distinguished citizens who fought for the city were heroized (in a religious sense) after their death and honoured with an elaborate burial mound by the citizens of Olbia.[80] Such a practice would correspond with the idea of the soteric and apotropaic function of a dead hero, who was perceived as a guardian of the whole community.[81] If this is the case, the exceptional position of the person buried in barrow 55 at Koshary may have also been an example of the heroization of a distinguished member of a local community.[82]

---

75 Petersen 2010, 141 draws similar conclusions in her analysis of the early-4th-century BC kurgan complex in Kerkinitis: the expression of ethnicity by maintaining traditional burial customs appears less important to the local society than the expression of social status through various prestige markers, such as kurgan construction, derived from different cultural spheres.
76 Rusyaeva 2000, 106.
77 Rusyaeva 2000, 107.
78 Rusyaeva 2000, 107.
79 Redina 2007, 96.
80 Rysyaeva 2000, 108.
81 Rusyaeva 2000, 108.
82 At the necropolis of Koshary, a separate area of warrior burials (concentrated in the northern part of a kurgan) has been recorded, which suggests that warrior identity was widely recognizable for the local community; see Papuci Władyka & Redina 2011, 289; also Redina 2007, 96; who associates these warrior burials with the campaign of Zopyrion.

It is quite significant that the use of barrows was widespread among the Greeks at least as early as the time of Homer. Burial mounds were an important element of cultural landscapes, linked not only to the buried individual but also to the collective memory of a given community.[83] Also, the exaggerated size of the cremation pyre (measuring 100 square feet) for Patroklos, the long duration of the burial ceremonies for Achilles (18 days) and Hector (11 days), as well as the impressive size of the burial mound prepared for Achilles demonstrates the considerable effort expended on funerary rites, which expressed the high status of the dead.[84] In the case of Black Sea *apoikiai*, it seems possible that the existence of kurgan structures in the steppe, associated with the nomadic Scythian culture, prompted the re-introduction of this old Greek custom among the Olbian population, especially during the Hellenistic period, when kurgans started to be associated with the cult of heroes.[85]

VI. ADOPTION OF RELIGIOUS PRACTICES

The cultural link between the city and its *chora* is visible in the adoption of specific burial practices. The analysis of funerary structures found at rural necropoleis between the 6th and 3rd centuries BC and their comparison with the city's cemetery has demonstrated that more elaborate tombs, such as niche graves and underground chambers, as a rule appeared at rural cemeteries later than in the urban necropolis.[86] This may indicate that trends created in the city spread among rural communities gradually, which may have been caused not only by different economic developments, but also by the fact that rural communities are usually more conservative in their lifestyle. Such traditionalism of rural settlements can also be observed in the attachment to older customs, such as organic bedding in graves[87] and the use of open cult areas – the so-called zol'niks.

Their use has widely been discussed, particularly in connection to its non-Greek origin.[88] As far as the northern Black Sea region is concerned, the use of zol'niks was especially popular in the Lower Bug area as well as in the Kimmerian Bosporus.[89] They are usually described as open cult areas in the form of a heap

---

83   Crielaard 2003, 59.
84   Crielaard 2003, 60f.; Hom. *Od.* 63–84; *Il.* 24.664–666; 23.164.
85   Rusyaeva 2000, 110.
86   Snŷtko 2013, who has omitted the cemetery at Koshary for unknown reasons. A similar trend can be observed in the Crimea: anthropomorphic grave steles that were popular in Chersonesos have their analogies in the *chora*. They are less elaborate and schematic and therefore were associated for a long time with the Taurian population; see Zubar' 1982, 244.
87   E.g. Koshary, graves 111 and 107: Chochorowski 2008.
88   Nosova 2002; 2008; Rusyaeva 2006a. For the non-Greek origin of zol'niks, see Marchenko 2013, 69f.
89   They have been recorded in so-called small Bosporan towns only, not in the larger political centres of the region; see Nosova 2002, 64f.

made of ash and earth in which sacred objects like votive offerings and bones of sacrificed animals were deposited. They have been recorded at Koshary (4th to the early 3rd century BC),[90] Kozÿrka 2 (350–270 BC),[91] Kutsurub 1 (575–475 BC), Katelino 1 (4th and 3rd centuries BC),[92] Glubokaya Pristan' (5th to the middle of the 3rd century BC),[93] and Panskoye 1 (400–270 BC) in the North-West Crimea (see Figs. 1 and 2 above).[94] There are also nine zol'niks recorded in the Western Temenos in Olbia. The role of these zol'niks was to collect the sacred remains of offerings and ashes from nearby sanctuaries that functioned between 4th and 1st centuries BC. It is possible that six archaic heaps found at the Berezan settlement (Borysthenes) and a heap from Tendra Spit also served as zol'niks connected with local sanctuaries.[95] The ash heap at Tendra Spit was in use for several centuries, until the Roman period, during which steles with dedications to Achilles were erected.[96] The material found in the heap includes about 1200 coins issued in various cities between the 2nd century BC and the third quarter of the 3rd century AD.[97]

Notably, the description of Greek altars of burnt offerings given by Pausanias – an ashen altar of Samian Hera in Olympia (Paus. 5.13.8) and an open-air altar of Zeus Lykaios in Peloponnese (Paus. 8.38.7) – confirms the presence of open cult areas in Greek tradition. It seems probable that such cultic areas had a cross-cultural character and their use was widely comprehensible by North Pontic communities. The popularity of zol'niks in Olbian rural settlements may have been prompted by a similar, conceptually recognizable religious tradition that was long present among steppe communities. However, both traditions had their own unique character as a result of the different religious practices specific to a given community.[98]

It is possible that Greek zol'niks served as rural religious centres: such settlements as Kozÿrka 2, Koshary, Glubokaya Pristan' and Panskoye 1 may have marked the geographical expansion of the Olbian *chora* during the peak of its economic development.[99] Two zol'niks discovered at Glubokaya Pristan' on the right bank of the Dnieper Liman are also exceptional due to their size. Deposits of votive offerings, bones of animals, pottery and other objects such as arrow heads,

---

90   Papuci-Władyka & Redina 2011, 287; Nosova 2002.
91   Krÿzhtskiï et. al. 1990, 58f.; Marchenko 2013.
92   Rusyaeva 2006a, 98.
93   Buïskikh 1993.
94   The settlement was founded by Olbia; however, half a century later, it was taken over by Chersonesos; see Hannestad et al. 2002, 9.
95   Rusyaeva 2006a, 95–98; 2006b; the sanctuary of Achilles on Tendra Spit is attested by Strabo 7.23.19; see also Tunkina 2007. The cult of Achilles on Berezan is attested by votive graffiti, mostly dating to the 5th–4th centuries BC; see Ivantchik 2017a, 39.
96   *IOSPE* I² 328; 330 and most probably *IOSPE* I² 329, 331.
97   Rusyaeva 2006a, 99.
98   Rusyaeva 2006a, 100.
99   Nosova 2002, 66.

coins and oil lamps, were initially accumulated in ravines, which over time created hills. The thickness of the cultural layer of the two zol'niks reached 4 and 2.8 metres.[100] The relatively short time of their use may suggest that the open cult area at Glubokaya Pristan' was also visited by inhabitants of other settlements, who deposited their votive offerings there. The importance of such open cult areas for the development of a local identity of the inhabitants is also visible at Katelino 1. The zol'nik was situated on an older Bronze Age kurgan, which shows a conceptual link between the Greek tradition and the local cultural landscape.[101] Votive offerings found here include a lead *bucranium*, a limestone altar, a terracotta statuette (possibly that of Herakles), and a lead figurine with a representation of a chariot led by Achilles (or Helios).[102]

The cult of Achilles played an important role in the religious life of the Greek *apoikiai* in the whole North Pontic region and can be traced to the earliest periods of Greek settlement in this area.[103] Its strongest tradition was established in the north-west Black Sea region where the hero was worshipped in Olbia, Beïkush, Berezan, Tendra Spit, Kinburn Peninsula, Cape Hippolaos, Tyras, and the Panhellenic sanctuary on the island of Leuke, which was under Olbian protection (see *IOSPE* I^2 325; Fig. 3).[104] The close link between Achilles and the northern Black Sea region was noted by Alkaios around 600 BC, which is roughly contemporaneous with the early stages of Greek settlement in Olbian territory (Alkaios 14 Diehl = 354 Lobel-Page). The preserved lyric fragment refers to Achilles as the lord of Scythia: Ἀχιλλεὺς ὁ τᾶς Σκυθίκας μέδεις.[105] The myth associated with the island of Leuke, according to which Achilles was buried there, is likely to have been already established in the archaic period.[106]

It is noticeable that the cult of Achilles was particularly practiced in the Olbian *chora*, unlike other cults that seem to have been more popular in the city. The cults of Apollo (with the epithets Ietros and Delphinios), Zeus, Athena and other Olympian gods that were worshipped in Olbia are almost absent outside the city. This conceptual distinction between urban and rural cults may have related to the specific role that Achilles played in 'familiarising' the land by providing a strong mythological link with the territory. The evidence of the cult of Achilles can be found not only at sanctuaries, but also at rural settlements, such as the archaic

---

100 Rusyaeva 2006a, 98.
101 Rusyaeva 2006a, 98.
102 Krÿzhitskiï et al. 1990, 58; Rusyaeva 2006a, 98.
103 The cult had centres in the Bosporan Kingdom, Chersonesos and the Scythian Neapolis; see Rusyaeva 1979, 139; Bujskikh 2007, 201; Hupe 2006a; 2007.
104 The cult of Achilles is attested by numerous votive and dedicatory inscriptions, votive gifts, and literary sources; see Rusyaeva 1979, 125; Bujskikh 2007; Tunkina 2007; Okhotnikov & Ostroverkhov 1993.
105 The existence of such an epithet of Achilles may be supported by the dedicatory graffito at the bottom of an Attic black-figured vessel found at Tyras and dated to the last quarter of the 5th century BC; the graffito reads …]ΛΕΙ ΣΚΥ[…; see Bujskikh 2007, 209.
106 Braund 2007, 54 with n. 66; Ivantchik 2017a, 9–13.

settlement near Bolshaya Chernomorka, where a graffito with the name of Achilles has been discovered (Fig. 2, no. 3).[107]

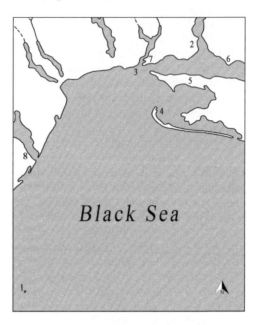

*Fig. 3: Geographic Locations of the Sanctuaries of Achilles, Adapted from Hupe 2006, Taf. 28. ©
Joanna Porucznik, Opole 2020. – 1. Leuke; 2. Olbia; 3. Berezan; 4. Tendra Spit; 5. Kinburn Peninsula; 6. Cape Hippolaos; 7. Beĭkush; 8. Tyras.*

Significantly, a similar distinction between the popularity of specific cults in the city and countryside at certain periods of time is visible in Chersonesos. The popularity of the worship of Herakles in the rural area has been demonstrated by finds of limestone reliefs representing the hero in a standing, reclining or feasting pose. Such artefacts have been discovered at Panskoye I and other sites in the northwestern Crimea (the so-called distant *chora*), including Chaĭka, Saki, the settlement at Lake Moĭnaki and Mezhvodnoe.[108] In contrast to the city, where freestanding statues and statuettes of Herakles were also common, the limestone reliefs seem very typical of the rural territory.[109] They date to the period from the late-4th to the early-3rd century BC, which was the time of the highest prosperity

107 Rusyaeva 1979, 137.
108 Stolba 2004, 60–62; for the Chaĭka settlement, see Popova &Kovalenko 1996.
109 Stolba 2004, 60.

of the *chora*.[110] A 4th-century-BC dedication to Herakles Soter found on a large black-glazed *kantharos* discovered at Panskoye I points to the protective meaning of his cult in the countryside.[111] Dedications to Herakles are well attested in the *chora*, whereas the dedications to Parthenos, who was the city's main deity, as well as other evidence of her worship, are limited to Chersonesos itself and its closest surroundings.[112]

Strikingly, the disappearance of the Chersonesitan coin types representing Herakles' head in the middle of the 2nd century BC was contemporaneous with the devastation of the city's *chora*.[113] This may suggest that the coin type representing Herakles reflected the situation in the *chora* of Chersonesos, whose prosperity guaranteed good fortune for the city itself, where the hero was worshipped as a protector and patron, whereas Parthenos appears to be merely the goddess and guardian of the city.[114]

This evidence may challenge the argument put forward by Irene Polinskaya that the city-countryside distinction is meaningful only in administrative and topographical contexts and that there was no distinction between urban and rural spheres in Greek religion.[115] She rightly points out that it is not possible to classify deities into those that were worshipped exclusively in the city or exclusively in the countryside.[116] Moreover, one must take into consideration the fact that the same individuals may have participated in religious rituals both in the city and the countryside. However, the evidence from Olbia and Chersonesos suggests that a distinction between the prominence of specific cults – which was a reaction to the socio-economic and political situation in the region – can be observed during certain periods. Also, the distance between the sanctuary and the *polis* did not have to reflect the importance of the sanctuary for either rural or urban communities. The peripheral location of Olbian extra-urban sanctuaries,[117] especially during the first period of Greek settlements in the Lower Bug region, played an important role in the demarcation of the psychological borders and the spatial structure of the city, which encompassed the surrounding rural and coastal areas.[118]

The Orphic-Dionysiac cult attested in Olbia as early as the 6th century BC played an important role in expressing the city's unity through the establishment of a close relationship between the territory and the citizens of the *polis*. It has been demonstrated through parallels in other 'colonial' areas of the Greek world that the Orphic-Dionysiac cult, which is associated with transcendence and utopi-

---

110 Stolba 2004, 61.
111 Stolba 1989; 2004, 63; see also Popova & Kovalenko 1996, 70.
112 Stolba 2004, 63f.
113 Stolba 2004, 60.
114 Stolba 2004, 63f.
115 Polinskaya 2006.
116 Polinskaya 2006, 65f.
117 See a discussion about the problematic use of the terms urban, sub-urban and extra-urban sanctuaries by Polinskaya 2006, 65–67.
118 Buïs'kikh 2004. For the location of sacred areas at border zones, see McInerney 2006.

an thinking, may have had an important impact on the development of group identity among settlers in the early stage of the city, as a response to their new living
conditions and the stress associated with the process of migration.[119] Moreover,
the Olbian cults of Demeter and Dionysos had a distinctive chthonic nature related
to agriculture, which may have stayed in conceptual opposition to the nomadic
traditions of the Scythians.[120] This was likely to have been reflected in Herodotos'
stories about the Scythian king Skyles and the philosopher Anacharsis, in which
such a juxtaposition of the rural and urban environments and their way of life is
expressed.

According to the story, Skyles inherited the kingship after his father's death;
however, he frequently rejected the Scythian way of life and preferred to stay in
Olbia. Due to the fact that his mother was Greek, Skyles was bilingual and accus-
tomed to living in a Greek way. He had two wives: the first one, Opoia, was a
native Scythian who lived in Scythian territory, whereas the second wife was
Greek and stayed in Skyles' house in the city. The king chose to be initiated into
the Bacchic mysteries of Dionysos, for which he was executed by the Scythians.
A similar fate was met by the philosopher Anacharsis, who was killed by his fellow Scythians for performing the rites of the Mother of the Gods.[121]

The cults of Demeter and Dionysos have a strong agricultural character which
is in opposition to the nomadic customs of the Scythians and their way of life as
defined by the Greek concept of civilisation, which is closely linked with agriculture.[122] This, too, may be reflected in Herodotos' tale of Skyles, who used to leave
his Scythian train outside, on the outskirts of Olbia, when coming to the city.
Once he crossed the city boundary, he took on Greek manners: Greek dress, language and cult. In this sense, the urban zone seems to be closed to outsiders, i.e.
the steppe population, who live in a Scythian, non-Greek way. In Herodotos' perception, the cultural boundaries seem to be expressed in the division between an
urban and a steppe community, which were the boundaries between the Greek and
non-Greek environments, that is, between Greek and non-Greek cultural identity.
The fact that Greek seasonal and stationary settlements were also established in
the steppe territory may have prompted the development of the concept of crossing the urban and steppe zones.[123]

The popularity of the use of lead *bucrania* in a funerary context, especially
during the Hellenistic period, can perhaps be perceived as the continuation of a

119 Guldager Bilde 2008. For a parallel from Athens, see Seaford 1994, 235–280.
120 Hinge 2008, 383–388.
121 See Hdt. 4.76f. on Anacharsis, with Corcella 2007, 636–638; Hdt. 4.78–80 on Skyles, with
    Corcella 2007, 638–40. On Anacharsis and Skyles, see also chapters III (Ph. Harland), IV (M.
    Oller Guzmán) and V (A.V. Podossinov) in this volume.
122 Hinge 2008, 383–388. For the ambivalent attitude to pasturage and its association with wild-
    ness and hinterland in Greek tradition, see McInerney 2006, 43–50. On the cult of Dionysos,
    see also chapter IV by M. Oller Guzmán in this volume.
123 10 settlements have been discovered in the area of Adzhigol'skaya Balka (Fig. 2, no. 5),
    which date to the archaic period (Kryzhtskiï et al. 1990, 23f.).

similar chthonic Dionysian cult. Such popular use of *bucrania* does not have analogies in other parts of the Black Sea and appears to be a local Olbian tradition.[124] *Bucrania* bear a representation of a bull head, associated with the cult of Dionysos.[125] Also, lead double axes (*labryes*) that sometimes occur together with *bucrania* can be connected to the same cultic practices. *Bucrania* and double axes have been found in large quantities in the Olbian necropolis as well as in the area of the Olbian temenos and at rural settlements, such as Berezan, Petukhovka, Chertovatoe 7 and Didova Khata (see Fig. 2, nos. 8 and 15).[126] They date to the time from the 4th to 2nd centuries BC,[127] which encompasses the period of the economic and political crisis in the northern Black Sea region in the second half of the 3rd century BC that was caused by multiple factors such as climate change, monetary crisis and movements of peoples in the steppe region. The effect of the crisis was a gradual decline of the city and a considerable reduction of the Olbian *chora*. The popularity of the use of *bucrania* may have been connected with the unstable economic and political situation in the North Pontic region during that period, which undoubtedly affected all spheres of social life. The development of a collective consciousness of anxiety and instability at that time probably prompted people to involve themselves again in mystic cults and religious-philosophical ideas of resurrection and the endless cycle of life and death,[128] ideas that were in vogue during the first phase of the development of the Olbian settlement.

VII. CONNECTING WITH THE PAST

The cult of Achilles and Apollo gained new importance in Olbia during the Roman period. As mentioned above, the city was destroyed by the Getai and Dacians in the middle of the 1st century BC. As a consequence, it was abandoned for several decades. Scholars assume, based on prosopographic evidence, that some citizens took refuge in other Greek cities as well as in local rural settlements.[129] Olbia started to be resettled and rebuild at the end of the 1st century BC.[130] One of the most significant changes during that time was the increase of a Sarmatian component in the city's community. It is assumed that the Sarmatian group was granted

---

124 Three *bucrania* dated to 360–320 BC have been discovered at the necropolis of Apollonia Pontike and they represent different stylistic features. The use of *bucrania* did not gain much popularity in Apollonia; see Zaïtseva 2004, 370.

125 For the connection between *bucrania* and the cult of Dionysos in its Orphic context, see Rusyaeva 1978; 1979, 88–90; Zaïtseva 1971.

126 Zaïtseva's catalogue consists of 280 objects (Zaïtseva 2004). Also, a considerable number of lead *bucrania* and *labryes* that are constantly being excavated remain unpublished.

127 Zaïtseva 2004; Kryzhtskiï et. al. 1989, 142; Rusyaeva 1979, 87f.

128 Kuzina 2007.

129 Krapivina 2007a, 161.

130 A newly published Olbian dedicatory inscription suggests that Olbia may have been rebuilt with the support of Rome; Ivantchik 2017b.

Olbian citizenship following a treaty between Olbia and the Sarmatian king Pharzoios (or his heir Inismeus), which is supported by the coinage of the Sarmatian king struck in Olbia about AD 50–80.[131] The great number of non-Greek, mostly Iranian names that occur at this time is attested in the dedicatory inscriptions for Achilles Pontarches and Apollo Prostates. These two cults were mainly associated with the most important political institutions and magistracies in Olbia.[132] This demonstrates that a certain number of citizens who preferred to follow a non-Greek onomastic tradition belonged to the city's elite. Since both epithets, Pontarches and Prostates, have not been attested before the Getic attack, the establishment of these cults appears to be a reaction to the new situation in the city. It appears probable that participation in the urban cults of Achilles Pontarches and Apollo Prostrates, which were rooted in the older Greek religious traditions of the region, strengthened the sense of cohesion and collective identity in the city through common ritual and a symbolic link between the past and the present.

The temple of Achilles Prostates is mentioned several times in the epigraphic record.[133] However, a considerable amount of the steles dedicated to him have been found outside the city, in such places as Beïkush and Berezan, which had been associated with the cult of Achilles for many centuries. Such a location for public dedicatory inscriptions may have had a symbolic meaning: they could mark the boundary of the Olbian territory, which seemed especially important during the revival of the Olbian *chora* at that time. The bond with older places of cult may have also carried an additional apotropaic function of the steles. It is possible to assume that the stele found at Sofievka marked the northern margins of the *chora* (Fig. 2, no. 1).[134]

The attachment to old traditions during the Roman period can be observed through the reoccurrence of the archaic tradition of child burials in pits, amphoras and other vessels under the floors of houses at rural settlements. The tradition of the so-called *enchytrismoi* is well attested in the archaic period, but is virtually absent in the Olbian necropolis of the time. It has been assumed that this funeral custom was most probably connected with a recurrence of the archaic cult of the dead. There have been 55 *enchytrismoi* found in the territory of the *chora*; this includes 46 burials at Kozÿrka, 3 burials in Petukhovka 2, and 6 burials in Zolotoï Mÿs (see Fig. 2).[135] The same burial tradition also occurs during the Roman period in the city and the *chora* of Chersonesos (Fig. 1). Several *enchytrismoi* have been discovered at the Chersonesitan necropolis: one burial dates to the 1st century AD, whereas the rest comes from a later period dating to the 5th–6th centuries AD.[136] The necropolis of Sovkhoz 10 (Sevastopol'skiï) (1st–5th centuries AD) is

131 Tokhtas'ev 2013. For the coins, see Krapivina 2007a, 163, fig. 23.
132 Hupe 2006b.
133 Krapivina 2007b, 592: *IOSPE* I² 80–115.
134 Rusyaeva 1979, 138.
135 Kryzhitskiï et al. 1989, 214f.; Burakov 1976, 138.
136 Zubar' 1982, 50f.

especially important, since two very archaic elements of funerary rites have been recorded there, namely children buried in amphoras in a contracted position.[137] Contemporary analogies can also be found in Tanaïs (Fig. 1).[138]

VIII. CONCLUSIONS

The material discussed above demonstrates that the traditional ethnic-oriented methodological approach, which is deeply rooted in culture-historical archaeology, has its serious limitations in the study of ancient societies. According to such an approach, material culture is assumed to provide clear ethnic markers that are used to demonstrate either the 'Greekness' of Olbia and its *chora* (assuming that they are perceived as culturally homogeneous), or a more 'mixed', Greek-barbarian character of the settlement. Instead, a distinction between ethnicity and cultural identity should be drawn, which allows for the analysis of cross-cultural objects and traditions without unnecessary attempts to attribute them to a given *ethnos*. The examination of cult and funerary practices in Olbia and its rural territory has demonstrated how Olbian society functioned in organised communities despite the potentially different ethnic, cultural and social backgrounds of the people living in its urban centre and the countryside, and how they created a common sense of identity based on shared experience, cultic rituals and beliefs, and a political and administrative order that was applied by the political elites living in the *asty*. Even though the cultural impact of Olbia on its *chora* is visible in the way in which rural communities followed trends that were characteristic of the urban area, it is important to notice that the *chora* was able to create its own local identity. This is apparent in such aspects as specific cults, local traditions, local history and conscious remembrance of the past.

The self-definition of the inhabitants of Olbia was created by a number of factors that changed over time. The common experience of migration, local religious and funeral practices developed by the settlers, the attachment to old traditions as well as close cultural and economic ties between the city, its *chora* and the steppe world were crucial aspects in the process of creating a local identity. Throughout the whole history of Olbia, the prosperity and condition of the city was reflected in the *chora*, which highlights a strong relationship between these two settlement zones. Rural sanctuaries that developed simultaneously with the expansion of the

---

137 Vÿsotskaya 2001/2, 271–273.
138 Shelov 1961, 88: 6 burials in amphoras plus 5 burials in which a child was partially covered by a sherd of an amphora (1st–4th centuries AD). Arsen'eva 1977, 115, 118 records 3 burials in amphoras: no. 165 (1st century AD), no. 144 and no. 185 (4th century AD); 2 burials of children covered with sherds: no. 191 (2nd century BC) and no. 194 (2nd/1st centuries BC); a child partially covered by a sherd: no. 198 (1st century AD)). During the excavations between 1981 and 1995, one child burial in an amphora was discovered: no. 9/1992/XVII, (second half of the 1st to the beginning of the 2nd century AD); in burial no. 35/1982, the child was covered with a sherd (2nd century AD); see Arsen'eva et al. 2001, 202.

rural area played an essential role in strengthening the link with the territory; they also served as important landmarks for both rural and urban communities. Similarly, zol'niks which were probably rural cult centres appear to have demarcated the territory of the Olbian *chora*. In the same way as kurgans, rural sanctuaries and zol'niks created a local cultural landscape and helped to shape and maintain collective memory. It seems reasonable to assume that Olbian kurgans and zol'niks were recognisable cross-culturally. This was particularly important in a 'colonial' milieu of the northern Black Sea region, in which cultural identity was going through a constant process of negotiation, reshaping and (re)invention.

The material discussed in this chapter has also demonstrated the importance of specific cults. The Orphic-Dionysiac cults popular in the first period of the existence of Greek *apoikia* seem to have helped settlers overcome the stressful process of migration. They also may have served to create a conceptual juxtaposition between urban and steppe communities through the imaginary border demarcated by the city's territory, which strengthened the sense of common identity based on a common lifestyle. Such a juxtaposition might have been especially prominent, when Olbia prospered the most as a political, administrative and economic centre, which resulted in the mass migration from rural to urban areas. During the Hellenistic period, the popularity of the mystic Orphic-Dionysiac cults is again visible and is likely to have been a reaction to the socio-economic crisis in the region and the growing anxiety connected to it. Again, the migration of the rural population to the city during that time is visible in the considerable shrinking of the *chora*.

In the post-Getic period, when a strong need for the preservation of cultural coherence is evident following the rebuilding of the city, a sense of collective identity based on older traditions appears to have played a crucial role. An example of this is the introduction of the cults of Achilles and Apollo with new epithets and new civic aspects. The re-introduction of the old burial custom of *enchytrismoi* around that time might have been another means of connecting with the past, which was caused by a strong need to preserve local Greek identity.

Acknowledgements

This research was supported by the National Science Centre, Poland, by a grant awarded to the research project 'The process of creating cultural identity in the North Pontic region in antiquity – the Greek polis and rural territories'. Project No. UMO-2015/17/N/HS3/02855.

Bibliography

Alekseev, A.Yu. 2003: *Khronografiya evropejskoï Skifii* (The Chronography of Europaean Scythia), Saint Petersburg.

Antonaccio, C.M. 2010: 'Redefining Ethnicity: Culture, Material Culture, and Identity', in S. Hales & T. Hodos (eds.), *Material Culture and Social Identity in the Ancient World*, Cambridge, 32–53.

Antonaccio, C. 2003: 'Hybridity and the Culture within Greek Culture', in C. Dougherty & L. Kurke (eds.), *The Cultures within Ancient Greek Culture: Contact, Conflict, Collaboration*, Cambridge, 57–74.

Antonaccio, C.M. 2013: Networking the Middle Ground? The Greek Diaspora, Tenth to Fifth Century BC, in W.P. van Pelt (ed.), *Archaeology and Cultural Mixture, Archaeological Review from Cambridge* 28.1, 241–255.

Arsen'eva, T.M. 1977: *Nekropol' Tanaisa* (The Necropolis of Tanaïs), Moscow.

Arsen'eva, T.M., Bezuglov, S.I. & Tolochko, I.V. 2001: *Nekropol' Tanaisa: raskopki 1981–1995 gg.* (The Necropolis of Tanaïs: Excavations in 1981–1995), Moscow.

Barth, F. 1969: *Ethnic Groups and Boundaries. The Social Organisation of Culture Difference*, Bergen.

Bhabha, H.K. 1994: *The Location of Culture*, London.

Bodzek, J. 2008: 'Koshary (Ukraine) – Coin finds in 2004–2005', in E. Papuci-Władyka (ed.), *Pontika 2006: Recent Research in Northern Black Sea Coast Greek Colonies*, Kraków, 12–23.

Bodzek, J. 2011: 'Remarks on Monetary Circulation in the Chora of Olbia Pontica – The Case of Koshary', in N. Holmes (ed.), *Proceedings of the XIVth International Numismatic Congress Glasgow 2009*, Glasgow, 58–64.

Braund, D. 2007: 'Greater Olbia: Ethnic, Religious, Economic, and Political Interactions in the Region of Olbia, c. 600–100 BC', in D. Braund & S.D. Kryzhitskiy (eds.), *Classical Olbia and the Scythian World*, Oxford, 37–77.

Bresson, A., Ivantchik, A. & Ferrary, J.-L. (eds.) 2007: *Une koinè pontique: cités grecques, sociétés indigènes et empires mondiaux sur le littoral nord de la mer Noire (VIIe s. a.C.–IIIe s. p.C.)*, Bordeaux.

Buĭskikh, A.V. 2015: 'Yuzhnyĭ temenos Ol'vii Pontiĭskoĭ (predvaritel'nŷe itogi izucheniya)' ('The Southern Temenos of Olbia Pontica (Preliminary Results)'), *VDI* 2, 6–21.

Buĭskikh, S.B. 1986: 'Nekotorŷe voprosŷ prostranstvenno-strukturnogo razvitiya Ol'viĭskoĭ khory' (Some Questions Concerning the Spatial and Structural Development of the Olbian Chora), in A.S. Rusyaeva (ed.), *Ol'viya i ee okruga. Sbornik nauchnykh trudov* (Olbia and Its Countryside. Collection of Scientific Papers), Kiev, 17–28.

Buĭskikh, S.B. 1993: 'Eskharŷ Glubokoĭ Pristani' (Escharae at Glubokaya Pristan'), *Drevnee Prichernomor'e: Kratkie Soobshcheniya Odesskogo Arkheologicheskogo Obshchestva*, 80–82.

Buĭs'kikh, S.B. 2004: 'Svyatilishcha *extra-urban* epokhi grets'koi kolonizatsii Nizhn'ogo Pobuzhzha' ('*Extra-Urban* Sanctuaries of the Greek Colonization Period in the Lower Bug River Region'), *Arkheologiya (Kiev)* 3, 3–14.

Bujskich, S.B. 2006: 'Die *chora* des pontischen Olbia: die Hauptetappen der räumlich-strukturellen Entwicklung', in P. Guldager Bilde & S.F. Stolba (eds.), *Surveying the Greek Chora. Black Sea Region in a Comparative Perspective*, Aarhus, 115–139.

Bujskikh, S. 2007: 'Der Achilleus-Kult und die griechische Kolonisation des unteren Bug-Gebietes', in Bresson, Ivantchik & Ferrary 2007, 201–212.

Buĭskikh, S.B. & Buĭskikh A.V. 2010: 'K khronologii arkhaicheskikh poseleniĭ khory Ol'vii Pontiĭskoĭ' ('To the Chronology of Archaic Settlements of Olbia of Pontos Chora'), *Bosporskie issledovaniya* 24, 3–64.

Bujskich, S.B. & Bujskich A.V. 2013: 'Zur Chronologie der archaischen Siedlungen in der Chora von Olbia Pontica', *Eurasia Antiqua* 19, 1–34.

Burakov, A.V. 1976: 'Kozŷrskoe gorodishche rubezha i pervŷkh vekov n.é.' (The Hillfort of Kozŷrka at the Turn of the Millennium and during the First Few Centuries AD), Kiev.

Bylkova, V.P. 1996: 'Excavations on the Eastern Boundary of the Chora of Olbia Pontica', *Échos du Monde Classique* 15, 99–118.

Bylkova, V.P. 2005: 'The Chronology of Settlements in the Lower Dnieper Region (400–100 BC)', in V.F. Stolba & L. Hannestad (eds.), *Chronologies of the Black Sea Area in the Period C. 400–100 BC*, Aarhus, 217–247.

Bylkova, V.P. 2011: 'Belozerskoe as a Settlement in a Greek-Barbarian *Contact Zone*', in Papuci-Władyka et al. 2011, 47–56.

Chochorowski, J. 2008: 'Social Aspects of Sacred Spatial Organization of Koshary Necropolis', in E. Papuci-Władyka (ed.), *Pontika 2006: Recent Research in the Northern Black Sea Coast Greek Colonies. Proceedings of the International Conference, Kraków, 18th March, 2006*, Kraków, 25–45.

Cohen, A. 1996: 'Ethnicity and Politics', in J. Hutchinson & A.D. Smith (eds.), *Ethnicity*, Oxford, 83–84.

Cojocaru, V. 2004: *Populaţia zonei nordice şi nord-vestice a Pontului Euxin în secolele VI–I a.Chr. pe baza izvoarelor epigrafice* (The Population of the Northern and North-Western Black Sea Region in the 6th–1st Century BC Based on Epigraphic Sources), Iaşi.

Cojocaru, V. 2008: 'Zum Verhältnis zwischen Steppenbevölkerung und griechischen Städten: Das *skythische Protektorat* als offene Frage', *Tyche* 23, 1–20.

Corcella, A. 2007: 'Commentary on Book IV', in D. Asheri, A.B. Lloyd & A. Corcella (eds.), *A Commentary on Herodotus: Books I–IV*, ed. by O. Murray & A. Moreno, with a contribution by M. Brosius, Oxford, 543–721.

Crielaard, J.P. 2003: 'The Cultural Biography of Material Goods in Homer's Epics', *Gaia: revue interdisciplinaire sur la Grèce Archaïque* 7, 49–62.

Dana, M. & Dana, D. 2007: 'Histoires locales dans le Pont-Euxin ouest et nord. Identité grecque et construction du passé', *Il Mar Nero* 5 (2001/3), 91–111.

Dolukhanov, P. 1995: 'Archaeology in Russia and Its Impact on Archaeological Theory', in P. Ucko (ed.), *Theory in World Archaeology: a World Perspective*, London, 342–372.

Dragadze, T. 1980: 'The Place of *Ethnos* Theory in Soviet Anthropology', in E. Geller (ed.), *Soviet and Western Anthropology*, London, 161–170.

Fless, F. & Lorenz, A. 2005a: 'Griechen, Skythen, Bosporaner? Zu den Problemen ethnischer Etikettierungen von Gräbern in den Nekropolen Pantikapaions', *Eurasia Antiqua* 11, 57–77.

Fless, F. & Lorenz, A. 2005b: 'Die Nekropolen Pantikapaions im 4. Jh v.Chr.', in F. Fless & M. Treister (eds.), *Bilder und Objekte als Träger kultureller Identität und interkultureller Kommunikation im Schwarzmeergebiet (Kolloquium in Zschorau/Sachsen vom 13.2–15.2.2003) (Internationale Archäologie. Arbeitsgemeinschaft, Symposium, Tagung, Kongress 6)*, Rahden, 17–26.

Foxhall, L., Gehrke, H.-J. & Luraghi, N. (eds.) 2010: *Intentional History: Spinning Time in Ancient Greece*, Stuttgart.

Gerasim, P.V. 2013: 'Greko-varvarskie étnokul'turnÿe protsessÿ perioda arkhaiki v Nizhnem Pobuzh'e: istoriografiya problemÿ' ('Greek and Barbarian Ethno-Cultural Processes in the Lower Buh River Region during the Antique Times: Historiography Issue'), *Teoriya i praktika obshchestvennogo razvitiya* 11, 303–306.

Gosden, Ch. 2004: *Archaeology and Colonialism. Cultural Contact from 5000 BC to the Present*, Cambridge.

Grechko, D. S. 2010: 'Kurganÿ kontsa VI – seredinÿ V vv. do n.é. Nizhnego Pobuzh'a: greki ili skifÿ?' ('The Barrows of the Lower Bug Littoral Zone of the End of VI – the Middle of V Century BC: the Greeks or the Scythians?'), *Drevnosti* 9, 116–135.

Guldager Bilde, P. 2008: 'Some Reflections on Eschatological Currents, Diasporic Experience, and Group Identity in the Northwestern Black Sea Region', in Guldager Bilde & Petersen 2008, 29–45.

Guldager Bilde, P. & Petersen, J.H. (eds.) 2008: *Meetings of Cultures in the Black Sea Region: Between Conflict and Co-Existence*, Aarhus.

Hall, S. 1996: 'Introduction: Who Needs *Identity*?', in S. Hall & P. du Gay (eds.), *Questions of Cultural Identity*, London, 1–17.

Hannestad, L., Stolba, V.F. & Shcheglov, A.N. (eds.) 2002: *Panskoye 1*. Vol. 1: *The Monumental Building U6*, Aarhus.

Harding, D.W. 2007: *The Archaeology of Celtic Art*, London.

Hinge, G. 2008: 'Dionysos and Herakles in Scythia – The Eschatological String of Herodotos' Book 4', in Guldager Bilde & Petersen 2008, 369–397.

Hupe, J. (ed.) 2006a: *Der Achilleus-Kult im nördlischen Schwarzmeerraum vom Beginn der griechischen Kolonisation bis in die römische Kaiserzeit. Beiträge zur Akkulturationsforschung*, Rahden.

Hupe, J. 2006b: 'Die olbische Achilleus-Verehrung in der römischen Kaiserzeit', in Hupe 2006a, 165–234.

Hupe, J. 2007: 'Aspekte des Achilleus Pontarches-Kultes in Olbia', in Bresson, Ivantchik & Ferrary 2007, 213–223.

Hurst, H. & Owen, S. (eds.) 2005: *Ancient Colonizations: Analogy, Similarity & Difference*, London.

Hutchinson, J. & Smith, A.D. (eds.) 1996: *Ethnicity*, Oxford.

Ivantchik, A. 2017a: 'The Greeks and the Black Sea: The Earliest Ideas about the Region and the Beginning of Colonization', in V. Kozlovskaya (ed.), *The Northern Black Sea in Antiquity: Networks, Connectivity, and Cultural Interaction*, Cambridge, 7–25.

Ivantchik, A. 2017b: 'A New Dedication from Olbia and the Problems of City Organization and of Greco-Barbarian Relations in the 1st century AD', *ACSS* 23, 189–209.

Jones, S. 1997: *The Archaeology of Ethnicity: Constructing Identities in the Past and Present*, London.

Krapivina, V.V. 2007a: 'Olbia and the Barbarians from the First to the Fourth Century AD', in D. Braund & S.D. Kryzhitskiy (eds), *Classical Olbia and the Scythian World*, Oxford, 161–172.

Krapivina, V.V. 2007b: 'Olbia Pontica in the 3rd–4th Centuries AD', in D.V. Grammenos & E.K. Petropoulos (eds.), *Ancient Greek Colonies in the Black Sea 2*, vol. 1, Oxford, 591–626.

Kryzhitskii, S.D. 2000: 'The Chora of Olbia Pontica: The Main Problems', *Échos du Monde Classique* 44.19, 167–178.

Kryžickij, S.D. 2006: 'The Rural Environs of Olbia: Some Problems of Current Importance', in P. Guldager Bilde & V.F. Stolba (eds), *Surveying the Greek Chora. Black Sea Region in a Comparative Perspective*, Aarhus, 99–114.

Krÿzhitskiï, S.D. (ed.) 1987: *Kul'tura naseleniya Ol'vii i ee okrugi v arkhaicheskoe vremya* (The Culture of the Population of Olbia and its Rural Area in the Archaic Period), Kiev.

Kryzhytskyy, S.D. & Krapivina, V.V. 2003: 'Olbian Chora', in D.V. Grammenos & E.K. Petropoulos (eds.), *Ancient Greek Colonies in the Black Sea*, vol. 1, Thessaloniki, 507–561.

Krÿzhitskiï, S.D., Buïskikh, S.B., Burakov, A.V. & Otreshko, V.M. (eds.) 1989: *Sel'skaya okruga Ol'vii* (The Rural Area of Olbia), Kiev.

Krÿzhitskiï, S.D., Buïskikh, S.B. & Otreshko, V.M. (eds.) 1990: *Antichnÿe poseleniya nizhnego Pobuzh'ya (arkheologicheskaya karta)* (Ancient Settlements of the Lower Bug Region (An Archaeological Map)), Kiev.

Kuzina, N.V. 2007: 'Dionisiïskaya simvolika v pogrebal'noï obryadnosti naselenya Severnogo Prichernomor'ya' ('Dionysus Symbolism in Burial Rituals of the North Black Sea Coastal Region Ancient Centers Population') *Bosporskie Issledovaniya* 16, 112–129.

Kuznetsov, V.D. 1999: 'Early Types of Greek Dwelling Houses in the North Black Sea', in G.R. Tsetskhladze (ed.), *Ancient Greeks West and East*, Leiden, 531–564.

Lordkipanidze, O. (ed.) 1979: *Problemÿ grecheskoï kolonizatsii severnogo i vostochnogo Prichernomor'ya. Materialy I Vsesoyuznogo simpoziuma po drevneï istorii Prichernomor'ya, Tskhaltubo – 1977* (Problems of Greek Colonisation of the Northern and Eastern Black Sea Region. Materials from the 1st All-Union Symposium on the Ancient History of the Northern Balck Sea Region, Tskhaltubo – 1977), Tbilisi.

Lordkipanidze, O. (ed.) 1981: *Demograficheskaya situatsiya v Prichernomor'e v period velikoï grecheskoï kolonizatsii. Materialy II Vsesoyuznogo simpoziuma po drevneï istorii Prichernomor'ya, Tskhaltubo – 1979* ('The Demographic Situation in the Black Sea Littoral in the Period of the Great Greek Colonization. Materials of the 2nd All-Union Symposium of the Ancient History of the Black Sea Littoral, Tskhaltubo – 1979'), Tbilisi.

Malkin, I. 2003: 'Networks and the Emergence of Greek Identity', *Mediterranean Historical Review* 18.2, 56–74.

Malkin, I. 2009: 'Foundations', in K.A. Raaflaub & H. van Wees (eds.), *A Companion to Archaic Greece*, Chichester, 373–394.

Marchenko, K.K. 1988: *Varvarÿ v sostave naseleniya Berezani i Ol'vii vo vtoroï polovine VII – pervoï polovine I v.do n.é. Po materialam lepnoï keramiki* (Barbarians in the Composition of the Population of Berezan and Olbia in the Second Half of the 7th – the First Half of the 1st Century BC), Leningrad.

Marchenko, K.K. 2013: 'Khora Ol'vii' (The Chora of Olbia), *Stratum Plus* 3, 17–130.

McInerney, J. 2006: 'On the Border: Sacred Land and the Margins of the Community', in R.M. Rosen & I. Sluiter (eds.), *City, Countryside, and the Spatial Organization of Value in Classical Antiquity*, Leiden, 33–59.

Mordvintseva, V. 2013: 'The Sarmatians: The Creation of Archaeological Evidence', *Oxford Journal of Archaeology* 32.2, 203–219.

Nosova, L.V. 2002: 'O kul'tovÿkh zol'nikakh antichnÿkh poseleniï Severo-Zapadnogo Prichernomor'ya (v svyazi s razkopkami Kosharskogo arkhelologicheskogo kompleksa)' (On the Cultic Zol'niks from Ancient Settlements in the South-West Black Sea Region (Regarding Excavations of the Archaeological Complex at Koshary)), in M.Yu. Vakhtina, V.Yu. Zuev, S.V. Kashaev & V.A Khrshanovskiï (eds.), *Bosporskiï fenomen: pogrebal'nye pamyatniki i svyatilishcha*, vol. 2, Saint Petersburg, 62–68.

Nosova, L.V. 2008: 'Antichnÿe zol'niki Severnogo Prichernomor'ya – zhertvenniki-eskhary ili musornÿe svalki?' (Ancient Zol'niks of the Northern Black Sea Region – Altars-Escharae or Waste Dumps?), *Kratkie soobshcheniya odesskogo arkheologicheskogo obshchestva* (Brief Reports of the Odessa Archaeological Society), 114–136.

Novikova, O.V. 2012: 'Problema vzaimootnosheniï grekov i varvarov Zapadnogo Krÿma v dorimskuyu épokhu v sovetskoï istoriografii poslevoennogo perioda (do kontsa 50-kh gg. XX v.)' (The Problem of the Relationship between the Greeks and Barbarians in West Crimea before the Roman Period in the Soviet Historiography of the Post-War Period (until the End of the 50s of the 20th Century), in A.R. Kemalova (ed.), *Istoriko-kul'turnoe nasledie Tarkhankuta: Novÿe tendentsii razvitiya, novÿe vozmozhnosti* (Historical-Cultural Heritage of Tarkhankut: New Tendencies for Development, New Possibilities), Simferopol, 112–121.

Ober, J. 2003: 'Postscript: Culture, Thin Coherence, and the Persistence of Politics', in C. Dougherty, L. Kurke (eds.), *The Cultures within Ancient Greek Culture: Contact, Conflict, Collaboration*, Cambridge, 237–255.

Okhotnikov, S.B. & Ostroverkhov, A.S. 1993: *Svyatilishche Achilla na ostrove Levke (Zmeïnom)* (The Sanctuary of Achilles on the Island of Leuke (Snake Island)), Kiev.

Papanova, V. 2006: *Urochishche Sto Mogil (Nekropol' Ol'vii Pontiïskoï)* (The Place of Hundred Burials (The Necropolis of Olbia Pontike)), Kiev.

Papuci-Władyka, E., Redina, E.F., Bodzek, J. & Machowski, W. 2009: 'The Koshary Project (Ukraine, Odessa Province), Seasons 2004–2006', *Recherches Archéologiques* 1, 487–507.

Papuci-Władyka, E., Redina, E.F., Bodzek, J. & Machowski, W. 2010: 'Polish-Ukrainian Koshary Project, Seasons 2007–2008', *Recherches Archéologiques* 2, 257–275.

Papuci-Władyka, E. & Redina, E.F. 2011: 'Ten Years of the Polish-Ukrainian Koshary Project (1998–2008)', in Papuci-Władyka et al. 2011, 283–297.

Papuci-Władyka, E., Vickers, M., Bodzek, J. & Braund, D. (eds.) 2011: *Pontika 2008: Recent Research on the Northern and Eastern Black Sea in Ancient Times. Proceedings of the International Conference, 21st–26th April 2008, Kraków*, Oxford.

Parovich-Peshikan, M. 1974: *Nekropol' Ol'vii éllinisticheskogo vremeni* (The Olbian Necropolis of the Hellenistic Period), Kiev.

Patterson, T.C. 2003: *Marx's Ghost: Conversations with Archaeologists*, Oxford.

Petersen, J.H. 2010: *Cultural Interactions and Social Strategies on the Pontic Shores. Burial Customs in the Northern Black Sea Area c. 550–270 BC*, Aarhus.

Polinskaya, I. 2006: 'Lack of Boundaries, Absence of Oppositions: The City-Countryside Continuum of a Greek Pantheon', in R.M. Rosen, I. Sluiter (eds), *City, Countryside, and the Spatial Organization of Value in Classical Antiquity*, Leiden, 61–92.

Popova, E.A. & Kovalenko, S.A. 1996: 'On the Cult of Herakles in the North-Western Crimea: Recent Finds from the Chaika Settlement', in G.R. Tsetskhladze (ed.), *New Studies on the Black Sea Littoral*, Oxford, 63–71.

Redina, E.F. 2007: 'Antichnaya pogrebal'naya traditsiya v formirovanii sakral'nogo prostranstva grecheskogo nekropol'ya IV–III vv do r. Chr. u s. Koshary' (Ancient Burial Tradition in the Greek Spatial Development of the Necropolis of the 4th–3rd century BC Near the Village of Koshary), in P. Guldager Bilde, M.Yu. Vakhtina, V.Yu. Zuev, S.V. Kashaev, O.Yu. Sokolova & V.A. Khrshanovskiĭ (eds.), *Bosporskiĭ fenomen: sakral'nȳĭ smȳsl regiona, pamyatnikov, nakhodok. Materialȳ Mezhdunarodnoĭ nauchnoĭ konferentsii*, vol. 2, Sankt Petersburg, 95–99.

Rusyaeva, A.S. 1978: 'Orfizm i kul't Dionisa v Ol'vii' (Orphism and the Cult of Dionysos in Olbia), *VDI* 1, 87–104.

Rusyaeva, A.S. 1979: *Zemledel'cheskie kul'ty v Ol'vii grecheskogo vremeni* (Agrarian Cults in Olbia in the Greek Period), Kiev.

Rusyaeva, A.S. 1992: *Religiya i kul'ty antichnoĭ Ol'vii* (Religion and Cults of Archaic Olbia), Kiev.

Rusyaeva, A.S. 2000: 'Kurganȳ Ol'vii kak simvol ee slavȳ i sakral'noĭ okhranȳ' ('Barrows of Olbia as a Symbol of its Glory and Sacral Protection'), *Arkheologicheskie vesti* 7, 106–111.

Rusyaeva, A.S. 2006a: 'Drevnegrecheskie sakral'nȳe zol'niki v Nizhnem Pobuzh'e' ('Ancient Greek Sacral Cinder-Heaps in the Lower Bug Region'), *Rossiĭskaya Arkheologiya* 4, 95–102.

Rusyaeva, A.S. 2006b: 'Svyatilishche Akhilla na Tendre v kontekste istorii i religii Ol'vii Pontiĭskoĭ' ('The Sanctuary of Achilleus on the Tendra in the Context of History and Religion of Olbia Pontica'), *VDI* 4, 98–123.

Rusyaeva, A.S, Krȳzhitskiĭ, S.D, Krapivina, V.V. et al. 2006: 'Drevneĭshiĭ temenos Ol'vii Pontiĭskoĭ' (The Oldest Temenos of Olbia Pontike), *MAIET Supplementum* 2.

Seaford, R. 1994: *Reciprocity and Ritual: Homer and Tragedy in the Developing City-State*, Oxford.

Sewell, W. 1999: 'The Concept(s) of Culture', in V.E. Bonnell & L. Hunt (eds.), *Beyond the Cultural Turn: New Directions in the Study of Society and Culture*, Berkeley, 35–61.

Shelov, D.B. 1961: *Nekropol' Tanaisa (raskopki 1955–1958 gg.)* (The Necropolis of Tanaïs (Excavations in 1955–1958)), Moscow.

Shennan, S.J. 1991: 'Some Current Issues in the Archaeological Identification for Past Peoples', *Archaeologia Polona* 29, 29–37.

Skundova, V.M. 1988: *Archaicheskiĭ nekropol' Ol'vii* (The Archaic Necropolis of Olbia), Leningrad.

Snitko, I.O. 2011: 'Pidkurganni pokhovannya nekropoliv khorȳ Ol'vii VI–V st. do n. e.' ('Burials under the Barrows in the Necropolei at Olbian Chora of the 6th and the 5th Centuries BC'), *Arkheologiya (Kiev)* 1, 11–22.

Snȳtko, I.A. 2013: 'Pogrebal'nye sooruzhenija sel'skikh nekropoleĭ Ol'vii v kontekste drevnegrecheskoĭ ritual'noĭ praktiki naseleniya Nizhnego Pobuzh'ya pozdnearkhaicheskoĭ, klassicheskoĭ i éllinisticheskoĭ épokh (VI–III vv. do n.é.)' ('Funerary Structures at Rural Necropoleis of Olbia in the Context of the Ancient Greek Ritual Practices among the Population of the Lower Bug Region in the Late Archaic, Classical and Hellenistic Periods (6th–3rd Centuries BC)'), *Arkheologicheskie Vesti* 19, 91–104.

Stolba, V.F. 1989: 'Novoe posvyashchenie iz Severo-Zapadnogo Krȳma i aspektȳ kul'ta Gerakla v Khersonesskom gosudarstve' ('A New Dedication from the North-Western Crimea and Aspects of the Cult of Herakles in the Chersonese State'), *VDI* 4, 55–70.

Stolba, V.F. 1996: 'Barbaren in der Prosopographie von Chersonesos', in B. Funck (ed.), *Hellenismus: Beiträge zur Erforschung von Akkulturation und politischer Ordnung in den Staaten des hellenistischen Zeitalters: Akten des Internationalen Hellenismus-Kolloquiums, 9.–14. März 1994 in Berlin*, Tübingen, 439–466.

Stolba, V.F. 2004: 'Guderne i Chersones: Parthenos og Herakles' (The Gods in Chersonesos: Parthenos and Heracles), in J. Højte & P. Guldager Bilde (eds.), *Mennesker og guder ved Sortehavets kyster*, Aarhus, 53–64.

Stolba, V.F. 2011: 'Multicultural Encounters in the Greek Countryside: Evidence from the Panskoye I Necropolis, Western Crimea', in Papuci-Władyka et al. 2011, 329–340.

Stolba, V.F. 2012: 'La vie rurale en Crimée antique: Panskoye et ses environs', in P. Burgunder (ed.), *Études pontiques: histoire, historiographie et sites archéologiques du basin de la mer Noire*, Lausanne, 311–364.

Tokhtas'ev, S.R. 2000: 'Iz onomastiki Severnogo Prichernomor'ya: X–XVII' (From the Onomastics of the Northern Black Sea Region: X–XVII), *Hyperboreus* 6.1, 124–131.

Tokhtas'ev, S.R. 2002: 'Ostrakon s poseleniya Kozÿrka XII ol'viïskoï khorÿ' (An Ostracon from the Settlement of Kozÿrka 12 in the Olbian Chora), *Hyperboreus* 8.1, 72–98.

Tokhtas'ev, S.R. 2013: 'Iranskie imena v nadpisyach Ol'vii I–III vv. n. é' (Iranian Names in the Inscriptions from Olbia of the 1st–3rd Century BC), in S.R. Tokhtas'ev & P. Luria (eds.), *Commentationes Iranicae Vladimiro f. Aaron Livschits nonagenario donum natalicium*, Saint Petersburg, 565–607.

Trigger, B. 1989: *A History of Archaeological Thought*, Cambridge.

Tsetskhladze, G.R. 2004: 'On the Earliest Greek Colonial Architecture in the Pontus', in C.J. Tuplin (ed.), *Pontus and the Outside World. Studies in Black Sea History, Historiography, and Archaeology*, Leiden, 225–278.

Tunkina, I.V. 2007: 'New Data on the Panhellenic Achilles' Sanctuary on the Tendra Spit (Excavation of 1824)', in Bresson, Ivantchik & Ferrary 2007, 225–240.

Vinogradov, Yu.G. 1981: 'Varvary v prosopografii Ol'vii VI–V vv. do n. é.' ('Barbarians in the Prosopography of Olbia of the 6th–5th Centuries BC'), in Lordkipanidze 1981, 131–148.

Vinogradov, Yu.G. 1989: *Politicheskaya istoriya ol'viïskogo polisa VII–I vv. do n. é. Istorikoépigrafícheskoe issledovanie* ('The Political History of Olbia of the 7th–1st Century BC. A Historical-Epigraphic Study'), Moscow.

Vinogradov, Y.G. 1994: 'New Inscriptions on Lead from Olbia', *ACSS* 1.1, 103–111.

Vÿsotskaya, T.N. 2001/2: 'Pogrebal'nÿe sooruzheniya i obryad pogrebenniï mogil'nika "Sovkhoz no. 10" (Sevastopol'skiï)' ('Burial Structures and Funerary Customs of the Necropolis *Sovkhoz 10* (Sevastopol'skiï)'), *Stratum Plus* 4, 270–277.

White, R. 1991: *The Middle Ground. Indians, Empires, and Republics in the Great Lakes Region 1650–1815,* Cambridge.

Zaïtseva, K.I. 1971: 'Ol'viïskie kul'tovÿe svintsovÿe izdeliya' (Olbian Cultic Lead Objects), in G.D. Belov (ed.), *Kul'tura i iskusstvo antichnogo mira* ('The Culture of the Ancient World'), Leningrad, 84–106.

Zaïtseva, K.I. 2004: 'Svintsovÿe izdeliya v vide golovok bÿkov, baranov i sekir iz Ol'vii' (Lead Objects from Olbia in the Form of Bull and Ram Heads, and Axes), *Bosporskie Issledovaniya* 7, 356–390.

Zubar', B.M. 1982: *Nekropol' Khersonesa Tavricheskogo I–IV vv. n.é.* ('The Necropolis of Chersonesos Taurike of the 1st–4th Century AD'), Kiev.

'THE MOST IGNORANT PEOPLES OF ALL'

Ancient Ethnic Hierarchies and Pontic Peoples

Philip A. Harland

Abstract: Attention to ethnic hierarchies in the ancient context can clarify nuances in Greek approaches to other ethnic groups. Alongside ideologies that sought to justify hegemonic rankings of non-Greek peoples, including Scythians and other Pontic peoples, were some limited attempts to attenuate negative evaluations of certain ethnic groups. After providing some background on sociological theories regarding group hierarchies, the ancient mainstream views are outlined. They are best represented by the 'Father of History' Herodotos and the Hippocratic author of *On Airs, Waters, and Places* in the 5th century BC: distance from the centre of the ingroup society is defined as the crucial factor, besides climatic features as hot versus cold and dry versus moist. A century later, the philosopher Aristotle seemed to be agreeing with these criteria on the one hand, but identified 'heart' / 'spirit' and 'intelligence' / 'skill' as the factors that decided on the virtue of citizens and their value as an ethnic. His contemporary, Ephoros put much more emphasis on individual virtue and intelligence rather than allowing geographical factors to determine characteristics. In the 3rd century BC, the geographer Eratosthenes vociferously echoed Ephoros in opposition to the prevailing view. In the largest extant work of ancient geography composed at the end of the 1st century BC, Strabo drew on both traditions generously. While leaning more towards the dichotomic model of Herodotos and the Hippocratic corpus, he failed to present a consistent synthesis.

Абстракт: «*Самые невежественные люди из всех*» – *понтийские народы и древние этнические иерархии*: В этой главе показано, как обращение внимания на этнические иерархии в древнем контексте может прояснить нюансы в отношении греков к другим этническим группам. Наряду с идеологиями, которые пытались обосновать преимущество негреческих народов, в том числе скифов и других понтийских народов, древними авторами были также предприняты ограниченные попытки смягчить негативные оценки определенных этнических групп. После представления научных основ социологических теорий, касающихся иерархии групп, автор текста изложил основные древние взгляды на эту тему. Наиболее верно они были описаны в трудах «отца истории» Геродота, а также в Гиппократовском трактате V века до н.э.: решающим фактором в определении иерархии, помимо климатических признаков, таких как жара – холод и засуха – влажность, является расстояние данной группы людей от центра. Спустя сто лет философ Аристотель с одной стороны согласился с этими критериями, но также идентифицировал «сердце» / «дух» и «ум» / «умение» как решающие факторы, говорящие о достоинстве граждан и их значении как этнической группы. Именно современный Эфор уделял гораздо больше внимания индивидуальному достоинству и интеллекту, не позволяя географическим факторам определять характер человека. За громким протестом против преобладающего мнения последовал географ Эратосфен в III веке до н.э. В крупнейшем из сохранившихся трудов древней географии, созданном в конце I-го века до н.э., Страбон в значительной степени опирался на обе традиции. Склоняясь больше к дихотомической модели Геродота и Гиппократовского корпуса, он не смог представить единозначного синтеза.

I. INTRODUCTION

> The area of the Euxine Sea to which Darius was leading his army is, except for the Scythians, inhabited by the *most ignorant peoples of all* (ἔθνεα ἀμαθέστατα). For we cannot cite the wisdom (σοφίη) of any people there, other than the Scythian people, nor do we know of any man noted for wisdom in the Pontic region other than Anacharsis. The Scythian kinship group (γένος) is *most clever* (σοφώτατα) of all in making the most important discovery we know of concerning human affairs, though I do not admire them in other respects. They have discovered how to prevent any attacker from escaping them and how to make it impossible for anyone to overtake them against their will (*Histories* 4.46.1f.).[1]

Herodotos is notorious for his description of northern peoples around what is now the Black Sea. Among his most extensive characterizations is that concerning those labelled 'Scythians' specifically. Herodotos' comparative approach to Pontic and other peoples, with his superlative comments, would be echoed in subsequent ethnographic writing throughout the Hellenistic and Roman eras. There is a widespread proclivity to rank peoples in relation to one another. Attention to these more specific categorizations may help us to move beyond the frequently reiterated idea that the Greek literary elites tend to construct an ethnic self-understanding primarily in juxtaposition to a generalized inferior 'other', the 'barbarians', as the works of François Hartog, Edith Hall, and Jonathan M. Hall emphasize.[2] An approach to ancient ethnic rivalries and interactions that moves beyond 'the other' may also help us to follow both ancient historians who emphasize nuances in Greek perceptions of other peoples and postcolonial scholars, such as Robert J.C. Young, who call us to abandon the category of 'the other' altogether.[3]

Instead, Herodotos' comment instantiates the concern for more particularity in grading other non-Greek peoples in relation to one another, in this case placing those identified generally as 'Scythians' above all other Pontic peoples with respect to their intelligence as manifest in military skill. More scholarly work remains to be done on how Greeks like Herodotos position supposedly 'lesser' peoples in relation to one another (rather than merely on a somewhat simplistic Greek-barbarian dichotomy) and on what shared legitimizing ideologies often accompanied such categorizations.

Rather than studying Greek ethnography in isolation, this paper places these ancient phenomena within the framework of social scientific theories and findings regarding intergroup conflict and prejudice within societies generally, particularly with respect to the concept of 'ethnic hierarchies'. What may at first glance seem peculiar to ancient ethnographic traditions could in some important respects be another example of commonly attested intergroup phenomena, I would argue. Although considerable scholarly work has been done on Greek or Roman perceptions of other peoples, seldom have the results been studied in terms of intergroup

---

1   Translation Strassler & Purvis 2009, with adaptations.
2   Hartog 1988 [1980]; E. Hall 1989 (on tragedians of the 5th century); J.M. Hall 2002, 172–188 (on a shift from aggregative to oppositional constructions of Hellenicity in the 5th century). On this see also Thomas 2000, 43–45 and Vlassopoulos 2013.
3   Keim 2018 (on this shift among ancient historians); Young 2012, 36–39.

prejudice (affective, negative attitudes or evaluations), stereotypes (external cate-gorizations based, in part, on prejudice), and discrimination (behaviours following from prejudice and stereotypes), on the one hand, and in terms of the ideologies that justify such hierarchies, on the other.[4] While the present contribution aims at understanding some attitudes and stereotypes that accompanied a low position for Pontic peoples in Greek representations of ethnic hierarchies, in another study I explore Greek inscriptional evidence for peoples from the Black Sea area settled in Greek city-states.[5] Such evidence for Pontic diasporas provides an opportunity to consider implications of hierarchies (that are our concern in this chapter) for social interactions between Greeks and Pontic peoples in local settings. Social identity theory and social dominance theory may help to further understand both ideological and social dimensions of the ancient situation.

II. INSIGHTS FROM THE SOCIAL SCIENCES

It is important to begin by briefly clarifying that most social scientific theories that help to frame the present study of ethnic groups and hierarchies owe some-thing to the important theories of Frederik Barth on ethnicity (since 1969) and Henri Tajfel and John C. Turner on social identity and intergroup conflict (since 1979). Barth's seminal anthropological study on ethnic groups replaces primordial notions of ethnicity with a more fluid and situational understanding of how mem-bers of such groups – with beliefs that they have distinctive cultural customs and a shared common ancestry – formulate boundaries and define themselves in relation to other peoples on a situational basis.[6]

Within the discipline of social psychology, Turner and Tajfel's social identity theory is concerned with the value members of social groups (including ethnic groups) attach to membership in a given group.[7] Specifically, the theory is fo-cused on how members seek a positive self-image for the group. They do this by, in part, favouring members of the ingroup and, most importantly here, by repre-senting outgroups in a negative or ambivalent manner.

Drawing on both Barth and Tajfel, Richard Jenkins' study (1994) of how eth-nic groups (as social groups) formulate and maintain a sense of belonging together emphasizes two interrelated factors: internal identifications by members of the group and external categorizations or stereotypes formulated by outsiders. It is the interplay between the self-categorizations of group members and reactions to the viewpoints of those who belong to outgroups that make the process of identifica-tion and the development of self-understanding so dynamic, as Jenkins shows.[8]

4   See Snellman & Ekehammar 2005; Snellman 2007.
5   Harland 2020.
6   Barth 1969. Also see Jenkins 1994.
7   Tajfel & Turner 1979 = Tajfel & Turner 1986; Tajfel 1981 and 1982. Also see the research review by Howard 2000.
8   Jenkins 1994. Cf. Harland 2009. Brubaker helpfully problematizes categories which reify 'identity', proposing a more fruitful set of concepts relating to processes of 'identification', a

It is the latter of the two factors explained by Jenkins – external categorizations or stereotypes – that are so instrumental in understanding socially shared representations within a particular society or community that result in rankings of
specific ethnic groups. These representations are what Louk and Roeland Hagendoorn and their colleagues (since 1989) call an 'ethnic hierarchy', or what we
might also express using the image of an ethnic ladder.[9] As Louk Hagendoorn
explains:

> In a multi-ethnic context, each group will have stereotypes about several outgroups accentuat-
> ing negative differences from the ingroup. Outgroups will be placed further away from or fur-
> ther below the ingroup, the larger and more important these differences are. This means that
> the process of differentiation unavoidably entails a rank-ordering. In this way stereotypes
> generate an ethnic hierarchy.[10]

Replicated studies of ethnic hierarchies in Holland have found that (within various
subgroups, including minorities) a consensual stratification had North Europeans
at the top, East and South Europeans below that, and Africans and Middle Eastern
groups at the bottom.[11] Alexandra Snellman and Bo Ekehammar's study of hierarchies among various groups in Sweden found a relative consistency: subjects
ranked Swedes first, Italians second, and Latin Americans third, with Somalians,
Iranians, and Syrians most often ranked (in that order) further below.[12]

In some respects, the rankings reflect the degree to which members of one
people choose to maintain social distance from members of another. Social distance here pertains to the acceptance or rejection of members of outgroups as marriage partners, neighbours, friends, classmates, or workmates.[13] Contact with
members of ethnic groups that are placed lower on hegemonic hierarchies than
one's own group would be considered undesirable in this way and contact with
those higher would be desirable for status implications.[14] So, such representations
and ideologies have a direct impact on social relations and discrimination.

A particular ethnic ladder may reflect priorities of the upper echelons of a culturally hegemonic or politically powerful group. Yet studies by Hagendoorn,
Snellman, and others show that the process of intergroup interactions sometimes
results in a similar or common hierarchy being taken on by subordinated ethnic
groups, even though such minorities would be placed low in the hegemonic ranking. A result is that 'ethnic groups at the bottom of the ethnic hierarchy are rejected by dominant ethnic groups as well as by other ethnic minorities'.[15] In some
cases, both hegemonic and subordinated groups in a particular society may thus

---

     terminological approach that I also adopt here. See Brubaker 2004, 28–63 = Brubaker and
     Cooper 2000.
9   Hraba, Hagendoorn & Hagendoorn 1989; Hagendoorn 1993 and 1995; Hagendoorn et al.
     1998. Also see Snellman & Ekehammar 2005.
10   Hagendoorn 1993, 36.
11   Hraba, Hagendoorn, and Hagendoorn 1989; Hagendoorn 1993; Hagendoorn et al. 1998.
12   Snellman & Ekehammar 2005.
13   Hagendoorn 1995.
14   Hagendoorn 1995, 205.
15   Hagendoorn 1995, 222.

have highly similar if not consensual rankings of specific ethnic groups or peoples. Simultaneously, certain ethnic groups still struggle with one another for a more favourable position on lower rungs of the ladder. This tendency to adopt and justify current hierarchies (i.e. the status quo) even by members of disadvantaged groups is also a central proposition in system justification theory, as I discuss below.

However, in other cases, minorities may construct their own alternative hierarchies of peoples in a way that benefits their own people's status (i.e. favouring the ingroup). This phenomenon is more in line with social identity theory and social dominance theory in their original expressions. In a study of Judaeans specifically, I have explored how ethnic minorities in the Roman era reflect both approaches.[16] Sometimes cultural minorities in the ancient context adopt or adapt hegemonic hierarchies, on the one hand, and sometimes they formulate alternative hierarchies that challenge those in a socially or culturally dominant position, on the other. Furthermore, a certain member of an ethnic minority group, like Philo or Josephus, may also reflect both approaches, depending on the social or rhetorical situation.

Roughly contemporary with the work of Hagendoorn and his colleagues and focussing on the United States, Jim Sidanius and Felicia Pratto (since 1999) developed a more wide-ranging theory which they designate 'social dominance theory'. This theory draws on social identity theory, realistic group conflict theory, Marxist theory, and other approaches in order to examine processes (at institutional, group, and individual levels) that lead to and maintain societal group-based inequalities, namely what they label 'social hierarchies', a broader concept which envelops racial or ethnic hierarchies.[17] Particularly important for understanding ancient Greek or Roman elite perspectives on other peoples is social dominance theory's attention to the role of commonly shared attitudes and 'legitimizing myths' or ideologies. These contribute to and justify processes of discrimination, thereby maintaining hierarchies that favour dominant groups.[18] There are also attempts to consider the degree to which specific individuals have an investment in commonly shared hierarchies and in ideological justifications for such hierarchies. A high score in such a 'social dominance orientation' indicates a high degree of support for current stratification and, conversely, a low score an orientation that favours more equality and attenuates an existing hierarchy.

Recent contributions in social psychology that focus on some of the blindspots of both social identity theory and social dominance theory may help to provide balance in theoretical perspectives. John T. Jost, Mahzarin R. Banaji, and colleagues (since around 1994) have developed 'system justification theory' to

---

16 Harland 2019.
17 Sidanius & Pratto 1999, with clarifications in Sidanius et al. 2004 and Pratto, Sidanius & Levin 2006. Cf. Snellman and Ekehammar 2005; Snellman 2007.
18 For a brief summary, see Pratto, Sidanius & Levin 2006, 275f. For a more extensive explanation, see Sidanius & Pratto 1999, 103–126. For possible weaknesses of the theory, see, e.g., Turner & Reynolds 2003. Cf. Jost, Banaji & Nosek 2004.

address other dimensions of intergroup interactions.[19] These social psychologists
focus attention on cases of outgroup (rather than ingroup) favouritism where
members of subordinated groups show a motivation to justify a status quo or hierarchy that involves their own domination by more powerful groups.[20] Similarly,
intergroup emotions theory as developed by Diane M. Mackie, Eliot R. Smith, and
colleagues (since 1993) explores a range of possible attitudes from positive to
negative and ambivalent responses in intergroup relations.[21] So while social identity theory and social dominance theory shed light on how ingroup favouritism
leads to stereotypes and intergroup conflict, system justification theory and intergroup emotions theory notice variations in responses to both the ingroup and outgroups. The latter also tend to observe cases where outgroup favouritism on the
part of subordinated groups leads to the perpetuation of existing social arrangements or hierarchies. These are among the social dynamics that we will now witness in a case study of ancient ethnic hierarchies with a focus on the position of
Pontic peoples.

III. ETHNIC HIERARCHIES AND LEGITIMIZING IDEOLOGIES

One very important factor in social dominance theory is the role of commonly
shared legitimizing ideologies aimed at enhancing a particular hierarchical arrangement within a given society.[22] On the other hand are attenuating ideologies
that may work against existing hierarchies. While ideologies which serve to maintain current arrangements are usually held by the dominant group and those which
work against the status quo are sometimes held by subordinated groups, Sidanius
and Pratto emphasize the tendency for both hegemonic and subordinated groups
to maintain largely consensual ideologies and attitudes that enhance rather than
undermine established hierarchies.[23] System justification theory and Hagendoorn's ethnic hierarchies research likewise posit that both dominant and subordinated groups may tend to the former ideologies, those that maintain the status
quo. When held by subordinated groups, these legitimating ideologies are akin to
the Marxist concept of 'false consciousness', as Jost and Banaji clarify.[24]

Here I argue that Greek ethnographic writings provide glimpses into both of
these ideological tendencies but especially elite ideologies that enhance currently
established hierarchies. This can be witnessed in cases where elite authors indicate
theoretical justifications for rankings they presume. These became part of a tradition of explaining the inferiority of particular peoples in specific, largely shared,
though slightly varying, stratified representations. Ethnographic authors like Eph-

19   Jost & Banaji 1994; Jost, Banaji & Nosek 2004.
20   See Harland 2019 for examples involving Philo and Josephus.
21   Smith 1993; Mackie & Smith 2002; Mackie, Smith & Ray 2008; Mackie & Smith 2015.
22   Sidanius & Pratto 1999, 103–126.
23   Sidanius & Pratto 1999, 123–126.
24   Jost & Banaji 1994; Jost 1995.

oros and Eratosthenes, however, seem to offer cases of elites attenuating certain aspects of common categorizations.

Unfortunately, our evidence for the perspectives of subordinated groups themselves is quite limited, since we often lack literary or inscriptional evidence to assess these standpoints. Nonetheless, attenuating ideologies are also clearly evident among one of the few minorities whose perspectives have been preserved in a substantial corpus of ancient writings, namely Judaeans.[25] Such limitations notwithstanding, some indications of the perspectives of other peoples, including Pontic peoples, have survived within elite ethnographic writing by Greeks, as we will soon see.

There are two main sets of principles first evident in the latter half of the 5th century (and repeated thereafter) that sought to enhance or justify dominant hierarchies, explaining why certain peoples should be considered higher or lower in a ranking of peoples, with Greeks (and later Romans) at the top. The first is more in line with realities of ethnocentrism and the second ostensibly offers a 'rational' basis for such categorizations. Nonetheless, the second dovetails closely with the first and, it seems, primary factor. It is important to briefly set out these two interrelated ideologies before turning to a more in-depth look at sources concerning the position of northern peoples in the hierarchies these ideologies sought to bolster.

On the one hand, there is the *distance-from-centre justification* of ethnic rankings which is already clearly evident in and known to Herodotos, who composed his work in the second half of the 5th century BC (likely between 450 and 420).[26] This factor would, quite readily, give a low position to peoples at the reaches of the known world, including inhabitants north of the Black Sea. The further away an ethnic group was from a cultural centre (whether that be Persepolis, Athens, or Rome), the more likely such a group was to be placed lower on the scale in a particular ethnic hierarchy. Herodotos himself, who employs the language of honour, shows an awareness of this concept when outlining the customs of the Persians (1.131–140). However, he attributes such a stratifying approach to the Persians (who would therefore place Greeks and others low in their own ethnic hierarchies) without necessarily recognizing how this principle informed Greek assessments of non-Greek peoples, including his own assessments:

> [The Persians] honour (τιμῶσι) most highly those who live closest to them, next those who are next closest, and so on, assigning honour by this reasoning. Those who live farthest away they consider least honorable of all. For they think that they are the best of all people in every respect and that others rightly cling to some virtue (ἀρετή) until those who live farthest away are the worst (κακίστους εἶναι). The Medes were under the influence of a similar principle ...[27]

Having noted this tendency to assume that the farther away peoples were, the more barbaric or inferior they would be, it is important to observe cases where

---

25 See Harland 2019.
26 On this, also see Isaac 2004, although I do not subscribe to his particular view of the ancient origins of racism.
27 Hdt. 1.134.2. Translation adapted from Godley 1920 (LCL).

this was challenged or inverted by Greek-speaking elites. As the cases of Ktesias of Knidos and Ephoros show, far-off peoples (e.g., Scythians, Egyptians, Ethiopians, Indians) might be considered particularly noble or wise, a reverse ethnocentrism of sorts.[28] This was, in part, a way of critiquing one's own society but it also served to attenuate existing consensually held hierarchies, I would suggest.

A second main factor is that, quite often, Greek ethnographic rankings of peoples came to be informed by ideologies first attested in the 5th century in philosophical and health-related discussions, theories regarding the four elements in nature and the four humours in humans.[29] The theory of the humours is first clearly outlined in the Hippokratic writings on *The Nature of Man* and *On Airs, Waters, and Places*, both likely from the final decades of the 5th century BC and roughly contemporary with Herodotos.[30] Because the attributes of hot, cold, dry, and moist were so fundamental to both the elements and the humours – with earth / black bile thought to be cold and dry, water / phlegm cold and moist, fire / yellow bile hot and dry, and air / blood hot and moist – there was also a close correspondence not only with climatic changes of the seasons but also with movements of the planets or stars.

So, in many respects, this was a thorough-going environmental theory that came to be applied as an explanatory device in a variety of contexts, in this case to bolster Greek categorizations of other ethnic groups. The Greek (or Roman) elites who constructed or modified such theories tended to imagine that their own location was relatively balanced in terms of temperatures and seasons and, therefore, balanced in a combination of the elements or humours. For this reason, humoral theory often dovetails quite closely with the distance-from-centre factor. For, as one went further north, south, east or west of a cultural centre such as Athens (or Rome), increasingly extreme or unchanging climates or environments were imagined to exist. These climates were thought to negatively shape the character of populations living in those conditions, creating inferior populations (cf. Diodoros of Sicily, *Library of History* 3.34). This situation means that, once again, there was a tendency to affirm a low place for peoples north of the Black Sea, as well as other northern peoples (e.g., Gauls, Germans or Britons) as they became known.

Although these propositions or ideologies are presented by ancient authors as explanations for the relative inferiority or superiority of particular peoples, I would argue that, in large part, these were justifications for a pre-existing, widely-shared categorization of peoples as inferior to Greeks. This said, the theorizing itself when adopted by subsequent authors could influence how a particular writer categorized specific peoples within such widely-held hierarchies. Like astrological reasoning, the theory itself left considerable room for variant or even opposite

---

28   Photios' summary of Ktesias' work on India frequently emphasizes that the various peoples of India, including the supposed pygmies and dog-heads, were 'very just'. For Ktesias, there were no humans settled beyond India. See Nichols 2008, 111–116. Cf. Wells 1999, 99–121; Romm 1992, 45–81.

29   Also see Thomas 2000, 47–74 on the humours and ethnography. For a general discussion of the environmental theory, see Isaac 2004, 55–109.

30   Jouanna 2012, 335–360.

interpretations of the same ostensible data. This goes along with my point that these were often justifications for negative categorizations more so than the cause of negative categorizations.

IV. HIERARCHIES AND PONTIC PEOPLES IN ANCIENT ETHNOGRAPHY

1. Herodotos' Greek Perspective on Northern Barbarians

Now that we have a general sense of these two ideological factors, we can move on to the relative positions of Pontic and other peoples in common ethnic hierarchies, as well as justifications of, or challenges to, such rankings. Before proceeding, it is important to note that Herodotos' conception of the inhabited world, much like his predecessors (e.g., Hekataios of Miletos) and authors who followed, is often divided in three: Europe, Asia, and Libya. Scythians and Pontic peoples would be the northern-most (non-legendary) peoples of Europe in this case. In his description of northern peoples, Herodotos sometimes reveals his own rankings of these peoples or, less often, the supposed rationale for his approach.

It is in the passage that opens this chapter (4.46) that Herodotos most directly reveals the position of Pontic peoples generally within the ethnic hierarchy he presumes, and it is a very low position. Herodotos' primary criterion for the low position of all Pontic peoples, presumably in relation to other Europeans at least, seems to be their relative lack of 'wisdom' (σοφίη) or intelligence (demonstrated by a dearth of wise men beyond one well-known Scythian, Anacharsis). As we will soon see, this is a criterion of evaluation that is echoed later in Aristotle's notion that a superior people would possess a balance of both 'intelligence' or 'skill'(διάνοια, τέχνη) and 'heart' / 'spirit' (θυμός). Furthermore, as Brent Shaw details, another important factor in Herodotos' ranking of peoples is whether or not they were considered settled agriculturalists, which for Herodotos suggests a less inferior level, or pastoral nomads, who were most inferior or barbaric in his view.[31] It is noteworthy that this bears some resemblance to apparent grading of barbarians later on in the work of Strabo, the geographer of the Augustan age. Strabo's account is based, in part, on concepts about the progression of human societies Plato's *Laws*, where the development is from simple forms of life (e.g., nomadism and banditry) to agricultural forms to organized cities on the model of the Greek city-state (*polis*).[32]

What is clear is that Herodotos ranks the Scythians above all other Pontic peoples based on their exceptional intelligence in one important area similarly valued by a Greek like Herodotos, namely military skill. As Sven Rausch's study of northern peoples explores in detail, this evaluation is often repeated in subsequent authors, with Thracians or Gauls or Celts often placed alongside Scythians

---

31 Shaw 1982.
32 Plat. *Nom.* 3.676a–683a; cf. Almagor 2005, 51–55, citing L.A. Thompson.

as superior or persistent (if wild) fighters.[33] Yet, while the Scythians are placed above other northern Europeans, Herodotos places Persians (categorized as Asians for Herodotos) above Scythians, at least that seems to be the implication in his portrayal of Persia's effective (almost Greek-like) fighting against Scythians.[34] Herodotos' tendency to differentiate (rather than mix together) sub-groups of 'barbarian' peoples is also evident in his description of the peoples of India, where one people of India kills no living thing whatsoever (3.100) and another people (Padaians) kills anyone who is ill and feasts on his or her flesh (3.99).

This concern to grade distinguished peoples continues within subsequent ethnography both in the Hellenistic period and in the Roman era, as Diodoros' discussion of Libyan tribes (3.49) and Tacitus' discussion of Germanic peoples (*Germania* 28–46) illustrate well. Tacitus makes the broad generalization that Germans are better than 'worthless' (29.3) and 'spiritless' (28.4) Gauls (= Celts). But then he details that, among Germans, Batavians were most outstanding in valour (29.1) and Chattians in mental abilities (30.1).[35] Fosians were inferior even in the best conditions (36.2).

The fact that Herodotos has an ongoing concern to rank other ethnic groups below Greeks but also to place sub-groups of supposedly inferior peoples within a hierarchy is also clear in his discussion of the Thracians, on the western coast of the Black Sea.[36] Overall, as Matthew Sears states, Herodotos description of the Thracians 'leaves the reader with an impression of contempt mitigated by curiosity'.[37] In discussing Thracians succumbing to the advance of Darius, Herodotos claims that the Getians (*Getai*) were 'the most manly and just (ἀνδρηιότατοι καὶ δικαιότατοι) of the Thracians' (4.93). Yet even they could be relatively foolish and become enslaved. That the 'just' quality of the Getians is relative to a very low view of all Thracians generally becomes clear when Herodotos chooses to zero in on the Getians' custom of sacrificing a human to their god Salmoxis (4.94). This leads Herodotos into stories concerning this god, and he chooses to relate the perspective of Greek settlers in the area (perhaps in a settlement such as

---

33   Scythians as dangerous warriors: Thuk. 2.96 (as archers); Isok. *Paneg.* 4.67 (with Thracians and Persians); Plat. *Laches* 191a; Arist. *Pol.* 7.1324b (with Thracians and Celts); *Rhet.* 1.9. See Rausch 2013.

34   Hartog 1988, 44–50, 258f. points out that when the Persians attack Greeks, they are portrayed as people who do not know how to fight. Whereas when the Persians attack Scythians, they figure as well-organized fighters, as if they were Greek hoplites attacking inferior people. For Hartog, this becomes a further instance of dichotomous thinking on Herodotos' part. Instead, I am drawing attention to how this helps us see Herodotos' tendency to rank specific barbarians in relation to one another.

35   Even among the Suebian tribe, some sub-groups were more savage than others, so he claims (38.1–45.6). For an overview of ethnographic traditions regarding Germans, see Rives 1999, 11–41. Tan 2014 unfortunately says almost nothing concerning the environmental theory as it pertains to Tacitus' discussion of the geography of *Germania*; she seems to have a primordial definition of ethnicity (e.g., 'uncompromised Germanic ethnicity': p. 183 n. 15; cf. p. 191 n. 67).

36   On Athenian or Greek perceptions of Thracians, see Sears 2015, 314–316; also Sears 2013.

37   Sears 2015, 315.

Histria). These Greeks are said to regard Thracians as 'crude' (κακόβιοι) and 'somewhat stupid' (ὑπαφρονέστεροι) for their belief in Salmoxis, with Greek settlers asserting that Salmoxis was merely a famous person, not a god at all (4.95).[38] Herodotos, however, is hesitant to take on fully the viewpoint of the colonists that he relates (4.96). Here, with a picture of Greek settlers' attitudes, we are gaining a glimpse into how ethnic rivalries and stereotypes might manifest themselves in actual social relations between Greeks and other peoples in and around the cities of the Black Sea region. While recognizing common negativity towards Thracians, it should also be noted that there were signs of attraction to Thrace and Thracian culture on the part of some among the Athenian elites, as Aristophanes' playful reference to 'Thrace-frequenters' suggests (Θρᾳκοφοῖται).[39]

There is not enough space to deal with the epigraphic evidence for Pontic diasporas here, evidence which I discuss at length in another study.[40] Still, it is important to note that inscriptions from Athens, Rhodes, and other locales demonstrate the ongoing presence of Pontic peoples (Thracians, Scythians, Maiotians, Sindians, Sarmatians, Kolchians, and others), including but not limited to imported slaves, in Greek cities from the 5th century and on into the Hellenistic era.[41] So, instances of Pontic peoples ranked within hierarchies pertain not only to distant populations. Rather, these rankings and stereotypes have real implications for social relations within Greek cities and for discrimination against such settlers. Greeks like Herodotos would encounter northern peoples in everyday interactions in places like Athens. In addition, there is also evidence in Herodotos' own narrative regarding interactions between Greek colonists and indigenous peoples north of the Black Sea, which is pertinent to the issue of alternative hierarchies that work against Greek ones.[42]

---

38   Braund 2008a, 357–359.
39   Aristoph., cited by Ath. 12.75; cf. Sears 2015, 316–318.
40   Harland 2020.
41   Examples from Athens: *IG* I[3] 421, lines 34–49 (six slaves: three Thracian women, one Thracian man, one Scythian man, and one Kolchian, ca. 415 BC); *IG* II[2] 1283 (free or freed Thracians, 240/39 BC and earlier); *SGDI* II 1992, 2163 (Maiotians, 2nd BC); *IG* II[2] 8430, 10243, 10244, 12061 (Sarmatians, 2nd BC); *IG* II[2] 9049f. (graves of Kolchians, 2nd–1st centuries BC). One inscription from the island of Rheneia (*SEG* XXIII 381 = *IG* IX 1[2].4.1778, 100 BC) attests to a master who owned 22 slaves, one third of which are identified as coming from the Black Sea area (four male Maiotians, three male Thracians). Graves from 2nd- or 1st-century-BC Rhodes attest to at least three Scythians; three Kolchians; five Maiotians; three Sarmatians; one Sindian and three Thracians (*IG* XII 1.526f.; *I.RhodM* 421; *SEG* LI 1015; Hatzfeld 1910, 243, no. 8; *IG* XII 1.514; Jacopi 1932, 232, no. 122; *I.Lindos* 683; *MDAI(A)* 23 (1898) 394, no. 64; *IG* XII 1.525; Jacopi 1932, no. 95; *SEG* XXXVIII 789; *IG* XII 1.1385; *I.RhodM* 217f.; *I.Lindos* 695). As Lewis 2018 shows, approximately one third of the 179 Delphic manumissions that identify ethnicity pertain to Pontic slaves (including Thracians). On the Pontic slave-trade, see Finley 1962; Braund & Tsetskhladze 1989; Gavriljuk 2003; Avram 2007; Tsetskhladze 2008; Braund 2008b.
42   See Podossinov, chapter II in this volume on the Scythian police force; Oller Guzmán, chapter I in this volume on early interactions between Greeks and Scythians.

2. Northern Perspectives on Greeks as Reflected in Herodotos' Work

Herodotos' work provides, first and foremost, a glimpse into his own categorizations of specific outgroups. Nonetheless there are hints of other peoples' perspectives and categorizations within Herodotos' narrative. Although the extent of Herodotos' own travels are debated, it is reasonable to suggest that he spent some time at Borysthenes (Olbia), where he would have opportunity to witness interactions between Greeks and indigenous peoples.[43] He goes into greater detail about this Ionian (Milesian) colony and refers to a conversation with Tymnes – a Greek-speaking official of the Scythian king Ariapeithes – which may have taken place there, for instance (4.76). Some of Herodotos' stories seem to reflect ethnic tensions between Greek settlers and Pontic peoples in places like Borysthenes, whether Herodotos went there or not. And, in some cases, Herodotos reports what he claims are Scythian perspectives on Greek peoples and their customs, including that Scythians 'avoid foreign customs at all costs, especially those of the Greeks' (4.76).

Herodotos relates two main incidents that serve to illustrate Scythian disdain for Greek cultural customs and, it seems, for Greeks generally, who are thereby cast as inferior to Scythians in alternative ethnic hierarchies.[44] Herodotos also mentions local informants in both cases. On the one hand, the story goes that when Anacharsis imported from Kyzikos rites for the Mother of the Gods, he was observed by another Scythian. That Scythian informant reported this to the king, who then shot Anacharsis dead with an arrow for engaging in Greek customs (4.76). Herodotos also reports that Scythians even rejected the existence of Anacharsis because Anacharsis had adopted such Greek practices.

A second story of Scythian negativity regarding inferior Greeks and their customs involves the Scythian king Skyles, son of Ariapeithes, which occurred 'many years later'. According to Herodotos (4.78–80), king Skyles' mother was a Greek from Histria and this king 'was not at all content to live as the Scythians did but, because of his education [by his mother], was much more inclined to practice Greek customs'. Beyond dressing in Greek attire when not visible to other Scythians, he also 'set up sanctuaries to the gods in accordance with Greek customs', particularly spending time at the Greek settlement of Borysthenes. The king's demise (by beheading) came when other leading Scythians witnessed him participating in the rites of Dionysos. Herodotos' expresses a local Greek's supposed taunt, which suggests that the Scythians were known to laugh at Greek settlers for, among other things, engaging in rites for a god that was reputed to induce madness.[45]

---

43 See Braund 2008a. Cf. Skinner 2012, 164f. For a more pessimistic view regarding Herodotos' travels to the Pontic region, see Armayor 1978.

44 On the interaction of Greeks and Pontic peoples in such narratives, see Podossinov 1996; Podossinov 2019; Braund 2008a; Vlassopoulos 2013.

45 See Oller Guzmán and Podossinov, chapters I and II in this volume, for more on Anacharsis and Skyles.

Such ethnic rivalries manifested within social encounters like these could re-sult in competing hierarchies. We have already cited Herodotos' outline of a Per-sian hierarchy which would place Greeks low on the scale of honour because of their distance from Persia, for instance (1.134.2). Furthermore, Herodotos' discus-sion of Persian alliances with Ionians (Ionian tyrants) against Scythians (ca. 513 BCE) attributes to the Scythians a disdain for Ionian Greeks such as those settled at Olbia, placing Ionians low down in a ranking of peoples.[46] Herodotos states: 'the Scythians judged the Ionians as free men to be the worst and most unmanly (κάκιστοι καὶ ἀνανδρότατοι) of all humanity; but as slaves, to be the most fond of servility and the least likely to flee from their masters. Such were the insults cast at the Ionians by the Scythians' (4.142). Even if this viewpoint is expressed in Herodotos' words,[47] this at least shows an awareness that Greek ethnic hierarchies could be actively challenged by other peoples, including northern peoples ('Scyth-ians') who rejected Greek categorizations.[48]

3. Medical Theories Contemporary with Herodotos

More accessible than these alternative rankings are the ideological justifications for the structures Herodotos and other Greek elites presume. Although not dealing with ethnic hierarchies, Deborah Thomas' work demonstrates Herodotos' aware-ness of medical theories at some length.[49] Herodotos, like other contemporaries, seems to believe the environment determines the character and relative quality of a people (e.g., 1.142; 9.122). He is aware of medical theories (e.g., 2.77) like the ones I discuss below in connection with Hippokratic literature. These commonali-ties may suggest widespread legitimizing ideologies among the Greek elites at least by the mid- to late-5th century. In Herodotos' case, for instance, Egyptians (who were viewed as southern Asians) are described as having paradoxical cus-toms that are precisely the opposite of what Herodotos considers normal, and this is linked to 'the contrary nature of Egypt's climate and its unique river' (2.35). Egyptians are nonetheless regarded as the second healthiest people next to Liby-ans (2.77) and praised for being exceedingly pious (2.37), and some Greek cus-toms regarding deities are then traced back to Egyptians (e.g., 2.49). This suggests ways in which Egyptians might be considered superior to at least some other peo-ples from Herodotos' perspective.

Such ambivalent attitudes – negative alongside positive evaluations – towards an outgroup are quite common, as Mackie and Smith's study of emotions in inter-group relations shows.[50] However, Herodotos does not clearly express such envi-ronmental theories in his discussion of the northerners, even though the cold may

---

46 Cf. Braund 2008b, 4–6.
47 Cf. Braund 2008b, 7.
48 See Harland 2019 for an example involving Judaean perspectives.
49 Thomas 2000, 28–101.
50 Mackie & Smith 2002, 2f.

explain how tough the Scythians were considered to be. This holds true especially in light of the fact that, elsewhere, Herodotos cites the view that 'soft places tend to produce soft men' (9.122). These ideological developments became more widespread, and eventually began to play a key role in justifying rankings within hierarchies.

The Hippokratic work *On Airs, Waters and Places*, which is usually dated to the final decades of the 5th century BC, seems to be the earliest extensive explanation of these ideologies regarding the correspondence between climate (with its four elements), the four humours within humans, and the relative inferiority or superiority of peoples.[51] *On Airs, Waters, and Places* (abbreviated as *On Airs* here) presents itself as a guide-book for travelling physicians. The author details what a successful physician must take into account regarding seasons, climate, environment, and the movement or position of the sun, moon, and stars. The author's emphasis on 'the contribution of astronomy' is echoed later in theories that accompany ethnic hierarchies of Vitruvius (*On Architecture* 6.1.3–12; late-1st century AD) and Claudius Ptolemy (*Tetrabiblos* 2.2.1–7; 2nd century AD), where the environmental theory is combined with astrological reasoning.[52] Although not expressly stated, the argument in *On Airs* presumes the theory of the four humours in human health, a theory that was first clearly expressed in the roughly contemporary Hippokratic work titled *The Nature of Man*.

Most importantly, *On Airs* offers a comparison of Asia and Europe (with the boundary being the Phasis or Maiotian Lake, it seems) regarding 'how the peoples of one differ entirely in form (μορφή) from those of the other' (*On Airs* 12). This is an issue that is mirrored in the roughly contemporary work on *Regimen*, where Pontic peoples are briefly contrasted to the southern Libyans (2.37), but *On Airs* goes deeper. What is most important in this case is the way in which *On Airs* justifies evaluations of inferior or superior characteristics of peoples in Europe and Asia, illustrating the *rhetoric* used to express and legitimate rankings of ethnic groups.

The Hippokratic author's views do not seem entirely consistent, but his main emphasis in discussing Asian peoples is that the climate in most parts of Asia is moderate and consistently hot, with very little variations from one season to the next (*On Airs* 12–16). This results in good agricultural production and healthy animals, especially in Egypt and Libya (the southernmost portions). At the same time, this also results in a homogeneously inferior population that is 'weak' (ἄναλκις), 'unmanly' (ἄνανδρος), 'lacking in heart' (ἄθυμος), and 'unwarlike' (ἀπόλεμος) (16). This is explained as going along with a tendency towards a lack of independence and susceptibility to rule by despots. A lack of variation in seasons in most of Asia is thought negative for the disposition of people. According to the author, there are 'no mental shocks nor extreme physical variation' that could have brought 'passion and arrogance' (τὴν ὀργὴν ἀγριοῦσθαί τε καὶ τοῦ

---

51   Cf. Thomas 2000, 86–101 regarding the continents in *On Airs*.
52   Cf. Galen, *The Soul's Dependence on the Body* 805. Cicero, who maintains the environmental theory, seems more hesitant in respect of combining it with astrology (*On Divination* 2.96).

ἀγνώμονος μετέχειν) or 'a share in a higher level of spirit' (θυμοειδέος μετέχειν μᾶλλον ἢ ...)' (16). This notion has affinities with Aristotle's emphasis on 'spirit' or 'heart', as I discuss below. Furthermore, passages such as this in *On Airs* continued to influence authors into the Roman era, as illustrated in Galen's discussion of *On Airs* centuries later.[53]

Still, there is mention of other (especially north-eastern) portions of Asia with more variations in seasons and temperatures. Moreover, the author anticipates a comparative internal ranking among Asian peoples when he states that you 'will find that Asians also differ from one another, some being superior (βελτίονες), others inferior (φαυλότεροι)'. And variations in seasons are once again the basis of these supposed differentiations (16).

The Hippokratic author then turns to European peoples (17–24). The author begins in the extreme north, contrasting Egyptians of the south to Scythians of the north, each population relatively homogeneous due to a lack of variation in seasons, the one hot, the other cold (18). A consistently cold and moist climate is the factor that renders a people inferior (soft, impotent, fat, and lazy), presumably in comparison with other Europeans. The Scythians are so undifferentiated that even the distinction between men and women is blurred (22). It seems that the author is justifying a placement of Scythians at the bottom of a hierarchy of European peoples.

The author then argues that European peoples are far more varied than Asians because of the variability of seasons in different parts of Europe (23). The more varied European peoples will therefore be superior to the largely homogeneous Asians. This notion is encapsulated in the assertion that Europeans are 'more courageous' (εὐψυχότεροι) than Asians because they live in colder conditions (23). The discussion of variations in European environments which concludes the work presumably facilitates a positive evaluation of other Europeans. Greeks, who would likely be placed at the pinnacle of all European peoples, are likely in mind, although this is not as expressly stated as it is by Aristotle. The implication is that, for this Hippokratic author, Scythians are at the bottom of the European hierarchy and Greeks are at the top, with most European peoples being superior to Asians. The fact that the author directly compares Egyptians and Scythians as the most extreme peoples in the most extreme environments suggests that both would be placed together at the bottom of an ethnic hierarchy.[54]

4. Ethnic Hierarchies in the Works of Aristotle

I have given considerable space to a discussion of *On Airs* in part because it is among the earliest and most extensive works concerned with theorizing the relative position of peoples. But I have also done so because the concepts it expresses

53 Galen, *Soul's Dependence on the Body* 8f. = 798–805; Strohmaier 2004; Cf. Isaac 2004, 85–87.
54 Cf. Diod. 3.33f.

came to influence subsequent ethnographic works that are concerned with justifying rankings of northern and other peoples in certain hierarchies. It is notable that neither Herodotos nor the Hippokratic author clearly explains the position of the Greeks at length, although both seem to presume the superiority of Greek peoples in relation to others that are subordinated.

In *Politics* (about a century after the above works), Aristotle is less hesitant to express his stratifications and the ideological concepts that may, in some respects, inform the positions of both Herodotos and the Hippokratic author regarding northern and southern peoples. Aristotle reveals the overall ethnic hierarchy that underlies his discussion when he states that 'barbarians' generally are 'more servile in character than Greeks, Asians more servile than Europeans' (*Politics* 7.1285a; cf. *On Airs* 16). So Greeks are at the top, other Europeans below that, and then Asians at the bottom, at least with respect to slave-like characteristics that lead to domination by others. Peoples around the Black Sea specifically seem to be placed particularly low by Aristotle, despite the fact that, in theory, they might possess the 'spirit' that cold climates foster.

Elsewhere in *Politics* (7.1338b), Aristotle puts northern peoples forward as an instance of the 'most savage' (ἀγριώτατοι) people – though lacking in true courage (ἀνδρεία) – comparing them to animals and highlighting the case of cannibalism. The claim that these people of the north lack true courage seems to clash with Aristotle's theory regarding conditions favourable to a high degree of 'spirit', which I discuss soon. Such theories and the categorizations they reflect are quite fluid and open to varying interpretations when used in connection with particular peoples. If we can assume some overlap between lower European peoples and higher Asian ones, then Aristotle's ethnic hierarchy seems to match both Herodotos and *On Airs*, though neither of these makes a statement as clear as Aristotle's on the position of Greeks.

Aristotle also combines what appears to be a theory of the humours and climate with the notion that Greece is the centre of the known world, reflecting both elements of widely shared legitimizing myths. Although Aristotle is certainly critical of contemporary Athenian societal arrangements in *Politics*, he nonetheless goes into more detail regarding the superiority of Greece, with geography and climate central to the reasoning. Aristotle outlines the positions of northern Europeans and Asians within a hierarchy of peoples. Yet he also reveals his supposed criteria for determining the inferiority or superiority of different Greek peoples (without naming specific city-states), namely, a balance of both 'heart' / 'spirit' (θυμός) and 'intelligence' (διάνοια) or 'skill' (τέχνη) which leads to virtuous citizens:

> The peoples inhabiting the cold places and those of Europe are full of *spirit* but inferior with regard to *intelligence* and *skill*, so that they continue to be comparatively free, but lack civic organization and the ability to rule their neighbours (θυμοῦ μέν ἐστι πλήρη, διανοίας δὲ ἐνδεέστερα καὶ τέχνης, διόπερ ἐλεύθερα μὲν διατελεῖ μᾶλλον, ἀπολίτευτα δὲ καὶ τῶν πλησίον ἄρχειν οὐ δυνάμενα). The peoples of Asia, on the other hand, are intelligent and skilful in temperament, but lack spirit, with the result that they continue to be subjected and enslaved (τὰ δὲ περὶ τὴν Ἀσίαν διανοητικὰ μὲν καὶ τεχνικὰ τὴν ψυχήν, ἄθυμα δέ, διόπερ ἀρχόμενα καὶ

δουλεύοντα διατελεῖ). But the Greek kinship group (γένος) participates in both characters, just as it occupies the middle position geographically, for it is both spirited and intelligent. For this reason, it continues to be free, to have the best civic institutions, and – if it attains a united civic constitution – to have the ability to rule everyone. The same variety also exists among Greek peoples (ἔθνη) in comparison with one another: while some have a singular nature, others have a good combination of both these qualities [i.e. spirit and intelligence]. So it is clear that those who are likely to be guided to virtue by the lawgiver must be both intellectual and spirited in their nature.[55]

Aristotle, like Herodotos and the Hippokratic author, leaves room for more specific rankings of Europeans and others, including the evaluation of which Greek peoples (or *poleis*) would be considered superior to other Greeks. The tradition of evaluating the degree to which a people possessed 'spirit' or 'intelligence' continues in subsequent ethnography and can be seen clearly, for instance, in Tacitus' evaluation of Germanic tribes, where environmental theories also play a role.[56]

5. Questioning Hegemonic Ethnic Hierarchies: The Approach of Ephoros

Justifying the subordination of other peoples with reference to theories of the humours and climate continues in subsequent ethnography into the Roman era.[57] Yet only certain dimensions of Herodotos' account of the Scythians specifically seem to remain prominent in discussions in the Hellenistic and Roman periods.[58] While Herodotos speaks of the Scythians as superior to other northern peoples and makes further distinctions regarding other peoples, later authors were less attentive to Herodotos' specifics. In particular, Herodotos claims that 'beyond [the Scythian farmers] dwell the Man-eaters (ἀνδροφάγοι), who are in no way Scythian, but a completely distinct people' (4.18; cf. 4.106). Yet subsequent mentions of the Scythians confuse the situation with the Scythians generally depicted as savage cannibals who engage in human sacrifice.[59] Herodotos' report that Scythians drink the blood of the first man they slay in battle (4.64f.) may contribute to these later simplified characterizations. This confusion was also, in part, because the Taurians, who were said to engage in sacrificing Greek sailors to a goddess, were sometimes subsumed under the general designation 'Scythians'. They were

---

55 Arist. *Pol.* 7.1327b. Translation adapted from Rackman 1932 (LCL). Cf. *Household Management* (*Oikonomika* 1.5.5 = 1344b), which reflects similar thinking in advice for choosing slaves.

56 Tac. *Germ.* 28–46, esp. 28.4; 29.2; 30.2. Cf. *Agricola* 11, where he also reflects environmental theories (based on the humours) in stating that 'shared climatic conditions produce the same physical appearance' (translation Birley 1999, 10).

57 E.g., Polyb. 4.21; Diod. 3.33f. (contrasting Ethiopian Trogodytes and Scythians as the extremes); Vitruvius, *On Architecture* 6.1; Strabo, *Geogr.* 2.5.26 (126f.C); Tac. *Agricola* 11; Galen, *Soul's Dependence on the Body*, 798–805; Lucan, *Civil War* 8.294–308. See the summary discussion by Isaac 2004, 82–101.

58 On the continuing influence of Herodotos in the Hellenistic era, see Priestley 2014, 109–156.

59 E.g., Apollodoros as reported by Strabo in *Geogr.* 7.3.6 (298C); Plut. *On Superstition* 13; Apollod. *Library* E 6.26 and 2.8.

thought to inhabit land within Scythian territory, whereas Herodotos located them just west of the Kimmerian Bosporos on the Crimean peninsula.[60] So, the supposed Man-eaters and Taurians readily stood in as representative of 'Scythians' after Herodotos. Some authors were concerned to reassert distinctions, including Ephoros. Cases such as this provide glimpses into elites who held ideologies that attenuated consensual rankings and worked against widely held justifications for such hierarchies.

Ephoros, whose work was likely composed around 350 BC, seems to have had a noteworthy discussion of Pontic peoples in his *Histories* (a work that only survives in citations by others, primarily Strabo and Diodoros of Sicily).[61] More importantly here, Ephoros also seems to have known and sought to pull the rug out from under common ideologies that sought to support Greek hegemonic ethnic hierarchies, hierarchies that placed Scythians at or near the bottom. As Frances Pownall convincingly argues, cases where citations can be checked in parallel sources suggest that Strabo's summaries faithfully reproduce Ephoros' work.[62] According to geographer Strabo, who cites Ephoros in order to defend Homer against the works of Eratosthenes of Cyrene's *Geography* (ca. 246–218 BC) and Apollodoros of Artemita (ca. 100 BC), the European section of Ephoros' work stressed that the lifestyle of Sauromatians and other Scythians varied considerably.[63]

There is reference to some engaging in cannibalism but others abstaining from killing any living thing, for instance. Strabo's point is that Ephoros critiqued other ethnographic writers for focusing on the most savage (ὠμότης) examples while ignoring contrary, positive evidence about distant peoples. Instead, these distant peoples, including Scythian nomads, could be put forward as positive moral examples of the 'most just' (δικαιότατοι) mode of life. In this case of positive attitudes towards outgroups, Ephoros proposed that the Scythians were 'the most straightforward', 'frugal' and 'independent' people.[64] Ephoros also argued that it was contact with inferior Greek customs that tainted what was originally superior. The phrasing in Strabo does not clearly indicate that Ephoros denied the existence of very particular northern peoples who engaged in cannibalism, but the direction of the argument does seem to indicate that Ephoros did reject this.

Strabo seems quite concerned to refute both Eratosthenes and Apollodoros who claimed that Scythians 'sacrificed strangers, ate their flesh, and used their

---

60 E.g., Hdt. 4.103; Eur. *Iph. Taur.* 72, 276–278, 775f. (see E. Hall 1989, 110–112); Diod. 4.44.7; Paus. 1.43.1; *Orphic Argonautica* 1075; Ov. *Pont.* 3.2.45–58; Amm. Marc. 22.8.3. On stories of Taurian human sacrifice (which likely circulated as early as the 6th century BC), see Rives 1995.

61 Hudak 2009, 5–8.

62 Pownall 2010, 116; Hudak 2009, 42.

63 Strab. *Geogr.* 7.3.9 (302f.C). Strabo does not believe that Ephoros was consistently truthful in other respects, however; see Roller 2010, 118–122.

64 Strab. *Geogr.* 7.3.8 (301C); cf. Pownall 2010, 127f. See also Diod. 1.9.5, where Ephoros is said to believe that barbarian peoples could claim greater antiquity than Greeks, which would imply superiority to Greeks (something that Diodoros dismisses).

skulls as drinking-cups', and that Homer was ignorant of these 'facts'.[65] Over a century after Strabo, Aulus Gellius was far more direct in dismissing as 'disgusting' and 'worthless' writings that claimed that the most remote of the Scythians engaged in cannibalism (ἀνθρωποφάγοι). However, he may well have had Herodotos himself (i.e. the Man-eaters passage) in mind, along with others who followed or misconstrued Herodotos, perhaps Eratosthenes and Apollodoros (Gell. *Attic Nights* 9.4). Whatever the case may be regarding the charges of human sacrifice and cannibalism, what is important here is that Ephoros attenuates common ideologies and begins to construct alternative hierarchies that place superior Scythians above other peoples, including Greeks, it seems.[66] Ethnography is here used as a means of critiquing customs of the author's own ethnic group, something that is also clearly evident later on in Tacitus' *Germania*, for instance.[67]

6. Attenuating Ideologies and Strabo's Response

Further signs of ideologies that attenuate widespread hierarchies are evident in sources cited or summarized by Strabo. In his citation of Ephoros and his refutation of Eratosthenes, he seeks to defend Homer. The principle concern seems to be that Eratosthenes did not value Homer as an accurate source, while Strabo did.[68] Eratosthenes may have accepted widespread, strongly negative categorizations of Scythians, although there are other signs that he himself worked against common elite representations and rankings in his own geographical work.[69] In particular, he seems to have challenged the normal Greek-barbarian dichotomy. Instead, he emphasized the measure of 'virtue' (ἀρετή) or 'vice' (κακία) independent of distance from a centre or independent of climate in describing the relative position of different peoples in the grand scheme of things, a viewpoint that Strabo is hesitant to adopt.[70]

Thukydides, who himself had Thracian ancestry, likewise tends to downplay the distinction between Greeks and 'barbarians' in his *Peloponnesian War*, emphasizing similarities between the lifestyles of those labelled 'barbarians' and of

---

65   Strab. *Geogr.* 7.3.6 (298C); cf. 7.3.7 (300C); cf. Gardiner-Garden 1986, 222–224.
66   In the discussion of peoples in Asia Minor, Strabo does not agree with Ephoros' tendency to break down the usual Greek-barbarian dichotomy, at least when Ephoros speaks of some peoples (γένη) as 'mixed' (μιγάδη), a category that does not exist for Strabo (*Geogr.* 14.5.23–25 [678f.C]). See Almagor 2005, 43f.
67   Gruen 2011. This function can also be seen, at times, in barbarian characters of comedy. See Long 1986, 165–167.
68   Cf. Roller 2018, 9f.
69   For a recent translation of Eratosthenes, see Roller 2010.
70   Strab. *Geogr.* 1.4.9 (66f.C). On Strabo's understanding of Greek and barbarian, see Almagor 2005; van der Vliet 2003; Dueck 2000, 58, 75–84. Closely related are traditions regarding the superiority of barbarian wisdom, which are also reflected in fictional narratives and writings. Cf. Philostr. *VA*; Letters of Anacharsis in the *Cynic Epistles*; cf. Harland 2011; also see Podossinov, chapter II in this volume.

earlier Greeks.[71] It should be clarified that even a challenger of the ethnocentric approach to Pontic peoples – like Eratosthenes or the author of the so-called letters of Anacharsis (*Cynic Epistles*) – was assuming an ethnic ladder, albeit one with different criteria for positioning peoples on the rungs. Eratosthenes' reconfiguration of ethnic hierarchies and rejection of the usual barbarian categorization happens to place Indians, Romans and Carthaginians on a high rung as 'refined' or 'urbane' peoples (ἀστεῖοι), rather than 'bad' peoples (κακοί). Eran Almagor convincingly argues that Strabo rejects Eratosthenes' alternative and is concerned to maintain the Greek-barbarian dichotomy, just as Daniela Dueck sees the Greek-barbarian dichotomy as central to Strabo's overall work.[72]

Strabo himself does seem concerned with ranking peoples even while maintaining a Greek-barbarian dichotomy. First of all, it seems clear that he places most Europeans above both Egyptians and Libyans.[73] Notwithstanding its coldest limits (e.g., the Tanaïs River), Europe tends to create superior peoples in comparison with Egypt and Libya. This is because Europe is 'both varied and most naturally suited for excellence in men and civic organization (πολυσχήμων τε καὶ πρὸς ἀρετὴν ἀνδρῶν εὐφυεστάτη καὶ πολιτειῶν)' (2.5.26). Beyond its mountains, Europe's climate is varied and temperate and therefore conducive to peace and independence. This region has therefore been the source of leading nations, such as the Greeks, Macedonians and Romans, who help to subdue any war-like inhabitants of mountainous or cold regions of Europe (2.5.26).

Dueck shows that Strabo sometimes adopts the Roman perspective, with Greeks and Romans grouped together as a civilized 'us' in contrast to barbarians, such as Britons and Germans.[74] However, in other respects, Strabo firmly places Greeks above Romans in his own hierarchy, with the Greeks being more ancient and both intellectually and culturally superior. So, Greeks are at the top of his ladder and Romans are near the top (similar to Eratosthenes' notion of Romans as refined 'barbarians'). Conversely, various barbarian peoples are placed on lower rungs, with Europeans generally above Egyptians (as southern Asians) and Libyans. Furthermore, both Patrick Thollard and Almagor demonstrate that Strabo assumes a 'scale' of barbarity, a scale that incorporates various 'civilizing' factors, including the distinction between primitive peoples (e.g., those engaged in a nomadic lifestyle or in banditry), more developed peoples (e.g., those engaged in a settled, agricultural lifestyle), and most developed peoples, namely Greeks with an organized civic constitution.[75]

There are signs that Strabo, like Herodotos and the Hippokratic author, shares the notion that climate and environment explain both the lifestyles and the relative

71 Thuk. 1.5f.; cf. Sears 2015, 315.
72 Strab. *Geogr.* 1.4.9 (66f.C) on Eratosthenes. See Almagor 2005, 49f.; Dueck 2000, 75–84.
73 Strab. *Geogr.* 2.5.26–33 (126–131C).
74 Dueck 2000, 75–84. For Strabo's explanation of the category 'barbarians', see *Geogr.* 1.4.9 (67C) (the Eratosthenes debate) and 14.2.28 (662C) (on the origin and meaning of the term).
75 Strab. *Geogr.* 17.3.24 (839C). See Thollard 1987 (dealing with *Geogr.* 3–4 only); Almagor 2005, 51–55. Cf. van der Vliet 2003.

inferiority of specific barbarian peoples, as exemplified with the Ethiopians.[76] Further factors that differ from Greek customs pertain to food-manufacturing, eating, bathing, clothing, and trading: all of these serve as criteria that justify a relative grading of peoples as more or less inferior.[77] Thus while the people of Britain are 'more simple and more barbaric' than Celts, still other peoples are 'completely barbarians (τελέως βάρβαροι)' or 'semi-barbarian' (ἡμιβάρβαροι).[78] The latter is not dissimilar to ideas attributed to Eratosthenes, despite the fact that Strabo critiques that author precisely on the 'barbarian' issue.

As Almagor points out, there is a lack of uniformity in Strabo's perspective on 'barbarians', and I would suggest that Strabo may not always adopt a strict Greek-barbarian dichotomy in his grading of peoples. Instead, he is using as his model some variation on commonly shared ethnic hierarchies, with different barbarian peoples being graded differently. The difficulty is that Strabo does not provide us with a consistent explanation of where exactly he places each specific barbarian people on the ethnic ladder. It is noteworthy that he also sees ways in which 'softness' or 'luxury' (τρυφή) associated with Greek and Roman lifestyles comes to have a negative influence on barbarian peoples.[79] Overall, though, the degree to which Strabo himself attenuates largely consensual categorizations of other peoples (as did Ephoros and Eratosthenes in more emphatic ways) remains debatable. In many respects, Strabo's approach, like others we have investigated here, serves to enhance and bolster widely held ethnic hierarchies.

V. CONCLUSION

While it seems that the majority of Greek intellectuals continued to legitimize hegemonic rankings of other peoples, placing Pontic peoples low on the ladder, there were a few others such as Ephoros who actively challenged such approaches and offered alternatives. These attempts to attenuate commonly held views would have affinities with certain subordinated or colonized peoples who actively sought to challenge their own low position within hegemonic hierarchies, assuming quite different arrangements on an ethnic ladder.

Acknowledgments

I would like to thank Altay Coşkun & Maia Kotrosits, who provided helpful feedback on the paper.

---

76   Strab. *Geogr.* 17.1.3 (786f.C); cf. Dueck 2000, 78f.
77   Cf. Dueck 2000, 78; Shaw 1982, 29f.
78   Strab. *Geogr.* 2.5.32 (130C); 4.6.4 (203C); 13.1.58 (611C); cf. 4.5.2 (200C): ἀπλούστερα καὶ βαρβαρώτερα.
79   Strab. *Geogr.* 7.3.7 (300C). For more on this, see Podossinov, chapter II in this volume.

Abbreviations

I.Rhod.M.   Maiuri, A. (ed.) 1925: *Nuova silloge epigrafica di Rodi e Cos*, Florence.
SGDI   Collitz, H., Bechtel, F. et al. 1894–1915, *Sammlung der griechischen Dialekti-schriften*, Göttingen.

Bibliography – Ancient Sources

Birley, A.R. 1999: *Agricola and Germany*. Oxford World's Classics, Oxford.
Godley, A.D. 1920: *Herodotus*. LCL, Cambridge, MA.
Rackman, H. 1932: *Politics*. LCL, Cambridge, MA.
Strassler, R.B. & Purvis, A.L. 2009: *The Landmark Herodotus: The Histories*, New York.

Bibliography – Modern Scholarship

Almagor, E. 2005: 'Who is a Barbarian? The Barbarians in the Ethnological and Cultural Taxonomies of Strabo', in D. Dueck, H. Lindsay & S. Pothecary (eds.), *Strabo's Cultural Geography: The Making of a Kolossourgia*, Cambridge, 42–55.
Armayor, O.K. 1978: 'Did Herodotus Ever Go to the Black Sea?', *HSCP* 82, 45–62.
Avram, A. 2007: 'Some Thoughts about the Black Sea and the Slave Trade before the Roman Domination (6th–1st Centuries BC)', in V. Gabrielsen & J. Lund (eds.), *The Black Sea in Antiquity. Regional and Interregional Economic Exchanges*, Aarhus, 239–251.
Barth, F. 1969: *Ethnic Groups and Boundaries*, Oslo.
Braund, D. 2008a: 'Scythian Laughter: Conversations in the Northern Black Sea Region in the 5th Century BC', in P.G. Bilde & J.H. Petersen (eds.), *Meetings of Cultures in the Black Sea Region: Between Conflict and Coexistence*, Aarhus, 347–367.
Braund, D. 2008b: 'Royal Scythians and the Slave-Trade in Herodotus' Scythia', *Antichthon* 42, 1–19.
Braund, D. & Tsetskhladze, G.R. 1989: 'The Export of Slaves from Colchis', *CQ* 39, 114–125.
Brubaker, R. 2004: *Ethnicity without Groups*, Boston.
Brubaker, R. & Cooper, F. 2000: 'Beyond "Identity"', *Theory and Society* 29, 1–47.
Dueck, D. 2000: *Strabo of Amasia: A Greek Man of Letters in Augustan Rome*, London.
Finley, M. I. 1962: 'The Black Sea and Danubian Regions and the Slave Trade in Antiquity', *Klio* 40, 51–59.
Gavriljuk, N.A. 2003: 'The Graeco-Scythian Slave-Trade in the 6th and 5th Centuries BC', in P.G. Bilde, J.M. Hojte & V.F. Stolba (eds.), *The Cauldron of Ariantas: Studies Presented to A.N. Ščeglov on the Occasion of His 70th Birthday*. Black Sea Studies 1, Aarhus, 75–85.
Hagendoorn, L. 1993: 'Ethnic Categorization and Outgroup Exclusion: Cultural Values and Social Stereotypes in the Construction of Ethnic Hierarchies', *Ethnic and Racial Studies* 16, 26–51.
Hagendoorn, L. 1995: 'Intergroup Biases in Multiple Group Systems: The Perception of Ethnic Hierarchies', *European Review of Social Psychology* 6, 199–228.
Hagendoorn, L., Drogendijk, R., Tumanov, S. & Hraba, J. 1998: 'Inter-Ethnic Preferences and Ethnic Hierarchies in the Former Soviet Union', *International Journal of Intercultural Relations* 22, 483–503.
Hall, E. 1989: *Inventing the Barbarian: Greek Self-Definition through Tragedy*, Oxford.
Hall, J.M. 2002: *Hellenicity: Between Ethnicity and Culture*, Chicago.
Harland, P.A. 2009: *Dynamics of Identity in the World of the Early Christians: Associations, Judeans, and Cultural Minorities*, New York.

Harland, P.A. 2019: 'Climbing the Ethnic Ladder: Ethnic Hierarchies and Judean Responses', *JBL* 138, 665–686.

Harland, P.A. 2020: 'Pontic Diasporas in the Classical and Hellenistic Eras', *ZPE* 214, 1–19.

Hartog, F. 1988: *The Mirror of Herodotus: The Representation of the Other in the Writing of History*, transl. by J. Lloyd, Berkeley.

Hatzfeld, J. 1910: 'Inscriptions de Rhodes', *BCH* 34, 242–248.

Howard, J.A. 2000: 'Social Psychology of Identities', *Annual Review of Sociology* 26, 367–393.

Hraba, J., Hagendoorn, L. & Hagendoorn, R. 1989: 'The Ethnic Hierarchy in the Netherlands: Social Distance and Social Representation', *British Journal of Social Psychology* 28, 57–69.

Isaac, B. 2004: *The Invention of Racism in Classical Antiquity*, Princeton.

Jacopi, G. 1932: 'Nuove epigrafi dalle Sporadi meridionali', *Clara Rhodos* 2, 165–256.

Jenkins, R. 1994: 'Rethinking Ethnicity: Identity, Categorization and Power', *Ethnic and Racial Studies* 17, 197–223.

Jost, J.T. 1995: 'Negative Illusions: Conceptual Clarification and Psychological Evidence Concerning False Consciousness', *Political Psychology* 16, 397–424.

Jost, J. T., & Banaji, M. R. 1994: 'The Role of Stereotyping in System-Justification and the Production of False Consciousness', *British Journal of Social Psychology* 33, 1–27.

Jost, J.T., Banaji, M.R. & Nosek, B.A. 2004: 'A Decade of System Justification Theory: Accumulated Evidence of Conscious and Unconscious Bolstering of the Status Quo', *Political Psychology* 25, 881–919.

Jouanna, J. 2012: *Greek Medicine from Hippocrates to Galen: Selected Papers*, ed. by P. van der Eijk, transl. by N. Allies, Leiden.

Keim, B. 2018: 'Communities of Honor in Herodotus' *Histories*', *AHB* 32, 129–147.

Lewis, D. 2017: 'Notes on Slave Names, Ethnicity, and Identity in Classical and Hellenistic Greece', *U Schyłku Starożytności: Studia Źródłoznawcze* 16, 183–213.

Mackie, D.M., & Smith, E. R. 2002: *From Prejudice to Intergroup Emotions: Differentiated Reactions to Social Groups*, New York.

Mackie, D.M. & Smith, E. R. 2015: 'Intergroup Emotions', in M. Mikulincer (ed.), *APA Handbook of Personality and Social Psychology*, vol. 2, Washington, DC, 263–293.

Mackie, D.M., Smith, E.R. & Ray, D.G. 2008: 'Intergroup Emotions and Intergroup Relations', *Social and Personality Psychology Compass* 2, 1866–1880.

Nichols, A. 2008: 'The Complete Fragments of Ctesias of Cnidus: Translation and Commentary with an Introduction', Ph.D., Gainesville, FL.

Podossinov, A.V. 1996: 'Babarisierte Hellenen – hellenisierte Barbaren: Zur Dialektik ethnokultureller Kontakte in der Region des Mare Ponticum', in B. Funck (ed.), *Hellenismus: Beiträge zur Erforschung von Akkulturation und politischer Ordnung in den Staaten des hellenistischen Zeitalters*, Tübingen, 415–425.

Podossinov, A.V. 2019: 'Nomads of the Eurasian Steppe and Greeks of the Northern Black Sea Region: Encounter of Two Great Civilisations in Antiquity and Early Middle Ages', in C. Hao (ed.), *Competing Narratives between Nomadic People and Their Sedentary Neighbours*, Szeged, Hungary, 237–251.

Pratto, F., Sidanius, J., & Levin, S. 2006: 'Social Dominance Theory and the Dynamics of Intergroup Relations: Taking Stock and Looking Forward', *European Review of Social Psychology* 17, 271–320.

Priestley, J. 2014: *Herodotus and Hellenistic Culture: Literary Studies in the Reception of the Histories*, Oxford.

Rausch, S. 2013: *Bilder des Nordens: Vorstellungen vom Norden in der griechischen Literatur von Homer bis zum Ende des Hellenismus*, Berlin.

Rives, J.B. 1995: 'Human Sacrifice Among Pagans and Christians', *JRS* 85, 65–85.

Rives, J.B. (ed.) 1999: *Tacitus: Germania*. Clarendon Ancient History, Oxford.

Romm, J.S. 1992: *The Edges of the Earth in Ancient Thought*, Princeton.

Sears, M.A. 2013: *Athens, Thrace, and the Shaping of Athenian Leadership*, Cambridge.

Sears, M.A. 2015: 'Athens', in J. Valeva, E. Nankov & D. Graninger (eds.), *A Companion to An-
    cient Thrace*, London, 308–319.
Shaw, B.D. 1982: '"Eaters of Flesh, Drinkers of Milk": The Ancient Mediterranean Ideology of
    the Pastoral Nomad', *AncSoc* 13/4, 5–31.
Shaw, B.D. 2000: 'Rebels and Outsiders', in A.K. Bowman, P. Garnsey, & D. Rathbone (eds.),
    The Cambridge Ancient History, vol. 11, 2nd ed., Cambridge, 361–404.
Sidanius, J. & Pratto, F. 1999: *Social Dominance: An Intergroup Theory of Social Hierarchy and
    Oppression*, Cambridge.
Sidanius, J., Pratto, F., van Laar, C. & Levin, S. 2004: 'Social Dominance Theory: Its Agenda and
    Method', *Political Psychology* 25, 845–880.
Skinner, J. 2012: *The Invention of Greek Ethnography: From Homer to Herodotus*, Oxford.
Smith, E.R. 1993: 'Social Identity and Social Emotions: Toward New Conceptualizations of Pre-
    judice', in D.M. Mackie & D.L. Hamilton (eds.), *Affect, Cognition and Stereotyping*, San Di-
    ego, CA, 297–315.
Snellman, A. 2007: *Social Hierarchies, Prejudice, and Discrimination*, Uppsala.
Snellman, A. & Ekehammar, B. 2005: 'Ethnic Hierarchies, Ethnic Prejudice, and Social Domi-
    nance Orientation', *Journal of Community and Applied Social Psychology* 15, 83–94.
Strohmaier, G. 2004: 'Galen's Not Uncritical Commentary on Hippocrates' *Airs, Waters, Places*',
    in P. Adamson (ed.), *Philosophy, Science and Exegesis in Greek, Arabic and Latin Commen-
    taries*, vol. 2, London, 1–9.
Tajfel, H. 1981: *Human Groups and Social Categories: Studies in Social Psychology*, Cambridge.
Tajfel, H. 1982: *Social Identity and Intergroup Relations*, Cambridge.
Tajfel, H., & Turner, J.C. 1979: 'An Integrative Theory of Intergroup Conflict', in W.G. Austin &
    S. Worchel (eds.), *The Social Psychology of Intergroup Relations*, Monterey, CA, 33–47.
Tajfel, H., & Turner, J.C. 1986: 'The Social Identity Theory of Intergroup Behaviour', in S.
    Worchel & W.G. Austin (eds.), *Psychology of Intergroup Relations*, Chicago, 7–24.
Tan, Z.M. 2014: 'Subversive Geography in Tacitus' "Germania"', *JRS* 104, 181–204.
Thollard, P. 1987: *Barbarie et civilisation chez Strabon. Étude critique des livres III et IV de la
    Geographie.*, Paris.
Thomas, R. 2000: *Herodotus in Context: Ethnography, Science, and the Art of Persuasion*, Cam-
    bridge.
Tsetskhladze, G. 2008: 'Pontic Slaves in Athens: Orthodoxy and Reality', in P. Mauritsch (ed.),
    *Antike Lebenswelten: Konstanz, Wandel, Wirkungsmacht. Festschrift für Ingomar Weiler zum
    70. Geburtstag*, Wiesbaden, 309–320.
Turner, J.C., & Reynolds, K.J. 2003: 'Why Social Dominance Theory Has Been Falsified', *British
    Journal of Social Psychology* 42, 199–206.
van der Vliet, E.Ch.L. 2003: 'The Romans and Us: Strabo's Geography and the Construction of
    Ethnicity', *Mnemosyne* 56, 257–272.
Vlassopoulos, K. 2013: 'The Stories of the Others: Storytelling and Intercultural Communication
    in the Herodotean Mediterranean', in E. Almagor & J. Skinner (eds.), *Ancient Ethnography:
    New Approaches*, London, 49–75.
Wells, P.S. 1999: *The Barbarians Speak: How The Conquered Peoples Shaped Roman Europe*,
    Princeton.
Young, R.J.C. 2012: 'Postcolonial Remains', *New Literary History* 43, 19–42.

FACING THE GREEKS

Some Responses of Local Populations to Greek Settlers

Marta Oller Guzmán

Abstract: In their colonial endeavours, the ancient Greeks encountered a huge variety of populations along the coasts of the Mediterranean and the Black Sea: some of them responded in a peaceful and friendly fashion to the arrival of newcomers, others were reluctant to receive foreigners and used violence against them. Interactions were of course much more complex, and the relations that developed between the communities were not necessarily consistent over time and space or throughout the different social strata. As most of our sources do, scholarship has largely focused on the Greek perspective. The present chapter approaches our Greek sources in an attempt to gain a better understanding of the perspectives of the locals and the motivations that guided their actions. The presence of the Greeks offered many new opportunities to them, though always combined with challenges and sometimes even threats. Many members of the local elites were inclined to cooperate with the settlers or at least to take over some of their cultural elements. There is more evidence for resentment against the Greeks among the common people. Repeatedly, conflicts escalated only with some delay after the initial settlement, perhaps because Greek intrusion often resulted in gradually destabilizing the social and political relations within neighbouring communities or among neighbouring tribes. The examples are drawn from the archaic and classical periods. They cover the appropriation of territory, trading with natural resources including slaves and produce as well as the transfer of cultural and cultic elements. Discussions will emphasize the attested or presumed perception of such actions on the local population.

Абстракт: Перед лицом греков: Реакции коренных народов на греческих поселенцев: В своих колониальных начинаниях древние греки столкнулись с огромным разнообразием населения вдоль побережья Средиземного и Черного морей: некоторые из них мирно и дружелюбно отреагировали на вновь прибывших, другие неохотно принимали иностранцев и применяли насилие к ним. Взаимоотношения конечно были гораздо более сложными и отношения, которые развивались между жителями, не всегда на протяжении времени или в разных социальных слоях были неизменными. Как и большинство античных источников, так современные исследования в значительной степени сосредоточены на греческой точке зрения. Настоящая глава обращается к греческим источникам, пытаясь лучше понять точку зрения местных жителей и мотивы, которыми они руководствовались. Присутствие греков открыло для них много новых возможностей, хотя они всегда связаны были с проблемами, а иногда и с угрозами. Многие представители местной элиты были склонны сотрудничать с поселенцами или, по крайней мере, перенять некоторые из элементов их культуры. Существует больше доказательств неприязни к грекам среди простых людей. Неоднократно конфликты обострялись чуть позже чем возникновение первых греческих поселений, возможно, потому что греческое вторжение часто приводило к постепенной дестабилизации социальных и политических отношений в соседних общинах или между соседними племенами. Примеры показанные в главе взяты автором из архаического и классического периодов. Они охватывают присвоение территории, торговлю природными ресурсами, в том числе рабами и продуктами, а также передачу культурных и культовых элементов. В дискуссии автор подчеркивает подтвержденное источниками или предполагаемое восприятие таких действий местным населением.

I. ON HOSPITABLE AND INHOSPITABLE PEOPLE

Throughout ancient Greek history, from the Mycenaean period to Roman times, migration is a characteristic feature of Greek civilisation. The archaic colonisation or diaspora of the Greeks is probably the best example of this phenomenon, as it made possible the extensive spread of Greek culture in large areas of the Mediterranean and Black Sea.[1] For those Greeks established abroad, contact with other peoples and different social, political and cultural realities was one of the major challenges they had to face.[2] According to the ancient sources, some of those peoples were peaceful and friendly, whilst others were clearly reluctant to accept foreigners into their lands and used violence to prevent them from settling there. However, the Greeks managed to found *apoikiai* in many places while interacting with a wide range of different local groups.

Indeed, the relationships between Greek colonists and locals are fundamental to an understanding of the development of many colonies. We may assume that good connections with the indigenous population were essential at the very time of settlement, since they could determine either the success or failure of the colonial enterprise. Therefore, knowing the degree of hospitality and the conditions upon which to achieve and maintain it certainly was an important part of organising a colonial expedition. Already in the archaic period, stories about distant peoples were circulated at least orally. Besides their entertaining character, they also provided traders and sailors with some (more or less) useful data for navigation.

---

1   In recent decades, the use of the word 'colonisation' to refer to the Greek migration of the archaic period has been criticised by a number of scholars, since it is associated with modern phenomena of imperialism and colonialism; see Garland 2014, 34 and Malkin 2016, 28–31. The word 'diaspora' has progressively gained acceptance as a more suitable term, although it is not a completely satisfactory alternative either; see Martinez-Sève 2012. In this paper I will use both words indiscriminately.

2   From the large and quickly growing bibliography, I mention the following: already in the 1960s, the *Atti del convegno di studi sulla Magna Grecia* (Taranto) provided an initial space for debate on the relationship between Greeks and locals, mostly centred around the archaeological finds. The publication of the *Atti* continued without interruption until 2019 and today the collection runs to 55 volumes. Advances in research are well represented by the collaborative volumes of Descœudres 1990; Tsetskhladze 1999; Petropoulos & Grammenos 2007. For Black Sea studies, see in particular Tsetskhladze 1998; Cojocaru 2005; Bresson, Ivantchik & Ferrary 2007; Cojocaru, Coşkun & Dana 2014. For the Iberian Peninsula, see Dietler & López-Ruiz 2009. Tréziny 2010 shows a particular interest in Greek and indigenous peoples around the Mediterranean Sea.

In a way, the *Odyssey* inaugurates this tradition of travel literature,[3] which would be developed further by ethnographers such as Hekataios of Miletos and, above all, Herodotos.[4] Some passages of his *Histories* offer splendid examples of these kinds of stories of foreign lands and native customs. Often spiced with legendary elements, those descriptions of exotic and little-known territories provided an initial portrait of indigenous populations that could warn of any dangers those who were planning to sail there.[5]

Let us take, as an example, two completely opposed stories about indigenous populations, as they appear in Herodotos' work. The first involves the king of Tartessos, whose realm was located in the southern part of the Iberian Peninsula. When the Phokaians arrived there, they became good 'friends' (προσφιλέες) with the local king Argantonios. The term φιλία ('friendship') refers not only to personal affection, but was also used to designate treaties and alliances between communities from archaic times onwards.[6] The 'friendship' between Argantonios and the Phokaians thus had a political and diplomatic dimension, which explains the king's subsequent invitation of the Greeks to settle in his *chora*. Certainly, the granting of land to foreigners was a rare privilege in archaic Greece,[7] which further emphasises the generosity of Argantonios.[8] In contrast to this idyllic vision of

---

3 See Gómez Espelosín 1994, 20f.; 2009, 283–285.

4 On Herodotos as the 'first travel writer', see Casson 1974, 95–111. A useful anthology of ancient Greek travel literature, with Spanish translations, can be found in García Moreno & Gómez Espelosín 1996.

5 On the ancient reception of Herodotos, see Rösler 2002, 85–91, who suggested that Herodotos wrote his work 'with future readers in mind' rather than thinking of a contemporary audience. However, it is generally accepted that Herodotos' *Histories* achieved wide popularity immediately after their publication; cf. Flory 1980.

6 Hdt. 1.163. The Greek word φιλία – φιλότης in epic poetry – refers to a wide range of reciprocal relationships: friendship, sexual love, hospitality and alliances or deals between states. Indeed, Taillardat 1982, 11, following E. Benveniste, argues that its primary meaning was 'le pacte en général', from which the idea of 'friendship' or 'hospitality' was developed. For its use in Homer, see Adkins 1963, 36f.; Mitchell 1997, 12–14 and Santiago Álvarez 2013, 94f. An inscription from Olympia dated c. 550–500 BC attests φιλότης in the context of a long-lasting agreement between the people from Sybaris and their allies with the Serdaioi, probably an indigenous people from Magna Graecia; see Santiago Álvarez 2013, 101–103. The term is 'un hapax nella documentazione epigrafica dei trattati', according to Giangiulio 1992, 35. Hdt. 2.181 uses φιλότης alongside συμμαχία to name the agreement between the Egyptian Amasis and the Kyrenaians. Thuk. 6.34 matches φιλία with συμμαχία in the speech of Hermokrates to the Syracusans.

7 See Piñol Villanueva 2013, 114–131, who collects and analyses examples of the granting of land to foreigners in Homer, Hesiod and Herodotos, together with the earliest epigraphic attestations; cf. Zurbach 2017, 228–236.

8 Argantonios is the only character identified by name in this story. His idealised portrait contributes to the idea of the blessed land: the wealth of his kingdom and his own longevity – he is claimed to have lived for 120 years – are illustrative examples of this notion. He is described as a fatherly monarch and rich benefactor who surprised the Greeks with his extraordinary generosity. Gómez Espelosín 2009, 289 is probably right when he states that Herodo-

Tartessos, the description of the Taurians, who inhabited the Crimean peninsula on the north coast of the Black Sea, seems to hail from a horrendous nightmare. Herodotos says that they performed human sacrifices and then brutally mutilated the corpses of the victims. They also killed any enemies they captured and decapitated them in order to use their heads as protective emblems for their houses. He adds, in conclusion, that they lived on piracy and looting.[9]

If we compare these two accounts, it is clear that Tartessos is depicted as a far safer and more desirable destination. However, the Greeks settled successfully in both places, so that one might want to know how much of these stories related to the realities encountered in place and how much was simply drawn from literary or fabulous traditions on exotic countries. At any rate, such narratives shaped the imagination of Greek merchants and settlers and alerted them of the possible dangers that might have been awaited them on their ways or at their destinations.

Most of such stories provide information on the Greeks' perceptions of the locals, but what of the locals' perceptions of the Greeks? How did the Greek presence impact these peoples? How did their perspective on the settlers develop over time? Written sources for this kind of questions are rare and were, for the most part, composed by Greek and Latin authors. Obviously, their views tend to be heavily biased and show little interest in the impact that the Greeks had on others. However, some of those texts seem to concur with respect to the discomfort and uneasiness of indigenous populations when facing the Greeks. On the following pages, I shall discuss some of these responses. A first group of examples (II) will focus on recurring patterns in the narratives of settlement histories (Kyrene, Lampsakos, Massalia, Emporion and New Kryassa), before a second part will explore various aspects of Hellenisation, such as the roles of the local elites, the introduction of new cults as well as the exploitation of natural and human resources (III).

II. SETTLEMENT NARRATIVES AND THE OCCUPATION OF LAND

Land and territory form a major area of study when investigating ancient Greek colonisation. An adequate introduction to this subject would require twice the length of this paper, so I shall confine myself to presenting some examples that are particularly revealing. The focus will be on five striking cases of changing attitudes: certain indigenous peoples initially welcomed Greek settlers, but realised over time that the Greek presence was a threat to their communities and then

tos' depiction of the king of Tartessos was 'more likely influenced by the model of Alkinoos' in the *Odyssey*; cf. Plácido 1995/6, 27f. On Argantonios and Tartessos, see also Gangutia Elícegui 1998, 242–249.

9   Hdt. 4.103. On this passage and its possible influence on the characterisation of the Taurians in Euripides' *Iphigenia among the Taurians*, see Oller Guzmán 2008.

reacted aggressively. Perhaps the most ancient account of this kind of 'broken' relationship can be found in the history of Kyrene.

1. KYRENE

Herodotos' *Histories* provide a detailed narrative of the foundation of Kyrene, which can roughly be dated to the last third of the 7th century BC.[10] It is striking that the whole process took several years. First, the settlers had difficulties to understand the indications of the Pythia from Delphi correctly, and then their lack of familiarity with the territory caused further problems. They were very much dependent on the intervention of the indigenous people to find a suitable place.[11] The latter could be obliged to accept the settlement of foreigners in their territory, but they deceived the Greek settlers and gave them inferior land.

After the foundation of Kyrene, we learn from Herodotos that the number of inhabitants remained stable for 56 years. When Battos the Happy (c. 583–570 BC)[12] succeeded to the throne, an oracle was uttered according to which all Greeks were invited to settle in Kyrene: a new distribution of land was announced (ἐπὶ γῆς ἀναδασμῷ). As a result of this call, many Greeks arrived in northern Africa.[13] This put an end to the peaceful relations with the locals and led to a war. The uneasiness of the Libyans can be clearly understood: they were 'deprived of their land and suffered violence by the Kyrenaians' (τῆς τε χώρης στερισκόμενοι καὶ περιυβριζόμενοι ὑπὸ τῶν Κυρηναίων). As they were unable to stop the expansion of Kyrene, they asked their Egyptian neighbours for help, but the alliance was in vain and they were thoroughly defeated. It is usually accepted that the γῆς ἀναδασμός was generally applied to colonial territory belonging to the city but not yet distributed,[14] often located in outlying areas and with lesser agricultural value. This territory was available for newcomers in case of need.[15] However, the new distribution of land in Kyrene, as announced by the oracle, was, in fact, a covert way of enlarging the civic body and expanding the territory at the expense of neighbouring indigenous people.[16]

---

10   Hdt. 4.150–167. For the date of the foundation, see Boardman 1995, 195; Austin 2008, 192; Zurbach 2017, 655.
11   Cf. Calame 1996, 128–135, esp. 135: 'Les indigènes prennent donc le relais du dieu de Delphes et finissent par conduire les colons de Théra … C'est alors au *Destinateur* indigène qu'il appartient de prononcer la *sanction* de l'action finalement parvenue à son achèvement.'
12   Chamoux 1953, 210 (*non uidi, apud* Zurbach 2017, 656).
13   On the arrival of additional settlers or ἔποικοι in Pontic colonies, see Avram 2012.
14   Asheri 1966, 10f.
15   Nevertheless, many inscriptions contain clauses that explicitly prevent the new distribution of land with a double objective: to maintain the same number of lots of land as was established at the foundational moment, and to preserve the inalienability of property, see Asheri 1966, 21–24.
16   Piñol Villanueva 2015, 52; Zurbach 2017, 656f.

This story illustrates how problematic the occupation of land by Greek settlers could be, particularly when such colonial expansion clashed with the interests of the locals, who, in the end, bore the brunt of the conflict. Indeed, the victory of Kyrene ensured that their territory was reinforced, and the city consolidated its sovereignty over the land, most likely overcoming its initial dependence on the locals.[17] It is not clear what historical veracity can be attributed to the account of Herodotos, but it is worth noting that the violent reaction of the Libyans to their abuse by the Kyrenaians is not a unique case in ancient written sources.

2. Lampsakos

The foundation myth of Lampsakos provides a good parallel for abusive behaviour by the Greek settlers and a violent response by the locals.[18] According to Plutarch,[19] who cites Charon of Lampsakos as his source,[20] Mandron, king of the Bebrykes,[21] promised his friend and guest (φίλος καὶ ξένος) Phobos of Phokaia that he would be granted 'part of his territory and his city' (μέρος τῆς τε χώρας καὶ τῆς πόλεως) as a reward for helping his people to defeat their neighbours. Sometime later, a group of Greek colonists arrived from Phokaia and acquired the promised land. However, these colonists began making incursions into the territory of the neighbouring indigenous peoples, wresting considerable spoils from them (ὠφελείας δὲ μεγάλας καὶ λάφυρα καὶ λείας ... λαμβάνοντες). These plundering expeditions provoked the discomfort of the Bebrykes and thus they decided to eradicate the Greeks. However, Mandron, who was a fair man, disapproved of the plan, and so a plot was hatched to kill the colonists during his absence. According to the story, it was the king's daughter Lampsake who warned the Greeks of the trap and thus became the eponymous heroine of the new city, Lampsakos.

At a first glance, Lampsake's act seems to amount to treason, but Plutarch clearly states that, before warning the Greeks, she tried to persuade her friends and relatives (τοὺς φίλους καὶ οἰκείους) not to perpetrate such a crime against the Phokaians, arguing how terrible a crime it was 'to kill their benefactors, allies and now also citizens' (εὐεργέτας καὶ συμμάχους ἄνδρας νῦν δὲ καὶ πολίτας ἀποκτιννύντες). The reluctance of the king and his daughter are, therefore, presented as justified: attacking the Greeks would constitute not only a grievous act

---

17 Zurbach 2017, 656f.
18 Lampsakos was founded in 654/3 BC by Phokaians; see Avram 2004, 986f.; Zurbach 2017, 576.
19 Plut. *Mul. Vir.* 18 (255a–c); Charon, *FGrH* 262 F 7.
20 Charon was allegedly a contemporary of Herodotos and wrote a book on the local history of his hometown; see Boulogne 2002, 305, n. 206; Fowler 2013, 641–643.
21 On this people, see Moret 2006, 59–64.

against their allies, but indeed a crime against these newly incorporated citizens.[22] However, despite being naturalised, Greek settlers abused the hospitality of the Bebrykes and probably endangered their relations with their neighbours by raiding their lands.

Certainly, the situation was awkward and the Bebrykes disagreed about how best to react to this threat, while sharing a sense of uneasiness at the abuses committed by the Greeks. According to Plutarch, the Bebrykes were at first envious of the gains made by the Greeks, but later they began to fear the Phokaians (Φωκαεῖς ... ἐπίφθονοι τὸ πρῶτον εἶτα καὶ φοβεροὶ τοῖς Βέβρυξιν). The question is: What were they afraid of? Various possibilities come to mind: first, they may have been afraid that the Greeks, with the wealth they brought, might become a rival power in political or economic terms; second, they perhaps suspected to suffer the same attacks as their neighbours; third, they probably expected to be held accountable for the raids by the Greeks whom they had invited to settle. One way or another, the network of relations amongst the indigenous peoples were destabilised and war was a likely result, just as in the case of Kyrene. Perhaps their fear stemmed from a mixture of these considerations, but it is worth stressing that envy and fear among the indigenous people are also specified as driving motivations in other accounts of Greek settlements.

3. Massalia

As regards envy, the foundation of Massalia offers an interesting parallel to the situation of Lampsakos. According to Justin's *Epitome* (2nd/4th century AD) of Pompeius Trogus' *Philippic Histories* (1st century AD),[23] a group of Phokaians arrived at Massalia under the command of two men, Simos and Protis. As in the case of Tartessos, the Greeks sought out Nannos, the king of the Segobriges,[24] in order to establish friendly relations with him, as a first step towards achieving the land they wanted (*regem Segobrigiorum, Nannum nomine, in cuius finibus urbem condere gestiebant, amicitiam petentes conveniunt*). Nannos welcomed them as guests (*hospites*) and invited them to participate in the wedding ceremony of his daughter Gyptis. By chance, Protis was chosen by Gyptis as her husband and thus became a member of the royal family. This incident conveys a legendary touch to the story, as similar marriages with foreign guests are well known from Greek mythology. But there may be more than mere symbolism behind this narrative,

---

22 The naturalisation of the Phokaians is mentioned here for the first time and must be interpreted as a consequence of land allocation made by King Mandron, thus confirming the close relation between citizenship and land tenure in the ancient Greek world, see Piñol Villanueva 2015.

23 Just. 43.3.8–13. On this passage, see Hermary, Hesnard & Tréziny 1999, 36–67.

24 This ethnic group is only mentioned by Justin, see Urso 2016, 175, n. 14.

given the widely known phenomenon of dynastic marriages to seal alliances or interstate treaties.[25]

After the city had been founded, its extraordinary development began to provoke resentment among the local population. Justin mentions the 'envy' of the Ligurians (*Ligures incrementis urbis invidentes*), whose discomfort induced them to permanent attacks on the Phokaian settlement. We can infer that there were also tensions between the indigenous peoples themselves, since the marriage agreement had been established with the tribe of the Segobriges, whereas the attacks are attributed to all the Ligurians, with no distinction drawn between the different tribes. In other words, the Greek presence and the prosperity of their colony seem to have destabilised the indigenous world and triggered conflicts that lasted for many years.

Justin continues the story as follows: Komanos, the son of Nannos, decided to put an end to the expansion of Massalia, as he feared that the Greeks might become masters of the land instead of tenants (*Non aliter Massilienses, qui nunc inquilini videantur, dominos quandoque regionum futuros*).[26] So, he devised a plot to take the city during the celebration of the *Floralia*,[27] but an anonymous indigenous woman, who was in love with a Greek, came to ruin his stratagem and helped the Massaliotes consolidate their position. Her role reminds us of Lampsake, as has been noted previously by other scholars.[28] Such similarities seem to constitute a pattern of foundation legends widely known in antiquity.

4. Emporion

Strabo informs us on yet another Phokaian settlement: Emporion (Empúries) on the east coast of the Iberian Peninsula.[29] After having occupied a little island offshore for a while, the small commercial settlement grew into a real city on the mainland.[30] This urban development appears to have had the blessing of the Indicetes, as the local inhabitants were called. According to the Geographer, the city of Emporion was initially divided in two by a wall because some of the Indicetes

---

25 Cf. the examples of Bellerophon, the Corinthian hero who married the daughter of the king of Lykia, Yobates, and of Odysseus, who was offered Nausikaa as his wife, although he declined since he was already married. On the exchange of women as a way to establish bonds of solidarity between the Greek and indigenous aristocracies, see Nenci & Cataldi 1983, 591–594. Parallel cases of intermarriage in the archaic period are considered in Rougé 1970 and Coldstream 1993.
26 Just. 43.4.3–8.
27 This episode recalls an aetiological myth explaining the reason why the city of Massalia closed its gates during this festivity. The stratagem of Komanos is somewhat similar to the Trojan horse, perhaps a borrowed mythical theme, as Meulder 2004 argues.
28 See Meulder 2004 and Moret 2006.
29 Strab. *Geogr.* 3.4.8 (159f.C).
30 On the different phases of its development, see Oller Guzmán 2013a with bibliography.

wanted to be surrounded by a common enclosure with the Greeks, for the sake of safety (ἀσφαλείας χάριν). In other words, outside the enclosure they felt threatened, so that they sought shelter inside the perimeter of the Greek city. After another unspecified period, so Strabo goes on to tell us, the two communities grew into one with 'mixed' customs and institutions. Although we have many indications for the existence of such 'double communities' and the Geographer points out that he knows several instances, it is quite unique that a literary source specifies the physical division within such an artificial unity.[31]

5. New Kryassa of Karia

The narrative of the foundation story of New Kryassa in Karia, as told by Plutarch, recalls the structure of our earlier examples: a friendly beginning, rapid development and a delayed escalation of conflicts.[32] The Melians sent out colonists due to shortage of land. After a storm destroyed their ships, they arrived at Kryassa in Karia and the locals, moved by pity or intimidated by their audacity, invited them to remain and gave them a part of their territory (τῆς χώρας μετέδωκαν). Nevertheless, since the Melian settlement grew rapidly (πολλὴν ἐν ὀλίγῳ χρόνῳ λαμβάνοντας αὔξησιν ὁρῶντες), the Karians devised a plan to destroy it, but again they failed thanks to the intervention of an indigenous woman. Her name was Kaphene and she was in love with a Greek called Nymphaios. Later, the indigenous city was completely destroyed and replaced by New Kryassa.

6. Intermediate Conclusions

Despite some differences, most of these episodes share a tripartite scheme. At first, the Greeks arrived in an unknown land with the intention of establishing a settlement. They achieved their goal thanks to the co-operation of the locals, who accepted their presence and even willingly offered a part of their territory. Later, the good relations ended when the rapid development and prosperity of the Greek city aroused envy and fear among the indigenous people. On some occasions, it is clear that the Greeks acted in an abusive manner by depriving the locals of their own land and endangered stability among the indigenous tribes. These were thus forced to react. In three cases, an indigenous woman played a crucial role in the rescue of the Greeks. Next followed a confrontation between them and the locals, whether in open combat (as in Kyrene) or through a stratagem (as in Massalia). Either way, the defeat of the indigenous people was an important condition for the

---

31   See Roller 2018, 156; cf. Garland 2014, 50.
32   Plut. *Mul. Virt.* 7 (246d–e).

consolidation of the Greek settlement.[33] The case of Emporion deviates slightly in that the relation between the Phokaian settlers and the Indicetes remained cordial, to a degree that they even amalgamated into one community; however, the fear of other local tribes seems to confirm that the advent of the Greeks had a destabilizing effect on indigenous communities and triggered potentially violent reactions.

At any rate, how much trust can we put in these stories? It is tantalising to deny such stories any credibility, given their partly legendary and largely apologetic nature. They have a strong inclination to ascribe legitimacy to the Greek settlers while putting blame on envious or even treacherous indigenous neighbours. However, whilst setting aside some surely fictional aspects or moralizing layers of the narratives, these texts quite plausibly reflect the fragile balance of relationships between Greeks and locals during these foundational periods. Greeks and indigenous people took significant risks when approaching each other.

The case of Emporion shows that such fears could leave clear traces in the material remains. Another example for that can be adduced from archaic Sicily. According to Stefano Vassallo, the arrival of the Greeks at Himera and the subsequent rapid development of their colony undoubtedly constituted a high level of risk. As a result, the indigenous people in their neighbourhood 'concentrated in larger and more secure dwellings'.[34] We can therefore conclude that Greek colonisation introduced, along with new opportunities, also new dangers and potential instability in the daily life of the local populations. The fears among the locals, whether regarding the settlers or other indigenous tribes, are variously reflected in our sources.

Resentment and fear continue to play important roles in our next set of examples, which will focus on cultural and religious aspects construed as identity markers. These may have been relevant also in the conflicts unfolded so far, but they have not been brought to the fore in our sources, at least not as strongly as in the famous case of the Scythian king Skyles: he was forced to hide his taste for 'Greek life' inside the walls of Pontic Olbia and was finally murdered by his own people because of his philhellenism.

33 However, the expansion of a colony through looting and appropriation of land from neighbouring populations was not always successful. See, for example, the failed expansion of the Phokaians in Alalia, where the new Greek settlers were defeated by a coalition of Etruscans and Carthaginians around 540 BC; see Hdt. 1.165f., with Zurbach 2017, 647–650.

34 Vassallo 2010, 52: 'Quando i primi coloni sbarcarono alla foce dell'Imera Meridionale è verosimile che gruppi indigeni si trovassero sparsi lungo le vallate, occupando insediamenti di tipo per lo più rurale per meglio controllare e sfruttare le risorse economiche dell'area, e che soltanto in caso di pericolo la popolazione si concentrasse in siti più sicuri […]. Solo dopo l'arrivo dei Greci e il successivo rapido sviluppo della colonia, che indubbiamente costituirono un elevato fattore di rischio, gli indigeni si concentrarono in abitati più grandi e sicuri.'

III. FURTHER ASPECTS OF THE GREEK PRESENCE
AMONG LOCAL POPULATIONS

1. Skyles and the Attraction of the Greek Way of Life

Herodotos asserts that the Scythians were 'very reluctant to adopt foreign customs from any other people, but even less from the Greeks'.[35] As proof, he adds the story of the Scythian philosopher Anacharsis and that of the Scythian king Skyles. Since the former is dealt with in the chapter by Alexandr Podossinov in the present volume, I shall concentrate on the latter, who arguably provides the most illustrious example of how attracted an individual of the indigenous people could feel by the life style of the Greeks and thus incurred the wrath of his fellow tribesmen.[36] Skyles was the son of a Scythian king and a Greek woman. Despite belonging to the Scythian royal family, he was strongly seduced by Greek customs and, whenever he could, went to the Greek city of Pontic Olbia; once inside the walls, with the city gates closed, he took off his Scythian clothes, to dress and live there as a Greek following Greek customs. He even married a Greek woman and built a house in Borysthenes.

As we can see, in Skyles' story, the city wall was a true border between the Greek and the indigenous worlds.[37] In much the same way as the Indicetes did in Emporion, Skyles entered the perimeter of the city looking for security, because only inside Olbia could he live the life of his choice. Skyles may very well have realised the risk of dressing as a Greek and practicing foreign cults under the eyes of the Scythians, for which reason he led a double life, one that finally caused his death. Indeed, when his fellow Scythians learned that the young king had been initiated into the mystery cult of Dionysos, they killed him, believing that this god alienates the human mind.[38] Skyles' story provides good evidence of how difficult it was for some indigenous peoples to accept foreign customs and for the tensions that these might generate in the very heart of their communities. However, once we widen our perspective and consider the power struggle between Skyles and his brother Oktamasades, we may perhaps nuance our conclusion and realize that cherishing foreign customs could be construed as disloyalty in a hostile context. As we shall see, Greek style and art were in fact esteemed highly among the Scythians, quite in contrast to the conclusion that Herodotos drew.[39]

---

35 Hdt. 4.76.
36 See Hdt. 4.76f. on Anacharsis, with Corcella 2007, 636–638; Hdt. 4.78–80 on Skyles, with Heinen 2006, 23f.; 27 and Corcella 2007, 638–640.
37 Braund 2008, 352f. The wall is considered a symbol of civilisation in many Greek texts, Garlan 1989, 129–133.
38 Hdt. 4.78–80. Certainly, Pentheus' myth in the *Bacchae* has some similarities with this story. However, the roles are swapped: in Herodotos' story, it is the local king who wants to be initiated into the mysteries of Dionysos, while his people clearly reject it.
39 Cf. Corcella 2007, 636–638 for his emphasis on the dynastic struggle for power.

2. Local Elites and Greek Prestige Objects

Several studies have highlighted the active roles that local elites played in exchanges with the Greeks. Their privileged situations allowed for contacts in many
different ways: for example, Thukydides states that the Greek cities established in
Thrace payed 'tribute' (φόρος) to the Odrysian king Seuthes, along with many
other valuable 'gifts' (δῶρα),[40] probably in order to maintain peaceful relations
with him and protect their estates from being plundered. Such 'gifts' are well attested archaeologically and epigraphically in the Black Sea.[41] They appear to have
functioned as tribute-like payments that allowed Thracians and Scythians an easy
way to benefit from the wealth of the Greek cities, while still allowing them to
prosper.[42]

The fulfilment of these obligations, which was necessary for the survival of
some Greek settlement, was a burden for the settlers, but might also had another
consequence: according to Gocha Tsetskhladze, 'they were the means of spreading Greek culture to the local elites and societies'.[43] Indeed, the finding of Greek
luxury objects in many princely tombs in various Black Sea enclaves suggests an
interest amongst local elites in owning Greek products, probably as status symbols
that would emphasize their superior position within their indigenous society or
perhaps also among rivalling potentates.[44]

3. Greek Writing

The spread of the Greek alphabet is also a reliable sign of Hellenization, although
the phenomenon is more complex. Objects such as the ring of king Skyles with a
Greek inscription,[45] might be understood as an example of the Scythian elites'
access to a higher and more differentiated culture. However, Greek script could
have been used by other segments of the indigenous population for different pur-
poses, particularly by traders. As soon as the complexity of commercial transac-
tions began to require written records, knowledge of writing would have been an

---

40 Thuk. 2.67.
41 The decree of Protogenes (*Syll.*³ 495: Olbia, late-3rd or early-2nd century BC) attests 'severe
    barbarian pressure' (Austin 2006, no. 115, p. 218) on the Pontic Greek cities by indigenous
    kinglets. It contains a detailed chronicle of the city's challenges in which Protogenes gener-
    ously helped with private contributions. The city's financial needs were especially due to the
    demands of the locals, see lines A.10–11 (Saitaphernes asked for gifts), A.34–35 (the Saii
    came to collect gifts), A.44 (more gifts for Saitaphernes), A.84 (Saitaphernes asks for more
    favours). See Müller 2011 for a commentary.
42 Gabrielsen 2007, 301.
43 Tsetskhladze 2002, 84.
44 West 2002, 451–454; cf. Heinen 2006, 11–28. For comparison with the east coast of the Ibe-
    rian Peninsula, see Oller Guzmán 2013a, 189f. with further bibliography.
45 Dubois 1996, 11–14.

important advantage. In the Greek world, private commercial letters are attested from the end of the 6th century BC onwards, especially in peripheral colonial regions such as the northern coast of the Black Sea and the Iberian Peninsula. In both areas, some indigenous names are recorded in this kind of correspondence, together with Greek names, suggesting the involvement of locals in Greek commercial activity.[46] Whether or not these supposedly indigenous merchants were members of important local families is not clear, but the assumption does not appear far-fetched. Be that as it may, there is no doubt that indigenous populations – specifically their kings or elites – reaped the benefit of taxes imposed on Greek commercial activities in their territories, as proven by the inscription from the emporion Pistiros in inland Thrace and other Greek sources.[47]

4. The Introduction of Greek Cults

The role of extra-urban or peripheral sanctuaries in colonial areas as a space of contact between Greek and non-Greek communities has been widely discussed. Sanctuaries such as the Heraion of Kroton are considered to be spaces for 'mediation and sovereignty',[48] where the Greek presence in an area might be strengthened and, at the same time, Greek cultural practices spread. However, we should be open to the idea that influence was exerted in two (or even multiple) directions within most colonial contexts, so that the Greeks would have been exposed to indigenous practices and customs as well.[49]

At the same time, we should keep in mind that not all peoples shared the same degree of permeability when it came to adopting or integrating foreign cultic practices or religious beliefs. Likewise, not all rituals or mythical traditions were equally suitable for a transfer to a foreign society. As we have mentioned before, Skyles was put to death for being discovered in the midst of a Bacchic fervour.[50]

---

46 On the practice of Iberians to adapt and use the alphabet, see Oller Guzmán 2013b, 84–86. However, not all indigenous populations were equally permeable to the use of writing.

47 Pistiros: *SEG* XLIII 486; Chankowski & Domaradzka 1999, 248–51; cf. Santiago Álvarez & Gardeñes Santiago 2002, 21–25. And see Dana 2007, 83–85 on the letter of Apatorios (Kerkinitis, ca. 400 BC), a trader in salt-fish who encouraged Neomenios to inquire about who payed tax (τέλος) to the Scythians (ll. 6–8); cf. Alemany & Oller 2011, 341–343.

48 Polignac 2001, 16–18.

49 See, for example, the mixed composition – Greek and indigenous – of the votive materials in the sacred areas of Timmari and Garaguso in Magna Graecia, 'luoghi di culto ove i Greci entrano in contatto con le popolazioni locali, attivando così una serie di cerimoniali ... secondo un doppio codice... segmenti sovrapposti di pratiche greche e pratiche epicorie' (Osanna 2010, 610). On the risk of 'barbarisation of Greek populations', see Baralis 2014, 100f. On the joint use of rural sanctuaries, also see the examples of the Olbian Chora, as discussed by J. Porucznik, chapter II in this volume.

50 Osborne 2008, 334 argues that, for the Scythians, the problem with Skyles was not that of 'importing Greek cult practices', but of allowing a Scythian to take 'the initiative in Greek cult practice'.

The wise Anacharsis suffered a similar fate, after being discovered while performing rituals in honour of the Mother of the Gods. This might suggest that the Scythians had a particular aversion to mystery cults.[51]

Maybe Herodotos' digression on the city of Gelonos, presumably located
somewhere deep in the hinterland of the northern Black Sea coast,[52] lends further
support to such a view. The settlement had been established in the territory of the
Scythian Boudinoi by Greeks coming from trade-stations to its south. Gelonos is
described as a wooden city surrounded by high walls, hosting sanctuaries and altars of Greek gods (Ἑλληνικῶν θεῶν ἱρά) in its interior. Every other year, they
celebrated a festival of Dionysos, which included ecstatic mysteries. The rituals
were performed within the civic space, which may imply that this manifestation of
Hellenicity could not be freely expressed beyond the walls in 'barbarian' territory.

Herodotos mentions, together with the cult of Dionysos, the language of the
Gelonoi, which was a mixture of Greek and Scythian (γλώσσῃ τὰ μὲν Σκυθικῇ, τὰ
δὲ Ἑλληνικῇ χρέωνται),[53] as a clear indication that they actually had Greek ancestors.[54] The cult thus appears as a defining feature of Greek identity for those who
lived among Scythians, not only in Gelonos but also in other Greek colonies like
Olbia where the association between Dionysos and Orphics is well attested in epigraphic data from the 6th century BC.[55] Looking at the Dionysiac cult from this
angle, one might wonder whether negative responses to it by the Scythians were
in the first place directed against a Greek (or colonial) identity marker that could
easily be targeted as alien and hostile. Conversely, it is worth stressing that the
cult of Dionysos in Thrace seems to have reached wider acceptance. Herodotos
and other literary sources underline the peculiarities of this cult among the Thracians, granting it a prophetic aspect together with particular ritual practices far from
those known for the god in the Greek world.[56] In this case, the cult of Dionysos is
an ethnic identity marker to which the Thracians likely responded.

---

51 Hdt. 4.76.
52 Hdt. 4.108. The existence of this city has not been proven, but archaeological sites in the
    northern Pontic region, such as Belskoe, shed light on the presence and activity of Greek
    craftsmen at local settlements, see Tsetskhladze 1998, 50; West 2002, 454; Vachtina 2007,
    35.
53 On bilingualism in the ancient Black Sea area, see Oller Guzmán forthcoming.
54 The use of the Greek language was the main distinguishing feature between Greeks and bar-
    barians, from the first mentions of the word βάρβαρος, whose onomatopoeic origin simply re-
    ferred to the incomprehensible way in which foreigners spoke, without any pejorative impli-
    cations.
55 See some remarks on this Orphic connection in Osborne 2008, 335–337 and Tortorelli
    Ghidini 2013, 148f.; for epigraphic data, see Dubois 1996, 128f.; 154f. It is possible that the
    Greeks, during their travels into the *hinterland*, brought the cult of Dionysos with them, see
    Rusjaeva 1999, 77.
56 Hdt. 5.7; 7.111; Oller Guzmán 2018.

5. The Exploitation of Material and Human Resources

An important consequence of the colonial settlements was the exploitation of natural and human resources. Xenophon, in the *March of the Ten Thousand*, provides a lucid account of the goods that were particularly appreciated when choosing a territory in which to found a colony.[57] The coastal zone near Pontic Herakleia on the southern coast of the Black Sea had sufficient space for ten thousand people. There was a natural harbour, a fresh-water spring and a great deal of timber, especially well-suited for shipbuilding. The soil was fertile and produced 'barley, wheat, beans of all kinds, millet and sesame, an adequate quantity of figs, an abundance of grapes, which yield a good sweet wine'. It only lacked olives to produce oil, but, this inconvenience apart, the place was most suitable for settlement.

Polybios, when talking about the war between Byzantion and Rhodes in 220 BC, provides a list of the products coming from the Black Sea that were used in the rest of the world: cattle and slaves 'are supplied by the countries around the Black Sea, as is generally agreed, in greater quantity, and of better quality than those provided by any other country; and as far as luxuries are concerned, they supply us with honey, wax and salt-fish in abundance'.[58] As for the Pontic grain trade,[59] Polybios states that there was a balanced import-export exchange between the Black Sea and the rest of the Mediterranean world, most likely depending on the year and the specific situation – surplus or shortage – of each region.

We have already mentioned the fact that Greek activity could also be a source of wealth for indigenous populations, but not everyone benefited likewise, and violent conflicts could often be the result. Indeed, if the Greeks succeeded in imposing their might, indigenous peoples risked losing partial or total control over their territories and resources, as in the case of Kyrene; or worse still, as with Kryassa, they could be expelled or annihilated. In some cases, such as Akanthos in Thrace,[60] the local populations fled even before the arrival of the Greek settlers. They surely were afraid of suffering defeat and enslavement.[61]

Indeed, we have some evidence for the subjugation of indigenous populations by Greek colonists, such as the Mariandynoi at Pontic Herakleia, who, according to Strabo, were reduced to slavery by the first settlers coming from Miletos.[62] Most probably, they were rural slaves, whose main activity was farming.[63] Yet

---

57 Xen. *An.* 6.4.3–6.
58 Polyb. 4.38.1–10, see translation and commentary in Austin 2006, 216f.
59 From among the large bibliography, see, e.g., Braund 2007.
60 Plut. *Quaest. Gr.* 30 (298a–b); cf. Zurbach 2017, 572.
61 Garlan 1989, 75–77; Heinen 2006, 66–76; Avram 2007, 247.
62 Strab. *Geogr.* 12.3.4 (542C); cf. Roller 2018, 695f. Similar situations might have occurred in other colonial areas: the Kyllirians, in Syracuse, and the Bithynians, in Byzantium, most probably belonged to the same type of indigenous populations forced to work the land of the new Greek settlers; see Garlan 1984, 116–121.
63 Zurbach 2017, 583–589

slaves were also trafficked to all parts of the ancient Greek world, and the Black
Sea is usually considered a primary source for the ancient Mediterranean slave-
trade.[64] There is abundant – and constantly increasing – literary and epigraphic
evidence to support this view.[65] To give but one recently-found example: in a private lead letter from the Taman peninsula and dated to the late-5th or early-4th century,[66] a man called Pistos writes to Aristonimos about certain sums that must be
paid (ἀποτείσασθαι). The names of the debtors are then given, together with the
amounts to be paid. In line 3, in connection with a 'gold stater' (l. 2 στατῆρα
χρυσō), a 'slave' (ἀνδράποδον) is clearly mentioned as part of a payment from a
man called Sapasis (Σαπασιν), a name otherwise unknown, maybe an indigenous
person. It is worth noting that this slave is the only piece of merchandise specified
among the amounts to be paid, perhaps because he was included in a delivery as
part of a commercial agreement.

What impact did this kind of human trafficking have on indigenous populations? This question is difficult to answer, because the number of enslaved people
might vary depending on the area and the period upon which we focus. However,
locals were likely to have been involved in this commercial practice: Herodotos
says that some Thracian tribes sold their own children,[67] perhaps as a way of con-
trolling a surplus population. On the other hand, military victories provided an
important source of slaves, whether they came from conflicts between barbarian
peoples, between Greeks and barbarians or, to a lesser extent, between Greeks
themselves.[68] Although slavery was not a phenomenon introduced only by the
Greeks into the Black Sea region, one may well wonder if the high demand of
slaves in the Pontic and Mediterranean Greek cities and further the presence of
large slave markets in colonies such as Tanaïs at the mouth of the Don strongly
increased the slave trade. In this regard, slavery must have enhanced the sense of
insecurity among the locals.

IV. CONCLUDING REMARKS

To conclude, throughout this article we have aimed to understand the impact of
the Greek presence on the indigenous populations of colonial areas. The difficulty
we are confronted with is that we are nearly exclusively depending on a Greek
literary tradition that has filtered and distorted past events and perceptions. And,
yet, if the various stories are read against the grain and compared with each other,

---

64 On the formation and development of the slave-trade in the northern Black Sea in the 6th and
    5th centuries BC, see Gavrilyuk 2003.
65 Heinen 2001/6; Avram 2007.
66 Zavoïkina & Pavlichenko 2016.
67 Hdt. 5.6.1, see Avram 2007, 247, who also mentions the Phrygians as sharing the same prac-
    tice, according to Philostratos, VA 8.7.1.
68 Garlan 1984, 60–64.

we can still learn about different ways of interaction between the Greek settlers and the locals. It is worth stressing that, according to the ancient sources analysed above, many colonial enterprises started quite peacefully. At least initially, friendly relation and cooperation resulting in mutual benefit seem to have prevailed. Tensions and conflicts were often there as well, but, as we have seen, in many cases they arose years after the arrival of the Greeks, when their colonial endeavour turned out to be more successful than expected and the newcomers (or even their offspring) began to claim more resources than seemed to be their fair share.

In several instances, disputes concerned the control of land, which was a standard problem. But beyond this, aggressive intrusions could also pertain to other material or economic resources, ranging from the exploitation of quarries over enslavement to challenging pre-existing trade relations and alliances with other neighbouring peoples. Not even the Greek sources deny that Greek settlers often committed abuses in those strives, despite a certain tendency to justify their actions. Often those abuses escalated pre-existing tensions, sometimes up to the level of a full-scale war, although responses to the Greek threat varied a lot. One of the more interesting consequences is that, in various ways, Greek settlers tended to destabilise the relations between the indigenous populations, giving birth to oppositions between groups or individuals in favour and against the Greeks and their way of life. Most probably, the envy and fear experienced by local populations and often mentioned in the texts were manifestations of a complex reality that included a mixture of discomfort and fascination when facing the Greeks.

Acknowledgment

This paper has been produced within the framework of two Research Projects: Prosopographia Eurasiae Centralis Antiquae et Medii Aevi FFI2014-58878-P and Estudio diacrónico de las instituciones socio-políticas de la Grecia antigua y de sus manifestaciones míticas FFI2016-79906-P (AEI/FEDER, UE).

Bibliography

Adkins, A.W.H. 1963: '"Friendship" and "Self-Sufficiency" in Homer and Aristotle', *CQ* 13.1, 30–45.

Alemany, A. & Oller, M. 2018: 'Contactos entre griegos y escitas en el litoral septentrional del mar negro durante el s. IV aC', in J. Pascual, B. Antela-Bernárdez & D. Gómez Castro (eds.), *Cambio y pervivencia. El mundo griego en el el siglo IV a.C.*, Madrid, 335–347.

Asheri, D. 1966: *Distribuzioni di terre nell'antica Grecia*, Torino.

Austin, M. 2006: *The Hellenistic World from Alexander to the Roman Conquest. A Selection of Ancient Sources in Translation.* 2nd ed. Cambridge.

Austin, M. 2008: 'The Greeks in Libya', in G.R. Tsetskhladze (ed.), *Greek Colonisation. An Account of Greek Colonies and Other Settlements Overseas*, vol. 2, Leiden, 187–217.

Avram, A. 2004: 'The Propontic Coast of Asia Minor', in M.H. Hansen & Th.H. Nielsen (eds.), *An Inventory of Archaic and Classical Poleis,* Oxford, 975–999.

Avram, A. 2007: 'Some Thoughts about the Black Sea and the Slave Trade before the Roman Domination (6th–1st Centuries BC)', in Gabrielsen & Lund 2007, 239–251.

Avram, A. 2012: 'Le rôle des époikoi dans la colonisation grecque en mer Noire: quelques études de cas', *Pallas* 89, 197–215.

Baralis, A. 2014: 'Hellénisation et déshellénisation dans l'espace pontique: le passé antique à l'épreuve des constructions identitaires modernes', in H. Ménard & R. Plana-Mallart (eds.), *Contacts de cultures, constructions identitaires et stéréotypes dans l'espace méditerranéen*, Montpellier, 91–106.

Boardman, J. 1995: *Les Grecs outre-mer: colonisation et commerce archaïques*, Naples.

Boulogne, J. 2002: *Plutarque, Oeuvres morales, IV. Conduites méritoires de femmes. Étiologies romaines – Étiologies grecques. Parallèles mineurs*, Paris.

Braund, D. 2007: 'Black Sea Grain for Athens? From Herodotus to Demosthenes', in Gabrielsen & Lund 2007, 39–68.

Braund, D. 2008: 'Scythian Laughter: Conversations in the Northern Black Sea Region in the 5th Century BC', in P. Guldager Bilde & J. Hjarl Petersen (eds.), *Meetings of Cultures in the Black Sea Region: Between Conflict and Coexistence*, Aarhus, 347–367.

Bresson, A., Ivantchik, A. & Ferrary, J.-L. (eds.) 2007: *Une koinè pontique – Cités grecques, sociétés indigènes et empires mondiaux sur le littoral nord de la mer Noire (VIIe s. a.C.–IIIe s. p.C.)*, Bordeaux.

Calame, C. 1996: *Mythe et histoire dans l'Antiquité grecque. La création symbolique d'une colonie*, Lausanne.

Casson, L. 1974: *Travel in the Ancient World*, London.

Chamoux, F. 1953: *Cyrène sous la monarchie des Battiades*, Athènes.

Chankowski, V. & Domaradzka, L. 1999: 'Réédition de l'inscription de Pistiros et problèmes d'interprétation', *BCH* 123, 247–258.

Cojocaru, V. (ed.) 2005: *Ethnic Contacts and Cultural Exchanges North and West of the Black Sea from the Greek Colonization to the Ottoman Conquest*, Iaşi.

Cojocaru, V., Coşkun, A. & Dana, M. (eds.) 2014: *Interconnectivity in the Mediterranean and Pontic World during the Hellenistic and Roman Periods*, Cluj-Napoca.

Coldstream, J.N. 1993: 'Mixed Marriages at the Frontiers of the Early Greek World', *OJA* 12.1, 89–107.

Corcella, A. 2007: 'Book IV', in D. Asheri, A. Lloyd & A. Corcella (eds.), *A Commentary on Herodotus Books I–IV*, ed. by O. Murray & A. Moreno, with a contribution by M. Brosius, Oxford, 543–721.

Dana, M. 2007: 'Lettres grecques dialectales nord-pontiques (sauf IGDOP 23–26)', *REA* 109.1, 67–97.

Dana, M. 2011: *Culture et mobilité dans le Pont-Euxin*, Bordeaux.

Descœudres, J.-P. (ed.) 1990: *Greek Colonist and Native Populations. Proceedings of the First Australian Congress of Classical Archaeology held in honour of Emeritus Professor A.D. Trendall (Sydney 9–14 July 1985)*, Oxford.

Dietler, M. & López-Ruiz, C. (eds.) 2009: *Colonial Encounters in Ancient Iberia: Phoenician, Greek, and Indigenous Relations*, Chicago.

Dubois, L. 1996: *Inscriptions dialectales grecques d'Olbia du Pont*, Geneva.

Flory, S. 1980: 'Who Read Herodotus' *Histories*?', *AJP* 101.1, 12–28.

Fowler, R.L. 2013: *Early Greek Mythography*. Vol. 2: *Commentary*, Oxford.

Gabrielsen, V. 2007: 'Trade and Tribute: Byzantion and the Black Sea Straits', in Gabrielsen & Lund 2007, 287–324.

Gabrielsen, V. & Lund, J. (eds.) 2007: *The Black Sea in Antiquity. Regional and Interregional Economic Exchanges*, Aarhus.

Gangutia Elícegui, E. 1998: 'La Península Ibérica en los autores griegos: de Homero a Platón', in J. Mangas & D. Plácido (eds.), *Testimonia Hispaniae Antiqua* II A, Madrid.

García Moreno, L.A. & Gómez Espelosín, F.J. 1996: *Relatos de viajes en la literatura griega antigua*, Madrid.

Garlan, Y. 1984: *Les esclaves en Grèce ancienne*, Paris.

Garlan, Y. 1989: *Guerre et économie en Grèce ancienne*, Paris.

Garland, R. 2014: *Wandering Greeks. The Ancient Greek Diaspora from the Age of Homer to the Death of Alexander the Great*, Princeton.

Gavrilyuk, N.A. 2003: 'The Graeco-Scythian Slave-Trade in the Sixth and Fifth Centuries BC', in P. Guldager Bilde, J.M. Højte & V.F. Stolba (eds.), *The Cauldron of Ariantas: Studies Presented to A.N. Shcheglov on His 70th Birthday*, Aarhus, 75–85.

Giangiulio, M. 1992: 'La φιλότης tra Sibariti e Serdaioi (Meiggs-Lewis, 10)', *ZPE* 93, 31–44.

Gödde, S. 2011: '"Fremde Nähe". Zur mythologischen Differenz des Dionysos', in R. Schlesier (ed.), *A Different God? Dionysos and Ancient Polytheism*, Göttingen, 85–104.

Gómez Espelosín, F.J. 1994: 'Relatos de viajes en la Odissea', *Estudios Clásicos* 36 (106), 7–31.

Gómez Espelosín, F.J. 2009: 'Iberia in the Greek Geographical Imagination', in M. Dietler & C. López-Ruiz (eds.), *Colonial Encounters in Ancient Iberia. Phoenician, Greek, and Indigenous Relations*, Chicago, 281–297.

Heinen, H. 2001/6: 'Sklaverei im nördlichen Schwarzmeergebiet. Zum Stand der Forschung', in H. Bellen & H. Heinen (eds.), *Fünfzig Jahre Forschungen zur antiken Sklaverei an der Mainzer Akademie 1950–2000. Miscellanea zum Jubiläum*, Stuttgart 2001, 487–503, Taf. II = idem, *Vom hellenistischen Osten zum römischen Westen. Ausgewählte Schriften*, Stuttgart 2006, 520–538.

Heinen, H. 2006: *Antike am Rande der Steppe. Der nördliche Schwarzmeerraum als Forschungsaufgabe*, Stuttgart.

Hermary, A., Hesnard, A. & Tréziny, H. 1999: *Marseille grecque. La cité phocéenne (600–49 av. J.-C.)*, Paris.

Hupe, J. (ed.) 2006: *Der Achilleus-Kult im nördlichen Schwarzmeerraum vom Beginn der griechischen Kolonisation bis in die Römische Kaiserzeit. Beiträge zur Akkulturationsforschung*, Rahden, Westf.

Ivantchik, A.I. 2005: *Am Vorabend der Kolonisation. Das nördliche Schwarzmeergebiet und die Steppennomaden des 8.–7. Jhs. v.Chr. in der klassischen Literaturtradition: Mündliche Überlieferung, Literatur und Geschichte*, Berlin.

Malkin, I. 2016: 'Greek Colonisation: The Right to Return', in L. Donnellan, V. Nizzo & G.-J. Burgers (eds.), *Conceptualising Early Colonisation*, Brussels, 27–50.

Martinez-Sève, L. (ed.) 2012: *Les diasporas grecques du VIIIᵉ à la fin du IIIᵉ siècle av. J.-C.* = *Pallas. Revue d'Études Antiques* 89.

Mitchell, L.G. 1997: *Greeks Bearing Gifts: The Public Use of Private Relationships in the Greek World, 435–323 B.C.*, Cambridge.

Meulder, M. 2004: 'La prise de Marseille par les Ségobriges: un échec'. *DHA* 30.1, 11–32.

Moret, P. 2006: 'La formation d'une toponyme et d'une ethnonymie grecques de l'Ibérie: étapes et acteurs', in G. Cruz Andreotti, P. Le Roux & P. Moret (eds.), *La invención de una geografía de la Península Ibérica*. Vol. 1: *La época republicana*, Madrid, 39–76.

Müller, Ch. 2011: 'Autopsy of a Crisis: Wealth, Protogenes and the City of Olbia in c. 200 BC', in Z.H. Archibald, J.K. Davies & V. Gabrielsen (eds.), *The Economies of Hellenistic Societies. Third to First Centuries BC*, Oxford, 324–344.

Nenci, G. & Cataldi, S. 1983: 'Strumenti e procedure nei rapporti tra greci e indigeni', in *Forme di contatto e processi di trasformazione delle società antiche*, Pisa, 581–604.

Oller Guzmán, M. 2008: 'Ifigenia ξενοκτόνος', *Faventia* 30.1–2, 223–240.

Oller Guzmán, M. 2013a: 'Griegos e indígenas en *Empórion* (s. VI–IV a.C.): un estado de la cuestión', in Santiago Álvarez & Oller Guzmán 2013, 187–202.

Oller Guzmán, M. 2013b: 'Quelques réflexions autour du commerce grec au littoral septentrional de la Mer Noire d'après l'épigraphie (VIᵉ–IVᵉ siècles av. J.-C.)', in G.R. Tsetskhladze, S. Atasoy, A. Avram, S. Dönmez & J. Hargrave (eds.), *The Bosporus: Gateway between the An-*

cient West and East (1ˢᵗ Millennium BC–5ᵗʰ Century AD). Proceedings of the 4ᵗʰ International Congress on Black Sea Antiquities, Istanbul, 14–18 September 2009, Oxford, 83–87.

Oller Guzmán, M. 2018: 'What was Thracian in the Cult of Dionysos in Roman Thrace?', in L. Vagalinski, M. Raycheva, D. Boteva & N. Sharankov (eds.), *Proceedings of the First International Roman and Late Antique Thrace Conference "Cities, Territories and Identities" (Plovdiv, 3ʳᵈ–7ᵗʰ October 2016), Izvestiya na Natsionalniya Arkheologicheski Institut / Bulletin of the National Archaeological Institute* 44, 211–220.

Oller Guzmán, M. forthcoming: 'Langues en contact et commerce: les défis linguistiques dans le cadre de la colonisation ionienne', in G.R. Tsetskhladze (ed.), *Ionians in the East and West. Colloquia Antiqua 27*, Leuven.

Osanna, M. 2010: 'Greci ed indigeni nei santuari della Magna Grecia: i casi di Timmari e Garaguso', in H. Trézini (ed.), *Grecs et indigènes de la Catalogne à la Mer Noire*, Paris, 605–611.

Osborne, R. 2008: 'Reciprocal Strategies: Imperialism, Barbarism and Trade in Archaic and Classical Olbia', in P. Guldager Bilde & J.H. Petersen (eds.), *Meetings of Cultures between Conflicts and Coexistence*, Aarhus, 333–346.

Petropoulos, D.V. & Grammenos, E.K. (eds.) 2007: *Ancient Greek Colonies in the Black Sea*, vols. 1–2, Oxford.

Piñol Villanueva, A. 2013: 'Acceso de extranjeros a bienes inmuebles: primeros testimonios (s. VIII–V a.C.)', in Santiago Álvarez & Oller Guzmán 2013, 113–145.

Piñol Villanueva, A. 2015: *El extranjero en la Grecia arcaica: acceso a la tierra y a la justicia*, PhD Bellaterra. URL: https://ddd.uab.cat/record/164364.

Plácido, D. 1995/6: 'La imagen simbólica de la Península Ibérica en la Antigüedad', *Studia Historica. Historia Antigua* 13/4, 21–35.

Polignac, F. de 2001: 'Mediation, Competition and Sovereignty: The Evolution of Rural Sanctuaries in Geometric Greece', in S.E. Alcock & R. Osborne (eds.), *Placing the Gods. Sanctuaries and Sacred Spaces in Ancient Greece*, Oxford, 3–18.

Roller, D.W. 2018: *A Historical and Topographical Guide to the Geography of Strabo*, Cambridge.

Rösler, W. 2002: 'The *Histories* and Writing', in E.J. Bakker, I.J.F. de Jong & H. van Wees (eds.), *Brill's Companion to Herodotus*, Leiden, 79–94.

Rougé, J. 1970: 'La colonisation grecque et les femmes', *Cahiers d'histoire* 15, 307–317.

Rusjaeva, A.S. 1999: 'Les *temene* d'Olbia à la lumière de son histoire au VIᵉ siècle av. n.è.', in O. Lordkipanidze, P. Lévêque, A. Fraysse & E. Geny (eds.), *Religions du Pont-Euxin*, Paris, 75–84.

Santiago Álvarez, R.-A. 2013: 'De hospitalidad a extranjería', in Santiago Álvarez & Oller Guzmán 2013, 89–111.

Santiago Álvarez, R.-A. & Gardeñes Santiago, M. 2002: 'Interacción de poblaciones en la antigua Grecia', *Faventia* 24.1, 15–30.

Santiago Álvarez, R.-A. (coord.) & Oller Guzmán, M. (ed.) 2013: *Contacto de poblaciones y extranjería en el mundo griego antiguo. Estudio de fuentes*, Bellaterra.

Taillardat, J. 1982: 'Φιλότης, πίστις et foedus', *REG* 95, 1–14.

Tortorelli Ghidini, M. 2013: 'Dionysos versus Orpheus', in A. Bernabé, M. Herrero de Jáuregui, A.I. Jiménez San Cristóbal & R. Martín Hernández (eds.), *Redefining Dionysos*, Berlin, 144–158.

Tréziny, H. (ed.) 2010: *Grecs et indigènes de la Catalogne à la Mer Noire*, Paris.

Tsetskhladze, G.R. (ed.) 1998: *The Greek Colonisation of the Black Sea Area: Historical Interpretation of Archaeology*, Stuttgart.

Tsetskhladze, G.R. (ed.) 1999: *Ancient Greeks West and East*, Leiden.

Tsetskhladze, G.R. 2002: 'Ionians abroad', in G.R. Tsetkhladze & A.M. Snodgrass (eds.), *Greek Settlements in the Eastern Mediterranean and in the Black Sea*, Oxford, 81–96.

Urso, G. 2016: 'Marsiglia e l'Occidente nelle *Storie Filippiche*', in A. Galimberti & G. Zecchini (eds.), *Studi sull' Epitome di Giustino*. Vol. 3: *Il tardo ellenismo. I Parti e i Romani,* Milan, 171–191.

Vachtina, M.J. 2007: 'Greek Archaic Orientalising Pottery from the Barbarian Sites of the Forest-Steppe Zone of the Northern Black Sea Coastal Region', in Gabrielsen & Lund 2007, 23–37.

Vassallo, S. 2010: 'L'incontro tra indigeni e Greci di Himera nella Sicilia centro-settentrionale (VII–V sec. a.C.)', in H. Tréziny (ed.), *Grecs et indigènes de la Catalogne à la Mer Noire,* Paris, 41–54.

West, S. 2002: 'Scythians', in E.J. Bakker, I.J.F. De Jong & H. Van Wees (eds.), *Brill's Companion to Herodotus*, Leiden, 437–456.

Zavoĭkina, N.V. & Pavlichenko, N.A. 2016: 'Pis'mo na svintsovoĭ plastine iz Patreya' (A Letter on a Lead Plate from Patrey), in V.L. Kuznetsova (ed.), *Fanagoriya. Rezul'taty arkheologicheskikh issledovaniĭ* (Phanagoria. Results of Archaeological Studies). Vol. 4, Moscow, 230–249.

Zurbach, J. 2017: *Les hommes, la terre et la dette en Grèce c. 1400 – c. 500 a.C.*, Bordeaux.

GRAUSAME UND EDLE SKYTHEN

Paradoxe Barbarenprojektionen
bei antiken Autoren (besonders Strabon)

Alexandr V. Podossinov

Abstract: Cruel and Noble Scythians. Paradoxical Projections of Barbarians in Ancient Authors (Especially Strabo): Two contradictory images of the Scythians were commonplaces in ancient literature. On the one hand, those 'barbarians' of the North were well-known for their cruelty and lack of human emotions, an image probably shaped in 5th-century-BC Athens. On the other hand, the simplicity of their nomadic life style, for which they were known since the days of Homer, allowed for representing them as models of ethical and uncorrupted humans, a contrast foil for the decadent Greek society, as unfolded in much detail by Ephoros. Interestingly, not only have these contradictory descriptions survived from the literature of Late Antiquity, but also the theme of the influence that other societies, especially the Greeks and Romans, supposedly exerted on them. Here, too, we see the same paradox: their impact could either be positive and thus civilize the savage Scythians, or negative and thus infect them with the greed of the seafaring Greeks or the debauchery of Mediterranean cultures. Especially popular was the *topos* of the drunken Scythian, attested since the poetry of Anakreon in the 6th century BC, but even this could be turned upside down, with the Scythian philosopher Anacharsis posing also as a teacher of sobriety. The construed nature of these characterizations is apparent, and the choice would depend on the predilection of an author or perhaps only the genre he was writing. The Augustan geographer Strabo provides the largest number of examples, quoting the two opposing traditions side by side. The essay ends with quotations from Christian authors, whose cosmopolitan ideology gradually dissolved the opposition of Greeks (or Romans) and Scythian barbarians.

Абстракт: Жестокие и благородные скифы: парадоксальные проекции варваров у античных авторов (особенно у Страбона): В античной литературе существовало два противоречащих друг другу образа скифов. С одной стороны, северные «варвары» были хорошо известны своей жестокостью и отсутствием человеческих эмоций – образ, вероятно, сформировавшийся в V в. до н. э. в Афинах. С другой стороны, простота их кочевнического образа жизни, известная еще со времен Гомера, позволяла представлять их как образец не испорченных цивилизацией людей, резко контрастирующий с деградацией греческого общества, как это было подробно показано Эфором. Интересно, что не только эти противоречивые описания дожили до поздней античности, но и также тема влияния, которое на них предположительно оказали другие общества, особенно греки и римляне. Здесь мы также усматриваем некий парадокс: их влияние могло быть или положительным, цивилизующим диких скифов, или, наоборот, отрицательным, заражающим их жаждой наживы мореплавателей греков или развратом средиземноморских культур. Особенно был распространен топос пьянствующего скифа, известный с поэзии Анакреонта в VI в. до н. э. Но даже и он мог быть опровергнут, если вспомнить скифского философа Анахарсиса, который выступал как проповедник трезвости. Искусственное конструирование этих характеристик и их выбор зависел, очевидно, от предпочтений автора или возможно даже от того жанра, в котором он творил. Наибольшая часть свидетельств в статье взято из «Географии» писателя августовского времени Страбона, который демонстрирует оба противоположных мнения один за другим. Анализ этого явления заканчивается цитированием христианских авторов, чья космополитическая идеология постепенно снимала оппозицию «греки (римляне) – скифы-варвары».

I. AMBIVALENTE SKYTHENBILDER

Die Skythen als ‚Barbaren' *par excellence* besaßen in den Augen der antiken Autoren alle Merkmale, die in einer Gegenüberstellung von ‚Zivilisation und Barbarei' auftreten: auf der einen Seite waren sie kriegerisch, wild, heimtückisch, primitiv und ungebildet, auf der anderen Seite enthaltsam, uneigennützig, friedlich, naiv und weise, außerdem tapfer und furchtlos.[1] Das jeweilige Bild des skythischen Barbaren ist dabei abhängig von den ideologischen Vorstellungen des Autors und von dem Genre, in dem er schrieb. Es ist mithin nicht verwunderlich, dass sowohl die xenophoben als auch xenophilen Strömungen, die reichlich (nicht nur in antiken) Gesellschaften vertreten waren, auf die Ausprägung des Skythenbildes eingewirkt haben.[2]

Das grundlegende Bild eines skythischen Nomaden basiert auf Herodot, der einen großen Teil des vierten Buches seiner Historien dem ‚Skythenlogos' widmete.[3] Es muss dabei besonders auf die ‚Toleranz' der herodoteischen Beschreibungen hingewiesen werden, da Herodot die Skythen an keiner Stelle als ‚Barbaren' bezeichnet.[4] Dafür wurde er scharf von Plutarch kritisiert, der ihn als ‚Barbarenfreund' (φιλοβάρβαρος) bezeichnete.[5] Herodot zeigt Verständnis dafür, dass jedes Volk seine eigenen Traditionen besitzt und diese respektiert werden

---

1   Über die Wahrnehmung der Barbaren seitens antiker Autoren gibt es umfangreiche Literatur; hier sei nur eine Auswahl aus den wichtigsten Arbeiten zum Thema genannt: Zahn 1896; Eichhorn 1904; Thompson 1921; Jüthner 1923; Lovejoy & Boas 1935; Bengtson 1954, 25–40; Schmitt 1957/8; Bacon 1961; Schwabl u.a. 1962; Christabel 1976; Balsdon 1979; Dauge 1981; Lévi 1984, 5–14; Long 1986; Cunliffe 1988; Hall 1989a; Nippel 1990; Georges 1994; Dihle 1994; Timpe 1996, 34–50; Bichler 1996, 51–74; Coleman 1997; Bäbler 1998; Motta 1999, 309–315; Harrison 2002; Mitchell 2007; Gruen 2011; Rausch 2013, 35–43; Vlassopoulos 2013; Podossinov, Jackson & Konovalova 2016.
2   S. hierzu Riemer & Riemer 2005.
3   Zur Beziehung von Herodot zu den Barbaren siehe Kothe 1969, 15–88; Hartog 1980; Bichler 1988, 117–128; Giuseppe & Reverdin 1988; West 1999, 76–86; Bichler 2000; West 2004, 73–89; Nesselrath 2009, 307–330. Einen einschlägigen Kommentar zum ‚Skythen-Logos' hat Aldo Corcella in Asheri, Loyyd & Corcella 2007, 543–721 vorgelegt.
4   Weiß 2015, 20.
5   Plut. *De malign. Herod.* 12 (857a). Zu Plutarchs Haltung zu Herodot s. Nesselrath 2009, 307–312; vgl. auch allgemein hinsichtlich seiner Beziehung zu Barbaren Schmidt 1999. In ihrer Studie lateinischer Autoren, welche die Skythen als ‚Barbaren' betrachten, führt Kuhnert 2012, 72–90 unter anderem folgende Stellen an: Rut. Lup. 2.9; Cic. *Nat. deor.* 2.88; Prop. 3.16; Tib. 3.4.91; Curt. 7.6; 7.8; Ov. *Trist.* 4.6.47; Pomp. Trog. *Prol.* 1.8; 2.4; Sen. *Cons. Helv.* 7.1; *Phaedr.* 166–168; Val. Max. 5.4.5; Gell. *NA* 9.4.6.

müssen.[6] An einer Stelle seiner *Historien*, in der Erzählung von den frevelhaften Handlungen des persischen Königs Kambyses in Ägypten, formuliert er die Idee einer nationalen Toleranz, die auch heute noch aktuell klingt:

> Mir ist völlig klar, dass Kambyses gänzlich wahnsinnig war; sonst hätte er sich nicht an Tempeln und Bräuchen zum Spott vergriffen. Wenn man alle Völker der Erde aufforderte, sich unter all den verschiedenen Sitten die trefflichsten auszuwählen, so würde jedes nach genauer Untersuchung doch die eigenen allen anderen vorziehen. So sehr ist jedes Volk davon überzeugt, dass seine Lebensformen die besten sind (οὕτω νομίζουσι πολλόν τι καλλίστους τοὺς ἑωυτῶν νόμους ἕκαστοι εἶναι). Es ist also ganz natürlich, dass nur ein Wahnsinniger über so etwas spotten kann.[7]

Weiter verbreitet in der griechisch-römischen Antike war aber die Meinung, dass die Skythen-Barbaren grausam, militant, sittenlos, heimtückisch und ignorant seien. Zum Beispiel war der unter Augustus wirkende Historiker und Geograph Strabon sehr empfänglich für die Darstellung von entfernt lebenden Fremden als grausamen und kriegerischen Wilden.[8] Kannibalismus (4.5.5 [201C]; 7.3.6f. [298–301C]), die Opferung von Fremden (5.3.12 [239C]; 7.3.6f.), die Verwendung menschlicher Schädel als Trinkschalen (7.3.6 [300f.C]), die Grausamkeit und brutale Erbarmungslosigkeit der Iberer, Thraker und Skythen (3.4.17f. [164f.C]) sowie die Militanz der Skythen und Sarmaten (11.3.3 [500C]) – all dies spiegelt die traditionell negative Haltung gegenüber den Barbaren wider.

Im Folgenden will ich versuchen, einige Besonderheiten des griechisch-römischen Skythenbildes zu beschreiben. Nach einigen einführenden Beispielen zum Bild des grausamen Skythen sowie den Griechen und Römern als Kulturbringern werde ich weitere Belege für die gegenteilige Sicht von der gerechten und edlen Gesinnung dieses Nordvolkes anführen, gefolgt von einigen Stimmen, welche den Sittenverfall der Skythen analog mit dem Einfluss der dekadenten Griechen erklären. Nach einem kurzen Exkurs zur sprichwörtlichen skythischen Trunkenheit werde ich einen Ausblick auf die Spätantike geben, in welcher sich der traditionelle Kontrast zwischen einerseits Griechen oder Römern und andererseits skythischen Barbaren vor allem unter christlichem Vorzeichen auflöste.

II. SKYTHEN ALS GRAUSAME BARBAREN

Die Athener hatten sehr früh die Möglichkeit, die Skythen aus der unmittelbaren Nähe zu beobachten. Einige Berichte attischer Redner bezeugen,[9] dass skythische

---

6 Vgl. Braund 2008, 348: ‚The point is he shows an unusual openness to other cultures which was unusual and remarkable …‘

7 Hdt. 3.38. Übersetzung des Herodot hier und im Folgenden von J. Feix.

8 Zum Barbarenbegriff bei Strabon s. Thollard 1987; Almagor 2005, 42–55, bes. 42: ‚The *Geography* of Strabo, the most comprehensive ethnographic work to survive from classical antiquity, is one of the main sources for a study of the ancient attitude towards other races and nations, viz. the barbarians‘. Laut Almagor benutzt Strabon das Wort *barbaros* ca. 150 mal in seiner *Geographie*.

9 Andokides 3.5; Aischines 2.173. S. auch: *Schol. Aristoph. Acharn.* 54; *Suda s.v.* Τοξόται.

Bogenschützen nach 480 v.Chr. öffentlich Polizeifunktionen in Athen ausführten. Ihre Zahl soll 300 betragen haben. Sie lebten in ihren Wagen auf der Agora und später auf dem Ares-Hügel.[10] Sie stellten beliebte Ziele für Spott in der athenischen Komödie dar. Etwa in den *Acharnern, Rittern, Thesmophoriazusen* sowie der *Lysistrata* des Aristophanes begegnen sie als dumme, lächerliche Menschen, die gebrochenes Griechisch sprechen.[11] Die Athener konnten sich also direkt ein Bild von den aus der Ferne stammenden Skythen machen, die vielen anderen Griechen nur aus der literarischen Produktion etwa auch des Aischylos oder Herodot bekannt waren.[12]

Selbst im täglichen Leben war es üblich, Unhöflichkeit, schlechte Manieren oder Sittenlosigkeit mit skythischen Gesichtszügen zu verbinden. In Senecas Tragödie *Phaedra* (660) sagt die Stiefmutter zu Hippolyt, der ihre Liebe zurückwies: ‚Auf deinem griechischen Gesicht ist skythische Härte (*Scythicus rigor*) zu sehen‘. Phaedras Vergleich setzt also den Kontrast skythischer Härte mit griechischer Freundlichkeit und Großzügigkeit voraus. Ähnlich sagt Philostrat in *Brief* 5 zu einem jungen Mann:

> Sag mir, woher kommst du, ein junger Mann, so hart mit der Liebe verbunden? ... Du scheinst mir ein Skythe und ein Barbar zu sein, der von jenem berühmten Altar und der unwirtlichen Schau kam (Σκύθης μοι δοκεῖς καὶ βάρβαρος εἶναι ἀπ᾽ ἐκείνου τοῦ βωμοῦ καὶ τῶν ἀξένων θεαμάτων).[13]

Die Vorstellung von der Wildheit der Barbaren hielt sich über viele Jahrhunderte und begegnet auch bei frühchristlichen Autoren. So beschreibt Tertullian (ca. 160–230 n.Chr.), einer der ältesten christlichen Autoren, bei der Verurteilung seines theologischen Gegners Markion aus Sinope die Sitten der Einwohner des Pontos-Gebiets als durchweg negativ:

> Der sogenannte Pontos Euxeinos, das gastliche Meer, ist eine Negation seiner Natur und ein Hohn auf seinen Namen. Schon infolge seiner Lage wird niemand den Pontus für gastlich halten, er ist zu weit von unseren milden menschlicheren Gestaden entfernt, fast aus einem gewissen Schamgefühl über seine Barbarei. Ganz wilde Völkerschaften umwohnen ihn, sofern man auf einem Wagen hausen überhaupt wohnen nennen kann. Ihre Wohnsitze sind nicht ständig, ihre Lebensweise roh, die Befriedigung des Geschlechtstriebes geschieht ohne Schranken und meistens ohne alle Scham. Auch wenn sie sich dabei der Öffentlichkeit ent-

---

10 Zu literarischen und materiellen Zeugnissen für skythische Polizisten in Athen s. Vos 1963; Lissarrague 1990, 125–149; Braund 1997, 48–56; Skržinskaja 1998, 209–261; Frolov 1998, 135–152; 2004, 195–220; Bäbler 1998, 165–181; Ivantchik 2005b, 100–114; Jatsenko 2006, 47–83; Podossinov, Jackson, Konovalova 2016, 174–195.

11 Plassart 1913, 151–213; Hall 1989b, 38–54; Sier 1992, 63–83; Bäbler 2005, 114–125; Sinizyn 2008, 269–292.

12 S. z.B. Bäbler 1998, 165–167. Von den Griechen, die während der Großen Griechischen Kolonisation mit den Skythen an der nördlichen Schwarzmeerküste zusammentrafen, gibt es praktisch keine schriftlichen Zeugnisse, so dass wir ihre Einstellung zu den Skythen nicht beurteilen können. Nur manchmal können wir annehmen, dass sich in diesen oder anderen Mitteilungen zu den Skythen in den Schriften der ‚mediterranen‘ Autoren einige Informationen der pontischen Griechen widerspiegeln.

13 Dies bezieht sich auf den Altar der Artemis in der skythischen Tauris (Krim), auf dem Fremde geopfert wurden.

ziehen, hängen sie zum Anzeichen an einem Joche ihre Köcher auf, damit sich nicht unverse-
hens jemand nähere. Die Leichname ihrer Eltern fressen sie mit Tierfleisch zusammengehackt
bei ihren Gastmählern. Ist jemand einer Todesart erlegen, die ihn ungenießbar macht, so gilt
dies als ein Fluch. Selbst die Weiber sind nicht etwa entsprechend der Eigenart ihres Ge-
schlechts milder und gesitteter; Kinder stillen ist ihre Sache nicht, statt Wolle zu spinnen,
hantieren sie mit Äxten, sie wollen lieber Kriegsdienste tun als heiraten. Der Himmelsstrich
ist rau, das Tageslicht ist niemals vollkräftig, die Sonne niemals ganz frei, die ganze Atmo-
sphäre ein Nebel, das ganze Jahr Winter, alle Winde, die wehen, kommen von Norden. Die
Getränke müssen durch Feuer erst wieder flüssig gemacht werden, die Ströme sind durch eine
Eisdecke gefesselt und auf den Gebirgen lagern mächtige Schneemassen. Alles ist träge, alles
starr; nichts ist dort feurig als die Wildheit, jene Wildheit nämlich, welche die Opfer der Tau-
rier, die Liebeshändel der Kolcher und die Kreuze der Kaukasier als Bühnenstoffe geliefert
hat. Allein nichts ist so befremdlich für uns und so traurig für Pontos, als dass dort Markion
geboren wurde, der abschreckender ist als ein Skythe, unsteter als ein Hamaxobier, un-
menschlicher als ein Massagete, verwegener als eine Amazone, dunkler als der Nebel, kälter
als der Winter, spröder als Eis, trügerischer als die Donau, gefahrvoller als der Kaukasus.[14]

III. GRIECHEN UND RÖMER ALS ZIVILISATOREN

Doch in vielen Fällen, in denen Griechen mit fremden Kulturen konfrontiert wurden, seien es die Äthiopier, Ägypter, Inder, Perser, Skythen, Thraker, Germanen
oder Kelten, die sie als ‚unzivilisiert‘ betrachteten, bemerkten sie ‚Fortschritte‘ in
Richtung Zivilisation. Solche Entwicklungen seien dem Einfluss soziopolitischer,
wirtschaftlicher, kultureller, religiöser oder ethischer ‚Errungenschaften‘ der griechischen Gesellschaft zuzuschreiben. Als sich etwa Medeia in der gleichnamigen
Tragödie des Euripides bei Jason über seine Undankbarkeit beklagt, obwohl sie
um seinetwillen ihr Vaterland und ihre Familie verlassen habe, erwidert er, dass
sie ihm dafür dankbar sein solle, dass er sie – eine Barbarin – mit den griechischen Bräuchen vertraut gemacht habe (*Med.* 529–531).

Die Milderung der rohen Lebensart gewalttätiger und kriegerischer Barbaren
unter dem Einfluss der Griechen oder der Römer ist einer der häufigsten Topoi in
der Gegenüberstellung von griechisch-römischer Zivilisation und barbarischer
Welt.[15] In den folgenden Beispielen aus Strabons *Geographie* sind es indes
durchweg die Römer, die als Kulturbringer in Erscheinung treten:[16]

(2.5.26 [127C]): ‚Und die Römer, die viele Völker übernommen haben, die ihrer Natur nach
unzivilisiert waren, infolge ihrer Umwelt […] haben die Kontaktlosen miteinander in Kontakt
gebracht und den Wilden beigebracht, in einem geordneten Gemeinwesen zu leben‘ (καὶ τοὺς
ἀγριωτέρους πολιτικῶς ζῆν ἐδίδαξαν).

(4.1.12 [186C]) über die gallischen Cavaren: ‚die auch gar keine Barbaren mehr sind, sondern
zum größten Teil den Stil der Römer, sowohl in der Sprache, als auch in der Lebensweise,
manche auch in der Staatsordnung, übernommen haben‘ (οὐδὲ βαρβάρους ἔτι ὄντας, ἀλλὰ

14 Tert. *Adv. Marc.* 1.1. Übersetzung in Anlehnung an K.A. Heinrich Kellner.
15 Herodot hat als erster die Militanz als charakteristisches Merkmal der Skythen beschrieben
    (4.6–66).
16 Alle Strabon-Stellen folgen – mit gelegentlichen leichten Anpassungen – der Übersetzung
    Stefan Radts.

μετακειμένους τὸ πλέον εἰς τὸν τῶν Ῥωμαίων τύπον καὶ τῇ γλώττῃ καὶ τοῖς βίοις, τινὰς δὲ καὶ τῇ πολιτείᾳ).

(3.3.8 [155f.C]) über die Völker im nördlichen Teil des spanischen Iberiens: ‚Das Schwerbe-zähmbare und Wilde (τὸ δὲ δυσήμερον καὶ ἀγριῶδες) ist nicht nur eine Folge ihrer Kriege, sondern auch ihrer Entlegenheit … und dadurch, dass sie kaum mit anderen in Kontakt kommen, haben sie den Sinn für Gemeinschaft und Menschlichkeit (ἀποβεβλήκασι τὸ κοινωνικὸν καὶ τὸ φιλάνθρωπον) verloren. Heute ist das bei ihnen weniger stark dank des Friedens und des Aufenthalts der Römer (διὰ τὴν εἰρήνην καὶ τὴν τῶν Ῥωμαίων ἐπιδημίαν); diejenigen aber, bei denen das nicht der Fall ist, sind umso widerspenstiger und wilder (χαλεπώτεροί εἰσι καὶ θηριωδέστεροι) <…> zum anderen ist es seinem (des Augustus) Nachfolger Tiberius gelungen, sie durch die Stationierung der von Caesar Augustus dazu bestimmten Truppen dreier Legionen in der Gegend nicht nur friedfertig, sondern manche von ihnen schon soziabel (οὐ μόνον εἰρηνικοὺς ἀλλὰ καὶ πολιτικοὺς) zu machen‘.

Einen ähnlichen Aspekt der römischen Herrschaftsideologie hat Plinius der Ältere ausgedrückt:

Wer nämlich dächte nicht, dass durch die Vereinigung der ganzen Erde unter der Hoheit des römischen Reiches auch das Leben aus dem Handelsverkehr und aus einem gemeinsamen glücklichen Frieden Vorteile erhalten habe und alles (d.h. Natur- und Kulturgüter), auch das, was früher verborgen war, zum allgemeinen Nutzen freigegeben sei?[17]

So verhält es sich nach Strabon (11.2.4 [494C]) auch mit den Barbaren in der nordpontischen Region: ‚…die [Maioten] in der Nähe von Tanaïs sind wilder (ἀγριώτερα), die, welche den Bosporos berühren, zahmer (χειροήθη μᾶλλον)‘. Hiermit will Strabon ausdrücken, dass sich Tanaïs – eine griechisch-barbarische Stadt – am entferntesten Rande der griechischen Welt befinde, an der Grenze zu den Steppennomaden,[18] der Bosporus aber ein zivilisierter griechischer Staat sei, der einen guten Einfluss auf die Barbaren ausübe.

IV. SKYTHEN ALS EDLE BARBAREN

Aber schon früh begannen Historiker und Philosophen, eben diese Skythen zu idealisieren und ihnen die Züge ‚edler Barbaren‘ (*noble savage*) zuzuschreiben. Ausgangspunkt hierfür war Homers *Ilias*, in der die Abier, ein sonst unbekanntes, wohl nördlich von den Thrakern verortetes Volk, als ‚wunderbare Stutenmelker, Milchesser und … gerechteste der Menschen‘ bezeichnet werden.[19] Herodot setzte diese Linie in gewisser Weise fort, aber zur vollen Entfaltung gelangte sie erst im Werk des griechischen Historikers Ephoros im 4. Jh. v.Chr. In einem von Strabon überlieferten Fragment präsentiert Ephoros die Skythen als Vorbilder in morali-

---

17 Plin. *NH* 14.1.2: *Quis enim non communicato orbe terrarum maiestate Romani imperii profecisse vitam putet commercio rerum ac societate festae pacis omniaque, etiam quae ante occulta fuerant, in promiscuo usu facta?* Übersetzung von R. König.

18 Vgl. Strab. *Geogr.* 11.2.3 (494C) über die Stadt Tanaïs: ‚Sie war der gemeinsame Handelsplatz der asiatischen und der europäischen Nomaden‘.

19 Hom. *Il.* 13.1–7. Es gibt viele Arbeiten, die diesem Thema gewidmet sind, s. z.B. Riese 1875; Motta 1999, 309–315; speziell zu dieser Homer-Stelle s. Ivantchik 2005a, 18–52.

scher und sozialer Hinsicht. Diese Nomaden hätten, da sie weit entfernt von der korrupten Zivilisation lebten, die hohen moralischen Prinzipien bewahrt:

> Die anderen Autoren nun, sagt er (Ephoros), sprechen nur von ihrer (der Skythen) Grausamkeit, da sie wissen, dass das Schreckliche und Erstaunliche Eindruck macht; man müsse aber auch von dem Entgegengesetzten sprechen und es als Vorbild hinstellen ... Dann geht er auf die Ursachen ein: dadurch dass sie in ihrer Lebensweise frugal seien und keinen Handel treiben, lebten sie erstens rechtlich miteinander, da sie alles, und besonders Frauen, Kinder und die ganze Verwandtschaft, gemeinsam besäßen, und seien sie außerdem für Auswärtige unüberwindlich und unbesiegbar, weil sie nichts hätten, um dessentwillen sie sich in Knechtschaft begeben würden.[20]

Es ist interessant, dass sich Ephoros in seinem Skythen-Diskurs von den Meinungen der ‚anderen Autoren‘ löst (Herodot soll hier nicht zu jenen gerechnet werden), die wohl mit Absicht die schreckliche Seite (δεινόν) des skythischen Nomadentums zu verschärfen versuchten, um ihre Werke interessanter (θαυμαστόν) zu gestalten. Beachten wir jedoch, dass Ephoros nicht der erste war, der die Nomaden idealisierte: es ist bekannt, dass bereits Choirilos aus Samos im 5. Jh. v.Chr. die Nomaden als Menschen beschrieb, welche die Gesetze und das Recht befolgten (ἄνθρωποι νόμιμοι).[21]

Die positive Charakterisierung der Skythen durch Ephoros übte einen starken Einfluss auf die nachfolgenden Beschreibungen der Skythen aus. Menschen, die weit weg von der Zivilisation lebten und deren Sitten deshalb unverdorben waren, die nicht raffgierig waren, Gerechtigkeit gegenüber allem und allen zeigten und einen milden Charakter besaßen, die quasi eine ‚kommunistische‘ Lebensweise pflegten – all diese Eigenschaften wurden seit dem 4. Jh. v.Chr. zu dominierenden Motiven in der Idealisierung der Skythen.[22]

Die Frömmigkeit und Gerechtigkeit der Nomaden infolge des Fehlens von Privateigentum betont auch Pseudo-Skymnos, wobei er offenbar Ephoros folgt:

> Sie (die Nomaden) sind sehr gottesfürchtig – kein einziger von ihnen würde je einem Lebewesen ein Unrecht zufügen (εὐσεβῆ πάνυ, ὧν οὐδὲ εἷς ἔμψυχον ἀδικήσαι ποτ' ἄν) –, tragen, wie Ephoros sagt, ihre Häuser mit sich und ernähren sich von Milch, indem sie, wie bei den Skythen üblich, ihre Stuten melken. Sie führen ihr Leben gemäß dem von ihnen aufgestellten Grundsatz, der Besitz und jeglicher Verkehr seien die gemeinsame Angelegenheit aller. Ephoros sagt, auch der weise Anacharsis stamme aus diesen so überaus gottesfürchtigen Hirtenvölkern.[23]

Ähnliche Aussagen finden sich bei Strabon wieder:

> Und es ist nicht verwunderlich, dass er (sc. Homer), wegen des bei uns weit verbreiteten Rechtsbruchs, bei geschäftlichen Verträgen, die Leute, in deren (sc. der Skythen) Leben nicht nur geschäftliche Verträge und Geldverdienen (ἐν τοῖς συμβολαίοις καὶ τῷ ἀργυρισμῷ) gar

---

20 Strab. *Geogr*. 7.3.9 (302C) = Ephoros, *FGrH* 70 F 42.
21 Strab. *Geogr*. 7.3.9 (303C) = Choirilos F 5.
22 Bäbler 2005, 122 äußert die vorsichtige Vermutung, dass das Bild von gerechten und nicht korrupten Skythen, die im 5. und 4. Jh. v.Chr. in Athen Polizeifunktionen ausübten, die Idealisierung der Skythen in der griechischen Philosophie und Geschichtsschreibung beeinflusst haben könnte.
23 Ps.-Skymnos 850–860. Übersetzung von M. Korenjak.

keine Rolle spielen, sondern die auch alles außer Schwert und Trinkbecher gemeinsam besitzen und an erster Stelle die Frauen und Kinder in platonischer Weise gemeinsam haben, die gerechtesten genannt hat? Auch Aischylos gibt zu erkennen, dass er dem Dichter beipflichtet, wenn er von den Skythen sagt: ‚Doch Skythen, Esser von Stutenkäs', ein rechtliches Volk (εὔνομοι Σκύθαι)'. Und diese Ansicht herrscht bei den Griechen auch heute noch: halten wir sie doch für die aufrichtigsten und am wenigsten arglistigen Menschen und für frugaler und genügsamer, als wir selber sind (ἁπλουστάτους τε γὰρ αὐτοὺς νομίζομεν καὶ ἥκιστα κακεντρεχεῖς εὐτελεστέρους τε πολὺ ἡμῶν καὶ αὐταρκεστέρους).[24]

Der griechische Historiker Nikolaos von Damaskus gibt im 1. Jh. v.Chr. eine Quintessenz des idealisierten Skythen:

> Die Galaktophagen (Milchesser), ein skythisches Volk, haben keine Häuser, wie die überwiegende Mehrheit der Skythen, und sie ernähren sich nur mit Stutenmilch, die sie trinken und aus der sie Käse machen, den sie essen, und deshalb ist es sehr schwierig, gegen sie zu kämpfen, da sie überall ihr Essen dabei haben. Sie haben auch Dareios in die Flucht geschlagen. Sie sind auch sehr gerecht, weil sie ihre Güter und Frauen gemeinsam haben, so dass sie die Älteren als Väter betrachten, die Jüngeren als Söhne und die Leute gleichen Alters als Brüder ... Man sagt, ihnen seien Eifersüchtige oder Hassende oder Fürchtende unbekannt, dank des gemeinsamen Lebens und der Gerechtigkeit (οὐδὲ εἷς οὔτε φθονῶν, ὥς φασιν, οὔτε μισῶν οὔτε φοβούμενος ἱστορήθη διὰ τὴν τοῦ βίου κοινότητα καὶ δικαιοσύνην). Bei ihnen sind Frauen nicht weniger kriegerisch als Männer, und im Notfall führen sie zusammen mit ihnen Krieg.[25]

Einen bemerkenswerten Vergleich zwischen der griechischen Kultur und den Sitten der ‚ursprünglichen' Barbaren lesen wir in Justins *Epitome* der von Pompeius Trogus verfassten *Historiae Philippicae*:

> Diese Anspruchslosigkeit ihrer Lebensart (*continentia*) ist auch der Grund ihrer Rechtschaffenheit, denn sie sind nicht begehrlich nach fremdem Gut; ... Ach wenn doch auch alle übrigen Sterblichen ebenso gemäßigt wären (*moderatio abstinentiaque*) und sich von fremdem Besitz fernhielten! Wahrhaftig, dann gäbe es nicht so viele Kriege, alle Jahrhunderte durch, in allen Ländern, ohne Unterlass, und dann würden nicht mehr Menschen durch Schwert und Waffengewalt als nach der Spielregel des natürlichen Schicksals dahingerafft; und ganz und gar wundersam muss es doch scheinen, dass jenen Leuten die schlichte Natur das schenkt, was die Griechen selbst durch die lange Lehre der Weisen und durch die Gebote der Philosophen nicht erlangen können, sondern dass vielmehr ihre hochkultivierte Sitte im Vergleich mit diesem kulturlosen Barbarenvolk (*inculta barbaria*) den Kürzeren zieht. So viel mehr schafft bei jenen die Unkenntnis der Laster (*uitiorum ignoratio*) als bei diesen die Kenntnis des Guten (*cognitio uirtutis*).[26]

Römische Dichter setzten diese Linie der idyllischen Haltung gegenüber den Skythen fort, ja sie steigern sogar die moralisierende Tendenz. So schreibt Horaz:

(9)   Lebt doch freier der Steppe Sohn,
             Der das schweifende Haus noch auf dem Karren zieht.

(17)   Kinder, denen die Mutter starb,
             Hegt die zweite Gemahlin lauteren Sinnes dort,

---

24 Strab. *Geogr.* 7.3.7 (300C), mit Aisch. F 198.
25 Nikolaos, *Sammlung seltsamer Völkersitten* 3 = Stob. *Anth.* 3.1.200.
26 Just. 2.2.10–15. Übersetzung in Anlehnung an O. Seel.

Dort missachtet, der Mitgift stolz,
Nicht das Weib ihren Mann, gleißender Buhlschaft froh.

Elterntugend ist dort (*magna parentium virtus*)
Und die Keuschheit (*castitas*), die flieht, ihrem Gelöbnis treu,
Selbst die Blicke des fremden Manns;
Untreu gilt als Vergehn, welches der Tod nur sühnt.[27]

Es ist offensichtlich, dass die Idealisierung der Skythen vor allem die Folge einer Enttäuschung der antiken Gesellschaft in Bezug auf ihre eigenen Werte war, die zuvor als Merkmal der ‚richtigen' Lebensweise wahrgenommen wurden. Diese Enttäuschung war der Grund dafür, dass in der antiken Literatur die Gestalten der skythischen Weisen Anacharsis und Abaris so populär wurden. Sie versinnbildlichten die Schlichtheit der Nomaden, welche dem korrupten und in Übel verstrickten Leben der ‚zivilisierten' Griechen entgegengehalten werden konnte. Strabon verband die Popularität dieser und anderer skythischer Weisen mit den idealisierten Eigenschaften der Skythen:

Deshalb waren auch Anacharsis, Abaris und etliche andere Männer dieser Art bei den Griechen geschätzt, weil sie einen ihrem Volk eigentümlichen Stempel von Freundlichkeit, Einfachheit und Gerechtigkeit trugen (ἐθνικόν τινα χαρακτῆρα ἐπέφαινον εὐκολίας καὶ λιτότητος καὶ δικαιοσύνης).[28]

Besonders die kynischen Philosophen nutzten Anacharsis als Projektionsfläche für ihre eigene Lehre.[29] Die meisten der unter seinem Namen zirkulierenden *Briefe*[30] reichen zurück bis an das Ende des 4. oder den Anfang des 3. Jhs. v.Chr.[31] Als ein Beispiel dafür möchte ich *Brief* 6 des Anacharsis zitieren, der an den Sohn eines gewissen Königs adressiert ist und in dem der Skythe den Luxus des ‚zivilisierten' Lebens mit der Einfachheit und Angemessenheit der ‚barbarischen' Lebensweise kontrastiert:

Du hast Flöten und einen Beutel voll Geld, ich Pfeil und Bogen. Darum bist du auch ein Sklave, ich aber bin ein freier Mensch. Du hast viele Feinde, ich keinen. Wenn du aber das Geld wegwirfst, wenn du Bogen und Köcher tragen und mit den Skythen leben willst, so stehen dieselben Vorzüge auch für dich bereit.[32]

Die Weisheit des Anacharsis war in den Augen Ciceros die Folge der Lebensweise des ‚natürlichen' Barbaren, die weit entfernt von den Versuchungen der Zivilisation sei:

…Oder konnte der Skythe Anacharsis Geld für Nichts erachten (*pro nihilo pecuniam ducere*), unsere Philosophen aber nicht? Ein Brief von jenem wird mit folgenden Worten überliefert: ‚Anacharsis grüßt Hanno. Ich besitze als Obergewand ein skythisches Fell, als Fußbekleidung

27 Hor. *Carm.* 3.24.9f., 17–24. Meine Übersetzung.
28 Strab. *Geogr.* 7.3.8 (301C).
29 Über Anacharsis in der antiken literarischen Tradition s. Heinze 1891, 458–468; Kindstrand 1981; Ungefehr-Kortus 1996; Schubert 2010; Kath & Rücker 2012. Zu Leben und Tod des Anacharsis s. auch Harland und Oller Guzmán, Kapitel III and IV in diesem Band.
30 Sie sind gesammelt bei Kindstrand 1981. Siehe auch Reuters 1963.
31 Reuters 1963, 1–5.
32 Übersetzung von F.H. Reuters.

die Schwiele der Fußsohlen, als Bett die Erde, als Kost den Hunger, ich lebe von Milch, Käse und Fleisch. Daher ist es Dir vergönnt, in mir einen gelassenen Menschen anzutreffen (*quare ut ad quietum me licet venias*). Jene Geschenke aber, über die du dich gefreut hast, gib' entweder deinen Bürgern oder den unsterblichen Göttern.'[33]

V. SITTENVERFALL AUCH BEI DEN SKYTHEN

Die zuletzt zitierte Stelle aus Strabons *Geographie* spielt nach dem Lob der skythischen Sitten nochmals auf den ethisch-kulturellen Einfluss der Griechen und Römer an, allerdings diesmal unter verkehrten Vorzeichen:

> Und das obwohl unsere Lebensweise (ὅ γε καθ' ἡμᾶς βίος) nahezu der ganzen Welt eine Verschlechterung gebracht hat, indem sie Luxus und Genüsse und tausende von Ränken zu deren Beschaffung einführte.[34]

Tatsächlich seien die Skythen nicht vom schädlichen Einfluss der griechisch-römischen Zivilisation verschont geblieben. Auch ihre – grundsätzlich ethische – Gesellschaft leide unter Verfallserscheinungen:

> Viel von dieser Schlechtigkeit (πολὺ οὖν τῆς τοιαύτης κακίας) ist daher auch nicht nur zu den übrigen Barbaren, sondern auch zu den Nomaden gedrungen. Denn als sie an das Meer gekommen waren (θαλάττης ἁψάμενοι), sind sie durch Räuberei und Fremdentötung schlechter geworden, und durch die Kontakte mit Vielen übernehmen sie deren Luxus und Handelsgeist (μεταλαμβάνουσι τῆς ἐκείνων πολυτελείας καὶ καπηλείας), Dinge die zwar als Mittel zur Zivilisierung gelten, aber den Charakter verderben (διαφθείρει δὲ τὰ ἤθη) und die vorhin genannte Aufrichtigkeit (ἀντὶ τῆς ἁπλότητος) durch Wendigkeit ersetzen.[35]

Noch negativer äußert sich Athenaios in seinem Werk *Deipnosophistai* über die Dekadenz der Skythen, wobei er die Worte des Klearchos von Soloi so wiedergibt:

> Klearchos erzählt folgendes: zuerst benutzte nur das skythische Volk allgemeine Gesetze; dann wurden sie wieder ‚unglücklichste aller Menschen' wegen ihrer Gewalttaten (διὰ τὴν ὕβριν): sie gaben sich dem Luxus hin wie niemand anderer, als Folge des vielen Glücks in allem, des Reichtums und der übrigen Arten von Wohlergehen ... Da sie als die ersten aller Menschen Luxus genossen, erreichten sie eine solche Grausamkeit (ὕβρις), dass sie bei allen Menschen, mit denen sie Verkehr hatten, die Enden ihrer Nasen zu kürzen begannen ...[36]

33 Cic. *Tusc.* 5.89f. Übersetzung aus Gerstacker u.a. 2015, 45.
34 Strab. *Geogr.* 7.3.7 (301C). Vgl. auch Pompeius Trogus, der in seinen für die Skythen sehr positiven Äußerungen die Unbesiegbarkeit der Skythen so erklärte (Just. 2.3.7): ‚Der ganze Stamm zeichnet sich durch Ausdauer in Arbeit und Krieg sowie durch außerordentliche körperliche Stärke aus; sie erwerben nichts, was zu verlieren sie fürchten könnten, und wenn sie Sieger sind, wünschen sie nichts anderes als Ruhm' (*Gens laboribus et bellis aspera, vires corporum inmensae; nihil parare, quod amittere timeant, nihil victores praeter gloriam concupiscunt*).
35 Strab. *Geogr.* 7.3.7 (301C).
36 Ath. 12.27. Meine Übersetzung.

Eine Variante findet sich beim christlichen Geschichtsschreiber Paulus Orosius, der im 5. Jh. n.Chr. die Grausamkeit und Militanz der Skythen fremdem Einfluss zuschreibt:

> 1300 Jahre vor der Gründung Roms verwüstete und eroberte der König der Assyrer Ninos ..., der aus dem Süden und vom Roten Meer erschien, im äußersten Norden den Pontos Euxeinos und lehrte die skythischen Barbaren, die bis dahin unkriegerisch (*inbelles*) und harmlos (*innocentes*) waren, und unfähig, ihre Grausamkeit (*saevitiam*) zu zeigen, ihre Stärke zu erkennen und schon nicht die Milch der Tiere, sondern das menschliche Blut zu trinken, und endlich durch Niederlagen zu gewinnen.[37]

Die oben zitierte Textstelle Strabons macht deutlich, dass er die Nähe zum Meer als gefährlich und schlecht für die Sittlichkeit der lokalen Stämme betrachtete. Und das ist verständlich. Wie wir wissen, beruht die griechische Zivilisation weitgehend auf Seefahrt. Schon Homer ließ Odysseus auf seiner Irrfahrt an die Grenzen der Oikumene gelangen. Während der Großen Griechischen Kolonisation beherrschten die Griechen einen Großteil der Mittelmeerküste. Dadurch wurden viele indigene Völker, die sie ‚Barbaren' nannten, in den Prozess der ‚Zivilisation' einbezogen. Mithin erhielt die Nähe der Barbaren zum Meer – im Rahmen dieser Theorie – eine negative Bewertung und wurde zu einem Faktor, der zur ‚Korruption' der Barbaren beitrug.[38] So sagt Strabon in seiner Beschreibung der nordpontischen Barbaren:

> Die Nomaden nun sind mehr Krieger als Räuber, und Krieg führen sie um die Abgaben… Die Ackerbauern werden […] als zivilisierter und sozialer betrachtet; da sie aber Geldverdiener und in Berührung mit dem Meer sind (θαλάττης ἁπτόμενοι), lassen sie sich nicht von Räuberei und anderen solchen Rechtsbrüchen und Übervorteilungen abhalten.[39]

Die Einschätzung von der Destruktivität der Nähe zum Meer teilten auch die früheren Philosophen, die über den idealen Staatsaufbau schrieben, wie wir etwa in Platons *Gesetzen* nachlesen können:

> Die Nähe des Meeres (ἐγγύτερον μέντοι τοῦ δέοντος κεῖται τῆς θαλάττης) bietet zwar Tag für Tag ihre süßen Reize dar, in Wahrheit aber ist es eine salzige und bittere Nachbarschaft. Indem sie nämlich die Bürger mit Handelsgeist und krämerischer Gewinnsucht erfüllt (ἐμπορίας γὰρ καὶ χρηματισμοῦ διὰ καπηλείας ἐμπιμπλᾶσα αὐτήν) und ihren Seelen einen trügerischen und unzuverlässigen Charakter einflößt, so entfremdet sie sie der Treue und dem Wohlwollen gegeneinander sowie gegen andere Menschen.[40]

Nicht zufällig sind Beschreibungen des ‚Goldenen Zeitalters' frei von Schifffahrt, eine Voraussetzung dafür, dass die Göttin der Gerechtigkeit noch in Ehren gehalten wurde.[41] Die Nähe zum Meer bot auch den Indigenen die Möglichkeit, sich

37 Oros. 1.4.1–3. Meine Übersetzung.
38 Siehe Ivantchik 2005a, 44f., der glaubt, dass diese Idee wahrscheinlich Strabon selbst gehört.
39 Strab. *Geogr.* 7.4.6 (311C).
40 Plat. *Nom.* 4.705. Übersetzung von Fr. Susemihl.
41 Z.B. Aratos von Soloi, *Phaen.* 110–114 (vgl. Horden & Purcell 2000; Schulz 2005, 210f.:
Χαλεπὴ δ' ἀπέκειτο θάλασσα,
καὶ βίον οὔπω νῆες ἀπόπροθεν ἠγίνεσκον,
ἀλλὰ βόες καὶ ἄροτρα καὶ αὐτὴ πότνια λαῶν

durch Piraterie am Wohlstand antiker Städte und Siedlungen zu bereichern.[42] So weist Strabon in der Beschreibung der Nordküste des Schwarzen Meeres darauf hin, dass die östlich von Gorgippia wohnenden Achaier, Zyger und Heniocher ,von der Seeräuberei' lebten. Wohl nicht zufällig handelte es sich nach Auffassung griechischer Schriftsteller bei den Achaiern und Heniochern um barbarisierte Griechen.[43]

VI. SKYTHISCHE TRUNKENHEIT

Ein gutes Fallbeispiel für die ambivalente Wahrnehmung des skythischen Barbaren bietet das Thema der Trunkenheit, das zu einer Art Klischee in der antiken Literatur und bildenden Kunst wurde. Schon der griechische Dichter des 6. Jhs. v.Chr. Anakreon warnte seine Freunde, sich nicht wie Skythen zu verhalten:

> Nun, Freunde, lasst uns nicht
> Mit solchem Lärm und Geschrei (πατάγῳ τε κἀλαλητῷ)
> Die Sauferei der Skythen (Σκυθικὴν πόσιν) imitieren
> Beim Weintrinken, vielmehr wollen wir ruhig
> Zum Klang süßer Hymnen trinken.[44]

Den Skythen wurde der Brauch zugeschrieben, reinen (ἄκρατος) Wein zu trinken, ihn also nicht mit Wasser zu verdünnen, wie es die Griechen taten. Herodot (6.84) erzählt eine in dieser Hinsicht erhellende Anekdote darüber, wie der Spartanerkönig Kleomenes (Ende des 6. bis Anfang des 5. Jhs. v.Chr.) dem Wahnsinn verfiel:[45]

> Die Spartaner selbst meinen, an der Krankheit des Kleomenes sei überhaupt keine Gottheit schuld; er habe sich durch seinen Umgang mit Skythen angewöhnt, ungemischten Wein (τὴν ἀκρητοποσίην) zu trinken, und sei dadurch wahnsinnig geworden … Seit dieser Zeit sagen sie, wie sie selbst erzählen, wenn sie den Wein einmal kräftiger trinken wollen: ,Reich ihn auf skythische Art!' (Ἐπισκύθισον).[46]

Man kann vermuten, dass der Allgemeinplatz der saufenden Skythen über die pontischen Griechen, die den Skythen viel Wein verkauften, wie zahlreiche Funde von Weingefäßen aus den skythischen Gräbern zeigen, ins griechische Mutterland gelangte. Dieser Wein kam von der nördlichen Ägäis wohl im Austausch für Getreide, das von der nördlichen Schwarzmeerküste nach Griechenland exportiert

μυρία πάντα παρεῖχε Δίκη δώτειρα δικαίων.
Τόφρ' ἦν ὄφρ' ἔτι γαῖα γένος χρύσειον ἔφερβεν.
42 Zur antiken Piraterie s. Wendt 2016, 79–91.
43 Strab. *Geogr.* 11.2.12 (496f.C). Einer gewissen Vorliebe für Mord und Anthropophagie gemäß beschuldigt Aristoteles die pontischen Achaier und Heniocher dieser Verbrechen (*Pol.* 7.1338b 17–28). Zur Wahrnehmung der Achaier und Heniocher in der antiken Literatur s. Asheri 1998, 265–285.
44 Anak. 57. Meine Übersetzung.
45 Zur Unzuverlässigkeit dieser Tradition s. Pečatnova 2007, 128–131.
46 Hdt. 6.84. Das gleiche Verb wird von Athenaios (10.29) für diejenigen verwendet, die weniger verdünnten Wein trinken möchten.

wurde.[47] Im Gegensatz zu den Griechen, für die verdünnter Wein ein alltägliches Getränk war, tranken die Skythen den Wein so, wie man es heutzutage (zumindest in nördlichen Breitengraden) tut, unverdünnt, aber nicht zu den täglichen Hauptmahlzeiten, wozu sie andere Flüssigkeiten wie Wasser, Milch und sogar Stutenmilch als Durstlöscher nutzten.[48]

Übrigens wurde der Brauch, Wein unverdünnt zu trinken, auch anderen ‚Barbaren' der Oikumene unterstellt. So schreibt Platon den Skythen, Persern, Karthagern, Kelten, Iberern und Thrakern diese Eigenschaft zu.[49] Aber selbst die berühmte skythische Trunksucht konnte in einem bestimmten Zusammenhang geleugnet werden. Eustathios zitiert in den Kommentaren zur *Ilias* das Sprichwort des Anacharsis:

> Auf die Frage, warum die Skythen keine Flötisten haben, sagte er: weil es keine Weinreben gibt ... Es läge näher zu sagen, dass die Skythen nicht einmal Trunkenheit und Wein haben.[50]

Diogenes Laertios, der dem Skythen Anacharsis ein ganzes Kapitel in seinem Buch *Leben der Philosophen* widmet, zitiert seine ‚antialkoholische' Rede:

> Anacharsis sagte, dass die Weinrebe drei Weintrauben bringe: die erste Vergnügen, die zweite Rausch, die dritte Ekel ... Als er gefragt wurde, wie man kein Trinker werden könne, sagte er: ‚Wenn du eine Schande der Betrunkenen (τὰς τῶν μεθυόντων ἀσχημοσύνας) vor deinen Augen hast ...'[51]

Allerdings konnte selbst Anacharsis wegen seiner skythischen Herkunft als Träger dieses Lasters betrachtet werden. Claudius Aelianus sagt von ihm in den *Bunten Erzählungen*:

> Man sagt, dass auch Anacharsis sehr viel bei Periander[52] ausgetrunken habe; diese Fähigkeit brachte er aus seiner Heimat mit: in der Tat neigen die Skythen dazu, reinen Wein zu trinken.[53]

VII. DIE NEUDEFINITION DES BARBARENBEGRIFFS UNTER CHRISTLICHEM VORZEICHEN

Die Opposition der griechisch-römischen und der barbarischen Kultur unterlag einem signifikanten Wandel bei frühchristlichen Autoren. An mehreren Stellen spricht der Apostel Paulus die universale, ethnische Grenzen überwindende Ausrichtung des Christentums an. Der *Römerbrief* sammelt – unter Rückgriff auf zahlreiche alttestamentliche Verse – die Hinwendung Gottes zu den Heiden, oder,

---

47   Vgl. Braund 2008, 354: ‚The trade in wine brought from the Greek world was a key feature of interaction with the Scythians.'
48   Ausführlicher darüber Porucznik 2013, 710–714.
49   Plat. *Nom.* 1.637d–e. Meine Übersetzung. Dasselbe erzählt Tacitus (*Germ.* 22) über die Germanen.
50   Eust. *Il.* 1.9. Meine Übersetzung.
51   Diog. Laert. 1.8.101–105, hier 103. Meine Übersetzung.
52   Der Tyrann von Korinth, 625–585 v.Chr.
53   Ail. *VH* 2.41. Meine Übersetzung.

der hebräischen Terminologie genauer folgend, zu den ‚Völkern' (*Röm* 15.9–12). Ein wenig älter ist der Galaterbrief, in dem Paulus seine Vision so formuliert (*Gal* 3.28): ‚Es gibt nicht mehr Griechen und Juden, nicht Sklaven und Freie, nicht Mann und Frau; denn alle seid einer in Christus Jesus!' Noch ausführlicher predigt er im Kolosserbrief, nachdem Paulus die Überwindung menschlicher Laster wie Schamlosigkeit, Gier, Zorn und Lüge zur Bedingung für die Erneuerung des Menschen erklärt hat (3.5–10), von der Aufhebung folgender Trennlinien (3.11):

> ‚Wo das geschieht, gibt es nicht mehr Griechen oder Juden, Beschnittene oder Unbeschnittene, Barbar (oder) Skythe, Sklave (oder) Freier, sondern alles und in allen (ist) Christus'.[54]

Der Barbar und der Skythe sind hier wohl einander entgegengesetzt, entsprechend den vorhergehenden und nachfolgenden gepaarten Gliedern der Aufzählung.[55] In jedem Fall verlässt der Skythe, der das Christentum annimmt, die Barbarei und alle inhärenten Laster[56].

Das Bild eines nördlichen Barbaren, der seine Wildheit ablegt, nachdem er die christliche Botschaft angenommen hat, wird in der Literatur noch vielfach begegnen, so etwa bei Hieronymus, der um die Wende vom 4. zum 5. Jh. n.Chr. schrieb:

> Die wilden Bessen und die große Zahl der mit Tierfellen bekleideten Völker, die einst bei ihren Totenopfern Menschen schlachteten, haben ihre rohe Sprache umgebrochen und zum Preisgesang auf das Kreuz veredelt. In der ganzen Welt gilt jetzt nur ein Wort: Christus.[57]

Etwa zeitgleich zu dem lateinischen Kirchenvater betont auch der Dichter Prudentius, dass alle Nationen in der Oikumene auf demselben Boden und unter demselben Himmel leben sowie vom selben Ozean umflossen werden:

> Denique Romanus, Daha, Sarmata, Vandalus, Hunnus,
> Gaetulus, Garamans, Alamannus, Saxo, Gaulala,
> una omnes gradiuntur humo, caelum omnibus unum est,
> unus et oceanus, nostrum qui continet orbem.[58]

So überwanden christliche Denker der (Spät-) Antike allmählich die Zweiteilung zwischen griechisch-römischer Zivilisation und barbarischen Randvölkern. Andererseits entstanden damals neue Trennlinien zwischen der als zivilisiert geltenden rechtgläubigen christlichen Welt der Römer und den als Barbaren betrachteten Häretikern oder Heiden. Doch dies führt uns bereits über unser Untersuchungsfeld hinaus, denn die gefürchteten nördlichen Barbaren hatten – nach den Einbrüchen an der Rhein- und Donaugrenze im 3. und 4. Jh., vor allem nach der Katastrophe von Adrianopel (378 n.Chr.) – ganz andere Namen als die der Skythen. Der Skythenlogos Herodots hatte den nördlichen Barbaren fast ein Jahrtausend lang ein sehr lebhaftes und zugleich hochgradig flexibles Gesicht im Bewusstsein der

---

54 ὅπου οὐκ ἔνι Ἕλλην καὶ Ἰουδαῖος, περιτομὴ καὶ ἀκροβυστία, βάρβαρος, Σκύθης, δοῦλος, ἐλεύθερος, ἀλλὰ [τὰ] πάντα καὶ ἐν πᾶσιν Χριστός.

55 Vgl. Weiß 2015, 23–26 (mit weiterer Literatur), der hier die Begriffe ‚skythisch' und ‚barbarisch' ebenfalls nicht als Synonyme, sondern Antonyme versteht.

56 Siehe ausführlicher Neutel 2015.

57 Hieron. *Ep.* 60.4. Übersetzung von L. Schade.

58 Prudent. *C. Symm.* 2.808–811. Meine Übersetzung.

griechischen (und römischen) Welt gegeben. Die von ihm geschaffenen Allgemeinplätze blieben, aber Namen und Konfigurationen änderten sich in Zeiten eines Umbruchs von welthistorischem Ausmaß.[59]

Danksagung

Ich danke Philipp Köhner für die Verbesserung der deutschen Version meines Beitrages.

Bibliographie

Almagor, E. 2005: ‚Who is a Barbarian? The Barbarians in the Ethnological and Cultural Taxonomies of Strabo‘, in D. Dueck, H. Lindsay & S. Pothecary (eds.), *Strabo's Cultural Geography. The Making of a Kolossourgia*, Cambridge, 42–55.

Asheri, D. 1998: The Achaeans and the Heniochi. Reflections on the Origins and History of a Greek Rhetorical Topos, in G.R. Tsetskhladze (ed.), *The Greek Colonisation of the Black Sea Area. Historical Interpretation of Archaeology*, Stuttgart, 265–285.

Asheri, D., Loyyd, A. & Corcella, A. 2007: *A Commentary on Herodotus Books I–IV*, ed. by O. Murray & A. Moreno, with a contribution by M. Brosius, Oxford.

Bäbler, B. 1998: *Fleissige Thrakerinnen und wehrhafte Skythen. Nichtgriechen im klassischen Athen und ihre archäologische Hinterlassenschaft*, Stuttgart.

Bäbler, B. 2005: ‚Bobbies or Boobies? The Scythian Police Force in Classical Athens‘, in D. Braund (ed.), *Scythians and Greeks: Cultural Interactions in Scythia, Athens and the Early Roman Empire (Sixth Century B.C. – First Century A.D.)*, Exeter, 114–125.

Bacon, H. 1961: *Barbarian in Greek Tragedy*, New Haven.

Balsdon, John P.V.D. 1979: *Romans and Aliens*, London.

Bengtson, H. 1954: ‚Hellenen und Barbaren: Gedanken zum Problem des griechischen Nationalbewusstseins‘, in K. Rüdinger (ed.), *Unser Geschichtsbild. Wege zu einer universalen Geschichtsbetrachtung*, München, 25–40.

Bichler, R. 1988: ‚Der Barbarenbegriff des Herodot und die Instrumentalisierung der Barbarentopik in politisch-ideologischer Absicht‘, in I. Weiler (ed.), *Soziale Randgruppen und Aussenseiter im Altertum*, Graz, 117–128.

Bichler, R. 1996: ‚Wahrnehmung und Vorstellung fremder Kultur. Griechen und Orient in archaischer und frühklassischer Zeit‘, in M. Schuster (ed.), *Die Begegnung mit dem Fremden. Wertungen und Wirkungen in Hochkulturen vom Altertum bis zur Gegenwart*, Stuttgart, 51–74.

Bichler, R. 2000: *Herodots Welt. Der Aufbau der Historie am Bild der fremden Länder und Völker, ihrer Zivilisation und ihrer Geschichte*, Berlin.

Braund, D. 1997: ‚Scythian Archers, Athenian Democracy and a Fragmentary Inscription from Erythrae‘, in *Antičnyj mir i Vizantija. K 70-letiju V.I. Kadeeva* (Die antike Welt und Byzanz. Zum 70-jährigen Jubiläum von V.I. Kadeev), Charkov, 48–56.

Braund, D. 2008: ‚Scythian Laughter: Conversations in the Northern Black Sea Region in the 5th Century BC‘, in P. Guldager Bilde & J. Hjarl Petersen (eds.), *Meetings of Cultures in the Black Sea Region: Between Conflict and Coexistence*, Aarhus, 347–367.

Christabel, L. 1976: *The Greek View of Barbarians in the Hellenistic Age*, Diss. Colorado.

Coleman, J. 1997: *Greeks and Barbarians: Essays on the Interactions between Greeks and Non-Greeks in Antiquity and the Consequences for Europocentrism*, Bethesda, MD.

---

59 Über Skythenbild und Skythenname im Mittelalter s. ausführlich Podossinov, Jackson & Konovalova 2016, 196–290.

Cunliffe, B.W. 1988: *Greeks, Romans and Barbarians: Spheres of Interaction*, London.

Dauge, Y.A. 1981: *Le Barbare. Recherches sur la conception romaine de la barbarie et de la civilization*, Brüssel.

Dihle, A. 1994: *Die Griechen und die Fremden*, München.

Eichhorn, A. 1904: *Barbaros quid significaverit*, Diss. Leipzig.

Frolov, E.D. 1998: 'Skify v Afinakh' (The Skythians in Athens), *VDI* 1, 135–152.

Frolov, E.D. 2004: 'Politsejskaja služba v demokratičeskom polise: skify v Afinakh' (Police Work in a Democratic City: The Scythians in Athens), in E.D. Frolov, *Paradoksy istorii – paradoksy antičnosti* (Paradoxes in History – Paradoxes in Antiquity), Sankt-Petersburg, 195–220.

Georges, P. 1994: *Barbarian Asia and the Greek Experience: from the Archaic Period to the Age of Xenophon*, Baltimore.

Gerstacker, A., Kuhnert, A., Oldemeier, F. & Quenouille, N. (eds.) 2015: *Skythen in der lateinischen Literatur. Eine Quellensammlung*, Berlin.

Giuseppe, N. & Reverdin, O. (eds.) 1988: *Hérodote et les peuples non grecs*, Genf.

Gruen, E. 2011: *Rethinking the Other in Antiquity*, Princeton.

Hall, E. 1989a: *Inventing the Barbarian: Greek Self-Definition through Tragedy*, Oxford.

Hall, E. 1989b: ‚The Archer Scene in Aristophanes' Thesmophoriazusae‘, *Philologus* 133, 38–54.

Harrison, T. 2002: *Greeks and Barbarians*, Edinburgh.

Hartog, F. 1980: *Le miroir d'Hérodote: essai sur la représentation de l'autre*, Paris.

Heinze, R. 1891: ‚Anacharsis‘, *Philologus* 50, 458–468.

Horden, P. & Purcell, N. 2000: *The Corrupting Sea. A Study of Medieval History*, Oxford.

Ivantchik, A.I. 2005a: *Am Vorabend der Kolonisation. Das nördliche Schwarzmeergebiet und die Steppennomaden des 8.–7. Jhs. v. Chr. in der klassischen Literaturtradition: mündliche Überlieferung, Literatur und Gesellschaft*, Berlin.

Ivantchik, A. 2005b: 'Who were the 'Scythian' Archers on Archaic Attic Vases?', in D. Braund (ed.), *Scythians and Greeks: Cultural Interactions in Scythia, Athens and the Early Roman Empire (Sixth Century B.C. – First Century A.D.)*, Exeter, 100–114.

Jatsenko, S. 2006: *Kostjum drevnej Evrazii (iranojazyčnye narody)* (Kostüm des alten Eurasiens [iranisch-sprechende Völker]), Moskau.

Jüthner, J. 1923: *Hellenen und Barbaren*, Leipzig.

Kath, R., Rücker, M. 2012: *Die Geburt der griechischen Weisheit oder: Anacharsis, Skythe und Grieche*, Halle.

Kindstrand, J.E. 1981: *Anacharsis. The Legend and the Apophtegmata*, Uppsala.

Kothe, H. 1969: ‚Skythenbegriff bei Herodot‘, *Klio* 51, 15–88.

Kuhnert, A. 2012: ‚Der Skythe: Friedliebender Krieger und weiser Dummkopf. Das Bild der Skythen in der lateinischen Literatur‘, in L. Prager (ed.), *Nomadismus in der ‚Alten Welt‘. Formen der Repräsentation in Vergangenheit und Gegenwart*, Berlin, 72–90.

Lévi, E. 1984: ‚Naissance du concept de barbare‘, *Ktèma* 9, 5–14.

Lissarrague, F. 1990: *L'autre guerrier. Archers, Peltastes, Cavaliers dans l'imagerie attique*, Paris.

Long, T. 1986: *Barbarians in Greek Comedy*, Carbondale, IL.

Lovejoy, A.O. & Boas, G. 1935: *Primitivism and Related Ideas in Antiquity*, Baltimore.

Mitchell, L.G. 2007: *Panhellenism and the Barbarian in Archaic and Classical Greece*, Swansea.

Motta, D. 1999: 'Scythae iustissimi barbarorum? Notazioni sulla fortuna de Erodoto nella cultura Greco-Romana', in P. Anello, (ed.), *Erodoto e l'Occidente*, Roma, 309–315.

Nesselrath, H.-G. 2009: ‚Fremde Kulturen in griechischen Augen – Herodot und die „Barbaren"‘, *Gymnasium* 116, 307–330.

Neutel, K.B. 2015: *A Cosmopolitan Ideal: Paul's Declaration 'Neither Jew nor Greek, neither Slave nor Free, nor Male and Female' in the Context of First Century Thought*, London.

Nippel, W. 1990: *Griechen, Barbaren und ‚Wilde‘*, Frankfurt.

Pečatnova, L.G. 2007: *Spartanskije tsari* (The Kings of Sparta), Moskau.

Plassart, A. 1913: ‚Les archers d'Athènes‘, *REG* 26, 151–213.

Podossinov, A.V., Jackson, T.N. & Konovalova, I.G. 2016: *Skifija v istoriko-geografičeskoj traditsii antičnosti i srednevekovja* (Scythia in the Historical and Geographical Tradition of Antiquity and the Middle Ages), Moskau.

Porucznik, J. 2013: ‚The Image of a "Drunken Scythian" in Greek Tradition', *Proceedings of the 1st Annual International Interdisciplinary Conference, AIIC 2013, 24–26 April, Azores, Portugal,* Lisbon, 710–714.

Rausch, S. 2013: *Bilder des Nordens. Vorstellungen vom Norden in der griechischen Literatur von Homer bis zum Ende des Hellenismus*, Berlin.

Reuters, F.H. 1963: *Die Briefe des Anacharsis*, Berlin.

Riemer, U. & Riemer, P. (eds.) 2005: *Xenophobie – Philoxenie. Vom Umgang mit Fremden in der Antike*, Stuttgart.

Riese, A. 1875: *Die Idealisierung der Naturvölker des Nordens in der griechischen und römischen Literatur,* Frankfurt.

Schmidt, T.S. 1999: *Plutarque et les Barbares. La rhétorique d'une image*, Brüssel.

Schmitt, H.H. 1957/8: *Hellenen, Barbaren und Römer*, Aschaffenburg.

Schubert, Ch. 2010: *Anacharsis der Weise. Nomade, Skythe, Grieche*, Tübingen.

Schulz, R. 2005: *Die Antike und das Meer,* Darmstadt.

Schwabl, H., Diller, H., Reverdin, O., Peremans, W., Baldry, H.C. & Dihle, A. 1962: *Grecs et Barbares*, Genf.

Sier, R. 1992: ‚Die Rolle des Skythen in den Thesmophoriazusen des Aristophanes', in C.W. Müller u.a. (eds.), *Zum Umfang mit fremden Sprachen in der griechisch-römischen Antike*, 63–83.

Sinitsyn, A.A. 2008: ‚Sofokl i skifskij logos Gerodota' (Sophokles und der Skythische Logos des Herodot), *Arkheologija Vostočno-Evropejskoj stepi* (Archaeologie der osteuropäischen Steppe) 6, Saratov, 269–292.

Skržinskaja, M.V. 1998: *Skifija glazami éllinov* (Skythien in Augen der Hellenen), Sankt-Petersburg.

Thollard, P. 1987: *Barbarie et civilisation chez Strabon. Étude critique des livres III et IV de la Géographie*, Paris.

Thompson, J.A.K. 1921: *Greeks and Barbarians*, London.

Timpe, D. 1996: ‚Rom und Barbaren', in M. Schuster (ed.), *Die Begegnung mit dem Fremden. Wertungen und Wirkungen in Hochkulturen vom Altertum bis zur Gegenwart*, Stuttgart, 34–50.

Ungefehr-Kortus, C. 1996: *Anacharsis, der Typus des edlen, weisen Barbaren*, Frankfurt.

Vlassopoulos, K. 2013: *Greeks and Barbarians*, Cambridge.

Vos, M.F. 1963: *Scythian Archers in Archaic Attic Vase-Painting*, Groningen.

Weiß, A. 2015: ‚Einführung – Die Skythen als paradigmatische Nomaden', in A. Gerstacker, A. Kuhnert, F. Oldemeier & N. Quenouille (eds.), *Skythen in der lateinischen Literatur. Eine Quellensammlung*, Berlin, 17–35.

Wendt, Ch. 2016: ‚Piraterie als definitorisches Moment von Seeherrschaft', in E. Baltrusch, H. Kopp, Ch. Wendt (eds.), *Seemacht, Seeherrschaft und die Antike*, Stuttgart, 79–91.

West, S. 1999: ‚Introducing the Scythians. Herodotus on Koumiss (4.2)', *Museum Helveticum* 56, 76–86.

West, S. 2004: ‚Herodotus and Scythia', in V. Karageorghis & I. Taifacos (ed.), *The World of Herodotus. Proceedings of an International Conference Held at the Foundation Anastasios G. Leventis, Nicosia, Sept. 18–21, 2003*, Nicosia, 73–89.

Zahn, R. 1896: *Die Darstellung der Barbaren in griechischer Litteratur und Kunst der vorhellenistischen Zeit*, Diss. Heidelberg.

**B. Studies in the Royal Dynasties
of Pontos and the Bosporan Kingdom**

THE BOSPORAN KINGS AND THE GREEK FEATURES
OF THEIR CULTURE IN THE BLACK SEA
AND THE MEDITERRANEAN

Madalina Dana

Abstract: The Bosporan rulers held the title of *archontes* of the Greek cities and *basileis* of the local populations in the inscriptions of their kingdom, but often called 'tyrants' or depicted with disdain in literary sources produced outside of their realm. But at least one historiographic tradition shows them in a much more positive light: there must have been some local chronicles or historiography that was perhaps initiated by them and portrayed them mostly favourable. Glimpses of other literary sources, if read against the grain, reveal the king's concern to be viewed as patrons of the arts. In addition, documentary evidence attests to the public honours they received from cities like Athens, which responded to their benefactions with the erection of statues. Other opportunities for royal munificence or the display of Philhellenism were offered by the Panhellenic sanctuaries (Didymai, Delos, Delphi) or on the occasion of Panhellenic games, besides the recognition of *asylia* for the *Asklepieia* of Kos. Even before the Hellenistic period, the rulers of the Bosporus made a considerable effort to be accepted as Greek kings by their Aegean trade partners, and so they were. Philhellenism began to be combined with diplomatic friendship with the Romans in the 1st century BC. The Bosporan rulers found a place in the largely pacified world of the Roman Empire, not least by willingly adhering to the ideals of the Second Sophistic.

Абстракт: Боспорские цари и греческие особенности их культуры в Понте и Средиземноморье: В надписях Боспорского Царства Боспорские правители носили титулы «архонтес» греческих городов и «басилеис» местного населения, но часто их называли «тиранами» или изображали с пренебрежением в литературных источниках, написанных за пределами их царства. Однако, по крайней мере, одна историографическая традиция показывает их в гораздо более позитивном свете: вероятно, существовали какие-то местные хроники или историография, которые возможно были инициированы правителями, и изображали их особенно благоприятно. Некоторые литературные источники, если их прочитать более внимательно, показывают, что правители заботились о том, чтобы их считали покровителями искусств. Кроме того, документальные источники свидетельствуют о публичных почестях, получаемых Боспорскими правителями в таких городах, как Афины, которые отплатили за их благотворительность возведением статуй. Разных возможностей для проявления царской благотворительности или филэллинизма было множество: в панэллинских святилищах (Дидима, Делос, Дельфы) или при проведении панэллинских игр, включая признание *asylia* для игр Асклепия на Косе. Еще до эллинистического периода правители Босфора предпринимали значительные усилия, чтобы их торговые партнеры из Эгейского моря считали их греческими правителями. Так это и было. В I веке до н.э. их филэллинизм начали связывать с дипломатической дружбой с римлянами. Боспорские правители нашли свое место в значительной степени безвоенном мире Римской империи, в основном благодаря добровольной приверженности идеалам второй софистики.

I. INTRODUCTION

As an apt introduction to our reflection on the Greek culture of the Bosporan kings, I would like to adduce a programmatic passage from an article by Heinz Heinen:

> La position périphérique du royaume du Bosphore et les difficultés d'une bibliographie abondante en langue russe ne devraient pas nous décourager d'aborder le littoral nord de la Mer Noire. Après tout, il serait dommage si nous perdions de vue ces dynastes à la lisière du monde scythe, alors que leurs images les rendaient tellement présents aux yeux des Athéniens fréquentant les marchés de la cité et du Pirée.[1]

This quotation expresses an interest in the culture of the 'peripheral' Greek community in the Bosporan Kingdom by those at its very 'centre'.[2] It encourages us to revisit the actions of the Bosporan dynasts, attested both in their own kingdom and elsewhere in the Greek world with a particular view to their cultural and ethnic implications. In contrast to other accounts which emphasize the 'barbarian' or 'ethnically mixed' nature of those kings from Pantikapaion, I shall try to show the coherence of our evidence for more than half a millennium. Systematically comparing the Bosporan rulers with their Bithynian counterparts will further strengthen the results of the present study.

As is well known, these rulers are designated as *archontes* of the Greek cities and as *basileis* of the indigenous peoples in the inscriptions of the northern Black Sea region.[3] For instance, they figure as 'archon of the Bosporus and Theodosia' as well as 'king of the Sindoi, Toretoi, Dandaroi and Psessoi' or 'king of the Sindoi and all the Maiotai'.[4] In contrast, the external literary evidence either calls them kings without distinctions or even 'tyrants', such as the hostile author Deinarchos in his speech *Against Demosthenes* (*Or.* 1.43: τοὺς ἐκ τοῦ Πόντου τυράννους). Their political programme anticipates the conditions of the Hellenistic kingdoms in more than one way, especially with a view to the symbiosis between Greek cities, which partially preserved their civic institutions, and the peoples living around. The political nomenclature of Tanaïs also mirrored its duality, in which two distinct peoples, the *Hellenes* and the *Tanaïtes*, lived side by side, led by *Hellenarches* and *archontes Tanaïton* respectively.[5]

The Spartokids, who ruled from 438 to ca. 108 BC, mostly bore Greek names (Satyros, Seleukos, Leukon, Gorgippos, Apollonios, Eumelos, Prytanis), besides some Thracian (Spartokos) and Iranian examples (Pairisades, Kamasarye). After the integration of the Bosporan kingdom into the realm of Mithradates VI Eupa-

---

1   Heinen 2006a, 292. He insists that the pro-Athenian policy of the Bosporan rulers was also relevant in the context of Athens' wars with Macedon. Grain supplies by the Spartokids and Athenian honours in return both before and after 338 thus represent an important continuity in the 4th and 3rd centuries amidst the several political vicissitudes.
2   See some of the conclusions of my book: Dana 2011a, 397–399.
3   See Vinogradov 1980, 82; Müller 2007, 148–150; Müller 2010, 39f.; Podossinov 2012.
4   *CIRB* 6 A, 7–11, 1037–1043; *SEG* LII 741.
5   The local agora was restored by the Hellenarch Basileides, son of Theoneikos: *IOSPE* II 430; *CIRB* 1245 (*ca.* 220 AD, under Rheskuporis III).

tor, we observe a partial continuity, though with a shift towards Persian. By the middle of the first century AD, the Bosporan kings adopted the Roman onomastic formula (Τιβέριος Ἰούλιος). In addition to that, most kings beginning with Asandros (48/7–20/19 BC) bore the Graeco-Roman epithet *Philorhomaios*, after most kings of the Mithradatid lineage from Pharnakes I (who died by 155 BC) to Pharnakes II (63–47 BC) had obtained the status of an *amicus populi Romani*.[6] Heinen judiciously remarked that the Roman period illustrates the results of a long symbiosis between Greek and Iranian (Scythian and Sarmatian) elements, in the context of subordination to the Roman domination of the Bosporan kings.[7]

These mixed traditions are in accordance with the intercultural exchange that is characteristic of the region.[8] My aim is to show that, despite their multiple ethnic backgrounds and diverse political power relations, the Bosporan dynasts persistently present themselves to the Greeks among their subjects and beyond as perfect Hellenic rulers. Obviously, cultural patronage and dedications to Greek gods are well attested in the classical era, when they did not have to justify their domination, but just wished to be regarded as cultivated monarchs – the tyrants of Herakleia or Syracuse may serve as parallels. In the Hellenistic period, such gifts and benefactions multiplied. It is during this time that rulers – heirs of Alexander, their descendants or minor dynasts – were in a state of permanent competition. They used especially the public space of major sanctuaries and Panhellenic festivals to affirm their close affiliation with Greek culture and their belonging to a politically fragmented, but culturally unified world.

II. EXPRESSIONS OF POLITICAL PIETY

Let us begin with a dedication by the king Leukon I (who ruled from 389 to 349 BC) to Apollo, dating from 360/55 BC. It is attested in an epigram phrased in elegiac couplets and carved in stone at Labrys (now Semibratnee, around 35 km from Gorgippia / Anapa):

> Having made a vow, Leukon, son of Satyros, erected this statue
> in honor of [Phoibos] Apollo, the valiant protector
> of this city, that of the Labrytai, (Leukon) who, as [*archon* of the] B[osporus] (=Pantikapaion)
> and of Theodosia, pushed with fight and might
> Oktamasades out of the land of the Sindoi, the son of Hekataios,
> king of the Sindoi, who after chasing his father
> out of power, [overran] this city.[9]

---

6   See Heinen 2006c; 2008 and Coşkun 2016 on Roman 'friendship' relations and so-called 'client kingdoms' in the eastern Mediterranean and especially of the Bosporan kingdom. On Pharnakes I, see also Payen, chapter VII, and on Pharnakes II, Ballesteros Pastor, chapter VIII in this volume.

7   Heinen 2006b, 334f.

8   Williams & Ogden 1994; Jacobson 1995; Fless & Treister 2005; Daumas 2009; Meyer 2013.

9   *SEG* XLIII 515; *BE* 1996, 306; *SEG* XLVIII 1027; *BE* 2003, 393; *SEG* LVI 885; Müller 2010, 359f., no. 1: Εὐξάμενος Λεύκων υἱὸς Σατύρ[ō τόδε ἄγαλμα] / [Φοίβωι] Ἀπόλλωνι

The epithet of the god Apollo (l. 2) was restored by the first editor as toponymic (*Labrytes*) according to the ethnic mentioned in line 3,[10] before Philippe Gauthier argued in favour of the more standardised epithet (*Phoibos*).[11] Nevertheless, it is obvious that the deity had a sanctuary in the city, whose name has been restored as *Labrys or *Labryta from the ethnic in the genitive plural Λαβρύτωμ.[12] This name, non-Greek but Hellenized,[13] is nowhere else attested directly. Apollo of Labrys is probably not identical with Apollo *Ietros* of the first Ionians, but a local manifestation, whom Leukon honours in a hexametric epigram. The dedication celebrates a military feat of the *archon* against Oktamasades, who is said here to be a usurper of his father, Hekataios, the king of the Sindoi. The latter is the same who in a stratagem attested by Polyainos (8.55) married queen Tirgatao of the Maiotai. It should be noted that both events, the confrontation of the two rivals around some local power issues and especially the dedication in Greek to a Greek god by the Spartokid, occurred in the middle of the 4th century. There is certainly a mark of the newly established domination of the dynast in the region, colonized by the Greeks from the 6th century BC, and a way of expressing his adherence to the Greek cultural values.

The Hellenistic period represents a significant change in the behaviour of the Bosporan kings, who, like other monarchs, appear as benefactors of Panhellenic sanctuaries. They imitate the Greek cities of Euxine Pontus: thus, the attestations of offerings in the sanctuary of Apollo of Didymai concerns a Sinopean, son of Theon, who dedicated a *phiale* in 224/3, and two sovereigns of the Bosporus. Around 180 BC, queen Kamasarye offered a golden object, and her son, Pairisades IV, a very precious golden *phiale* (177/6 BC). The Delphians thanked Kamasarye and Pairisades also for having followed the example of Spartokos VI, the father of Kamasarye, and for being generous toward the sanctuary and the people of Delphi.[14] One of their predecessors, Pairisades II, is known for having made an offering at Delos: five different accounts preserved epigraphically[15] mentioned a

---

στῆσε τῶι ἐνα[ρέτως] / τῆσδε πόλεως μεδέοντι Λαβρύτωμ, Β[οσπόρō ἄρχων?] / Θευδοσίης τε, μάχηι καὶ κράτει ἐξελ[άσας] /˙ Ὀκταμασάδεα γῆς ἐΞίνδων (= ἐξ Σίνδων), παῖδ᾿ Ἑκ[αταίō] /⁵ τοῦ Σίνδωμ Βασιλέως, ὃς πάτερα Q?[---] / ἐγβάλλων ἀρχῆς εἰς τήνδε πόλιγ κ[ατέθρεξεν?].

10 *SEG* XLIII 515.
11 *BE* 2003, 393.
12 Hansen & Nielsen 2004, no. 702.
13 For the details, see Müller 2010, 360.
14 Sinope: *I.Didyma* 447, l. 7. Kamasarye: *I.Didyma* 463. Pairisades IV: *I.Didyma* 464. Delphi: *FD* III 1, 453. See Giovannini 1997.
15 *IG* XI 2.287 B, l. 126 (ca. 250; mentioned together with the offering of Antigonos Gonatas); *I.Délos* 298 A, l. 95 (239); *I.Délos* 313 A, l. 74 (ca. 235); *I.Délos* 314 B, l. 82 (between 237 and 233 BC); *I.Délos* 320 B, l. 40 (229 BC). Cf. Vinogradov 1997, 47; for other donors from the Bosporan Kingdom in Delos and Rhodes, see Vinogradov 1997, 65. See also Bringmann & von Steuben 1995, no. 192: Pairisades II; no. 294: Pairisades IV; no. 295: Kamasarye, mother of Pairisades IV.

cup he had offered to the treasure of the sanctuary around 250 BC, just after the consecration of three silver *phialai* by their Chersonesitan neighbours.[16]

A first parallel with the Bithynian kings can be made for these offerings, intended to mark the place of these 'peripheral' dynasties in the multiple configurations of Hellenism. The Bithynian kings were active in supporting the sanctuaries and other places of culture of the Greeks,[17] as they consecrated *phialai* in Didymai,[18] a place closely connected with Milesian colonisation in the archaic period.

But the main argument of the prestige policy and of the quest for recognition of the Bosporan kings remains a fragmentary inscription of Kos. This is a response letter from an anonymous king who, at the request of the *theoroi* of Kos, recognized the *Asklepieia* as Panhellenic and penteteric games (242/1 BC):

> ... in accordance with their instructions, they made proclamation of the sacrifice and games, using their education (τῆι διδασκαλίαι χρώμενοι) so beautifully that wonder overcame us ... since the time when we acceded to our kingdom ... concerning these things ... so many and such long-standing benefactions toward you. As there are just now certain things standing in the way of these things, if for this the *theoroi* seem to find fault with us, we ask you nonetheless to judge that we are accommodating ourselves to the times, reckoning that indulgence is owed us, whenever ... we should be unable to send *theoroi* to the celebration of these games. I and my sister ... and our citizens accept both your proclamation that has been made for the god and the inviolability, and we also have accepted gladly the kinship as true and worthy of you and us, the best testimony being that of our father, which you have made clear he himself furnished. If any of other Greeks, making this a start of friendship, should first proclaim us kin, we would gladly approve them many times over, having full gratitude towards those who remember such kinship and blood ties and choose to preserve them. And now we approve your loyalty and in the future we shall try to maintain for your people the benefactions established in ancient times and now finely and suitably brought to the best renewal by you, loyal friends, and, heeding your requests, always to gratify you to the best of our ability. Farewell.[19]

---

16 *I.Délos* 298 A, l. 96 (239 BC); *I.Délos* 313 A, l. 74 (235 or 234 BC); *I.Délos* 314 B, l. 81 (between 237 and 233 BC); *I.Délos* 320 B, l. 40 (229 BC). See Dana 2011b, 49f.

17 On the *euergesia* of the 'peripheral kings', see Michels 2010.

18 Prusias II, between 182 and 149: the first time, one *phiale* of 300 Alexandrian drachms, and one hydria of 1490 dr.; the second time, one *phiale* (*I.Didyma*, no. 463, l. 10–16, 22–25; no. 469; 473, ll. 3f.; maybe no. 462, l. 5). Ps.-Skymnos v. 58f. (ed. Marcotte 2000) mentions the 'true cult' of Nikomedes (II or III) towards Apollo of Didyma.

19 I here quote the translation of Rigsby 1996, 121–124, no. 12. See also Curty 1995, 48–52, no. 24: ἀκολούθ[ως αἷς εἶχον ἐντολαῖς τὴν ἐπαγ-] /[8] γελίαν ἐποιή[σαντο τῆς τε θύσιας καὶ τῶν ἀγώ-] / νων τῆι διδασκαλίαι χρώμενοι οὕτω καλῶς ὥστε θαυμάζειν ἐπήει [ἡμεῖν ---] / ἀφ᾽οὗ τῆς βασιλεία[ς ---] /[12] .ίαν περὶ τούτων ὑσ[--- σ]θε τηλικούτων καὶ το[ιούτων φιλανθρώπων προ-] / ὑπαρχόντων ἡμεῖν πρὸς [ὑμᾶς· ὄντων δέ τινων νῦν] / τούτοις ἐμποδίων, εἰ κα[ὶ οἱ θεωροὶ τοῦτο ἡμεῖν] /[16] μέμψασ{α}θαι ἐφάνη{.}ν, πα[ρακαλοῦμεν ὅμως ὑμᾶς] / τοῖς καιροῖς ἴσως ἐπακ[ολ]ουθή[σαντας κρῖναι ἡ-] / μεῖν συγγνώμην ἐκτέον, ὅταν [---] / κότες μὴ δυνώμεθα τὰς τούτω[ν τῶν ἀγώνων συντελείας] /[20] θεωρεῖν· ἐγὼ δὲ καὶ ἡ ἀδελφ[ή] μ[ου ---] / καὶ οἱ ἡμέτεροι πολῖται τήν τε παρ᾽ [ὑμῶν ἐπαγγε-] / λίαν γινομένην τῶι θεῶι καὶ τὴν ἀ[συλία]ν δεχ[ό-] / μεθα καὶ τὴν συγγένειαν οὖσαν ἀλ[η]θινὴν καὶ ὑ[-] /[24] μῶν τε ἀξίαν καὶ ἡμῶν ἡδέως προσ[δε]δέγμεθα, / μαρτυρίας μεγίστης τῆς παρὰ το[ῦ ἡμ]ετέρου / πατρὸς προσγεγενημένης ἣν ἀπ[οπεφήν?]ατε / αὐτοῦ ἐκείνου ποιησαμένου· εἰ [δὲ καὶ τῶν λοιπῶν] /[28] τινες Ἑλλήνων ἀρχὴν φιλίας ποιούμενοι ταύτην / πρῶτον προσηγόρευον ἡμ[ᾶς συγγενεῖς, εὐ]λόγως / ἂν προσελαμβάνομεν τ[ούτους, τοιαύτη]ς φιλανθ- / ρωπίας ἡμεῖν

The first commentators identified the author of this royal letter as Gelon, son of Hieron II of Syracuse, since a decree of Gela was carved on the other side of the stele (Rigsby 1996, no. 49). Afterwards, most scholars, following the hypothesis of Jeanne and Louis Robert, agreed on identifying him as a king of the Bosporus.[20] The Roberts wondered if, since he expressed his happiness that some of the Greeks have acknowledged his kinship with them, we are to understand that he was not Greek or that his Greekness was doubtful. The king was most probably not Pairisades II, who came to power in 284/3 and was alive as late as 250 (cf. *IG* XI 2.287 B, l. 126), but one of his sons – the reference to the father indicates that he must have acceded to the throne only recently. Andrew Rigsby concludes: 'If rightly assigned, it is a precious document on the Hellenism of the Spartocid dynasty'.[21]

At any rate, the word *didaskalia* (l. 9) is to be translated better as 'lecture', having a similar meaning as *akroasis*. This would have allowed an orator to prove an ancestral connection between two cities. Perhaps the king was impressed by the claim of historical or mythical relations made in a speech delivered by a Koan ambassador. This admiration for the occasional speech pronounced by the Koan shows that the monarch was sufficiently cultivated to understand this flattery and to respond to it adequately. Even if the king does not seem able to send *theoroi*, he approves the Greeks' initiative and does not hide his pride at having been invited to attend the festival.

The comparison with the response of the Bithynian king Ziaelas not only reveals a similar use of language, but also some differences between two royal pronouncements.[22] Ziaelas' answer insists, on the one hand, on the continuity of the policy inaugurated by his father Nikomedes I and the friendship of Ziaelas with Ptolemy III, and, on the other hand, on the importance of the links not only with the Koans, but also with the other Greeks – that is those who frequented the sanctuary of Asklepios and were able to recognize his position.[23] The sentence 'we do

προϋπαρχ[ούσης πρό]ς γε τοὺς το[ι-] /³² αὐτὴν συγγένειαν καὶ [τη]λικ[αύτην ἀ]ναγκαιότητα / ἀναμιμνήσκον[τα]ς κα[ὶ] ταύτ[ην δι]αφυλάττειν προ- / αιρουμένους π[ολ]λαπλασίως· κ[αὶ νῦν] τὴν εὔνοιαν ἀπ[ο-] / δεδέγμεθα [κα]ὶ τὰ μετ[ὰ] ταῦτα [π]ε[ι]ρασόμεθα δια- /³⁶ [τ]ηροῦντε[ς τὰ ἐ]κ [π]αλαιῶν μὲν χρ[ό]ν[ω]ν συνεστη- / [κότ]α, ν[ῦν δὲ καλ]ῶς καὶ προσηκόντω[ς ε]ἰς τὴμ βελ- / [τίστην ἀνανέ]ωσιν ἠγμένα ὑφ' ὑμ[ῶν], εὖνοι φίλο[ι,] / [φιλάνθρωπα τῶι] δήμωι ὑπάρχε[ιν καὶ ὑπ]ακούοντε[ς /⁴⁰ τὰ ἀξιούμενα ἀεὶ χ]αρίζεσθαι εἰ[ς δύναμιν]. ἔρρωσθ[ε].

20   J. et L. Robert, *BE* 1953, 152.
21   Rigsby 1996, 123.
22   See Dana 2020a.
23   I here translate following Welles 1934, 118–120, no. 25 with minor adaptation: 'Ziaelas, king of Bithynia, to the council and the people of Kos, greetings. Diogeitos, Aristolochos, and Theudotos, your envoys, came and asked us to recognize as inviolable the temple of Asklepios in your city and to befriend the city in all other ways, just as our father Nikomedes was well disposed toward your people (καθόπερ καὶ Νικομήδης ὁ πατὴρ ἡμῶν εὐνόως διέκειτο τῶι δήμωι). We do in fact exercise care for all the Greeks who come to us as we are convinced that this contributes in no small way to one's reputation; especially do we continue to make much of our father's (other) friends and of you, because of his personal acquaintance

in fact exercise care for all Greeks who come to us, as we are convinced that this contributes in no small way to one's reputation' recalls the response of the king Theodoros and king Amynander of Athamania (a part of Epeiros, whose inhabitants were of Illyrian descent) to the city of Teos. The kings indeed recognize the city and its land as sacred, inviolable and tax-free in 205/1 BC: 'This we do because of our being in fact related to all the Greeks since we are related to the original himself of the common appellation of the Greeks'.[24] Concerning the letter that we have assigned to a Bosporan king, the insistence on 'kinship as true and worthy' and on gratitude for 'those who remember such kinship and blood ties' is crucial for understanding the text. This 'true kinship' was very likely entirely invented, but this practice was common even for cities whose Greek origin was beyond suspicion. Recognition of kinship was above all a rhetorical but coherent exercise, involving mutual harmonization of the narratives.

Both answers affirm the continuity of policy over generations, which implies *per se* the continuity of the dynasty and of the relationship. However, we have to distinguish between the expression of the Bosporan king willingly confirming the ambassador's claim, and Ziaelas' assertion. The diplomatic exchanges presented here do therefore not reflect examples of how a barbarian king was denied recognition by 'the Greeks', as assumed by Lise Hannestad, but rather attest to skilful networking among the representatives of different Greek or Hellenized communities.

III. CULTURAL PATRONAGE AND LOCAL HISTORIES

The Bosporan kings showed the same favourable attitude towards Greek culture in their own kingdom. Two famous musicians of antiquity, Stratonikos of Athens and Aristonikos of Olynthos, are attested to have stayed at the Bosporan court. Their presence has to be connected with the prestige policy deployed by the kings. Stratonikos is a well-known cithara-player who flourished between 375 and 350 BC.[25] After several trips to Greece, Macedon, Thrace, Asia Minor, Rhodes and all along the Black Sea coast, he was executed by the Cypriot king Nikokles, because

with your people, because king Ptolemy, our friend and ally, is friendly toward you, and still further because your envoys expressed with great enthusiasm the good-will which you have for us. In the future, as you may request, we shall try for each one individually and for all in common to favour you as much as lies in our power, and as for your seafaring citizens to take thought for those who happen to enter territory under our control, so that their safety may be assured, and in the same way also for those who are cast upon our coast because of an accident in the course of their voyage, we shall try to exercise all concern that may be injured by no one. We recognize also your temple as inviolable, as you have requested, and concerning these and our other wishes I have ordered Diogeitos and Aristolochos and Theudotos to report to you. Farewell.' Cf. Rigsby 1996, 118–121, no. 11; Hannestad 1996; Brodersen, Günther & Schmitt 1999, 12f., no. 409.

24   Translation adapted from Welles 1934, 152–156, no. 35.
25   Bélis 1999, 115–118; see Gilula 2000.

of his caustic ironies.[26] Stratonikos is mentioned by Athenaios as a guest of 'king Berisades' 'in the Pontos', actually Pairisades I of the Bosporus (ca. 344/3–311/10).[27] When the king was trying to hold back the musician who wanted to return to Greece, Stratonikos gave a very lively reply. According to Machon, Athenaios' source,

> Stratonikos the cithara-player once sailed to the Pontos at Berisades, who was then sovereign. A long time having already passed, Stratonikos wished to return to Greece. And as the king, it seems, did not allow him, it is said that he gave him the following answer: 'But you, do you intend to stay?'[28]

If this answer indicates a negative cliché on this peripheral zone, Stratonikos displayed the same irony in regards to cities of even higher importance. This type of replay had to be part of a collection of anecdotes that made their author famous. His words are mainly aimed at other musicians, composers, cithara-players and flute players, all pronounced in the middle of their recital. Another anecdote included in the work of Athenaios presents the musician criticizing a cithara player in the theatre of Byzantium.[29] Another time, Stratonikos committed slander by saying that he had come out of Herakleia 'as of a brothel'.[30]

Polyainos provides the second attestation of a famous artist at the Bosporan court, when he transmits a ploy of Memnon, the famous mercenary of Rhodes, during the preparation of a war against king Leukon I. Wishing to learn the number of inhabitants of the kingdom, the Rhodian sent to the 'tyrant of the Bosphoros' a man named Archibiades of Byzantium under the guise of an embassy. The 'ambassador' was accompanied by Aristonikos of Olynthos,[31] a cithara-player later known at the court of Philip II and Alexander, after the annexation of his homeland by the Macedonian king in 348. He was the cithara master of Alexander and followed the Conqueror to Asia. After he had died in an ambush in 328, Alexander raised a bronze statue for him at Delphi. This showed him standing, brandishing a spear in one hand and a cithara in the other.[32] His performances were highly appreciated because of his prestige, and the crowd was certainly eager to listen.[33] While Aristonikos performed in the theatres, Archibiades could easily estimate the number of the future enemies.

26 Ath. 8.352d.
27 Ath. 8.349d (the name of Pairisades was corrupted in Dein. *Or.* 1.43). The source is Machon of Corinth or Sikyon, an author of the New Comedy (ca. 360–250 BC), who wrote *Chreiai*, a collection of moralizing anecdotes; see von Wartensleben 1901, 125–138. The passage is quoted as Machon, section IX (Stratonikos) by Gow 1965, 42, who offers a commentary on pp. 89f.
28 Ath. 8.349d.
29 Ath. 8.350a; cf. Bélis 1999, 95f.
30 Ath. 8.351c.
31 Polyain. 5.44.1 and *Excerpta Polyaeni* 7.4 (mentioned only as κιθαρῳδὸν ἄριστον). See Bélis 1999, 219f. For the acquaintance of Aristonikos with the Macedonian court, see Momigliano 1992, 144 n. 27.
32 Arr. *Anab.* 4.16.6f.; Plut. *Mor.* 334f. See Bélis 1999, 223.
33 Polyain. 5.44.1.

One can imagine that both Stratonikos and Aristonikos were well received at the Bosporan court and that their reception let the kings appear as protectors of the arts and letters, just like Hellenistic rulers of later centuries. These are the best-known cases, but everything suggests that they were not the only ones. This royal patronage was certainly linked, at least in the collective imagination, to the kings' wealth. This wealth could be (and was) perceived negatively, as a mark of luxury, which was typical of tyranny. Athenaios' *Deipnosophistai* has more examples evoking the luxury of tyrants, especially that of the Cypriot kings.[34] As to the Bosporus, Leukon I figures as 'tyrant of the Pontos' in this work. When one of his flatterers had allegedly stolen his friends, Leukon shouts out: 'Gods! I should kill you if tyranny did not need men as miserable as you! (εἰ μὴ πονηρῶν ἀνδρῶν ἡ τυραννὶς ἐδεῖτο).[35]

However, this negative perspective on the Bosporan kings contrasts with a panegyric tradition inspired by some sources favourable to them. This would explain why the Stoic Chrysippos, for whom a good king is characterized by *euphyia* and *philomatheia*,[36] depicts the same Leukon I as wise and kind and why Dion Chrysostomos lists him in a line of 'good kings'.[37] The same Leukon is evoked by Aineias the Tactician (5.2), as giving up the services of bodyguards who were in debts because of games and other excesses. One may thus suppose the existence of a Bosporan *Lokalgeschichte* that conveyed positive images of its past rulers. By the end of the 1st century BC, Strabo still drew on this tradition, when mentioning the funerary monument of Satyros near Pantikapaion; the king died (in 389/8) 'one of the illustrious dynasts of the Bosporus'.[38]

An argument for the existence of such a favourable historiographical source is, in Rostowtzew's opinion, a long passage from Diodoros of Sicily (20.22–26), with abundant (especially chronological) details on several Bosporan dynasts as late as the *archon* Pairisades I (344/3–311/10). Diodoros uses this 'chronicle' together with other sources, most likely a *Lokalgeschichte* that continued the chronicle after Pairisades. This local history could have been inspired by Eumelos, one of the successors of Pairisades I, which would explain the eulogistic description of his reign. After the death of Pairisades, the succession was disputed by his three sons, Satyros, Eumelos and Prytanis. Satyros, as the elder brother, received the power from his father, but, dissatisfied with this arrangement, his brother Eumelos made war against him and seized the throne.[39] Three aspects in Diodoros' account clearly show the positive image of Eumelos, at least in the eyes of the Greeks: first, the victory and accession to the throne are explained as due both to his noble birth and virtue (καὶ κατὰ γένος καὶ κατ' ἀρετήν); next, his promise to the citizens of Pantikapaion to restore their ancestral constitution (πάτριος πολιτεία) and grant

34 See Raptou 1999, 215–217.
35 Ath. 6.257d–e.
36 Chrysippos, *Ethica*, *SVF* III 690–692 (Plut. *Mor.* 1061d); Strab. *Geogr.* 7.3.8 (301f.C).
37 Dio Chrys. *Or.* 2.77; see Rostowtzew 1931, 112f.
38 Strab. *Geogr.* 11.2.7 (494C).
39 See Rostowtzew 1931, 113–115; Vinogradov 1980, 65; Hind 1994, 480; Bosi 1986, 175.

other privileges is praised; finally, the historian insists on the generous attitude of Eumelos towards the Greeks of the Black Sea.[40] It is possible that Pompeius Trogus used such a favourable historiographical source and a Bosporan local history – and they may be one and the same, for his *Philippic Histories*.[41]

These local stories have to be seen in the broader context of the literary practices largely attested in Greek cities. The most pertinent parallel for a local historian is Syriskos of Chersonesos in the vicinity of Pantikapaion.[42] This *Bürgerhistoriker* is known to have reported the benefactions of a certain Parthenos to cities and kings in the 3rd century BC.

IV. ATHENIAN HONOURS FOR THE SPARTOKIDS

The most distinguished strategy of dynastic legitimation was undoubtedly maintaining a close relationship with Athens, but our analysis need not be limited to the well-documented benefactions – that is grain donations and tax privileges granted by the kings, to which the Athenians responded by granting public honours.[43] As we know, the close relations between Athens and the Bosporan kingdom are attested in several speeches of Demosthenes: *Against Phormion* (*Or.* 34, in 327/6), *Against Leptines* (*Or.* 20, in order to preserve the privileges granted to Leukon and his sons, in 355/4)[44], *Against Lakritos* (*Or.* 3, before 340, attesting the journey of some Athenian merchants from the Bosporus to Athens *via* Olbia and the left shore of the Black Sea).[45] The 16th speech of Lysias was written around 392/90 on behalf of Mantitheos of Thorikos, who was accused of collaboration with the Thirty. In his defence Mantitheos maintained that he had been in the Bosporus at the time, sent by his father to Satyros I (about 405/3).[46] Isokrates' *Trapezitikos* (*Or.* 17) provides some further background information. From this we learn that several compatriots of the young Bosporan man – traders or exiles that he is specifically accused of frequenting – were established in Athens, while many Athenian traders were sojourning in Pantikapaion.

The presence of the kings in Athens was of a symbolic nature: their statues and honorary decrees were displayed at strategic places of the city. More than that, the Athenians announced to Spartokos III the expulsion of the Macedonian garrison in 287, hence to a member of a dynasty that had nourished the city for generations.[47] Among the honours awarded to Spartokos for having responded

40   Asheri 1998, 274.
41   Pomp. Trog. *Prol.* 37: *dictaeque in excessu regum Bosporanorum et Colchorum origines et res gestae*; cf. Rostowtzew 1931, 115.
42   *FGrH* 807 T 1; *IOSPE* I² 344; Chaniotis 1988, E 7.
43   See Osborne 1983, 41–44, T 21–24; Braund 2003; Gallotta 2004; Moreno 2007, 144–208.
44   Dem. *Or.* 20.30; cf. Braund 1997, 132f.
45   Dem. *Or.* 3.10; Dem. *Lept.* 33; Isok. *Trapez.* 57.
46   Lys. *Or.* 16.4.
47   *IG* II² 653 = *Syll.*³ 370 (285/4 BC); Burstein 1978; Bringmann & von Steuben 1995, no. 34; Heinen 2005; Oliver 2007a, 253 n. 122; Oliver 2007b, 196f.; Müller 2010, 366–369, no. 8.

positively to the call of Athens is the erection of a bronze statue of the king in the agora near those of his ancestors (*para tous progonous*) and of another statue on the Acropolis. The bronze statues of the ancestors are not mentioned in an older Athenian decree, passed for the sons of Leukon I in 347/6, Spartokos, Pairisades and Apollonios,[48] but the statues were probably those for Pairisades I and his two sons, Satyros and Gorgippos. Demosthenes had proposed them between 330 and 324/3, as we can infer from a speech of Deinarchos, accusing Demosthenes of venality.[49] The decree for Spartokos III confirms the erection of statues of his ancestors not only in the agora, but also in the Piraeus. However, the second statue for Spartokos must have been erected elsewhere, since the port was still in the hands of Macedonians. The most likely location was thus on the Acropolis. Very important is the fact that the statue remained there even after the liberation of the Piraeus, given that the fragments of the inscription carved on its base were discovered *in situ*.[50]

The decrees and statues exhibited in the most frequented places – the agora, harbour and acropolis – reminded the Athenians and the foreign residents not only of what they owed to kings, but also of the fact that these 'barbarians' or 'tyrants' of the Bosporus were honoured according to democratic procedures. The Athenians were attentive to the protocol requiring the association of a sovereign with his sons, as Heinen has pointed out. It is in this way that the Spartokids refer to themselves in official documents issued in their kingdom, for example in a decree from Pantikapaion granting *proxenia* to an individual by 'Pairisades and his children (παῖδες)' (*CIRB* 1). The mention of children foreshadows the official practice of Hellenistic documents, in which children (τέκνα) are often mentioned after the king or the royal couple. Accordingly, the Athenians entirely respected this Spartokid etiquette, as is shown both by the texts and the visual representations, which highlight in particular the outstanding position of the eldest sons. On the relief accompanying the inscribed decree of 347/6, the two elder sons are seated on the same throne, sharing the power, while the youngest son is standing beside; an additional proposal granted him a crown.[51]

While the respect the Athenians paid those distant dynasts can be understood in view of their permanent need of the wheat, it is not as easy to explain the eagerness of the Boporan rulers to grant the Athenians preferential prices for the purchase of their produce, besides making large donations. In my book, I have explained this inclination as a response to their contested Greek origins and compromised legitimacy; close relations with democratic Athens would have countered this deficit.[52] I am less inclined to think so now. Dealing from a position of strength, the Spartokids did not need to request the approval of the Greeks for their form of government. Nevertheless, the kings were interested in their recogni-

48 *IG* II² 212 = *Syll.*³ 206. See Tuplin 1982; Müller 2010, 361–363, no. 4.
49 Dein. *C. Demosth.* (*Or.* 1), 43. Cf. Heinen 2006a.
50 Heinen 2006a, 291.
51 Heinen 2006, 287f.
52 Dana 2011a, 315f. Cf. Honigman 2007, 129–131.

tion and in receiving civic honours – decrees and statues – as an expression of their belonging to the cultural *koine*. These honours were as important for them as for the Athenians or the Greeks of the Black Sea region.

V. DIPLOMACY BETWEEN EGYPT AND THE BOSPORUS

Another example of the Bosporan kings' ability to integrate themselves culturally into the Greek world is provided by a letter Ptolemy Philadelphos' minister Apollonios sent to his agent Zenon of Kaunos. Dated 21 September 254 and kept in the archives of Zenon, this letter informs us that ambassadors of Pairisades II (284/3 to ca. 245) were present at the court of Ptolemy. A fairly recent discovery in the Bosporus proves the reciprocity of exchanges: a remarkable fresco in the sanctuary of Aphrodite of Nymphaion represents the warship labelled 'Isis', probably sent by the Lagids in the context of a diplomatic mission to the Bosporan kingdom, probably under the same Pairisades.[53] The abovementioned letter represents an urgent command to Zenon to arrange for means of transportation in Ptolemais at the entrance to the Fayum. They were necessary for the ambassadors of Pairisades and the *theoroi* of the city of Argos:

> 'Apollonios to Zenon greetings. As soon as you read this letter, send off to Ptolemais chariots and the other carriage-animals (?) and the baggage-mules for the ambassadors from Pairisades and the delegates from Argos whom the King has sent to see the sights of the Arsinoite nome. And make sure that they do not arrive too late for the purpose: for at the time of writing this letter they have just this moment sailed up. Farewell. Year 32, Panemos 26, Mesore 1.' (Addressed) 'To Zenon.' (Docketed) 'Year 32, Mesore 2, at the 10th hour. Apollonios about the animals for the envoys from Pairisades and Argos'[54]

Philadelphos had thus sent them on a sight-seeing tour of the Arsinoite nome. Theodore Skeat remarked that Pairisades is lacking the title *basileus* and wondered if Ptolemy II and his ministers 'did not condescend to recognize the Crimean kinglet as an equal'. In my opinion, however, the Bosporan king was too well known to need this specification. The reasons of the mission from Pairisades are unknown. A business co-operation between the two greatest grain-producers of (or for) the Near East would be a plausible assumption.[55] But there also remains

---

53 Bricault 2007, in particular 246f. and 247 n. 5 (bibliography); Müller 2010, 257–261. For the Pontic prosopography in Egypt, see Avram 2007.
54 *SB* 7263 = *P.Lond.* VII 1973: *recto*: Ἀπολλώνιος Ζήνωνι χαίρειν. ὡς ἂν ἀναγνῶις / τὴν ἐπιστολήν, ἀπόστειλον εἰς Πτολεμαίδα τά τε ἀρμάτια καὶ τὰ λοιπὰ βαδιστικὰ πορεῖα / καὶ τὰς νωτοφόρους ἡμιόνους ὥστε τοῖς παρὰ /⁵ Παιρισάδου πρεσβευταῖς καὶ τοῖς ἐξ Ἄργους / θεωροῖς οὓς ἀπέσταλκεν ὁ βασιλεὺς κατὰ θέαν / τῶν κατὰ τὸν Ἀρσινοΐτην. καὶ φρόντισον / ἵνα μὴ καθυστερήσῃ τῆς χρείας· ὅτε γὰρ/ ἐγράφομέν σοι τὴν ἐπιστολὴν ἀνεπεπλεύκεισαν ἤδη. /¹⁰ ἔρρωσο. (ἔτους) λβ, Πανήμου κς, Μεσορὴ α. *Verso*: (ἔτους) λβ, Μεσορὴ [[α]]β, Ἀπολλώνιο[ς] Ζήνωνι. / ὥρας ι. περὶ τῶν τοῖς / παρὰ Παιρισάδου καὶ / Ἀργείοις πρεσβευταῖς / πορείων. Cf. Skeat 1974, 62–66, whose translation is quoted above; also Samuel 1993, 172.
55 Archibald 2004.

the possibility that the Bosporan ambassadors were the *theoroi* who attended the *Ptolemaia*, a penteteric festival created by Philadelphos in honour of his father. For reasons of chronology, M. Bergmans thinks that the *theoroi* from Argos had come to announce the *Nemeia* of the following year,[56] but the chronology of the *Ptolemaia* is still debated. At any rate, the presence of ambassadors from Argos is surprising, since the city supported Antigonos Gonatas, a rival of the Ptolemies.

Preparing the festivals involved inviting *theoroi*, artists and athletes, but also organizing processions and military parades. This performance was destined to impress highly the guests, who had to leave in awe as if they had visited the most splendid capital of the world. For some delegations from kingdoms and cities alike, the trip to Egypt offered an opportunity to tour one of the most exotic countries. The Lagids were proud to show off their power and wealth, while visitors appreciated the antiquities along the Nile and the innovations of the Ptolemies in their vast territory.[57]

VI. THE BOSPORAN KINGS AS BENEFACTORS AND ADMIRERS OF GREEK CULTURE IN THE ROMAN WORLD

Royal *euergesia* towards Greek cities outside the Bosporan kingdom or Athens was practiced particularly under Roman imperial rule as a way for the monarchs to display their adhesion to Greek values. In this later period, various rulers were honoured by cities of the southern coast of the Black Sea:[58] thus Ti. Julius Sauromates I by the Bithynian city of Nikaia in AD 117 and in Sinope (at an uncertain date); Ti. Julius Rheskuporis III by Amastris in 221 as well as by Prusias on the Hypios in AD 222 and 223.[59] Sauromates I was honoured as *ktistes kai euergetes* not only of the city of Nikaia, but also of the college of the *neoi*, an institution that played a central role for Greek culture and civic identity:

> To good fortune! The great king Tiberius Julius Sauromates, descendant of kings, son of the king Rheskuporis, friend of Caesar and friend of the Romans, pious, high priest for life of the Augusti and benefactor and founder of our city, the college of the *neoi* of Nikaia honours him as their founder and benefactor for his piety, under the *epimeletes* Lucius Flavius Epitynchanon, in year 117, month Lôos.[60]

56 Bergmans 1979.
57 Cf. Herodas, *Mime* 1.23–33.
58 Lifshitz 1968, 28; Robert 1980, 80f. Lukian (*Alex.* 57) mentions an embassy sent to Bithynia by the Bosporan king Eupator (AD 154–173), when Lukian himself was visiting the region.
59 Sauromates I: *CIRB* 44 = *I.Nikaia* T 42a; *CIRB* 46 (Sinope, in Latin). Rheskuporis III: *CIRB* 54 (Amastris); *CIRB* 953 = *I.Prusias ad Hypium* T 3 (AD 222) and *CIRB* 55 = *I.Prusias ad Hypium* T 2 (AD 223). See also *IOSPE* II 44 on the Herakleot Julius Telesinos in AD 250.
60 *CIRB* 44: [Ἀγαθῆι] τύχηι. / [τὸ]ν ἐκ π[ρ]ογόνων [βα]σι- / [λέ]ων βα[σιλέα] μέγαν [Τιβέ-] / [ρι]ον Ἰού[λιον Σαυ]ρομάτ[ην], /⁵ υἱὸν βα[σιλέως] Ῥησκουπ[ό-] / ριδος, φιλο[κ]αίσαρα καὶ / φιλορώμα[ι]ον εὐσεβῆ, / ἀρχιερέα τ[ῶ]ν Σεβαστῶν / διὰ βίου κα[ὶ ε]ὐεργέτην /¹⁰ τῆς πατρίδος καὶ κτίστην, / Νεικαιέων νέων σύνοδος / τὸν ἑαυτῶν κτίστην καὶ/ εὐεργέτην εὐσεβείας χάριν, / ἐπιμεληθέντος Λουκίου Φλαουί- /¹⁵ ου Ἐπιτυνχάνοντος, / ἐν τῷ γιυ ἔτει καὶ μηνὶ Λώῳ.

An anecdote told by Philostratos presents a Bosporan king, probably Ti. Julius Sauromates I (AD 93–123) or his son Ti. Julius Kotys II (AD 123–131), as passionately interested in Greek culture.[61] The king, who went to Smyrna, fascinated by the reputation of this cultural hotspot, was humiliated by the sophist Polemon of Laodikeia, who refused to greet him:

> For, in truth Polemon, was so arrogant that he conversed with cities as his inferiors, Emperors as not his superiors, and the gods as his equals (…). And once when the ruler of the Bosporus, a man who had been trained in all the culture of Greece, came to Smyrna in order to learn about Ionia, Polemon not only did not take his place among those who went to salute him, but even when the other begged him to visit him, he postponed it again and again, until he compelled the king to come to his door with a fee of ten talents.[62]

This episode does not necessarily imply contempt of a man from the Black Sea region, because philosophers since the days of Alexander the Great made it a point of honour not to chase after the 'heroes of the day'. According to Philostratos, Polemon showed the same contemptuous attitude to another prominent individual from the Euxine, the sophist and advocate Marcus of Byzantium. Polemon finally recognized him from his Dorian accent and invited him to join his school.[63] Polemon was a disciple of Timokrates, from whom he had contracted his 'proud and haughty temper'. Timokrates, however, 'came from the Pontos and his birthplace was Herakleia, whose citizens admire Greek culture'.[64]

The Sophistic movement highly contributed to the prestige and the economic prosperity of the cities of the Greek East, which were proud of their intellectuals. As Glen Bowersock has put it, 'their fame attracted the rich and cultivated (the ones who could afford to travel) from distant regions of the empire'.[65] Among them was the young Bosporan king, who behaved in accordance with the practices of his time, though not much different either from the habits of his predecessors.

This was also the period in which the Greek cities incessantly claimed illustrious origins, proudly recalling their mythical ancestry or historical glory. Meanwhile, kings presumed to be descendants of royalty and heroes and were boasting friendship with the emperor(s). As a pertinent example, I here quote an inscription from Pantikapaion:

> To good fortune! Ulpios Antisthenes, son of Antimachos, chiliarch, honours king Tiberius Julius Rheskuporis, descendant of Herakles and Poseidon's son Eumolpos and of kingly an-

61 Cf. Skrzhinskaya 2016.
62 Philostr. *VS* 1.25, p. 535 ed. W. Cave Wright, whose translation is given above with adaptation: ὑπέφρων γὰρ δὴ οὕτω τι ὁ Πολέμων, ὡς πόλεσι μὲν ἀπὸ τοῦ προὔχοντος, δυνάσταις δὲ ἀπὸ τοῦ μὴ ὑφειμένου, θεοῖς δὲ ἀπὸ τοῦ ἴσου διαλέγεσθαι (…). ἀνδρὸς δέ, ὃς ἦρχε μὲν Βοσπόρου, πᾶσαν δὲ Ἑλληνικὴν παίδευσιν ἥρμοστο, καθ' ἱστορίαν τῆς Ἰωνίας ἐς τὴν Σμύρναν ἥκοντος οὐ μόνον οὐκ ἔταξεν ἑαυτὸν ἐν τοῖς θεραπεύουσιν, ἀλλὰ καὶ δεομένου ξυνεῖναί οἱ θαμὰ ἀνεβάλλετο, ἕως ἠνάγκασε τὸν βασιλέα ἐπὶ θύρας ἀφικέσθαι ἀπάγοντα μιστοῦ δέκα τάλαντα.
63 Philostr. *VS* 1.24, p. 528f. See Dana 2013, 34.
64 Philostr. *VS* 1.25, p. 536; cf. Goulet-Cazé 2016, who dates to the end of the first half of the 2nd century AD; also Dana 2020b.
65 Bowersock 1969, 17.

cestors, son of the great king Sauromates, friend of Caesar and friend of the Romans (*philo-kaisar kai philorhomaios*), pious, high priest for life of the Augusti, as his own saviour and master, in the year 512 (=216 AD) and the month of Loos, (day) 20.[66]

The dedicator, who bore his Greek name Antisthenes as cognomen, was also a Roman citizen, while holding a leading position in the Bosporan army.[67] Another text that has become available only recently also honours a high official in the Roman period, probably under Sauromates (I or II).[68] Its wording removes any doubts about the Greek culture that the Bosporan elites shared with their sovereigns. The poetic (often Homeric) lexemes and comparisons are remarkable.[69] Three keys are advised to the king for attaining perfection (ll. 32–35): endurance, war and persuasive speech, whereby *logos* ('word, speech, reason') is presented as a *pharmakon* ('medicine').[70] The author of the verse inscription had first been the educator of the young prince, as implied in the reference to the relationship between Chiron and the young Achilles (l. 23). Later, he became a military commander and a dignitary of the state. The reward to which the epigram refers (ll. 36–37) brings to mind the torque decorating the neck of Neokles, governor of Gorgippia, who was honoured by his son Moirodoros under Ti. Julius Sauromates in AD 186:

> Under the king Tiberius Julius Sauromates, friend of Caesar and friend of Romans (*philo-kaisar kai philorhomaios*), pious, Moirodoros, son of Neokles, governor of Gorgippia, has erected the statue of his own father (and benefactor?) Neokles, son of Moirodoros, ancient governor of Gorgippia, as a sign of honour, in year 483 (AD 186), the 15 Lôos.[71]

---

66 *IOSPE* II 41 = *CIRB* 53: [Ἀγαθῆι τύχηι]. / [τὸν] ἀφ᾽ Ἡρακλέου[ς] καὶ Εὐμόλ- / που τοῦ Ποσειδῶνος καὶ ἀπὸ / προγόνων βασιλέων βασιλέ- /⁵ α Τιβέριον Ἰούλιον Ῥησκούπο- / ριν, υἱὸν μεγάλου βασιλέως / Σαυρομάτου, φ[ι]λοκαίσαρα / καὶ φιλορώμα[ι]ον, εὐσεβῆ, ἀρ- / χιερέα τῶν Σεβαστῶν διὰ /¹⁰ βίου, Οὔλπιος Ἀντισθένης / Ἀντιμάχου χειλιάρχης / τὸν ἑαυτοῦ σωτῆρα καὶ / δεσπότην τειμῆς χάριν / ἐν τῶι · β·ι·φ · ἔτει /¹⁵ καὶ μηνὶ Λώωι κ. Translation adapted from Heinen 2006, 335.
67 See Heinen 2006b, 335f.
68 See Bowersock & Jones 2006, proposing Sauromates II (AD 194 or 195), supported by Heinen 2006c, 48. Vinogradov 1994, 73f. first reports this inscription, whose publication was delayed until 2005: Vinogradov & Shestakov 2005 and Saprykin 2005 (photo p. 64). The name of the king (l. 5) was restored by Vinogradov as Rheskuporis I (*PIR²* I 512, where his rule is dated – perhaps a bit early – to AD 67/8–91/2); Saprykin regards the king as Sauromates I (*PIR²* I 550: AD 93/4–123/4).
69 Vinogradov & Shestakov 2005, 43; see also Saprykin 2007, 315f.
70 Cf. Plut. *Mor.* 614c.
71 *CIRB* 1119 A: [---] /¹ [ἐπὶ βα]σιλέῳ[ς Τιβερίου Ἰουλίου] / [Σαυρομά]του φιλ[οκαίσαρος καὶ φιλο-] / ρωμα[ίου], εὐσεβο[ῦς, Νεοκλέα Ἡρο-] /⁴ δώρ<ου> τὸν πρὶν ἐπὶ [τῆς Γοργιππείας] / τὸν ἴδιον πατέρα κ[αὶ εὐεργέτην, Ἡ-] / ρόδωρος Νεοκλέο[υς ὁ ἐπὶ τῆς Γορ-] / γιππείας ἀνέστησ[εν τειμῆς χάριν]· /⁸ ἐν τῷ γ[πυ´ ἔτει Λώου ει´]. Cf. 1119 B: [---]ς? / [Νεοκλέα Ἡροδώρου] τὸν πρ[ὶν ἐπὶ τῆς] / [Γοργιππείας, τὸν ἴδιον] πατέρα [Ἡρόδω]ρος /⁴ [Νεοκλέους ὁ ἐπὶ τῆς Γο]ργιπ[πεί]ας ἀνέ- / [στησεν τ]ειμῆς χάριν / [ἐν τῷ γ]πυ´ Λώου ει´. Cf. Bowersock & Jones 2006, 126; Heinen 2006c, 42–48; also 1996, 90–95 and 2006b, 336–338 (all versions with photos).

Sergey Saprykin has established the reading Moirodoros, instead of Herodoros.[72] It relates our text to other inscriptions mentioning Moirodoros, son of Neokles, as a member of a very influential family in Gorgippia, close to the Bosporan king and engaged in maritime trade (*CIRB* 1134). The way in which Moirodoros has represented his father is very instructive for the Bosporan elite, whose king was the most important representative. Neokles is clad in a mantle (a *himation* and not a Roman *toga*), holding his right hand in a fold of his cloth, whereas the left hand descends towards a case with papyrus rolls. A better-preserved statue from the same city illustrates this type of representation, which had been traditional for intellectuals since the 4th century BC.[73] The neckband perhaps meant to emphasize his official function or to recall his military feat.

V. CONCLUSIONS

Both the Spartokid and the Aspurgian dynasties pursued a similar policy, attempting to appear as cultivated and generous kings, both in the Black Sea region and throughout the Greek world at large, irrespective of the marginal location of the Bosporan kingdom. Their strategies did not differ from those of other rulers who wished to manifest their belonging to the Hellenic community by embracing their cultural values. Prominent cities like Athens or sanctuaries like Delos offered the ideal stages to demonstrate these qualities, competing with other potentates for the appreciation of a global Greek audience. Gradually in the course of the 1st century BC, the adherence to the values of Hellenism was paired the display of friendly relations with the Roman people and the Emperor.[74] Indeed, it is in the period of the Second Sophistic that Hellenism finds its full expression in the revalorization of the Greek past, favoured by the expansion of the Empire. The Bosporan kings, as well as the cultivated elites of their kingdom, found their place in that complex world dominated by the most important of cultural values in the eyes of the Greeks: *paideia*.

Bibliography – Ancient Sources

FD   *Fouilles de Delphes*. III: *Epigraphie*, Paris, 1909–1985.
I.Délos   *Inscriptions de Délos*, Paris, I–VII, 1926–1972.
I.Didyma   Rehm, A. 1958: *Didyma*. II: *Die Inschriften*, Berlin.

---

72 Saprykin 1986.
73 Heinen 2006c, 47 fig. 19; cf. 1996, 93, fig. 4 and 2006b, 339, fig. 9. The sculpture is now in the Hermitage. The papyrus roll that he held in his hand has disappeared, but we can clearly see the place where it was placed originally, while five long papyri bound together are visible at the bottom right.
74 See above, n. 6.

I.Nikaia Şahin, S. 1979–1987: *Katalog der antiken Inschriften des Museums von Iznik (Nikaia)*, Bonn, I–II (IK 9–10).

I.Prusias ad Hypium Ameling, W. 1985: *Die Inschriften von Prusias ad Hypium*, Bonn (IK 27).

Bibliography – Modern Scholarship

Archibald, Z.H. 2004: 'In-Groups and Out-Groups in the Pontic Cities of the Hellenistic Age', in C.J. Tuplin (ed.), *Pontus and the Outside World*, Leiden, 1–15.

Asheri, D. 1998: 'The Achaeans and the Heniochi. Reflections on the Origins and History of a Greek Rhetorical Topos', in G.R. Tsetskhladze (ed.), *The Greek Colonization of the Black Sea Area. Historical Interpretation of Archaeology*, Stuttgart, 265–285.

Avram, A. 2007: 'L'Égypte lagide et la mer Noire: approche prosopographique', in A. Laronde & J. Leclant (eds.), *La Méditerranée d'une rive à l'autre: culture classique et cultures périphériques*, Paris, 127–153.

Bélis, A. 1999: *Les musiciens dans l'antiquité*, Paris.

Bergmans, M. 1979: 'Théores argiens au Fayoum (P.Lond. VII 1973)', *Chronique d'Égypte* 54, 127–130.

Bosi, F. 1986: 'La storia del Bosforo Cimmerio nell'opera di Strabone', in G. Maddoli (ed.), *Strabone. Contributi allo studio della personalità e dell'opera*, vol. 2, Perugia, 172–188.

Bowersock, G.W. & Jones, C.P. 2006: 'A New Inscription from Panticapaeum', *ZPE* 156, 117–128.

Braund, D. 1997: 'Greeks and Barbarians: The Black Sea Region and Hellenism under the Early Empire', in S.E. Alcock (ed.), *The Early Roman Empire in the East*, Oxford, 121–136.

Braund, D. 2003: 'The Bosporan Kings and Classical Athens: Imagined Breaches in a Cordial Relationship (Aisch. 3.171–172; [Dem.] 34.36)', in *The Cauldron of Ariantas. Studies Presented to A.N. Ščeglov on the Occasion of his 70th Birthday*, Aarhus, 197–208.

Braund, D. (ed.) 2005: *Scythians and Greeks. Cultural Interactions in Scythia, Athens and the Early Roman Empire (Sixth Century BC–First Century AD)*, Exeter.

Bricault, L. 2007: 'La diffusion isiaque en Mésie Inférieure et en Thrace: politique, commerce et religion', in L. Bricault, M.J. Versluys & P.G.P. Meyboom (eds.), *Nile into Tiber. Egypt and the Roman World. Proceedings of the IIIrd Confererence of Isis Studies, Faculty of Archaeology, Leiden University, May 11–14, 2005*, Leiden, 245–266.

Bringmann, K. & von Steuben, H. (eds.) 1995: *Schenkungen hellenistischer Herrscher an griechische Städte und Heiligtümer*, vol. 1, Berlin.

Brodersen, K., Günther, W. & Schmitt, H.H. 1999: *Historische griechische Inschriften in Übersetzung*, vol. III, Darmstadt.

Buraselis, K. 1993: 'Ambivalent Roles of Centre and Periphery. Remarks on the Relation of the Cities of Greece with the Ptolemies until the End of Philometor's Age', in P. Bilde, T. Engberg-Pedersen, L. Hannestad, K. Randsborg & J. Zahle (eds.), *Centre and Periphery in the Hellenistic World*, Aarhus, 251–270.

Burstein, S.M. 1978: 'I.G. II² 653, Demosthenes and Athenian Relations with Bosporus in the Fourth Century B.C.', *Historia* 27, 428–436.

Chaniotis, A. 1988: *Historie und Historiker in den griechischen Inschriften*, Stuttgart.

Coşkun, Altay 2016: 'Heinz Heinen und die Bosporanischen Könige – Eine Projektbeschreibung', in V. Cojocaru & A. Rubel (eds.), *Mobility in Research on the Black Sea. (Iaşi, July 5–10, 2015)*, Cluj-Napoca, 51–71.

Curty, O. 1995: *Les parentés légendaires entre cités grecques*, Geneva.

Dana, M. 2011a: *Culture et mobilité dans le Pont-Euxin. Approche régionale de la vie culturelle des cités grecques*, Bordeaux.

Dana, M. 2011b: Les relations des cités grecques du Pont-Euxin ouest et nord avec les centres cultuels du monde grec, *ACSS* 17, 47–70.

Dana, M. 2013: 'La cité de Byzance aux époques hellénistique et impériale: un centre culturel avant Constantinople', in G.R. Tsetskhladze, S. Atasoy, A. Avram, Ş. Dönmez & J. Hargrave (eds.), *The Bosporus: Gateway between the Ancient West and East (1ˢᵗ Millennium BC–5ᵗʰ Century AD). Proceedings of the Fourth International Congress on Black Sea Antiquities Istanbul, 14ᵗʰ–18ᵗʰ September 2009*, Oxford, 29–38.

Dana, M. 2020a: 'Local Culture and Regional Cultures in the Propontis and Bithynia', in M.P. de Hoz, J.L. Garcia Alonso & L.A. Guichard Romero (eds.), *Greek Paideia and Local Tradition in the Graeco-Roman East*, Leuven, 39–71.

Dana, M. 2020b: 'Peut-on être sophiste dans le Pont-Euxin? Philosophie, rhétorique et périphérie', forthcoming in A.-M. Favreau-Linder, S. Lalanne & J.-L. Vix (eds.), *Passeurs de culture. Transmission et circulation du savoir dans l'empire gréco-romain, II, Paris, 27–28 novembre 2015*, Turnhout.

Daumas, M. 2009: *L'or et le pouvoir: armement scythe et mythes grecs*, Paris.

Favreau-Linder, A.-M. 2012: 'Polémon (P. 218)', in R. Goulet (ed.), *Dictionnaire des philosophes antiques*, vol. Vb, Paris, 1194–1205.

Fless, F. & Treister, M. (ed.) 2005: *Bilder und Objekte als Träger kultureller Identität und interkultureller Kommunikation im Schwarzmeergebiet*, Rahden.

Gallotta, S. 2004: 'I rapporti tra Atene e il regno del Bosforo. Riflessioni sul tema', *Incidenza dell'antico* 2, 171–178.

Gilula, D. 2000: 'Stratonicus, the Witty Harpist', in D. Braund & J. Wilkins (ed.), *Athenaeus and His World. Reading Greek Culture in the Roman Empire*, Exeter, 423–433.

Giovannini, A. 1997: 'Offrandes et donations des souverains aux sanctuaires grecs', in *Actes du Xᵉ Congrès International d'Épigraphie Grecque et Latine*, Paris, 175–181.

Gow, A.S.F. 1965: *Machon. The Fragments*, Cambridge.

Goulet-Cazé, M.-O. 2016: *Timocratès d'Héraclée (T 155)*, in R. Goulet (ed.), *Dictionnaire des philosophes antiques*, vol. IV, 1206f.

Hannestad, L. 1996: '"This Contributes in No Small Way to One's Reputation": The Bithynian Kings and Greek Culture', in P. Bilde, T. Engberg-Pedersen, L. Hannestad & J. Zahle (eds.), *Aspects of Hellenistic Kingship*, Aarhus, 67–98.

Hansen, M.H. & Nielsen, T.H. (eds.) 2004: *An Inventory of Archaic and Classical Poleis*, Oxford.

Heinen, H. 1996: 'Rome et le Bosphore. Notes épigraphiques', *CCG* 7, 81–101.

Heinen, H. 2005: 'Athenische Ehren für Spartokos III. (IG II² 653)', in V. Cojocaru (ed.), *Ethnic Contacts and Cultural Exchanges North and West of the Black Sea. From the Greek Colonization to the Ottoman Conquest*, Iaşi, 109–125.

Heinen, H. 2006a: 'Statues de Pairisadès I et de ses fils érigées sur proposition de Démosthène (Dinarque, *Contre Démosthène* 43)', in H. Heinen, *Vom hellenistischen Osten zum römischen Westen. Ausgewählte Schriften zur Alten Geschichte*, Stuttgart, 283–294. First published in P. Carlier (ed.), *Le IVᵉ siècle av. J.-C. Approches historiographiques*, Paris, 1996, 357–368.

Heinen, H. 2006b: 'Greeks, Iranians and Romans on the Northern Shore of the Black Sea', in H. Heinen, *Vom hellenistischen Osten zum römischen Westen. Ausgewählte Schriften zur Alten Geschichte*, Stuttgart, 320–342. First published in G. Tsetskhladze (ed.), *North Pontic Archaeology. Recent Discoveries and Studies*, Leiden, 2001, 1–23.

Heinen, H. 2006c: *Antike am Rande der Steppe. Der nördliche Schwarzmeerraum als Forschungsaufgabe*, Stuttgart.

Heinen, H. 2008: 'Romfreunde und Kaiserpriester am Kimmerischen Bosporos. Zu neuen Inschriften aus Phanagoreia', in A. Coşkun (ed.), *Freundschaft und Gefolgschaft in den auswärtigen Beziehungen der Römer (2. Jh. v.Chr. – 1. Jh. n.Chr.)*, Frankfurt, 179–198.

Hind, J. 1994: 'The Bosporan Kingdom', *CAH²*, vol. 6, 476–511.

Honigman, S. 2007: 'Permanence des stratégies culturelles grecques à l'œuvre dans les rencontres inter-ethniques, de l'époque archaïque à l'époque hellénistique', in J.-M. Luce (ed.), *Identités ethniques dans le monde grec antique. Actes du Colloque International de Toulouse organisé par le CRATA. 9–11 mars 2006 = Pallas* 73, 125–140.

Jacobson, E. 1995: *The Art of the Scythians: The Interpenetration of Cultures at the Edge of the Hellenic World*, Leiden.

Lifshitz, B. 1968: 'Épigrammes grecques du Bosphore', *RhM* 111, 13–36.

Marcotte, D. 2000: *Les géographes grecs. I. Pseudo-Scymnos. Circuit de la terre*, Paris.

Meyer, C. 2013: *Greco-Scythian Art and the Birth of Eurasia. From Classical Antiquity to Russian Modernity*, Oxford.

Michels, Ch. 2010: 'Begrenzte Integration. Der Euergetismus der Könige von Bithynien, Pontos und Kappadokien in der griechischen Welt, in Ch. Antenhofer, L. Regazzoni & A. von Schlachta (eds.), *Werkstatt Politische Kommunikation*, Göttingen, 189–204.

Momigliano, A. 1992: *Philippe de Macédoine. Essai sur l'histoire grecque du IVᵉ siècle av. J.-C.*, Combas.

Moreno, A. 2007: *Feeding the Democracy. The Athenian Grain Supply in the Fifth and Fourth Centuries BC*, Oxford.

Müller, Ch. 2007: 'Insaisissables Scythes: discours, territoire et ethnicité dans le Pont Nord', in J.-M. Luce (ed.), *Identités ethniques dans le monde grec antique. Actes du Colloque International de Toulouse organisé par CRATA. 9–11 mars 2006, Pallas* 73, 141–154.

Müller, Ch. 2010: *D'Olbia à Tanaïs, Territoires et réseaux d'échanges dans la mer Noire septentrionale aux époques classique et hellénistique*, Bordeaux.

Oliver, G. 2007a: *War, Foot and Politics in Early Hellenistic Athens*, Oxford.

Oliver, G. 2007b: 'Space and the Visualization of Power in the Greek *polis*. The Award of Portrait Statues in Decrees from Athens', in P. Schultz & R. von den Hoff (eds.), *Early Hellenistic Portraiture. Image, Style, Context*, Cambridge, 181–204.

Osborne, M.J. 1983: *Naturalization in Athens*, Brussels, vol. 3.

Podossinov, A.V. 2012: 'Le royaume du Bosphore Cimmérien aux époques grecque et romaine: un aperçu', in P. Burgunder (ed.), *Études pontiques. Histoire, historiographie et sites archéologiques du bassin de la mer Noire*, Lausanne, 87–109.

Raptou, E. 1999: *Athènes et Chypre à l'époque perse (VIᵉ–IVᵉ siècles av. J.-C.). Histoire et données archéologiques*, Lyon.

Rigsby, K.J. 1996: *Asylia. Territorial Inviolability in the Hellenistic World*, Berkeley.

Robert, L. 1980: *À travers l'Asie Mineure: poètes et prosateurs, monnaies grecques, voyageurs et géographie*, Paris.

Rostowtzew, M. 1931: *Skythien und der Bosporus. I. Kritische Übersicht der schriftlichen und archäologischen Quellen*, Berlin.

Samuel, A.E. 1993: 'The Ptolemies and the Ideology of Kingship', in P. Green (ed.), *Hellenistic History and Culture*, Berkeley, 168–192.

Saprykin, S. 1986: 'Iz épigrafiki Gorgippii' (On the Epigraphy of Gorgippia), *VDI* 1986.1, 62–75.

Saprykin, S. 2005: 'Enkomiï iz Pantikapeya i polozhenie Bosporskogo tsarstva v kontse I–nachale II v.n.é' (An *enkomion* from Pantikapaion and the History of the Bosporan Kingdom at the End of the 1ˢᵗ and the beginning of the 2ⁿᵈ century AD), *VDI* 2005.2, 45–81.

Saprykin, S. 2007: 'The Kingdom of Bosporus at the Turn of the Common Era: Barbarian and Roman Impact', in A. Bresson, A. Ivantchik & J.-L. Ferrary (eds.), *Une Koinè pontique. Cités grecques, sociétés indigènes et empires mondiaux sur le littoral nord de la mer Noire (VIIᵉ s. a.C.–IIIᵉ s. p.C.)*, Bordeaux, 309–317.

Skeat, T.C. 1974: *Greek Papyri in the British Museum*. Vol. VII: *The Zenon Archive*, London.

Skrzhinskaya, M.V. 2016: 'Intellektual'nye zanyatiya borsporskoĭ élitẏ' (The Intellectual Occupations of the Bosporan Elite), in *AAVV: Elita Bospora i Bosporskaya élitarnaya kul'tura. Materialẏ mezhdunarodnogo Kruglogo stola (Sankt-Peterburg, 22–25 noyabrya 2016 goda)* (Bosporan Elite and its Culture. Papers of the International Round Table (Saint Petersburg, November 22–25, 2016)), Saint Petersburg, 16–20.

Tuplin, C. 1982: 'Satyros and Athens: IG II² 212 and Isocrates 17.57', *ZPE* 49, 121–128.

Vinogradov, Yu.G. 1980: 'Die historische Entwicklung der Poleis des nördlichen Schwarzmeergebietes im 5. Jahrhundert v.Chr.', *Chiron* 10, 53–100.

Vinogradov, Yu.G. 1994: 'Greek Epigraphy of the North Black Sea Coast, the Caucasus and the Central Asia (1985–1990)', *ACSS* 1, 63–74.

Vinogradov, Yu.G. 1997: *Pontische Studien. Kleine Schriften zur Geschichte und Epigraphik des Schwarzmeerraumes*, Mainz.

Vinogradov, Yu.G. & Shestakov, S.A. 2005: *Laudatio funebris iz* Pantikapeya (*Laudatio funebris from Pantikapaion*), *VDI* 2005.2, 42–44.

von Wartensleben, G. 1901: *Begriff der griechischen Chreia und Beiträge zur Geschichte ihrer Form*, Heidelberg.

Welles, C.B. 1934: *Royal Correspondence in the Hellenistic Period*, New Haven.

Williams, D. & Ogden, J. 1994: *Greek Gold: Jewellery of the Classical World*, London.

LES SUITES DE LA PAIX D'APAMÉE EN MER NOIRE

Germain Payen

Abstract: The Consequences of the Peace of Apameia for the Black Sea: The 2nd century BC began with a major geopolitical shift in the Mediterranean and Hellenistic world, one that has been studied extensively, though with a concentration on one of its aspects, i.e. the interconnectivity between the Roman and the Hellenic political spheres. In the aftermath of the settlement of the Peace of Apameia (188 BC) as regards the Black Sea area, many Anatolian kings and dynasts, freed from Seleukid control, actively reshaped the geopolitical order over the following decades. Their rivalries involved the rule of the Black Sea, whose southern shores belonged to the Pontic and Bithynian kingdoms, while control of the Sea of Marmara was disputed by the Attalids, Rhodians and Bithynians. These rulers conducted diplomatic relations with some cities of the Black Sea, such as Sinope, Herakleia or Tieion on the southern shore, but also Chersonesos Taurike and Mesambria on the northern and western shores. The war between the Attalid king and his Bithynian and Kappadokian allies against the Pontic king and his Galatian and Armenian allies, c.182–179, was of prime importance. This conflict ended with the victory of the Attalid Eumenes II, assuring a relative supremacy over Anatolia, as well as the conquest of Sinope and the conclusion of an alliance with Chersonesos for Pharnakes I, the king of Pontos (Polyb. 25.2). The conclusion of the war made clear that Roman hegemony was still a distant political factor, in spite of some secondary diplomatic accomplishments of the Senate. The kings of Pontos, Bithynia and Pergamon quite actively pursued their own agendas: Eumenes targeted Thrace and inland Anatolia, while opposing Bithynian expansion through diplomatic relations with cities opposed to Bithynia, particularly Herakleia. Pharnakes could keep Sinope after his defeat, which led to a new maritime orientation for his kingdom.

Абстракт: Последствия Апамейского договора в Причерноморье: II век до н.э. начался в средиземноморском и эллинистическом мире с крупного геополитического сдвига, который был тщательно изучен, однако сосредоточился на одном из его аспектов, а именно на взаимосвязи между римской и эллинской политическими сферами. В этой главе автор стремится обсудить последствия Апамейского договора (188 г. до н.э.) по отношению к Черноморскому региону. Освободившись от контроля Селевкидов, многие анатолийские цари и правители активно изменяли геополитический порядок в течение последующих десятилетий. Их соперничество включало контроль над Черным морем, южные берега которого принадлежали Понтийскому и Вифинскому царствам, в то время как контроль над Мраморным морем оспаривался Атталидами, родосцами и бифинийцами. Эти правители поддерживали дипломатические отношения с некоторыми городами черноморского региона, такими как Синопа, Гераклея или Тиеион на южном берегу, а также с Херсонесом Таврическим и Месамбрией на северном и западном берегах. Самой значительной является война, в которой атталидский царь и его союзники из Вифинии и Каппадокии выступили против понтийского царя и его союзников из Галатии и Армении, ок. 182–179 гг до н.э. Этот конфликт закончился победой атталидского царя Эвмена II, гарантирующей относительное превосходство над Анатолией, а также завоеванием Синопы и заключением союза между Херсонесом и Фарнаком, царём Понта (Полибий 25.2).

Завершение войны ясно дало понять, что несмотря на некоторые второстепенные дипломатические достижения Сената, римская гегемония все еще была отдаленным политическим фактором. Цари Понта, Вифинии и Пергама довольно активно выполняли собственные планы: Эвмен был нацелен на Фракию и внутреннюю Анатолию, одновременно выступая против экспансии Вифинии, устанавливая дипломатические отношения с городами, противостоящими Вифинии, особенно с Гераклеей. Фарнак, после своего поражения, был в состоянии удержать Синопу, что привело к новой морской ориентации его царства.

I. INTRODUCTION

Le début du IIe siècle avant notre ère fut marqué par plusieurs bouleversements géopolitiques à l'échelle du bassin méditerranéen. Dans la région anatolienne, l'événement fondateur fut la conclusion du traité d'Apamée. Ce traité fut conclu en 188 entre Rome et le royaume séleucide, après une victoire des premiers sur le plus puissant roi hellénistique du moment. Si ce triomphe consacrait le nouveau statut de grande puissance méditerranéenne de la république romaine, ses effets immédiats furent essentiellement ressentis en Anatolie. Cette péninsule, précédemment dominée par le roi séleucide, fut libérée de son emprise et laissée aux mains des dynastes et cités locales. Le recul séleucide libéra un large corridor courant de la Cilicie jusqu'au rivage thrace, sur la Propontide. Ces terres furent distribuées aux alliés de Rome, le roi attalide et la cité de Rhodes.[1] D'autres puissances se trouvèrent renforcées dans leur légitimité ou prirent leur indépendance, en particulier les royaumes de Bithynie, de Cappadoce et de Cappadoce pontique, plus communément appelée 'Pont', ainsi que l'Arménie qui fut divisée en deux royaumes reconnus par Rome.[2] En revanche, aucune source ne laisse supposer que les États des rives non-anatoliennes du Pont-Euxin furent touchés par le traité. Hormis en Thrace, à l'extrême ouest, l'influence séleucide ne semble pas avoir touchée les rivages du Pont-Euxin. C'est au cours de la décennie suivante que les conséquences sur cet espace prirent forme, lors de conflits entre puissances anatoliennes qui tentèrent de profiter de l'absence d'un *hegemon* régional.

1　Polyb. 21.42 et Liv. 38.38 (le traité d'Apamée); Polyb. 21.46.2–12 et Liv. 38.39.7–17 (le règlement territorial). Voir Bikerman 1937; Magie 1950, vol. 2, 758–764; McShane 1964, 151s.; Walbank 1979, 164–175; Will 1982, 226–228; Gruen 1984, 640–643; Sherwin-White 1984, 18–27; Eckstein 2008, 334s.; Payen 2020, 76–123.

2　Strab. *Geogr.* 11.14.5 (528C), 15 (531s.C). Sur la partition de l'Arménie, voir Schottky 1989, 139–145; Traina 1999/2000 59–63; Facella 2006, 183–198; Dédéyan 2007, 114–116; Ballesteros Pastor 2016; Coşkun 2016. Sur la situation de l'Arménie Majeure, voir Hewsen 1985, 57s.; Facella 2006, 190–192. Sur la Sophène, voir Frankfort 1963, 181–184; Hewsen 1985, 58–60; Syme 1995, 51–57.

II. DE LA PAIX D'APAMÉE À LA PAIX DE 179:
UNE HISTOIRE ANATOLIENNE DU PONT-EUXIN

1. La première mise à l'épreuve de l'ordre anatolien et ses conséquences
sur la mer Noire

À la suite du traité d'Apamée, les réseaux d'alliance évoluèrent dans la péninsule anatolienne.[3] En premier lieu, le dynaste attalide Eumène II dut faire face à une situation d'urgence, car il comptait de nombreux rivaux risquant de remettre en cause son nouveau statut. En effet, si la rivalité avec Rhodes avait été mise de côté le temps d'affronter un adversaire commun, elle refit surface dès 189 lors des pourparlers de paix à Rome.[4] Le dynaste bithynien Prusias Ier, quant à lui, s'était révélé comme le principal rival des Attalides en Asie Mineure septentrionale.[5] Le règlement territorial du traité d'Apamée avait laissé ce souverain lésé d'une partie de ses récentes conquêtes qui furent remises à Eumène.[6] Le dynaste bithynien et Rhodes pouvaient craindre un renforcement attalide sur la Propontide et le Pont-Euxin. Plus globalement, les Antigonides et les Séleucides avaient de bonnes raisons de rester hostiles envers ce nouveau pouvoir régional, en raison du problème des territoires thraces disputés entre Eumène et Philippe V et de celui de la Pamphylie soumise aux prétentions d'Eumène et d'Antiochos. Enfin, la nouvelle frontière entre l'Anatolie et le royaume séleucide était tenue par Ariarathe IV de Cappadoce, un roi lié à Antiochos III par alliance matrimoniale.[7] C'est de ce côté que la principale réorientation diplomatique eut lieu, par le truchement d'un nouveau mariage diplomatique entre dynasties ariarathide et attalide.[8] Ce renforcement vers le Sud-Est ne réglait pas les difficultés présentées par les Galates et le roi de Bithynie à l'Est et au Nord, tandis que l'équilibre des forces restait à clarifier sur les rives de la mer Noire.

En parallèle avec l'établissement de nouvelles alliances, les dynastes anatoliens ne tardèrent pas à mettre les décisions de 188 à l'épreuve des armes. Le premier test majeur fut imposé à Eumène par Prusias Ier de Bithynie. La guerre, commencée en 186 ou 185, semble avoir eu pour objet une portion de la Phrygie,

---

3   Sur le nouvel ordre anatolien, voir en dernier lieu Koehn 2007, 205–213.
4   Polyb. 21.18–24; Liv. 37.52.1–37.56.10; App. *Syr.* 44.229–231. Voir Walbank 1979, 111–116 sur le texte polybien. Sur le discours des Rhodiens, voir Ferrary 1988, 156s., qui notait le caractère anachronique du discours des ambassadeurs rhodiens qui fait davantage référence au thème de la liberté des cités grecques proclamée en 196 qu'à la situation de 189.
5   Voir Vitucci 1953, 37–65; Fernoux 2008, 225–236.
6   Liv. 37.25.8–14. C'est dans ce sens que doit être compris le passage: Μυσούς, οὓς <Προυσίας> πρότερον αὐτὸς παρεσκευάσατο, tiré de Polyb. 21.45.10. Liv. 38.39.15; Polyb. 21.22.14 et 46.10. Voir Habicht 1956, 92; Walbank 1979, 171s.; Will 1982, 228s.; Debord 2001, 142–144.
7   Sur le mariage d'Ariarathe avec Antiochis, dans les années 190: App. *Syr.* 5.18.
8   Liv. 38.39.6. Voir aussi Polyb. 21.44, cf. Walbank 1979, 153 et 164; Michels 2009, 122s.; Payen 2019, 281; McAuley 2019, 320s.

prise par le Bithynien à Eumène vers 196, et que Rome remit à ce dernier lors du règlement territorial de 188.[9] Dans un article récent, S. Dmitriev a mis en relation le siège d'Héraclée du Pont par Prusias Ier, évoqué par Memnon, avec la guerre entre le royaume bithynien et Eumène II. Ce rapprochement s'accompagnait d'une correction des dates du conflit *c.* 185–183 plutôt qu'en 186–184. S'il est convaincant dans la révision de la date du siège d'Héraclée après 188, en revanche il n'est pas nécessaire de nier toute implication de la portion de Mysie que se disputaient les deux rois, qui apparaît comme un mobile probable du conflit, quand bien même il ne fut pas le seul et peut-être pas l'élément déclencheur.[10] La cité autonome d'Héraclée du Pont subit donc l'assaut bithynien: Prusias désirait profiter de la vacance des autorités politiques supérieures pour reprendre sa politique de conquête, à l'encontre d'une puissance qui bloquait ses ambitions à l'égard du Pont-Euxin (voir Fig. 1).

Fig. 1 : La guerre de 185-183
(Germain Payen, 2020)

Après divers engagements mal connus, la guerre prit fin en 184, et, si le traité de paix consécutif n'est pas connu, il est clair qu'il était à l'avantage d'Eumène, qui s'assurait la possession de la partie de la Phrygie mise en jeu et de la cité de Tieion, et renforçait son influence politique sur les dynastes galates et sur les cités autonomes des Détroits.[11] La révision du contexte de l'attaque bithynienne contre Héraclée a pour conséquence que la cité de Tieion ne fut prise par Prusias que le temps de ce conflit, au terme duquel elle revint à Eumène. Cette acquisition représentait un ajout conséquent à la présence attalide sur le littoral pontique, même si la situation n'était sans doute pas modifiée outre-mesure. En effet, les Attalides avaient maintenu des liens forts avec cette cité tout au long du IIIe siècle.[12] Il n'y a pas de raison de penser que cette acquisition fut accompagnée de territoires faisant le lien entre Tieion et le reste du territoire d'Eumène: cette cité se trouvait simplement enclavée en territoire étranger, à l'instar de Telmessos.[13]

---

9   Liv. 38.39.15; Polyb. 21.22.14; 21.46.10. Voir Habicht 1956, 93–96; Will 1982, 228s.; Koehn 2007, 127–129; Fernoux 2008, 232s.

10   Dmitriev 2007, 135–137. Le passage mentionné est Memn. *FGrH* 434 F 19.1–3. Voir dernièrement Keaveney & Madden 2011, 61, n. 19.

11   Polyb. 25.2. Voir Hopp 1977, 43; Will 1982, 286; Petković 2012, 364.

12   Le fondateur de la dynastie Philétairos en était originaire pour Strab. *Geogr.* 12.3.8 (543C); Eumène (Ier) l'avait déjà tenu sous son autorité d'après Memn. *FGrH* 434 F 9.4.

13   Voir Magie 1950, vol. 2, 760, qui allait trop loin en niant l'obtention de Tieion par Eumène, assurée par sa récupération en 179, d'après Polyb. 25.2.7; Jones 1937, 420; McShane 1964, 162.

Le pouvoir attalide s'avéra ne pas être le seul à sortir de la guerre renforcé, même si les bénéfices principaux lui revenaient. Malgré la défaite bithynienne, ce conflit fut pour Prusias Ier l'occasion de marquer le territoire royal de son empreinte par l'intermédiaire de trois fondations civiques importantes.[14] En effet, les cités de Prousa de l'Olympe, Prousias de l'Hypios et Bithynion furent toutes trois fondées pendant ou au terme de la guerre, sans doute dans une optique de protection et de consolidation des acquis. Ainsi, Prousa de l'Olympe pouvait garder la frontière sud du royaume, face au territoire attalide, à la manière d'un puissant poste avancé.[15] Prousias de l'Hypios remplaça Kieros, un établissement héracléote conquis pendant la guerre et refondé à son terme, probablement pour s'assurer contre un retour de la cité ennemie.[16] Enfin, Bithynion fut établie plus à l'est, également à la limite du territoire d'Héraclée, mais aussi sur les contreforts des montagnes de Paphlagonie.[17] Ainsi Prusias assurait-il l'hellénisation physique et politique de son territoire, par la création de nouvelles cités, mais aussi sa défense grâce aux fortifications, et enfin la glorification de son nom et de sa dynastie. Malgré le recul bithynien en Phrygie et l'abandon du siège d'Héraclée, l'ancrage territorial de Prusias en direction de la mer Noire s'était renforcé.

Les enjeux de cette guerre annonçaient déjà la suivante: les statuts des peuples galates et des cités du littoral pontique s'affirmaient comme des clefs stratégiques dans la rivalité entre dynastes régionaux.[18] En définitive, le souverain sortait grandi de ce premier conflit post-apaméen, mais son renforcement prenait surtout forme du côté de la frontière galate, tandis que son influence sur les cités de la mer Noire se trouvait davantage confirmée que renforcée.[19]

2. LA GUERRE DE 182–179: LES SOURCES ET LEURS ENSEIGNEMENTS

Au lendemain de la guerre entre Eumène et Prusias, un nouveau coup fut porté à l'ordre anatolien établi à Apamée. Celui-ci fut plus sérieux et impliqua davantage de puissances anatoliennes.[20] L'événement déclencheur de cette crise fut l'attaque du roi du Pont, Pharnace, contre la cité de Sinope, en 183. Il s'agissait d'une réplique de la tentative du prédécesseur de ce roi, en 228 ou *c.* 220, lors de laquelle

---

14 Sur ces trois fondations, voir Fernoux 2004, 38–40.
15 Strab. *Geogr.* 12.4.3 (563C); Plin. *NH* 5.148. Voir Holleaux 1938, 114s.; Robert 1937, 228–235; Leschhorn 1984, 279–281.
16 Memn. *FGrH* 434 F 19.1. Voir Robert 1980, 61s.; Leschhorn 1984, 279; Ameling 1985, 3s.; Heinemann 2010, 217–219 et 228.
17 Voir Dörner 1952, 81; Jones 1940, 17; Robert 1980, 129–132; Marek 1933, 23.
18 Voir, en dernier lieu, Payen 2020, 148–161.
19 Sur les frontières galates au IIème siècle, voir en dernier lieu Coşkun 2019a, 617–622.
20 Sur cette guerre, voir McShane 1964, 161–163; Will 1982, 288–290; Petković 2012, 360–364.

Sinope avait fait appel à Rhodes pour repousser l'envahisseur.[21] Pharnace tentait de dépasser Mithridate III, toujours avec le même objectif géostratégique: prendre le contrôle du port le plus prospère du littoral sud du Pont-Euxin. Rhodes, dont les intérêts commerciaux en mer Noire se trouvaient menacés, n'était pas prête à intervenir directement et choisit d'envoyer une délégation au Sénat romain.[22] Cette initiative mit face-à-face les ambassadeurs de Pharnace et d'Eumène, lui aussi venu se plaindre du roi du Pont, ce qui amena ensuite les deux royaumes à s'affronter ouvertement. La réponse faite par le Sénat, qui accepta seulement d'envoyer des commissaires pour évaluer la dispute entre Pharnace et Eumène et la situation sinopéenne, suppose que les deux rois étaient alors déjà entrés dans une confrontation diplomatique. L'hypothèse la plus probable est que cette altercation avait pour motif principal un objet de rivalité autre que Sinope, à savoir la mainmise sur les peuples galates.[23] La suite du conflit fit intervenir toutes les principales puissances politiques et militaires de la péninsule, au gré des réseaux d'alliance créés ou consolidés à la suite du traité d'Apamée, et à nouveau modifiés au cours de la guerre.

La séquence des événements n'est que partiellement connue, surtout par l'intermédiaire de Polybe.[24] Celui-ci évoquait les ambassades à Rome des différents dynastes et cités impliqués, ce qui laisse penser qu'il se fondait sur des sources romaines pour les années 183/2, 182/1 et 181/80. Seules les dernières phases de la guerre et le traité conclu en 179 étaient évoqués avec plus de détails.[25] Le texte du traité de 179 constitue la pièce maîtresse dans la reconstitution des événements et des alliances. En ce qui concerne le Pont-Euxin, une inscription, transcrivant un traité de φιλία καὶ συμμαχία entre Pharnace et la cité de Chersonèse sous le patronage romain, semble entrer dans ce contexte, même si sa datation est débattue.[26] Enfin, diverses sources archéologiques diffuses permettent d'élaborer le contexte historique dans lequel se trouvait alors le Pont-Nord.

Le différend entre Eumène et Pharnace tourna au conflit armé en 182. Pour faire face, le roi attalide bénéficiait d'un allié, le roi Ariarathe IV de Cappadoce, qui participa d'autant plus activement aux opérations militaires que Pharnace monta une expédition contre la Cappadoce. Pharnace disposait également d'un

---

21 Sur l'attaque de Mithridate III: Polyb. 4.56.2s. Dan 2009, 94. Voir Petković 2012, 380, qui datait cette attaque en 228, plutôt que vers 220. Sur l'attaque de Pharnace en 183, et l'envoi de délégations pontique, attalide et rhodienne à Rome: Polyb. 23.9.1–3; Liv. 40.2.6. Cf. Petković 2012, 360; Roller 2020, 57 (Roller 2020 a été publié trop tard pour pouvoir être pleinement pris en considération dans cette communication).
22 Polyb. 23.9.2.
23 Polyb. 23.9.4.
24 Polyb. 24.1.1–3; 24.5.1–8; 24.14.1–11; 24.15.12s.; 25.2; Strab. *Geogr.* 12.3.11 (545f.C); Diod. 29 F 25–27.
25 Les opérations militaires en 180: Polyb. 24.14.1–11. La fin de la guerre et le traité de 179: Polyb. 25.2.
26 *IOSPE*, I² 402. Sur les problèmes de datation de cette inscription, les détails de cette question complexe seront évoqués plus bas.

allié: Mithridate d'Arménie Mineure ou de Sophène, dont le territoire souverain, difficile à préciser, bordait sans doute celui d'Ariarathe.[27] Après diverses confrontations mal connues, quelques trêves et des délégations sénatoriales impuissantes à régler le différend, celui-ci semble au contraire s'être élargi. En effet, outre la présence au côté d'Eumène de Prusias II de Bithynie, qui avait récemment succédé à son père, apparaissent au second plan des cités et des dynastes du Pont Nord et d'Arménie.

Avant d'en venir aux acteurs pontiques dont la situation dans ce conflit est peu claire, il convient d'évaluer les causes de l'entente entre Prusias et Eumène. Cette alliance était une véritable volte-face diplomatique qui mérite un examen, d'autant qu'elle concernait l'équilibre des forces sur le rivage du Pont-Euxin. La récente défaite de Prusias Ier contre le roi attalide, déjà en partie sur le sujet de la Galatie, pouvait faire craindre un encerclement à Eumène. La tournure des événements facilita un rapprochement: Pharnace choisit d'attaquer la Paphlagonie, à la frontière attalide et bithynienne, et son stratège Léokritos conquit Tieion malgré la présence d'une garnison de mercenaires attalides, menaçant du même coup le territoire bithynien.[28] Cette avancée dut rendre la perspective d'un danger pontique crédible, ce qui put contribuer à convaincre Prusias de choisir le camp attalide. Le roi attalide offrit un présent diplomatique important à Prusias, à savoir une promesse de remise de la cité de Tieion.[29] Ainsi, Prusias fit le choix de rejoindre le plus puissant royaume engagé, l'allié de Rome, tandis qu'Eumène consentit à abandonner une place-forte possiblement isolée en territoire montagneux entre la Bithynie et le Pont.

Ayant assuré l'alliance bithynienne, Eumène semble avoir usé de ses liens avec les grandes cités des rives de la Propontide et de la puissance de sa flotte pour tenter une approche indirecte, consistant à bloquer les détroits de l'Hellespont limiter les mouvements de ses ennemis. Rhodes, bien que n'ayant pas rejoint Pharnace, en empêcha Eumène car ses intérêts commerciaux étaient en jeu, ce qui aurait entamé une brouille entre les deux pouvoirs.[30] Il s'agissait d'une confrontation diplomatique importante pour les destinées géopolitiques de la mer Noire. Le roi attalide céda à la pression rhodienne et reconcentra ses efforts sur l'Anatolie

27 Polyb. 24.14.1s. L'identification de Mithridate d'Arménie et du territoire qu'il dirigeait reste sujet à débat: Mørkholm 1966, 29, n. 35 et Patterson 2001, 156, n. 6 y voient le dynaste de Sophène, tandis que Walbank 1979, 272s. et Ballesteros Pastor 2016, 276s. l'identifient au fils de Pharnace et prince du Pont Mithridate (IV), qui aurait alors été en charge d'une satrapie arménienne du royaume du Pont. Il peut aussi s'agir d'un dynaste local, indépendant de la Sophène comme du Pont: Payen 2020, 170s. Sur le royaume de Cappadoce dans ce conflit, voir Michels 2009, 123; McAuley 2019, 321s.; Payen 2020, 163f. et 167s.
28 Diod. 29 F 26. Voir Roller 2020, 60.
29 Polyb. 25.2.7, précisait la remise ultérieure de la cité par Eumène à Prusias, à titre de don. Voir Walbank 1979, 272, qui précisait qu'il s'agissait d'une parenthèse de Polybe dans le texte du traité; Petković 2012, 363s.
30 Polyb. 27.7.5. Voir Berthold 1984, 173, n. 17 et Gabrielsen 1999, 45, qui mettaient en parallèle cette action avec la guerre de Rhodes contre Byzance dans le thème de la protection du libre commerce maritime.

centrale, établissant ainsi une hiérarchie dans ses intérêts stratégiques. Il repoussa donc les troupes pontiques de Galatie, regagnant du même coup la soumission de certains dynastes locaux, avant de lancer une puissante offensive qui emporta la décision et qui lui permit de dicter ses conditions de paix.[31] En résumé, celles-ci consistaient, pour Pharnace, à renoncer à ses contacts diplomatiques avec les autorités galates, à rendre les territoires conquis en Paphlagonie et en Cappadoce et à payer des dédommagements à plusieurs de ses opposants.

À l'instar de ce qu'avait montré le conflit précédent, les cités aux frontières des territoires dynastiques, sujettes ou non, jouèrent le rôle d'objectifs politiques et territoriaux prioritaires pour les royaumes anatoliens. Les seules *poleis* autonomes à être citées parmi les *adscripti*, autrement dit en tant que garantes, étaient soit des alliées importantes d'Eumène (Cyzique et Héraclée), qu'il fallait protéger contre les ambitions pontiques et bithyniennes, soit des cités du Pont Nord (Mésambria et Chersonèse), éloignées du théâtre d'opération.[32] Il apparaît donc que la guerre connut une extension diplomatique sur la côte Nord de la mer Noire (voir Fig. 2).

Fig. 2 : La guerre de 182-179
(Germain Payen, 2020)

Pour analyser plus en profondeur cet aspect de la paix de 179, un document épigraphique est particulièrement utile, même si les difficultés liées à sa datation rendent son interprétation difficile, à savoir le traité entre Pharnace du Pont et la cité de Chersonèse.[33] Cette inscription retranscrit manifestement les serments

---

31 Polyb. 25.2. Voir Walbank 1979, 272.
32 Polyb. 25.2.13.
33 *IOSPE* I² 402 = *SEG* XXX 962. Voir les traductions de Burstein 1980, 4 (traduction anglaise); Højte 2005, 148s. (texte grec et traduction anglaise); en dernier lieu, Müller 2010, 93–95 et 379–381, a édité, traduit et commenté le texte.

échangés par les autorités du royaume du Pont et de la cité de Chersonèse lors de la conclusion d'une alliance entre ces deux entités politiques. Le point critique en ce qui concerne ce document s'avère être sa datation très discutée. La formule permettant de déterminer le contexte historique précis de l'épisode diplomatique en question est la référence à la 'cent cinquante-septième année (...) comme le compte le roi Pharnace' (l. 29–32). La seule ère dynastique utilisée par les rois du Pont de manière certaine, à savoir l'ère bithyno-pontique, ne peut correspondre à l'usage de Pharnace, puisqu'elle situerait ce texte en 140, date à laquelle il était mort depuis des années. L'identification d'une ère dynastique ou civique débutant en 337/6, bien que difficile à établir avec certitude, a amené de nombreux chercheurs à dater l'inscription de 179.[34] Cette hypothèse, adoptée ici, permet de faire correspondre le traité entre Pharnace et Chersonèse avec la fin de la guerre de 182–179.[35]

Néanmoins, aucune autre source ne soutient l'existence d'une ère pontique débutant en 337/6, date tirée de l'histoire du fondateur présumé de la dynastie mithridatide, telle que l'a transmise Diodore.[36] Depuis les années 1980, des objections ont été opposées à cette théorie. Ainsi, il existe une tradition différente au sujet de l'histoire des ancêtres de Pharnace, tradition qui ne supporte pas la reconstruction d'une ère débutée en 337/6.[37] De manière plus fondamentale, l'absence de sources corroborant l'existence d'une ère dynastique particulière avant la reprise de celle utilisée en Bithynie rend, au fait, hasardeux cette hypothèse.[38] Ainsi, il a été proposé de voir une référence à l'ère royale séleucide, dont la première année correspond à 312, ce qui signifierait que le traité aurait été conclu en 155.[39] La raison pour laquelle cette hypothèse reste douteuse doit être cherchée dans la chronologie du règne de Pharnace et de ses successeurs Mithridate IV et Mithridate V. En effet, si Mithridate IV put accéder au trône en 155 puisque la première attestation littéraire de son statut royal est datée de 155/4, il aurait eu un court règne, une autre inscription prêtant à son successeur Mithridate V le titre royal qu'il faut dater 152/1 si l'ère séleucide y était encore utilisée comme référence.[40]

---

34 En dernier lieu, voir Vinogradov dans *BE* 1990, 559; Heinen 2005, 37–42; Avram dans *BE* 2006, 298; Avram 2016, 216–223, en faveur d'une datation en 179. L'*editio princeps* de l'inscription prenait pour acquis cette datation, suivie en cela par la communauté scientifique jusqu'aux remises en cause de S.M. Burstein et B.C. McGing.
35 Polyb. 25.2.3–15. Outre le développement ci-avant, voir Walbank 1979, 271–274; Will 1982, 288–290.
36 Diod. 15.90.3. Voir Højte 2005, 140; Avram 2016, 114s. et 125.
37 Plut. *Dém.* 4.1.
38 Sur la diffusion de l'ère séleucide, voir Leschhorn 1993, 91–95; Savalli-Lestrade 2010, 63s.
39 Burstein 1980, 7; McGing 1986, 30; Ferrary 2007, 319; Arrayás Morales 2014, 938; Roller 2020, 62 et 65.
40 Reinach 1905, 114s. Il n'est pas impossible que cette inscription ait été datée selon l'ère bithyno-pontique, qui fut adoptée à une date inconnue par la dynastie pontique, ce qui décalerait sa datation jusqu'en 137/6. Dans ce cas, le témoignage d'App. *Mith.* 10.30 placerait le début du règne de Mithridate V en 149 au plus tard. Voir Højte 2005, 144.

Les sources amphoriques ne sont pas d'un grand secours, puisqu'elles sont datées en fonction de la tradition tirée de la datation haute de cette inscription.[41] En revanche, les monnaies frappées par Mithridate IV, nombreuses par rapport à celles aux noms de Pharnace ou de Mithridate V, portent à considérer que son règne dura plus de quelques années.[42] D'autres inscriptions pourraient aider à préciser la séquence des règnes si elles se trouvaient dater avec certitude, mais le même problème d'identification de l'ère utilisée apparaît.[43] De prime abord, la date haute suggérant une alliance conclue en 179 semble difficile à accorder avec le récit de la guerre de 182–179 fait pas Polybe, dans lequel Rome tint un rôle secondaire, puisque dans cette inscription la puissance romaine paraît influente sur les deux acteurs concernés. Néanmoins, il n'y a pas matière à sur-interpréter le rôle de Rome, que cette inscription caractérise comme un ami du roi pontique et de la cité de Chersonèse. Dans cet ordre d'idée, il faut noter que les articles récents d'H. Heinen et d'A. Avram ont réaffirmé la vraisemblance de la datation haute, en mettant en avant la relative modestie de l'influence romaine suggérée par le document, en critiquant certains arguments des tenants de la datation basse. Ainsi, A. Avram a formulé une hypothèse nouvelle concernant l'ère utilisée par Pharnace, suggérant une ère civique sinopéenne.[44]

En plus de ces cités, plusieurs dynastes apparaissent parmi les *adscripti*. Parmi ceux-ci, un dynaste d'Europe, Gatalos, qui a pour seul complément à son nom un ethnique, 'Sarmate'.[45] Bien que maigre, cet indice suppose qu'il s'agissait d'un chef nomade du littoral septentrional ou occidental du Pont-Euxin. La mention de ce dernier peut être un indice des déplacements de populations nomades dans le Pont-Nord qui se produisirent peut-être à partir de la fin du IVe siècle.[46] Cette hypothèse permet d'évoquer d'autres sources littéraires, épigraphiques et archéologiques, qui montrent qu'il convient de rester prudent en raison du manque de certitude sur ce sujet débattu. En effet, certains historiens ont pensé pouvoir mettre en rapport un passage de Diodore sur la dévastation du Bosphore par les Sarmates avec un autre extrait de ce même auteur, au sujet d'une guerre entre les

---

41 Finkielsztejn 2001a, 172–174 et Højte 2005, 146s., ont bien mis en relief le raisonnement circulaire ici en cause, ainsi que la fragilité de la datation traditionnelle des amphores rhodiennes, souvent utilisées comme références en la matière.
42 Waddington *et al.* 1925, 12s. Vinogradov dans *BE* 1990, 559, faisait de cette relative profusion de numéraires au nom de Mithridate IV une preuve d'un long règne, *contra* Højte 2005, 144.
43 *OGIS* 771 et Reinach 1905, 114s. Sur *OGIS* 771, voir McShane 1964, 142; Grainger 2002, 53–55, 109–111 et 116s.
44 Heinen 2005, 44–51; Avram 2016, 225–230. À l'inverse, Højte 2005, 140–147, a offert une synthèse commode des arguments en faveur de l'une ou l'autre datation, en précisant au final que 'the Seleucid calendar seems most the probable for determining the date of the inscription' (147). En dernier lieu, voir aussi Müller 2010, 93–95, qui restait prudente mais semblait favoriser la datation basse.
45 Polyb. 25.2.11–13.
46 Lebedynsky 2001, 92–98; Heinen 2005, 42–47; Müller 2010, 70–74; Coşkun, *APR s.vv.* Gatalos.

Scythes et le roi du Bosphore allié aux Sarmates, et, surtout, avec un décret de Chersonèse évoquant une attaque barbare au début du IIIe siècle et faisant sans doute référence à des Sarmates.[47] Ces mises en relation laisseraient penser qu'un mouvement sarmate de grande envergure déséquilibra l'équilibre régional dès la fin du IVe siècle et au début du IIIe siècle.[48] Néanmoins, d'autres chercheurs ont montré que les fouilles archéologiques ne témoignaient pas de l'installation de peuples sarmates, utilisant des kourganes, avant 150 a.C., tandis que des territoires de Crimée avaient été abandonnés dès le début du IIIe siècle, ce qui a mené ces savants à séparer les deux phénomènes et à dater l'arrivée des Sarmates du milieu du IIe siècle.[49] De fait, le décret de Chersonèse cité plus haut a été découvert dans un pauvre état de conservation et la mention des Sarmates y est incertaine et sans relation claire avec l'attaque évoquée. En tout état de cause, il convient d'appliquer la plus grande prudence dans l'identification et l'interprétation des sources grecques concernant cette partie du monde.

III. LE NOUVEL ORDRE PONTIQUE: ORIENTATIONS ET INCERTITUDES

1. La paix de 179 et le nouvel équilibre des pouvoirs en Anatolie-Pont Sud

L'étude de la scène géopolitique pontique au début du IIe siècle est rendue complexe par le biais des sources, qui ne permettent de l'évaluer qu'à l'aune des intérêts des puissances anatoliennes, et seulement par intermittence. Cette première décennie post-apaméenne permit d'éclaircir l'ordre anatolien et les rapports de force dans la péninsule, notamment entre les dynastes et les cités. La principale puissance maritime de la région, Rhodes, éprouva des difficultés face à la tâche impliquée par ses agrandissements territoriaux. Plus important, et ce qui en était peut-être la conséquence, la cité ne sut pas garantir les intérêts de son alliée principale du Pont-Euxin méridional, Sinope. Malgré le déclin des grands royaumes hellénistiques concurrents et l'augmentation fortuite des sources de revenu de Rhodes, celle-ci n'avait pas réellement su assurer son statut de thalassocratie pourtant annoncée et expérimentée depuis le milieu du siècle précédent.[50] La puissance rhodienne s'était montrée timorée et payait son manque d'engagement par

47 Diod. 2.43.7 (dévastation de la Crimée par les Sarmates); Diod. 2.20.22 (présence de Siraces, peuplade sarmate, dans l'armée du roi du Bosphore Eumèlos contre les Scythes de Satyros); *IOSPE* I² 343 (décret de Chersonèse). Sur cette dernière inscription, voir Vinogradov 1997.
48 C'était la thèse proposée par Rostovtzeff 1930, 573s., reprise par Shcheglov 1990, 186s. et Vinogradov & Shcheglov 1990, 361.
49 Khrapunov 2004, 126; Bylkova 2005, 142; Stolba 2005, 308.
50 Sur Rhodes, ses moyens et ses ambitions après la paix d'Apamée, voir Berthold 1984, 167–178, ancien mais utile; Gabrielsen 1999, 100–108 qui traitait aussi bien de l'organisation de la flotte rhodienne que des objectifs politiques et marchands poursuivis par la cité en direction des mers.

l'abandon de Sinope à Pharnace.[51] Il est tentant de faire le lien entre la mise de côté des intérêts rhodiens et le refus opposé au projet de blocage des détroits de l'Hellespont d'Eumène.[52] Rhodes semble avoir été dépassée par la mécanique des réseaux d'alliance impliquant les principales puissances de la région, mécanique dans laquelle s'insérait mal une cité marquée par une tradition de neutralité à l'égard des royaumes hellénistiques. Ses capacités militaires, représentées avant tout par sa flotte de guerre, avaient été retenues par la résistance de la confédération lycienne, qui n'acceptait pas l'état de sujétion politique et fiscale dans lequel elle se trouvait depuis 188.[53]

À l'inverse, Eumène II avait montré avec force qu'il était de taille à assurer le statut de grande puissance régionale que Rome lui avait confié, par son activité militaire et ses réussites diplomatiques. Le royaume attalide se trouvait être le seul rival de Rhodes dans la conquête des mers dont la puissance n'avait pas été récemment diminuée. En effet, si la puissance militaire et l'emprise territoriale des souverains de Pergame n'avaient jamais été à la mesure de celles des grandes dynasties, leur puissance navale s'était considérablement accrue à partir de la fin du règne d'Attale Ier.[54] Ainsi, les récentes recherches archéologiques menées à Elaia ont montré que le port de Pergame était bien plus étendu et développé qu'on a pu le penser jusqu'à un passé récent, ce qui aurait permis à Attale et à Eumène de mener leur jeu diplomatique du côté de la Grèce et de Rome, malgré les difficultés rencontrées en Anatolie même.[55] Les îles d'Andros et d'Égine, en Égée, avaient été récemment conquises lors des guerres menées en Grèce aux côtés de Rome et des Étoliens, puis des Achéens.[56]

Le traité d'Apamée avait considérablement modifié la situation. En effet, l'acquisition de terres étendues vers l'intérieur de la péninsule, en Phrygie et en Pisidie, devait obliger Eumène à orienter son action davantage vers le domaine continental, ce que montrèrent les deux guerres étudiées ici. Néanmoins, le roi avait aussi emporté en 188 la grande cité portuaire d'Éphèse, ainsi que des terres accessibles uniquement par la mer, à savoir la Chersonèse de Thrace en Europe, et Telmessos en Lycie.[57] Si la Galatie apparaît comme un objectif récurrent des conflits de la décennie 180, il est remarquable qu'Eumène ait également lutté sur mer dans les deux conflits. De fait, s'il ne fit pas de conquête majeure, puisque Tieion, prise en 185, fut rendue à Prusias en 179, en revanche ses avancées diplomatiques

---

51 Polyb. 23.9.1–3.
52 L'aide militaire apportée à Rhodes par Eumène: Polyb. 24.15.13; le refus opposé au roi par les Rhodiens au sujet du blocage des Détroits: Polyb. 27.7.5. Voir McShane 1964, 160; Ager 1991, 27s.
53 Polyb. 24.15.13: Le fait que Rhodes ait dû faire appel à Eumène en 180 prouve l'ampleur des difficultés rencontrées par les autorités rhodiennes dans les opérations terrestres.
54 Voir McShane 1964, 107–110 (sous Attale Ier), 158–161 et 180s. (sous Eumène II), 188–191 (sous Attale II); Ma 2013, 61s.
55 Voir Pirson 2004, 341–355.
56 Polyb. 22.8.10 (Égine); Liv. 31.45.1–8 (Andros).
57 Polyb. 21.46.9s; Liv. 38.39.14–16. Voir McShane 1964, 148–154.

furent remarquables, lui permettant de conserver la ligne politique de son prédécesseur en direction de la Grèce et de Rome, mais aussi de la Thrace et de la région des Détroits.[58] En revanche, la cession de Tieion à Prusias II et l'abandon de
Sinope à Pharnace marquèrent la faiblesse des ambitions attalides dans le Pont-
Euxin, où le maintien d'alliances avec des cités et dynastes influents devaient rester le principal aspect de la politique dynastique.

Comme le montre l'activité du roi attalide, les aspects maritimes de la géopolitique anatolienne furent marqués dans ces années par les ambitions diverses en
mer Noire et dans la zone des détroits de l'Hellespont. En l'occurrence, d'autres
dynastes y trouvèrent leur compte. Prusias et Ariarathe, les alliés d'Eumène, obte-
naient respectivement des compensations territoriales et financières, assurant leur
statut de dynastes majeurs dans la région. Le royaume de Cappadoce se trouvait
trop éloigné du rivage pontique pour y exercer une influence notable et se définis-
sait comme un État terrien, sans ambition maritime. Il en allait autrement de la
Bithynie. Prusias II, en dépit du relatif désintérêt bithynien à l'égard du front maritime, avait pu récupérer Tieion sur le Pont-Euxin, malgré les tentatives attalides
puis pontiques.[59] Ce retournement politique et diplomatique permettait à Prusias
de limiter l'échec de la politique menée par son père contre Héraclée quelques
années plus tôt, enclavant plus encore la puissante cité au sein de son propre territoire souverain. La solide alliance établie par les autorités héracléotes avec la dynastie attalide, mais aussi avec Rome, laissaient peu d'espoir de soumettre la cité,
mais l'établissement de colonies bithyniennes dans ses alentours donnait au roi
l'opportunité de faire peser son influence et de renforcer les échanges humains et
économiques avec elle. À plus long terme, le royaume de Bithynie se trouvait lui-
même encerclé par deux puissants royaumes, l'un exerçant une forte influence sur
les Détroits et en Anatolie centrale, l'autre semblant sur le point de réorienter ses
ambitions vers la mer Noire et la bordure Nord-Est de l'Anatolie. Le choix ulté-
rieur entre une stratégie maritime ou terrestre serait déterminant dans l'évolution
du réseau d'alliance bithynien.

Pour Pharnace, repoussé sur le plan continental, la défaite ne ruinait pas les
avantages récemment gagnés: le traité de paix restait peu sévère et lui laissait Sinope, qui devint résidence royale. Il obtenait aussi les anciennes colonies sino-
péennes, Cotyôra et Kérasous, et put ensuite fonder sur le site de cette dernière
une cité qu'il baptisa de son nom, Pharnakeia.[60] D'autre part, il est difficile de
préciser l'extension des terres qui lui furent concédées par la même occasion. Le
traité précise bien que la plupart de ses conquêtes en Anatolie continentale lui

58 *Syll.*[3] 627; Polyb. 25.2. Ma 2013, 57s., a souligné l'utilisation régulière des capacités de pro-
    jection de l'appareil militaire attalide par ces rois successifs, dans des guerres menées outre-
    mer qui constituaient un aspect très original de la politique extérieure d'Attale Ier et de ses
    successeurs. Voir aussi Kosmetatou 2003, 164; Michels 2019, 335s; Payen 2020, 180s.
59 Il n'y a pas de référence d'usage sur les rois de Bithynie et leur politique maritime, on se
    reportera donc essentiellement à Vitucci 1953 et Fernoux 2004, 30–111, qui s'y rapportaient
    incidemment.
60 Strab. *Geogr.* 12.3.16s. (548C); Arr. *PPE* 16.4. Cf. Counillon 1990, 496.

furent retirées, notamment en Paphlagonie et en Cappadoce, ainsi qu'en Galatie.[61] Néanmoins, le sort réservé aux terres de deux chefs galates, Gaezatorix et Kassignatos, dont la demande d'aide avait été repoussée par Eumène en 180,[62] n'est pas abordé directement dans le traité. Il est donc possible qu'elles soient restées entre les mains du roi du Pont. En l'état, si Kassignatos est connu comme le dirigeant des Tolistoboges, la tribu d'appartenance de Gaezatorix ne peut être que supputée : ayant fait sa démarche auprès d'Eumène en compagnie de Kassignatos, il est probable qu'il ait été lui-même le chef des Trocmes, étant donné que ces deux tribus partageaient des liens diplomatiques depuis longtemps. En ce cas, on peut envisager de situer son territoire souverain en Paphlagonie continentale, quelque part entre Sinope et Ankyra, précisément dans la zone touchée par les avancées de Pharnace en Paphlagonie.[63] L'absence de source évoquant à nouveau cet individu dans les années suivantes ainsi que son absence dans le traité de paix de 179 pourraient indiquer que ses terres furent laissées à Pharnace en complément de Sinope. Néanmoins, Kassignatos est mentionné par Tite-Live en tant que principal leader galate ayant combattu sous les ordres d'Eumène dans la troisième guerre de Macédoine, en 171.[64] Il semble donc que son autorité et ses territoires lui aient été rendus en 179, probablement au prix d'un accord avec le roi attalide.

Faute d'éléments supplémentaires, l'hypothèse la plus probable est que Gaezatorix partagea le même sort et put lui aussi récupérer ses terres, limitant ainsi les gains du roi du Pont. Repoussé hors de la Galatie et d'une partie de la Paphlagonie continentale, ce dernier fut porté à réorienter sa politique vers une nouvelle dimension maritime, ce qui transparaît dans la translation de la résidence royale vers Sinope.[65] L'importance accordée par la dynastie mithridatide à cette cité n'a rien pour surprendre, son rayonnement dans le Pont-Euxin apparaissant dans nombre d'œuvres d'auteurs classiques, hellénistiques et romains.[66] De manière plus concrète, la conquête de la région sinopéenne mettait aux mains de Pharnace non seulement des sources de revenu fiscal considérables, mais aussi un arrière-pays riche en ressources forestières à même de fournir le matériau de construction principal d'une flotte de guerre.[67] L'intégration d'une partie du littoral de la mer Noire dans

---

61 Polyb. 25.2.4s.
62 Polyb. 24.14.6s.
63 Sur les liens entre tribus galates et leur territoire respectif, à particulier pour les Trocmes et les Tolistoboges, voir Coşkun 2019a, 617–619; Payen ca. 2021. Sur Gaezatorix, voir aussi Coşkun, *APR s.vv.* Gaizatorix.
64 Liv. 42.57.9.
65 McGing 1986, 23, faisait de Sinope la plus importante addition au royaume pontique depuis sa création. Voir également Habicht 1989, 328–330. Plus récemment, Roller 2020, 58, soulignait l'importance de la réorientation territoriale et stratégique du royaume du Pont vers la mer Noire à l'issue de ce conflit, malgré sa préférence pour une datation du traité entre Pharnace et Chersonèse en 155 plutôt que 179.
66 Sur Sinope et sa situation de grande ville du Pont-Euxin, voir surtout Tsetskhladze 1999, 114; Doonan 2003, 1383–1385 (bibliographie: 1394–1399); Doonan 2004, 69–92; Heinen 2005; Barat 2006; Dan 2009, 69–84; Barat 2009, 353–358; Avram 2016, 226s.; Coşkun 2019b, 21s.
67 Strab. *Geogr.* 12.3.12 (546C).

la scène géopolitique régionale avait donc élaboré une nouvelle orientation politique pour les Mithridatides, un changement qui fera l'objet d'un examen plus poussé dans la section suivante.

En dernier lieu, les conflits de la décennie 180 clarifièrent les rôles tenus par les grandes puissances extérieures par rapport à la scène locale. La place du roi séleucide dans la guerre de 182–179 fait écho à celle que tenait le souverain antigonide dans le conflit précédent. Certes, Séleucos IV n'alla pas jusqu'à soutenir militairement son allié pontique, et la frontière du Taurus resta inviolée. Néanmoins, des pourparlers furent engagés, et l'intervention militaire décidée: dans le principe, la politique séleucide en Anatolie n'avait pas entièrement changé après 188, et cette région restait un axe géopolitique possible, comptant un certain nombre d'États alliés que le roi avait pour droit, pour devoir, de soutenir.[68] À l'image du roi de Macédoine, qui n'abandonna jamais ses ambitions en Thrace, le roi séleucide ne devait pas considérer le traité d'Apamée comme un point de non-retour, ce que les souverains anatoliens avaient manifestement compris, gardant ce roi encore puissant dans leur horizon diplomatique. Si le traité d'Apamée et le règlement des affaires anatoliennes par Rome n'avaient pas mis fin à une certaine influence des dynastes séleucides et antigonides sur la péninsule, les destinées du Pont-Euxin semblaient désormais hors d'atteinte, même si l'allié local des Séleucides se trouvait être le roi du Pont.

Rome, le plus important de ces États étrangers, mais aussi le plus éloigné, eut une attitude ambivalente dans les années 180, non sans une certaine cohérence politique. Après avoir tardé à intervenir diplomatiquement lors du conflit entre Prusias Ier et Eumène II, le Sénat envoya plus régulièrement des commissions sénatoriales pendant la guerre suivante, mais sans plus d'efficacité.[69] Si cette démarche semblait alors devenir la norme, le Sénat restait peu intéressé par les détails de la scène anatolienne aussi longtemps que ses ennemis historiques n'y tentaient pas un retour. Les autorités romaines semblent ainsi absentes du règlement militaire du conflit, comme du traité de paix, malgré leur présence dans l'établissement d'un traité entre Pharnace et Chersonèse.[70] La seule immixtion décisive de Rome a parfois été imputée à Flamininus, qui aurait convaincu Séleucos IV de ne pas intervenir.[71] Néanmoins, Flamininus n'était pas envoyé à la cour séleucide pour cette raison. Dans tous les cas, en l'absence de sources contredisant

---

68   Diod. 29 F 27; Polyb. F 96 Büttner-Wobst = *Souda s.v.* ἀκέραιος. Cette continuité des principes géopolitiques séleucides en Asie Mineure avant et après 188 a été soulignée par Chrubasik 2013, 105–116.

69   Sur la nouveauté diplomatique représentée par ces envois réguliers d'ambassades au Sénat, voir en dernier lieu Eckstein 2008, 352s.

70   Polyb. 25.2.

71   Diod. 29 F 27; Polyb. F 96 Büttner-Wobst = *Souda s.v.* ἀκέραιος. Voir Hopp 1977, 34; McGing 1986, 28; Will 1982, 289; Petković 2012, 362, faisaient de Flamininus l'agent romain du changement de plan de Séleucos. Chrubasik 2013, 109, restait prudent.

Polybe, l'avis de ce dernier garde tout son poids: Rome se fit remarquer par son manque d'entrain à faire cesser cette guerre.[72]

2. Le nouvel axe Pharnace – Chersonèse sous influence romaine, tournant géopolitique pour le Pont nord ou épisode circonstanciel?

L'aspect le plus surprenant et difficile à interpréter dans la guerre de 182–179 est la présence de cités et d'un dynaste du Pont-Nord parmi les *adscripti* du traité, ainsi que les liens éventuels avec le traité conclu par Pharnace avec Chersonèse taurique.[73] Le manque d'informations sur la configuration politique de la région rend toute tentative de reconstruction ardue et il est nécessaire de procéder par degré de plausibilité. L'étude des sources archéologiques ne permet pas de dater précisément les vagues de migration majeures, en particulier en ce qui concerne les Sarmates. Ce qui peut en être déduit, en revanche, est que l'ordre géopolitique de cette région était peu hiérarchisé, Chersonèse et les autres cités autonomes faisant reposer leur défense sur des alliances changeantes avec les rois du Bosphore cimmérien, installés à l'Est autour de la cité de Panticapée, les rois de Scythie Mineure, sur le Bosphore européen dans l'arrière-pays de Chersonèse, et différentes dynasties établies plus au Nord, qualifiées de Sarmates par les Grecs, peut-être en raison du caractère nomade ou semi-nomade de ces peuples. Le dynaste mentionné par Polybe dans le traité de 179 pourrait donc aussi bien être un allié qu'un ennemi de Chersonèse, et le terme 'Sarmate' qui lui est associé ne peut être considéré sans réserve comme un marqueur ethnique spécifique.[74] Dans l'immédiat, l'action de Gatalos put avoir un rapport avec la migration des populations bastarnes, qui arrivèrent à la frontière macédonienne, à la demande de Philippe V, au moment de la mort de ce dernier, en cette même année 179.[75] Il est notoire que le roi antigonide avait mené une politique agressive contre les peuples thraces, les attaquant en 184, 183 et 181, ce qui avait peut-être créé une vacance du pouvoir et des troubles locaux qui vinrent perturber les réseaux diplomatiques des belligérants anatoliens.[76]

Pour en revenir au traité conclu entre Pharnace et Chersonèse, il apparaît que sa conclusion venait d'une initiative des Chersonésitains, qui, sous la menace des 'barbares du voisinage', durent faire appel à un protecteur intéressé à la bonne

72 McShane 1964, 155–158 et 163; Will 1982, 289s.; Eckstein 2008, 355s.
73 Sur Chersonèse à l'époque hellénistique, voir Zolotarev 2003, 619–624, qui usait de sources archéologiques en premier lieu.
74 Müller 2010, 72, n. 78.
75 Polyb. 23.8.2–7; Liv. 41.19.3.
76 Sur la politique septentrionale de Philippe V après la guerre antiochique, Polyb. 23.8.2–7; Liv. 39.35.4; 39.53.12–16; 40.21.1–4; 40.22; 40.57s.; 41.19.3; Oros. 4.20; Just. 32.3.5. Voir Walbank 1979, 223–257; Will 1982, 252s.; Gruen 1984, 399–402; Hammond & Walbank 1988, 468–470.

tenue d'un port commercial et d'une tête de pont au Nord du Pont-Euxin.[77] L'identité de ces παρακείμενοι βάρβαροι n'est pas évidente, mais il pourrait s'agir des soldats du roi de Petite Scythie, étant donnée la proximité entre Néapolis et Chersonèse, ou des Sarmates. Dans le cas présent, la première option est à privilégier, puisque les Scythes étaient effectivement les voisins de Chersonèse.[78] Cela pourrait impliquer un conflit entre, d'un côté, le roi de Petite Scythie, et peut-être son allié le roi du Bosphore, de l'autre, Chersonèse et le dynaste sarmate. Ces alignements peuvent être suggérés par le témoignage de Polyen, faisant état d'une autre attaque scythe contre Chersonèse repoussée grâce à l'intervention de la reine sarmate Amagè.[79] Néanmoins, une autre possibilité (soit moins probable) n'est pas à écarter non plus: celle d'un affrontement entre Sarmates et Chersonésitains. Bien qu'isolé, ce document témoigne du rôle tenu par les populations 'scythiques' dans l'expansion de l'influence dynastique pontique au Nord du Pont-Euxin, quand bien même le roi du Pont n'eut pas l'initiative de cette alliance. En réalité, la démarche des Chersonésitains témoigne à la fois de l'instabilité de la scène politique locale et de l'absence d'une puissance dominante à laquelle s'adresser en priorité: Eumène s'était mis en retrait dans la scène pontique, Rhodes y avait perdu de sa superbe et exerçait une influence économique avant tout, Prusias restait tourné vers l'Anatolie centrale et les Détroits. Le cas de Rome mérite pour sa part un examen particulier.

Le texte du traité apporte la preuve d'une reconnaissance locale de l'influence de Rome, dont la *philia* est assurée aux deux puissances.[80] Il n'y a pas lieu de surestimer les clauses de maintien de l'*amicitia* romaine au point de faire de Rome la puissance hégémonique dans la région, une déduction qui ne correspondrait pas avec le développement de l'impérialisme romain, encore balbutiant et sans répercussions majeures en mer Noire.[81] En revanche, il est probable que les autorités sénatoriales aient joué un rôle dans les négociations de paix, profitant de l'occasion pour établir des relations avec les parties impliquées. Il s'agissait aussi d'une grande puissance neutre, plus à même de servir de garant pour Chersonèse sans froisser l'orgueil de Pharnace ou lui soumettre un choix stratégique, à l'inverse de ce qu'aurait représentée une intervention attalide dans le processus. Rome, pour sa part, restait une grande puissance lointaine et sans velléités d'intervention, mais son statut hégémonique prenait déjà forme dans une région qui n'avait pas subi sa puissance militaire.

En dernier lieu, on peut se poser la question de l'impact de cette mise en place géopolitique de 179 en mer Noire, ainsi que de son maintien dans les décennies suivantes. En effet, si la datation haute de l'inscription *IOSPE* I² 402 fait encore

---

77 *IOSPE* I² 402, l. 15.
78 Sur l'alignement des différents *adscripti* durant le conflit, voir Heinen 2005, 44–47.
79 Polyen 8.56. Le terme utilisé par Polyen correspond d'ailleurs en partie à celui que l'on peut trouver dans le texte du traité entre Chersonèse et Pharnace: οἱ παρακείμενοι Σκύθαι. Voir Müller 2010, 72s. et 94.
80 *IOSPE* I² 402, l. 3s. et 26s.
81 Heinen 2005, 43s.; Avram 2016, 218–220.

débat aujourd'hui, c'est en partie du fait de l'absence de témoignage d'une influence romaine ou mithridatide dans le Pont-Nord jusqu'au règne de Mithridate VI Eupatôr, dans les dernières années du IIe siècle. Si cela peut laisser entendre que l'établissement de nouveaux réseaux et d'un nouvel équilibre en 179 ne fut pas suivi de faits pendant plus d'un demi-siècle, il peut également s'agir d'un effet de source. De fait, l'histoire du Pont-Euxin comme celle du royaume du Pont avant la guerre entre Rome et Mithridate VI restent largement méconnus, faute d'œuvres littéraires ou d'inscriptions correspondantes. Pour ce qui est de Rome, il est en revanche probable que le Sénat ne démarcha pas beaucoup plus dans la région, faute d'agents locaux ayant régulièrement fait appel à lui.[82]

Les Rhodiens, de leur côté, ne semblent pas avoir tenté d'intervenir sur le plan politique, fragilisés qu'ils furent dans les décennies suivantes par les révoltes lyciennes et cariennes, puis par le retournement des faveurs romaines.[83] En revanche, le Pont-Nord, plus précisément Olbia et le royaume du Bosphore, restèrent des destinations commerciales importantes tout au long du IIe siècle pour les Rhodiens. Cette influence économique semble s'être répercutée sur la politique monétaire de certaines cités, qui adoptèrent des modèles iconographiques propres aux monnaies rhodiennes et autrement inconnus dans la région.[84] En revanche, il peut être noté que Chersonèse, la nouvelle alliée de Pharnace, ne faisait pas partie des cités sous influence économique rhodienne.

En tout état de cause, même sans avoir à présumer des liens non documentés par les sources entre la dynastie mithridatide et le Pont-Nord, le maintien de l'alliance est avéré par l'intervention de Mithridate VI vers 114.[85] Les conséquences de ce nouveau développement historique s'avérèrent rapides et de grande ampleur, puisque Mithridate en profita pour se constituer un empire sur le Pont-Euxin, qu'il utilisa comme un socle dans la guerre contre Rome.

82 Il ne peut cependant pas être exclu que des appels au sénat romain furent entrepris, sans avoir laissé de traces dans les sources. En effet, l'exemple de la Judée montre qu'en l'absence de certaines sources locales, judéennes dans ce cas précis, beaucoup d'échanges diplomatiques entre Rome et des puissances moyennes nous seraient inconnues. Voir la récente étude de ces sources par Coşkun 2019c.

83 Sur la Lycie et la Carie: Bresson 1999, 106–118. Sur Rome et Rhodes après la troisième guerre de Macédoine: Gruen 1975, 79–81; Gruen 1984, 569–572; Eckstein 1988, 415–424; Ager 1991, 29–37.

84 Shelov 1978, 170s.; Berthold 1984, 205–207; Frolova 1995, 9s.; Gabrielsen 1999, 64–71; Finkielsztejn 2001b, 187–190; Müller 2010, 263s.; Lund 2011, 288s.

85 Just. 37.3.1–3; Strab. Geogr. 7.3.17–4.4 (421–427C); IOSPE, I² 352 et 353. Sur cette guerre, voir Ballesteros Pastor 1996, 45–51; Müller 2010, 95–99, avec de nombreuses références bibliographiques.

Bibliographie – Sources anciennes

Diodore, *Bibliothèque historique. Fragments. Livres XXVII–XXXII*, texte établi et traduit par P. Goukowski, Paris 2012.

Justin, *Abrégé des histoires philippiques de Trogue Pompée et prologues de Trogue Pompée*, texte établi et traduit par É. Chambry & L. Thély-Chambry, vol. 2, Paris 1936.

Polybe, *Histoire*, texte traduit par D. Roussel, sous la direction de F. Hartog, Paris 2003.

Polybii *Historiae*, ed. Th. Büttner-Wobst, Leipzig, 5 vols., 2ème éd., 1893–1904.

Strabon, *Géographie, tome VIII (Livre XI)*, texte établi et traduit par F. Lasserre, Paris 1975 [réimpr. Paris, 2003].

Tite-Live, *Histoire romaine, Tome XXVIII, Livre XXXVIII*, texte établi et traduit par R. Adam, Paris 1982 [réimpr. Paris, 2003].

Bibliographie – Littérature moderne

Ager, S.L. 1991: 'Rhodes: The Rise and Fall of a Neutral Diplomat', *Historia*, 40, 10–41.

Ameling, W. 1985: *Die Inschriften von Prusias ad Hypium (IK 27)*, Bonn.

Arrayás Morales, I. 2014: 'La integración del Mar Negro en el mundo romano (ss. II–I a.C.)', *Latomus* 73, 938–967.

Avram, A. 2016: 'On the Date of the Treaty Between Pharnaces and Tauric Chersonese', in J.Ch. Couvenhes (ed.), *La* symmachia *comme pratique du droit international dans le monde grec. D'Homère à l'époque hellénistique*, Besançon, 213–237.

Ballesteros Pastor, L. 1996: *Mitrídates Eupátor, rey del Ponto*, Granada.

Ballesteros Pastor, L. 2016: 'The Satrapy of Western Armenia in the Mithridatid Kingdom', in V. Cojocaru & A. Rubel (eds.), *Mobility in Research on the Black Sea. (Iaşi, July 5–10, 2015)*, Cluj-Napoca, 273–287.

Barat, C. 2006: *Sinope dans son environnement pontique*, PhD Bordeaux.

Barat, C. 2009: 'Sinope et ses relations avec la péninsule anatolienne: réseaux, échanges des biens et des hommes', in H. Bru, F. Kirbihler & St. Lebreton (eds.), *L'Asie Mineure dans l'Antiquité: échanges, populations et territoires. Regards actuels sur une péninsule, Actes du colloque international de Tours 21–22 octobre 2005*, Rennes, 351–375.

Berthold, R.M. 1984: *Rhodes in the Hellenistic Age*, Ithaca, 203–207.

Bikerman, E. 1937: 'Notes sur Polybe I. Le statut des villes d'Asie après la paix d'Apamée', *REG* 50, 217–239.

Braund, D. (ed.) 2005: *Scythians and Greeks. Cultural Interactions in Scythia, Athens and the Early Roman Empire (Sixth Century BC–First Century AD)*, Exeter.

Bresson, A. 1999: 'Rhodes and Lycia in Hellenistic Times', in V. Gabrielsen (ed.), *Hellenistic Rhodes: politics, culture, and society*, Aarhus, 98–131.

Bresson, A., Ivantchik, A. & Ferrary, J.-F. (eds.) 2007: *Une koinè pontique. Cités grecques, sociétés indigènes et empires mondiaux sur le littoral nord de la mer Noire (VIIe s. a. C.–IIIe s. p. C.)*, Bordeaux.

Burstein, S.M. 1980: 'The Aftermath of the Peace of Apamea. Rome and the Pontic War', *AJAH* 5, 1–12.

Bylkova, V.P. 2005: 'Lower Dnieper Region as an Area of Greek-Barbarian Interaction', in D. Braund (ed.), *Scythian and Greeks. Cultural Interactions in Scythia, Athens and the Early Roman Empire (sixth century BC–first century AD)*, Exeter, 131–147.

Chrubasik, B. 2013: 'The Attalids and the Seleukid Kings, 281–175 bc', in P. Thonemann (ed.), *Attalid Asia Minor. Money, International Relations, and the State*, Oxford, 83–119.

Coşkun, A. (ed.), *APR: Amici Populi Romani* (Trèves 2007–2008 et Waterloo 2010–2019). URL: http://www.altaycoskun.com/apr.

Coşkun, A. 2016: ,Philologische, genealogische und politische Überlegungen zu Ardys und Mithradates, zwei Söhnen des Antiochos Megas (Liv. 33,19,9)', *Latomus* 75, 849–861.

Coşkun, A. 2019a: 'The "Temple State" of Phrygian Pessinus in the Context of Seleucid, Attalid, Galatian and Roman Hegemonial Politics (3rd–1st Centuries BC)', in G.R. Tsetskhladze (ed.), *Phrygia in Antiquity: From the Bronze Age to the Byzantine Period*, Leuven, 607–648.

Coşkun, A. 2019b: 'Pontic Athens. An Athenian Emporion in its Geo-Historical Context', *Gephyra* 18, 11–31.

Coşkun, A. 2019c: 'Triangular Epistolary Diplomacy with Rome from Judas Maccabee to Aristobulos I', in Coşkun & Engels 2019, 355–388.

Coşkun, A. & Engels, D. (eds.) 2019: *Rome and the Seleukid East. Selected Papers from Seleukid Study Day V, Université libre de Bruxelles, 21–23 Aug. 2015*, Bruxelles

Counillon, P. 1990: 'Arrien et Kérasous: un cas de toponymie rétroactive', *1er Congrès International sur la mer Noire. I. 1–3 juin 1988*, Samsun, 493–500.

Dan, A. 2009: 'Sinope, "capitale" pontique, dans la géographie antique', in H. Bru, F. Kirbihler & St. Lebreton (eds.), *L'Asie Mineure dans l'Antiquité: échanges, populations et territoires. Regards actuels sur une péninsule, Actes du colloque international de Tours 21–22 octobre 2005*, Rennes, 67–131.

Debord, P. 2001: 'Les Mysiens: du mythe à l'histoire', in V. Fromentin & S. Gotteland (eds.), *Origines Gentium*, Bordeaux, 135–146.

Dédéyan, G. (ed.) 2007: *Histoire du peuple arménien*, Toulouse.

Dmitriev, S. 2007: 'Memnon on the Siege of Heraclea Pontica by Prusias I and the War between the Kingdoms of Bithynia and Pergamum', *JHS* 127, 133–138.

Doonan, O. 2003: 'Sinope', in Grammenos & Petropoulos 2003, 1379–1401.

Doonan, O. 2004: *Sinop Landscapes. Exploring Connection in a Black Sea Hinterland*, Philadelphie.

Dörner, F.K. 1952: *Bericht über eine Reise in Bithynien ausgeführt im Jahre 1948 im Auftrage der Österreichischen Akademie der Wissenschaften*, Vienne.

Eckstein, A.M. 'Rome, the War with Perseus, and Third Party Mediation', *Historia* 37, 414–444.

Eckstein, A.M. 2008: *Rome enters the Greek East. From Anarchy to Hierarchy in the Hellenistic Mediterranean, 230–170 BC*, Oxford.

Facella, M. 2006: *La dinastia degli Orontidi nella Commagena ellenistico-romana*, Pise.

Fernoux H.-L. 2004: *Notables et élites des cités de Bithynie aux époques hellénistique et romaine (IIIe siècle av. J.-C.–IIIe siècle ap. J.-C.)*, Lyon.

Fernoux, H.-L. 2008: 'Rivalité politique et culturelle entre les royaumes de Pergame et de Bithynie', in M. Kohl (ed.), *Pergame. Histoire et archéologie d'un centre urbain depuis ses origines jusqu'à la fin de l'Antiquité*, Lille, 223–244.

Ferrary, J.-L. 1988: *Philhellénisme et impérialisme: aspects idéologiques de la conquête romaine du monde hellénistique, de la seconde Guerre de Macédoine à la Guerre contre Mithridate*, Rome.

Ferrary, J.-L. 2007: 'L'essor de la puissance romaine dans la zone pontique', in Bresson, Ivantchik & Ferrary 2007, 319–325.

Finkielsztejn, G. 2001a: *Chronologie détaillée et révisée des éponymes amphoriques rhodiens de 270 à 105 av. J.-C. environ. Premier bilan*, Oxford.

Finkielsztejn, G. 2001b: 'Politique et commerce à Rhodes au IIe s. a.C.: le témoignage des exportations d'amphores', in A. Bresson & R. Descat (eds.), *Les cités d'Asie Mineure occidentale au IIe siècle a.C.*, Bordeaux, 181–196.

Frankfort, Th. 1963: 'La Sophène et Rome', *Latomus* 22, 181–190.

Frolova, N.A. 1995: 'On the Monetary Circulation of the Bosporus in the 3rd Century BC', in N.A. Frolova (ed.), *Essays on the Northern Black Sea Region Numismatics*, Odessa, 2–16.

Gabrielsen, V. (ed.) 1999: *Hellenistic Rhodes: Politics, Culture, and Society*, Aarhus.

Giovannini, A. 1982: 'La clause territoriale du traité d'Apamée', *Athenaeum* 70, 224–236.

Grainger, J.D. 2002: *The Roman War of Antiochos the Great*, Leyde.

Grammenos, D.V. & Petropoulos, E.K. (eds.) 2003: *Ancient Greek colonies in the Black Sea*, Thessalonique.

Gruen, E.S. 1975: 'Rome and Rhodes in the Second Century BC: A Historiographical Inquiry', *CQ* n.s. 25, 58–81.

Gruen, E.S. 1984: *The Hellenistic World and the Coming of Rome*, Berkeley.

Habicht, Ch. 1956: ‚Über die Kriege zwischen Pergamon und Bithynien', *Hermes* 84.1, 90–110.

Habicht, Ch. 1989: 'The Seleukids and their rivals', *Cambridge Ancient History*, 2ème éd., vol. 8, 324–387.

Hammond, N.G.L. & Walbank, F.W. 1988: *A History of Macedonia*, III, Oxford.

Heinemann, U. 2010: *Stadtgeschichte im Hellenismus. Die lokalhistoriographischen Vorgänger und Vorlagen Memnons von Herakleia*, Munich.

Heinen, H. 2005: ‚Die Anfänge der Beziehungen Roms zum nördlichen Schwarzmeerraum. Die Romfreundschaft der Chersonesiten (*IOSPE* I², 402)', in A. Coşkun (ed.), *Roms auswärtige Freunde in der späten Republik und im frühen Prinzipat*, Göttingen, 31–54.

Hewsen, R.H. 1985: 'Introduction to Armenian Historical Geography IV: the boundaries of Artaxiad Armenia', *REArm* 19, 55–84.

Højte, J.M. 2005: 'The Date of the Alliance between Chersonesos and Pharnakes (*IOSPE* I², 402) and its Implications', in V.F. Stolba & L. Hannestad (eds.), *Chronologies of the Black Sea Area in the Period c. 400–100 BC*, Aarhus, 137–152.

Holleaux, M. 1938: *Études d'épigraphie et d'histoire grecques II. Études sur la monarchie attalide*, Paris.

Hopp, J. 1977: *Untersuchungen zur Geschichte der letzten Attaliden*, Munich.

Jones, A.H.M. 1937: *The Cities of the Eastern Roman Provinces*, Oxford.

Jones, A.H.M. 1940: *The Greek City: From Alexander to Justinian*, Oxford.

Keaveney, A. & Madden, J.A. 2011: 'Memnon', *Brill's New Jacoby* 434, 1–70.

Khrapunov, I.N. 2004: 'Etnicheskaya istoria Krÿma v rannem zheleznom veke', *Bosporskie issledovaniya. Vyp. VI.* (The Crimea in the Early Iron Age: an Ethnic History, BI 6), Simferopol.

Koehn, C. 2007: *Krieg – Diplomatie – Ideologie. Zur Außenpolitik hellenistischer Mittelstaaten*, Stuttgart.

Kosmetatou, E. 2003: 'The Attalids of Pergamon', in A. Erskine (ed.), *A Companion to the Hellenistic World*, Oxford, 159–174.

Lebedynsky, I. 2001: *Les Scythes*, Paris.

Leschhorn, W. 1984: *'Gründer der Stadt': Studien zu einem politisch-religiösen Phänomen der griechischen Geschichte*, Stuttgart.

Leschhorn, W. 1993: *Antike Ären. Zeitrechnung, Politik und Geschichte im Schwarzmeerraum und in Kleinasien nördlich des Tauros*, Stuttgart.

Lund, J. (2011): 'Rhodian Transport Amphorae as a Source for Economic Ebbs and Flows in the Eastern Mediterranean in the Second Century BC', in Z.H. Archibald, J.K. Davies & V. Gabrielsen (eds.), *The Economies of Hellenistic Societies, third to first Centuries BC*, Oxford, 280–295.

Ma, J. 2013: 'The Attalids: A Military History', in P. Thonemann (ed.), *Attalid Asia Minor. Money, International Relations, and the State*, Oxford, 49–82.

Magie, D. 1950: *Roman Rule in Asia Minor to the End of the Third Century after Christ*, 2 vols., Princeton.

Marek, C. 1993: *Stadt, Ära und Territorium in Pontus-Bithynia und Nord-Galatia*, Tübingen.

McAuley, A. 2019: 'L'ombre lointaine de Rome: la Cappadoce à la suite du traité d'Apamée', in Coşkun & Engels 2019, 309–332.

McGing, B.C. 1986: *The Foreign Policy of Mithridates VI Eupator King of Pontos*, Leyde.

McShane, R.B. 1964: *The Foreign Policy of the Attalids of Pergamum*, Urbana.

Michels, Ch. 2009: *Kulturtransfer und monarchischer 'Philhellenismus': Bithynien, Pontos und Kappadokien in hellenistischer Zeit*, Göttingen.

Michels, Ch. 2019: 'Unlike any Other? The Attalid Kingdom after Apameia', in Coşkun & Engels 2019, 333–352.

Mørkholm, O. 1966: *Antiochus IV of Syria*, Copenhague.

Müller, Ch. 2010: *D'Olbia à Tanaïs: territoires et réseaux d'échanges dans la mer Noire septentrionale aux époques classique et hellénistique*, Bordeaux.

Patterson, L. 2001: 'Rome's Relationship with Artaxias I of Armenia', *AHB* 15.4, 154–162.

Payen, G. 2019: 'L'influence séleucide sur les dynasties anatoliennes après le traité d'Apamée', in Coşkun & Engels 2019, 279–307.

Payen, G. 2020: *Dans l'ombre des empires. Les suites géopolitiques du traité d'Apamée en Anatolie*, Québec.

Petković, Ž. 2012: 'The Aftermath of the Apamean Settlement: Early Challenges to the New Order in Asia Minor', *Klio* 94, 357–365.

Pirson, F. 2004: 'Elaia, der maritime Satellit Pergamons', *MDAI(A)* 54, 197–213.

Reinach, Th. 1905: 'A Stele from Abonuteichos', *Numismatic Chronicles* 4 (5), 113–119.

Robert, L. 1937: *Études anatoliennes. Recherches sur les inscriptions grecques de l'Asie Mineure*, Paris.

Robert, L. 1980: *À travers l'Asie Mineure: poètes et prosateurs, monnaies grecques, voyageurs et géographie*, Paris.

Roller, D.W. 2020: *Empire of the Black Sea: the Rise and Fall of the Mithridatic World*, Oxford.

Rostovtzeff, M.I. 1930: 'The Bosporan Kingdom', *Cambridge Ancient History*, vol. 1, 561–589.

Savalli-Lestrade, I. 2010: 'Les rois hellénistiques, maîtres du temps', in I. Cogitore & I. Savalli-Lestrade (eds.), *Des Rois au Prince. Pratiques du pouvoir monarchique dans l'Orient hellénistique et romain*, Grenoble, 55–83.

Shcheglov, A.N. 1992: *Polis et chôra. Cité et territoire dans le Pont Euxin*, Besançon.

Shelov, D.B. 1978: *Coinage of the Bosporus VI–II Centuries BC*. BAR Int. Ser. 46, Oxford.

Sherwin-White, A.-N. 1984: *Roman Foreign Policy in the East, 168 BC to AD 1*, Londres.

Stolba, V. E. 2005: 'The Oath of Chersonesos (*IOSPE* I² 401) and the Chersonesean Economy in the Early Hellenistic Period', in Z.H. Archibald, J.K. Davies, V. Gabrielsen & G.J. Oliver (eds.), *Hellenistic Economies*, Londres, 298–321.

Syme, R. 1995: *Anatolica: Studies in Strabo*, Oxford.

Traina, G. 1999/2000: 'Épisodes de la rencontre avec Rome (IIe siècle av. J-C.–IIIe siècle ap. J.-C.)', *Iran & Caucasus* 3/4, 59–78.

Tsetskhladze, G. 1999: *Pichvnari and Its Environs, 6th c BC–4th c AD*, Paris.

Vinogradov, J.G. & Shcheglov, A.N. 1990: 'Obrazovanie territorial'nogo khersonesskogo gosudarstva' (The Formation of the territorial State of Chersonesos), in E.S. Golubtsova (ed.), *Éllinizm: ékonomika, politika, kul'tura* (Hellenism: Economics, Politics, Culture), Moscou, 310–371.

Vinogradov, J.G. 1997: 'Khersonesskiï dekret o nesenii Dionisa (*IOSPE* I² 343) I sarmatskoe vtorzhenie v Skifiyu' (The Chersonesian Decree on Carrying Dionysos, *IOSPE* I² 343, and the Sarmatian Invasion to Scythia), *VDI* 1997.3, 104–124.

Vitucci, G. 1953: *Il regno di Bitinia*, Rome.

Waddington, W.H., Babelon, E. & Reinach, Th. 1925: *Recueil général des monnaies grecques d'Asie Mineure*, 2ème éd., vol. I.1, Paris.

Walbank, F.W. 1979: *A Historical Commentary on Polybius*, vol. 3, Oxford.

Will, E. 1982: *Histoire politique du monde hellénistique*, vol. 2, 2ème éd., Nancy.

Zolotarev, M.I. 2003: 'Chersonesus Tauricus', in Grammenos & Petropoulos 2003, 603–644.

THE RETURN OF THE KING

Pharnakes II and the Persian Heritage

Luis Ballesteros Pastor

Abstract: Pharnakes II is mainly known for his war against Julius Caesar, but also for having been the instigator of the uprising which led his father Mithradates VI Eupator to suicide. Some sources even directly attribute the death of the great Pontic king to his son. As is well known, Pompey acknowledged Pharnakes as ruler of the Bosporus after the Mithradatic Wars in 63 BC, and he would reign over Kolchis as well at some point. In 48 BC, he went back to Anatolia and seized the old domains of his father. This chapter analyses how Pharnakes tried to be seen as a great Achaimenid ruler, following the official genealogy established in the court of Pontos which traced back the lineage of the Mithradatids to Cyrus and Darius the Great. Pharnakes' return to the realm of his ancestors represented an exaltation of the Iranian roots of this king. One may gain the impression that this particular emphasis on his genealogical and ideological connections with the ancient Persian kings was also intended to set him apart from his (uneasy) father.

Абстракт: Возвращение царя: Фарнак II и Ахеменидская традиция: Фарнак II известен в основном своей войной против Юлия Цезаря, а также тем, что подстрекал к восстанию, которое довело его отца Митридата VI Евпатора до самоубийства. Существуют даже некоторые источники, которые напрямую приписывают смерть великого понтийского царя его сыну. Как хорошо известно, Помпей признал Фарнака правителем Боспора после Митридатовых войн в 63 г. до н.э., и чуть позже он также начал править Колхидой. В 48 г. до н.э. он вернулся в Анатолию и захватил старые владения своего отца. В этой главе автор анализирует, как Фарнак пытался сделать так, чтобы его считали великим Ахеменидским правителем, следуя за официальной генеалогией, установленной при понтийском дворе, которая прослеживает происхождение рода Митридата до Кира и Дария Великого. Возвращение Фарнака в царство его предков представляло собой возвышение иранских корней этого царя. Может сложиться впечатление, что этот особый акцент на его генеалогические и идеологические связи с древними персидскими царями был также положен, чтобы отделить Фарнака от его (проблематичного) отца.

I. INTRODUCTION

With the death of Mithradates Eupator in 63 BC,[1] the Persian world seemed to be definitely subjugated by the West. As in the case of Antiochus III, the Pontic king

---

1   On Eupator's suicide, see App. *Mith.* 111.537–539; Plut. *Pomp.* 41.5; Cass. Dio 37.10.4; Flor. 1.40.26; Liv. *Per.* 102; Oros. 6.5.6f.; *Vir. Ill.* 76.8; Eutr. 6.12.3; Fest. *Brev.* 16.1; Just. 37.1.9; Gell. *NA* 17.16.5; Paus. 3.23.5; Val. Max. 9.2 ext. 3; Servilius Damocrates, *Theriaca* 101–106 (*Poetae Bucolici et Didactici* ed. Didot, vol. 3, p. 120); Galenos, *De Theriaca* ed. Kühn, vol.

had appeared before the eyes of the Romans as a reincarnation of the Achaimenids.[2] Pompey, as a new Alexander, emerged as the dominator of the hereditary enemies of western civilization. But immediately upon this success of the Republic, various Oriental monarchies were striving to extol their Persian heritage and, in certain cases, their Achaimenid descent. Paradoxically, the phenomenon of so-called 'Persianism' was reinvigorated around the same time: the rulers of Armenia, Kappadokia, Kommagene or Media Atropatene stressed their Iranian ancestry as a source of pride.[3] They were aware that Rome exerted a decisive influence over their respective territories, although it should also be borne in mind that the Parthians defeated the Roman Republic in 53 BC, a remarkable event that may well have changed some points of view.[4] In this context, Pharnakes II played an outstanding role, because he defied Rome and enjoyed an ephemeral glory, like a lightning flash, before being defeated by Caesar (48–47 BC). This king did not just reconquer the territories formerly ruled by his father in Anatolia,[5] but he was

16, 283f.; Suda *s.v.* Πομπήϊος; Ὦ φῶς; Φύσει; cf. Luc. *Phars.* 1.335f.; Mart. 5.76; Juv. *Sat.* 6.661f.; 14.252; *Schol. Juv. Sat.* 6.661, 10.273; Sidon. *Carm.* 7.79–82. Pharnakes is accused of having murdered his father by Cass. Dio 37.12.4; 37.14.2; App. *BCiv.* 2.13.92; Jos. *AJ* 14.3.4 (53); Zonar. 5.6; 10.5; *Schol. Bern. Lucan.* 1.336 (cf. Vell. 2.40.10); *Schol. Juv. Sat.* 14.252.2. Cassius Dio followed two different sources on Mithradates' end, because he reports that the king committed suicide and was murdered by his enemies: see Portanova 1988, 516 n. 794. On the revolt, see Reinach 1890, 406–410; Gajducevič 1971, 321f.; Ballesteros Pastor 1996, 279–281; Goukowsky 2001, 245–247, ns. 1036–1047; Abramzon & Kuznetsov 2011. On Pharnakes' image in literary sources, see in general Ballesteros Pastor 2005b.

2   On the Mithradatic conflict as an evocation of the Persian Wars, see Ballesteros Pastor 2011. In general, on the persistence of the Persian Wars in Roman memory, see Spawforth 1994; Hardie 2007; Russo 2014a; 2014b; Makhlaiuk 2015; Almagor 2019.

3   See Ballesteros Pastor 2018a, 148f. and *passim*. On these monarchies in general, see Hoben 1969; Sullivan 1990. On the debate around the concepts of 'Persianization' and 'Persianism', see Strootman & Versluys 2017. Regarding Persianism in the Mithradatid dynasty, see Canepa 2010, 11–13; 2014, 61; 2017, 217–222; Lerouge-Cohen 2017; Gatzke 2019; Strootman 2020, 205–210; cf. McGing 2014; Michels 2017. On the vindication of Achaimenid descent by several eastern royal houses, see further Ballesteros Pastor 2012; 2018a, 174–176; 2018b, 273, 276; Lerouge-Cohen 2016; 2017.

4   On the uncertain situation of the eastern dynasties in regards with Roman rule towards the middle of the 1st century BC, see Coşkun 2008; Strootman 2016, 225.

5   Pharnakes defeated Cn. Domitius Calvinus, who probably was a legate of Caesar: *BAlex.* 34–40; 65.4; 69.1; 74.3; Cic. *Deiot.* 14; 24; Liv. *Per.* 112; Plut. *Caes.* 50.1; Suet. *Jul.* 35.2; 36; App. *Mith.* 120.591; *BCiv.* 2.13.91; Cass. Dio 42.46.1f.; 42.47.1. Cf. Sweeney 1978, 183; Carlsen 2008, 74; Gaertner & Hausburg 2013, 88. On Pharnakes' war with Caesar, see *BAlex.* 34–40; 65; 69–77; Cass. Dio 47.1–3; Plin. *NH* 6.3.10; App. *Mith.* 120.593f.; *BCiv.* 2.13.91; Plut. *Caes.* 50.2; *Mor.* 206e; Liv. *Per.* 113; Flor. 2.3.11; Suet. *Jul.* 37; Strab. *Geogr.* 12.3.14 (547C); Frontin. *Str.* 2.2.3; Luc. *Phars.* 10.475f.; Sen. *Suas.* 2.22; Eutr. 6.22.2; Oros. 6.16.3; *Vir. Ill.* 78.7; Ampel. 34.1. Cf. Hoben 1969, 17–25; Gajducevič 1970, 322–324; Sherwin-White 1984, 299f.; Sullivan 1990, 156–158; Freber 1993, 81–83; Heinen 1994; Dobesch 1996; Gaertner & Hausburg 2013, 91f., 101–104, 120f.; Coşkun 2019, 132–135. On Pharnakes' rule over Kolchis, see Cass. Dio 42.45.3; Strab. *Geogr.* 11.2.17 (498C); cf. Braund 1994, 147–149. For a comprehensive discussion of Pharnakes' campaigns, see now Coşkun ca. 2020.

also eager to be regarded as an heir of the Persian tradition. He not only meant to claim the inheritance of the Mithradatids, but he also aimed more generally at being recognized as a continuator of the Achaimenids. As Eupator had done before, Pharnakes regarded them as the former legitimate masters of Asia Minor.[6]

Pharnakes' position with respect to his father is ambiguous. He had led the revolt that made Mithradates commit suicide, and he was recognized as a friend and ally of the Roman people in return. Nevertheless, Pharnakes did not disown his roots: to some extent, he decided to appear as a continuator of his father's policy, but, at the same time, seemed to emphasize perhaps even more his Iranian descent. This policy, however, did not imply denying the Hellenic features of his reign, which are likewise evident. Thus, Pharnakes' rule appears before us as a complex combination of influences, whose interpretation is very difficult, even more so when considering the scarcity of the extant evidence for this monarch. His ephemeral rule in Anatolia was accompanied by a strong reminder of some old Achaimenid practices. They were not only designed to convey legitimacy to Pharnakes' conquest of Pontos, but also emphasized more broadly the Iranian tradition in which he wanted his rule to be viewed. This may be seen as a matter of pride before the Romans, as had been the case with Mithradates, and furthermore as an element of competition with other eastern monarchies. The very name *Pharnakes* was related to the concept of **Xvarenah*, the 'glory', 'majesty' and 'legitimacy' of the ancient Persian rulers.[7] It is no abject idea that his own name thus further inspired Pharnakes to seek the charisma and grandeur of his most famous ancestors, perhaps trying to make forgotten his inglorious accession to the throne this way.

In short, the present chapter argues that Pharnakes went beyond claiming legitimacy as king of Pontos through his Mithradatid lineage, but evoked the return of Achaimenid government over Anatolia.

II. AN ACHAIMENID RULER

The kingship of Pharnakes II had common features with that of his father. As a son of Mithradates, Pharnakes had a double genealogy, Persian and Hellenic, and this was likewise reflected in several facets of his reign. As in Eupator's case, Pharnakes' coins bear a Hellenized face of the king, who appears without beard and wears a diadem.[8] Just as his father, however, he wore a tiara before his subjects in the Bosporos. As Appian tells us, when Pharnakes rose against Eupator, his men put upon his head a sheet of papyrus, as if it were a tiara: that is, the sol-

---

6   On Eupator's perspective on Achaimenid rule over Anatolia and the Black Sea, see Ballesteros Pastor 2013a; 2013c, 280; 2018a, esp. 156f.
7   Melikian-Chirvani 1993, 28f.; cf. Gnoli 1999; De Jong 1999.
8   On Pharnakes' coins, see Gajducevič 1971, 322f.; Golenko & Karyzskovsky 1972; Zograph 1977, 301f. On Eupator's coins, see above all De Callataÿ 1997. On the double genealogy of the Pontic dynasty, see below, n. 10.

diers recognized in the new king a sovereign with one of the symbols of Persian royalty.[9]

Pharnakes must have kept alive the memory of the official Mithradatid genealogy, which was connected with the Achaimenid branch of Darius.[10] As we know, Mithradates VIII of the Bosporus solemnly claimed descent from Achaimenes (Tac. *Ann.* 12.18.2) about a century later, thus echoing a tradition which probably had been cherished for many generations. The exaltation of Achaimenid ancestry is clearly reflected in the names Pharnakes chose for his own offspring, among whom we only know two sons (Darius and Arsakes)[11] and a daughter (Dynamis).[12] We can trace back this practice to his father, though not yet to the beginning of his reign, when he chose more typically dynastic names, such as Mithradates and Ariarathes. But later, possibly while he was at war with Rome and designed a more outspoken Achaimenid dynastic propaganda, he begot his sons Cyrus, Darius and Xerxes; we may well wonder whether there also was a Kambyses, who would have escaped recording in our sources.[13]

In a certain way, Pharnakes imitated his father in his attempt to be regarded as the founder of a new line of Persian kings. The ruler of the Bosporus was thus competing with other royal houses who also claimed to belong to the Achaimenid house, precisely because of their kinship with Mithradates Eupator. This would be the case with the rulers of Armenia,[14] Kappadokia,[15] Atropatene,[16] Judaea[17] and

9   App. *Mith.* 111.533; cf. Ritter 1965, 163. On the other symbols, see Briant 1996, 187, 239f.
10 On this genealogy of the Pontic dynasty, see Just. 38.7.1; App. *Mith.* 9.27; 112.540; Sall. *Hist.* F 2.73 ed. Maurenbrecher *apud* Ampel. 30.5; Tac. *Ann.* 12.18.2. Cf. Panitschek 1987/8; Ballesteros Pastor 2012; 2013b, 272–285; Lerouge-Cohen 2017. The Kappadokian Ariarathids, who claimed descend from Darius as well (Diod. 31.19.1–3), established kinship ties with the Pontic house in the 2nd century BC, see Just. 38.2.5; App. *Mith.* 9.29; 10.31; 12.39; cf. Ballesteros Pastor 2014. See also below, n. 34.
11 Darius, Pharnakes' son, ruled over a part of Pontos for a brief time after Philippi: see App. *BCiv.* 5.8.75; Hoben 1969, 34–39; Olshausen 1980, 909f.; Sullivan 1990, 160f.; Primo 2010, 162. Arsakes: Strab. *Geogr.* 12.3.38 (560C); cf. Marek 1993, 50; Syme 1995, 172; also see below, with n. 62.
12 It is debated whether Dynamis was the daughter Pharnakes offered to Caesar (App. *BCiv.* 2.13.91). See Rostovtzeff 1919, 98; Hoben 1969, 32; Braund 1984, 178f. n. 79; Coşkun & Stern, chapter IX in this volume for a positive conclusion.
13 Ballesteros Pastor 2015a, esp. 436f. The name of his son (and vice-regent in the Bosporus until 65 BC) Machares is of uncertain etymology. Other known sons of Eupator born before his first war with Rome were named after former satraps of western Anatolia: Pharnakes, Xiphares (Oibares) and Artaphernes. Taking into account that Persian adulthood began at 24 (Strab. *Geogr.* 15.3.18 [733C]), we may understand why Appian (*Mith.* 108.513) called Cyrus, Darius, Xerxes and Oxathres 'handsome children' in 64 BC. See further Ballesteros Pastor 1996, 321–323.
14 Tigranes II of Armenia (ruled ca. 95–55 BC) married Kleopatra, daughter of Eupator: Just. 38.3.2; 40.1.3; App. *Mith.* 13.44; 15.54; 104.487; Plut. *Luc.* 14.6; 22.1, 5; Memn. *FGrH* 434 F 1.29.6; 31.1f.; Cass. Dio 36.50.1; Ballesteros Pastor 2013b, 207f.
15 The wife of Archelaos I of Kappadokia (ruled 36 BC–17 AD) belonged to the Armenian royal house: Aug. *RG* 27.2; cf. Sullivan 1990, 185; Ballesteros Pastor 2013b, 29.
16 The wife of Mithradates I of Atropatene was a daughter of Tigranes II: Cass. Dio 36.14.2; Plut. *Luc.* 31.7; cf. Sullivan 1990, 101; 283; 447 n. 12; Ballesteros Pastor 2020. The identity

Kommagene.[18] As regards the name Arsakes, it could reflect a change in Pharnakes' policy, implying an affiliation with the Parthians, with whom we cannot discard some (otherwise unattested) kinship ties.[19]

III. PHARNAKES, GREAT KING OF KINGS

According to the preserved inscriptions, as well as to the gold staters issued between 55/4 and 51/50 BC, Pharnakes adopted the title 'Great King of Kings'.[20] Konstantin V. Golenko and Pyotr J. Karyszkowski suggested that this coinage appeared coinciding with the death of the Great King Tigranes II of Armenia, to whose throne Pharnakes would have aspired.[21] However, we consider that this title was probably related to other factors, since Tigranes had male descendants who guaranteed the dynastic succession.[22] In fact, Pharnakes was resuming a title formerly held by his father, namely 'King of Kings', while, at the same time, omitting any Greek epithet on his coins, such as 'Philorhomaios', which he had borne formerly.[23] At any rate, Pharnakes ruled over a conglomerate of peoples settled around the Maiotis and the Kimmerian Bosporus: the dynasts of these tribes were described as 'kings', so the sovereigns of the Bosporus could justify the adoption of the title 'King of Kings'.[24] Thus, this use of the imperial titulature

and regnal years of this Mithridates are hard to determine: see further the assumptions by Schippmann 2014.

17 Glaphyra, Archelaos' daughter, espoused Alexander, prince of Judaea, ca. 17 BC: Sullivan 1980, 1161–1164; 1990, 185; Wilker 2007, 30; 52 n. 17; 59; 72.

18 Mithradates III of Kommagene married Jotapa, princess of Atropatene: Sullivan 1990, 299; 326. On the role of royal women in the transmission of kingship in Hellenistic times, see Strootman 2016, 219–221.

19 On Mithradates Eupator and the Parthians, see Olbrycht 2009; Lerouge-Cohen 2014, 144–148. On the Parthian dynastic name Arsakes and its relationship with the Achaimenids, see Olbrycht 2019. On Arsakid marriage policy in Hellenistic times, see Dąbrowa 2018.

20 *CIRB* 28; Griffiths 1953, 146; Yaïlenko 1985, 619–627; Golenko & Karyszkovsky 1972; Primo 2010, 160; Muccioli 2013, 276; Ballesteros Pastor 2017.

21 Golenko & Karyszkovsky 1972, 32.

22 Tigranes II was succeeded by his son Artavasdes II: Aug. *RG* 27; Strab. *Geogr.* 11.14.10 (530C); Jos. *AJ* 15.4.3 (104); Cass. Dio 40.16.2; Moses of Chorene 2.22.

23 On Eupator as King of Kings, see Ballesteros Pastor 1995; 2018a; Muccioli 2013, 409–412. On Pharnakes' title 'Philorhomaios', see Heinen 2001, 360; 2008, 192; Muccioli 2013, 270f. with n. 688; 276. Primo 2010, 160 does not consider the change in the coin legends.

24 Ballesteros Pastor 2017, 299. On the rule of the Bosporan monarchs over these peoples, see also Gourova 2014. Some of those dynasts were 'kings', see, for instance, *Syll.*³ 709, ll. 7 and 23; Polyain. 8.56; Strab. *Geogr.* 7.4.3 (309C); 11.5.8 (506C); Tac. *Ann.* 12.15.2; 12.20.1. For the interpretation of 'King of Kings' as 'ruler over other kings' in the late Hellenistic period, see App. *Syr.* 48.247, referring to Tigranes II, and further Cass. Dio 63.4–6; Amm. Marc. 19.2.11; cf. Strootman 2014, 52; 2016, 219–221; Ballesteros Pastor 2017, 299 with n. 11; also Coşkun 2020b, section 3, who emphasizes that Pharnakes' title has no clear anti-Roman implication. Some scholars, however, propose that this title was just a claim of legitimacy: Wiesehöfer 1996, 29; 56; 121; 133; cf. Muccioli 2013, 401–409 (with further bibliography).

probably had something to do with certain achievements over peoples of the northern Black Sea coast, which we cannot specify.

Besides, we must bear in mind the peculiarity of the title 'Great King of Kings' borne by Pharnakes II, Mithradates II of Parthia, Artavasdes I of Atropatene and perhaps Mithradates Eupator.[25] As we can infer from the monument at Nemrud-Dağı, *Basileus Basileon Megas* was the Greek term used in the Late Hellenistic period to designate the titulature of the old Achaimenid emperors.[26] Therefore, by proclaiming himself 'Great King of Kings', the ruler of the Bosporus was trying to exalt himself as the legitimate successor of the glorious Persian monarchs of the past. It is noteworthy that, unlike Eupator, Pharnakes included this title on his coins. This would be a proof of Pharnakes' eagerness to proclaim his Achaimenid descent. Mithradates Eupator, even on his staters issued between 89 and 85 BC, which count after a new era, avoided the title 'King of Kings' on his coins. In other words, he refrained from spreading this Persian titulature among his Greek or Hellenized subjects. In contrast, Pharnakes chose this title for his issues, thus following a more recent trend that had been established by the Parthian and Armenian rulers.

Just as under the Achaimenids and Alexander the Great, the title 'King of Kings' implied the dominion over the two parts of the world, that is, Europe and Asia. This was one of the principles of the imperial idea which had been formerly developed by the Achaimenids and would continue during the Hellenistic period.[27] By crossing the Hellespont from Asia to Europe, Darius I had proclaimed his ecumenic rule, thus confirming the universalistic dimension of his empire. Analogously, Alexander the Great thought he was conquering the two parts of the world, not only when he arrived in Asia in 334 BC, but also when he crossed the Jaxartes, allegedly returning to Europe from Asia.[28] In the case of Pharnakes, who ruled the Kimmerian Bosporus, this sovereignty over lands on the two continents was easy to argue and contributed to promoting an image of grandeur of the

25 On Pharnakes, see above n. 20. On Mithradates II, see *I.Délos* 1581; Muccioli 2013, 405f.; on Artavasdes, De Callataÿ & Lorber 2011, 438; Ballesteros Pastor 2017, 297f. ns.1 and 5 (with bibliography). Eupator's title has been reconstructed on some inscriptions: Yaïlenko 1985, 618; Arsen'eva, Böttger & Vinogradov 1995, 205–207; Ivantchik & Tokhtas'ev 2011 (*SEG* XLV 1020–1023; *AE* 2009, 1225–1226; Avram, *BE* 2010, 471); Yaïlenko 2010, 199–204 (*AE* 2010, 1444).

26 See *OGIS* 388f.; 392, and two letters attributed to Hippokrates (Smith 1990, 48–51). Cf. Darius' title in the inscription at Behistun (*DB* I.1; Kuhrt 2007, 141): 'I am Darius, Great King, King of Kings'. For further remarks, see Engels 2014, 345; Ballesteros Pastor 2017, 297f. Alternatively, the title could appear as *Megas Basileus Basileon*: *CIRB* 29; cf. Muccioli 2013, 410.

27 On the ecumenic perspective among the empires of the Hellenistic period, see Walbank 1984, 66; Briant 2002, 178–183; Tuplin 2010, 290–292; Strootman 2014, 49; 2016, 221–226.

28 Tuplin 2010, 290f.; Ballesteros Pastor 2013b, 205; 2015b, 93. See also Just. 12.16.5: two eagles were seen when Alexander was born, which is interpreted as an omen of his future rule over both Europe and Asia; cf. Curt. 7.8.30; Ballesteros Pastor 2003, 34. On the dominion over the two parts of the world as a feature of Hellenistic royal charisma, see further Polyb. 11.34.14–16; Gehrke 1982, 254.

Bosporan kingdom: despite its limited extension, it could boast an exceptional strategic location at the very border between Asia and Europe.[29] This ecumenic perspective had also been spread by Mithradates Eupator, who was acclaimed as the master of the two parts of the world, and was glorified as *kosmokrator*.[30] Pharnakes' claim to be a 'Great King of Kings' thus followed in the footsteps of his father, although at a far more modest level. The use of a title reminiscent of the one borne by the Achaimenids was, once more, a proclamation of legitimacy and aspirations, rather than a mere geopolitical reality.

IV. THE EMASCULATION OF THE BOYS OF AMISOS

One of the most relevant episodes of Pharnakes' brief reign in Pontos was undoubtedly the castration of the young boys of Amisos, a punishment of the city for having sided with the Romans.[31] This measure recalled an old Achaimenid practice: as we know from Herodotos, the generals of Darius castrated the young Greeks in 494 BC after the Ionian revolt (Hdt. 6.6.9). By inflicting such a punishment, the Persians ensured that there would be no new generation of rebels. Moreover, the Achaimenid kings used to impose as a tribute the delivery of young eunuchs destined to functions in the court.[32] But, at the same time, the punishment of Amisos was evoking actions of the satraps of Daskyleion, ancestors of Pharnakes' family. As Arrian reports, the satrap Pharnabazos ordered the emasculation of the boys of Chalkedon during the Peloponnesian War, probably in 409 BC, to suppress a rebellion against the Persians. The memory of this dramatic event had remained alive in the city of the Propontis throughout the centuries. When Arrian

---

29 This idea may be expressed in *CIRB* 29: [Φαρνάκης(?) μέγας βασιλε]ὺς βασιλέων | [ὑποτάξας βαρβάρους τοὺς κα]τὰ τὴν Εὐρώπην | [καὶ τὴν Ἀσίαν? ἀνέθηκε] Διὶ Γενάρχηι. Cf. *SEG* XL 627 (1); Vinogradov, *BE* 1990, 580. On Pharnakes' dominion over both sides of the Bosporus, see further Strab. *Geogr.* 11.2.11 (495C); Luther 2002, 268f.; Heinen 2005; Engels 2014, 346.

30 Ballesteros Pastor 2013b, 205; 2018a, 142; Palazzo 2016, 336; cf. Flor. 1.40.3. On Mithradates as *kosmokrator*, see in particular Poseidonios, *FGrH* 87 F 36 *apud* Ath. 5.213b.

31 App. *BCiv.* 2.13.91; *BAlex.* 70.6: *Nam neque interfectis amissam vitam neque exsectis virilitatem restituere posse; quod quidem supplicium gravius morte cives Romani subissent.* Cf. 41.2: *bona civium Romanorum Ponticorumque diripuit, supplicia constituit in eos qui aliquam formae atque aetatis commendationem habebant ea quae morte essent miseriora*; Ballesteros Pastor 2013a, 189. Gaertner & Hausburg 2013, 104, 109, suppose that only Roman citizens were castrated, but Appian does not specify this aspect; in fact, *BAlex.*41.1 speaks of confiscations among both Romans and Pontics. On the circumstances of the resistance of Amisos and the time of its conquest, see Coşkun 2020b, section 5.

32 The annual tribute from Babylon included 500 young eunuchs (Hdt. 3.92.1). The castration of male children was repudiated by the Greeks (Hdt. 8.105.1). Cf. Bosworth 1997, 301 with n. 17; Fisher 2002, 215; Hornblower 2003; Kuhrt 2007, 229; 577; 591; 674; Ballesteros Pastor 2013a, 189. This punishment was also used by Periander of Corinth: Hdt. 3.48f.; cf. Desmond 2004, 35f.

was writing in the 2nd century AD, the anniversary day of this disastrous event was still a *dies nefastus*.[33]

By this cruel order, Pharnakes was therefore remembering his Persian ancestors, a dynasty of satraps that had ruled for much time over the south-western part of the Black Sea coast. It was, in short, a way to proclaim that the king had returned: Pharnakes was not only the son of Mithradates Eupator, but also a descendant of Artabazos, the valiant commander of Xerxes' army who saved his troops from the Battle of Plataia and was appointed to rule over Daskyleion by the Great King.[34]

V. THE GREAT KING AT WAR: PHARNAKES' SCYTHED CHARIOTS

One of the peculiarities of the army of Pharnakes II was no doubt the use of scythed chariots.[35] Although their invention could be traced back to the neo-Assyrian period, these chariots were probably associated with Persian kings, who had utilized them in their wars with the Greeks.[36] In the Hellenistic period, scythed chariots were also included in the armies of the Seleukids, and the Romans learned how to dodge them during the war with Antiochos III.[37] Mithradates Eupator employed scythed chariots against Ariarathes VII of Kappadokia and the legions of Sulla. In this way, the Pontic king was remembering his ancestors, the satraps of Daskyleion, who had used this type of weapon. Eupator, however, noticed its scarce effectiveness and probably avoided it in his combats with Lucullus and Pompey.[38] However, Pharnakes had scythed chariots with him on his Pontic campaign and deployed them against Caesar's army, despite the experience of Mithradates. This was certainly not due to tactical reasons, but in the first place of symbolic importance: an exaltation of the Iranian elements of his reign. Once more, the son of Mithradates showed off his lineage and demonstrated that the ancestral kings were ruling Pontos again.

33 Arr. *Bith. FGrH* 156 F 79f.; Bosworth 1997. According to Bosworth 1997, 299, a reference to this episode may also be found in Plut. *Cam.* 19.9, where the Carthaginians are mistakenly cited instead of the Chalkedonians.

34 The Mithradatids' descent from Artabazos is explicitly mentioned in Sall. *Hist.* F 2.85 ed. Maurenbrecher; Flor. 1.40.1; cf. Ballesteros Pastor 2012. See also above, n. 10.

35 *BAlex.* 75.2f.; Cass. Dio 42.47.5.

36 See Rop 2013; cf. Nefiodkin 2004; Ballesteros Pastor 2013b, 183.

37 Liv. 37.41.5–12; Baker 2003, 380; Bugh 2007, 278–280; Sekunda & De Souza 2007, 348.

38 Just. 38.1.8; Plut. *Sulla* 15.1; 18.2f.; *Luc.* 7.4; App. *Mith.* 18.66; 42.163f.; Frontin. *Str.* 2.3.17; Veget. *Mil.* 3.23.1; cf. Plut. *Luc.* 37.3; Sall. *Hist.* F 3.21 ed. Maurenbrecher; Couvenhes 2009, 421f.; Ballesteros Pastor 2013b, 183. On the scythed chariots and the satraps of Daskyleion, see for instance Xen. *Hell.* 4.1.17.

VI. PHARNAKES II AND THE UNFINISHED TOMB OF AMASEIA

The monarchs of Pontos had their monumental tombs in Amaseia by the river Iris. Amaseia had been the capital of the kingdom for some time, until the court was moved to Sinope, probably by Pharnakes I.[39] Altogether, there are five tombs, one of which was left unfinished. These burial monuments are carved into the cliffs above the city, and corridors provide access to them. The identification of the kings buried in each of these graves has been a matter of discussion. The traditional hypothesis maintains that Mithradates I Ktistes and the three kings who followed him – Ariobarzanes, Mithradates II and Mithradates III – were buried in Amaseia. The next ruler, Pharnakes I, left his tomb unfinished presumably because he decided to be buried in Sinope, his new capital. The fact that there is an inscription dedicated to the wellbeing of king Pharnakes not too far from this monument might support this view.[40]

But this reconstruction implies some problems. Neither the number nor the identity of the rulers buried in Amaseia can be determined: except for the one inscription mentioned (whose king is also hard to specify), there are no other epigraphic documents identifying the respective owner of each tomb; furthermore, each chamber may have kept more than one corpse.[41] There is no proof that the first monarchs of Pontos conquered lands east of the Halys. Mithradates I Ktistes first settled in Kimiata, on the slopes of Mount Olgassys, located in Paphlagonia. From there, the Mithradatids extended their realm combating the Galatians and taking also possession of Amastris.[42] Therefore, Amaseia, placed in Pontic Kappadokia, may have become the capital of the kingdom at the earliest in the time of Ariobarzanes, the successor of Mithradates Ktistes, or rather Mithradates II, the

---

39   On the royal tombs at Amaseia, see Fleischer 2009; 2017; Højte 2009b. We do not know which king moved the capital to Sinope. Rostovtzeff & Ormerod 1933, 218, thought it was Pharnakes I, who conquered this Greek city; see further Olshausen 1978, 436f.; Avram 2016, 225; Payen, chapter VII in this volume. McGing 1986b, 250f. doubts that this measure can be attributed to either Pharnakes I or Mithradates V with any certainty; cf. Barat 2012, 55. In general, on Amaseia, see Olshausen & Biller 1984, 112 (with further bibliography).

40   Rostovtzeff & Ormerod 1933, 217f.; Fleischer 2009, 111–115; 2017, 85–87. For the inscription, see *OGIS* 365; cf. Reinach 1890, 456; Anderson, Cumont & Grégoire 1910, 114f., no. 94; Fleischer 2017, 87f.: Ὑπὲρ βασιλέως / Φαρνάκου / [Μη]τρόδωρος / [...]ιου φρουραρ- / χήσ]ας [τὸ]ν βω- / [μ]ὸν καὶ [τ]ὸν / ἀνθεῶνα / θεοῖς. 'On behalf of King Pharnakes (dedicated) [Me]trodoros, son of [...]os, the commander of the garrison, an altar and a flower garden to the Gods.'

41   Cf. Højte 2009b, 126. For this discussion, see Reinach 1890, 293; Rostovtzeff & Ormerod 1933, 218; McGing 1986b, 250f.; Burstein 1980, 11 n. 35; Fleischer 2009; 2017, 85–88, 107–109; Ballesteros Pastor 2013a, 226 n. 23; Avram 2015, 115.

42   Memn. *FGrH* 434 F 1.11.4; 1.16.1; Ballesteros Pastor 2013a, 186, 196 n. 23. Galatians were nonetheless recruited as mercenaries in the Pontic army by Mithradates I: Apollon. Aphr. *FGrH* 740 F 14; Olshausen 1978, 404; Coşkun 2011, 88 with n. 11. We can see a similar circumstance in the case of Eupator, who seized Galatia and slaughtered the tetrarchs, but still had Galatians in his service: see Just. 37.4.6; 38.3.6; 38.5.3–6; Plut. *Mor.* 259a–d; App. *Mith.* 46.178f.; 54.218; 111.539; Liv. *Per.* 102.3; *Vir. Ill.* 76.8; Oros. 6.5.6; Ballesteros Pastor 2013a, 23, 85–87, 165–167, 215.

first Pontic king who established a marriage alliance with the Seleukids.[43] For his part, Højte proposed to ascribe the first tomb of the Pontic royal cemetery to Mithradates III.[44] The burial place of Mithradates Eupator has not yet been identified with certainty. Claire Barat's study has reached the conclusion that Mithradates Euergetes and his son Eupator were presumably buried in Sinope.[45] For our part, we therefore think that the construction of the unfinished tomb at Amaseia began in the brief reign of Pharnakes II.

Upon recovering Pontos, Pharnakes II sought above all to be seen as the heir of the ancestral kings of that country, in order to be recognized as a legitimate sovereign. The construction of a monumental tomb beside the ones of his royal ancestors would have endorsed this claim. This context notwithstanding, the design of the tomb for Pharnakes II differed from the other funerary monuments in Amaseia. Three of the tombs were built according to the Hellenistic model, with a portico formed by fronton and columns. In contrast, the unfinished grave has a portico without columns and with stone revetments, following the same pattern of the second grave built there.[46] Robert Fleischer noted this difference, and attributed the original design 'to a local tradition, unknown to us'.[47] One rational may have been to speed up the work, but perhaps the most likely explanation is that Pharnakes just wanted to imitate his ancestor in tomb B.

Apart from these considerations, the non-Hellenic design of Pharnakes' grave without a portico may reflect the aim to emphasize the king's Iranian roots. Even in the case of the monuments that followed the Hellenistic fashion, the conception of the Pontic royal tombs at Amaseia is essentially Persian. The burial is placed inside the rock, in order to prevent the dead body from polluting, in accordance with Iranian rules of purity.[48]

Pharnakes' monument was left unfinished after Caesar defeated the king in the Battle of Zela on 2 August 47 BC. Not much later, the king died fighting against Asandros in the Bosporan kingdom, and his burial place remains unknown.[49] We do not know how much time the construction of this funerary monument might have taken. At any rate, it is worth remembering that Darius, the son of Pharnakes, ruled over Amaseia sometime after the Battle of Philippi (42 BC) and may well have continued working on the tomb.[50]

---

43 On this alliance, see Petković 2009; Ballesteros Pastor 2013b, 240f.; D'Agostini 2016; Coşkun 2018, 226f., arguing for 244/2 BC. When the offspring of this marriage, princess Laodike, left home to marry Antiochos III, Polyb. 5.43.2 says that she departed from 'Pontic Kappadokia'.
44 Højte 2009b, 127.
45 Barat 2012, 58–60. See also Højte 2009b, 128.
46 Fleischer 2009, 111–116 and *passim*; Id. 2017.
47 Fleischer 2009, 118. For a detailed description of the tomb, see Fleischer 2017, 71–88.
48 Fleischer 2009, 115; 2017, 130; Canepa 2010, 12; 2014, 53, 61.
49 On Pharnakes' defeat at Zela, see above, n. 5. On his death that followed within not much more than a month, see App. *Mith.* 120.595; Strab. *Geogr.* 13.4.3 (625C); Cass. Dio 42.47.5; cf. Coşkun 2019, 132–137; 2020.
50 See above, n. 11.

VII. CONCLUSION

We can conclude that Pharnakes II represents the survival of the Persian conception of kingship in the Bosporan kingdom and, if only ephemerally, also in
Pontos. Despite some opposition to him, because his arrival meant a new confrontation with Rome, the native Anatolian population largely recognized Pharnakes
as the legitimate king, as heir to an ancestral royalty, which was basically Iranian.
Once the former Mithradatid kingdom of Pontos had been divided between Rome
and local dynasts most loyal to Rome (first Deiotaros Philorhomaios, then Polemon Eusebes),[51] this Iranian conception of kingship may have fallen into oblivion
in that region. However, in the Bosporan kingdom, the idea of Iranian royalty survived under the successors of Pharnakes, despite their political affiliation with the
Roman Emperors. The titles 'King of Kings' or 'Great King of Kings' remained
the most distinguished elements of royal titulature under the successors of Pharnakes until the 2nd century AD.[52]

Acknowledgments

*This paper has been drawn up within the Research Project: 'La aportación de las culturas locales
al helenismo de Asia Menor y a su expansión por el Mediterráneo oriental'. I am grateful to Altay
Coşkun for his helpful remarks.*

Bibliography – Ancient Sources

Goukowsky, P. 2001: *Appien. Histoire Romaine. Tome VII. Livre XII. La Guerre de Mithridate*,
    Paris.
Smith, W.D. 1990: *Hippocrates, Pseudepigraphic Writings: Letters, Embassy, Speech from the
    Altar, Decree*. Edited and Translated with an Introduction, Leiden.

Bibliography – Modern Scholarship

Abramzon, M.G. & Kuznetsov, V.D. 2011: 'The Rebellion in Phanagoria in 63 B.C. New Numis-
    matic Evidence', *ACSS* 17, 75–110.
Almagor, E. 2019: 'Echoes of the Persian Wars in the European Phase of the Roman-Syrian War
    (with an Emphasis on Plut., *Cat. Mai.* 12–14)', in A. Coşkun & D. Engels (eds.), *Rome and
    the Seleukid East. Selected Papers from Seleukid Study Day V, Brussels, 21–23 Aug. 2015*,
    Brussels, 87–133.
Anderson, J.G.C., Cumont, F. & Grégoire, H. 1910: *Studia Pontica*, vol. 3: *Recueil des inscrip-
    tions grecques et latines du Pont et de l'Arménie*, Brussels.

---

51 See Strab. *Geogr.* 11.2.18 (499C); 12.3.13 (546f.C); 12.3.29 (555f.C); 12.3.38 (560C); cf.
    Coşkun, chapter X in this volume.
52 See Funck 1998; Ballesteros Pastor 2017, 301 with n. 28.

Arsen'eva, T.M., Böttger, B. & Vinogradov, Y.G. 1995: 'Griechen am Don. Die Grabungen in Tanais 1994', *Eurasia Antiqua* 1, 213–263.

Avram, A. 2016: 'Sur la date du traité entre Pharnace et Chersonèse Taurique', in Couvenhes, J.-C. (ed.), *La symmachia comme pratique du droit international dans le monde grec*, Besançon, 213–237.

Ballesteros Pastor, L. 1995: 'Notas sobre una inscripción de Ninfeo en honor de Mitrídates Eupátor, rey del Ponto', *DHA* 21, 111–117.

Ballesteros Pastor, L. 1996: *Mitrídates Eupátor, rey del Ponto*. Granada.

Ballesteros Pastor, L. 2003: 'Le discours du Scythe à Alexandre le Grand', *RhM* 146, 23–37.

Ballesteros Pastor, L. 2005a: 'El reino del Ponto', in V. Alonso Troncoso (ed.), *ΔΙΑΔΟΧΟΣ ΤΗΣ ΒΑΣΙΛΕΙΑΣ. La figura del sucesor en la realeza helenística*, Madrid, 127–138.

Ballesteros Pastor, L. 2005b: 'Nekotorÿe aspektÿ obraza Farnaka II v antichnoï literature' (Some Aspects of Pharnaces II's Image in Ancient Literature), *Antiquitas Aeterna* 1, 211–217.

Ballesteros Pastor, L. 2011: '*Xerxes redivivus*. Mitrídates, rey de Oriente frente a Grecia', in J.M. Cortés et al. (eds.), *Grecia ante los Imperios. Actas de la V Reunión de Historiadores del Mundo Griego Antiguo (Carmona 2009)*, Sevilla, 253–262.

Ballesteros Pastor, L. 2012: 'Los herederos de Artabazo. La satrapía de Dascilio en la tradición de la dinastía Mitridátida', *Klio* 112, 366–379.

Ballesteros Pastor, L. 2013a: '*Nullis umquam nisi domesticis regibus*. Cappadocia, Pontus and the Resistance to the Diadochi in Asia Minor', in V. Alonso Troncoso & E.M. Anson (eds.), *After Alexander. The Time of the Diadochi (323–281 BC)*, Oxford, 183–198.

Ballesteros Pastor, L. 2013b: 'Pompeyo Trogo, Justino y Mitrídates. Comentario al *Epítome de las Historias Filípicas* (37,1,6–38,8,1)', Hildesheim.

Ballesteros Pastor, L. 2014: 'A Neglected Epithet of Mithridates Eupator (*IDélos* 1560)', *Epigraphica* 76, 81–85.

Ballesteros Pastor, L. 2015a: 'Los príncipes del Ponto. La política onomástica de Mitrídates como factor de propaganda dinástica', *REA* 117, 425–445.

Ballesteros Pastor, L. 2015b: 'Quinto Curcio ante sus fuentes: el episodio de Alejandro y los escitas del Tanais', *Gerión* 33, 91–110.

Ballesteros Pastor, L. 2017: 'Pharnaces II and his Title "King of Kings"', *AWE* 16, 297–303.

Ballesteros Pastor, L. 2018a: 'De rey del Ponto a rey de reyes. El imperio de Mitrídates Eupátor en el contexto del Oriente tardo-helenístico', in L.R. Cresci & F. Gazzano (eds.), De Imperiis. *L'idea di impero universale e la successione degli imperi nell'Antichità*, Rome, 137–170.

Ballesteros Pastor, L. 2018b: 'Les réseaux de Mithridate', *DHA* 44, 273–288.

Ballesteros Pastor, L. 2020: 'The Origins of the Ariobarzanid Dynasty', *Gephyra* 19 (forthcoming).

Baker, P. 2003: 'Warfare', in A. Erskine (ed.), *A Companion to the Hellenistic World*, Oxford, 373–388.

Barat, C. 2012: 'Représentations de la dynastie du Pont, images et discours', in E. Santinelli-Foltz & G.-C. Schwentzel (eds.), *La puissance royale*, Rennes, 45–61.

Bosworth, A.B. 1997: 'The Emasculation of the Chalcedonians: A Forgotten Episode of the Ionian War', *Chiron* 27, 297–313.

Braund, D. 1984: *Rome and the Friendly King: The Character of the Client Kingship*, London.

Braund, D. 1994: *Georgia in Antiquity. A History of Colchis and Transcaucasian Iberia 500 BC–AD 562*, Oxford.

Briant, P. 2002: *From Cyrus to Alexander. A History of the Persian Empire*, Winona Lake, IN.

Bugh, G.R. 2007: 'Hellenistic Military Developments', in G.R. Bugh (ed.), *The Cambridge Companion to the Hellenistic World*, Cambridge, 265–294.

Burstein, S.M. 1980: 'The Aftermath of the Peace of Apamea: Rome and the Pontic War', *AJAH* 5, 1–12.

Canepa, M.P. 2010: 'Achaemenid and Seleucid Funerary Practises and Middle Iranian Kingship', in H. Börm & J. Wiesehöfer (eds.), Commutatio et Contentio: *Studies in Late Roman, Sasanian and Early Islamic Near East. In Memory of Zeev Rubin*, Düsseldorf, 1–21.

Canepa, M.P. 2014: 'Topographies of Power: Theorizing the Visual, Spatial and Ritual Contexts of Rock Reliefs in Ancient Iran', in Ö. Harmanşah (ed.), *Of Rocks and Water. Towards an Archaeology of Place*, Oxford, 53–92.

Canepa, M.P. 2017: 'Rival Images of Iranian Kingship and Persian Identity in Post-Achaemenid Western Asia', in Strootman & Versluys 2017, 201–222.

Carlsen, J. 2008: 'Cn. Domitius Calvinus: a Noble Caesarian', *Latomus* 67, 72–81.

Coşkun, A. 2008: 'Das Ende der "romfreundlichen" Herrrschaft in Galatien und das Beispiel einer "sanften Provinzialisierung" in Zentralanatolien', in idem (ed.), *Freundschaft und Gefolgschaft in den auswärtigen Beziehungen der Römer (2. Jahrhundert v.Chr. – 1. Jahrhundert n.Chr.)*, Frankfurt, 133–164.

Coşkun, A. 2011: 'Galatians and Seleucids: a Century of Conflict and Cooperation', in K. Erickson & G. Ramsey (eds.), *Seleucid Dissolution. The Sinking of the Anchor*, Wiesbaden, 85–108.

Coşkun, A. 2018: 'The War of Brothers, the Third Syrian War, and the Battle of Ankyra (246–241 BC): a Re-Appraisal', in K. Erickson (ed.), *The Seleukid Empire, 281–222 BC. War within the Family*, Swansea, 197–252.

Coşkun, A. 2019: 'The Date of the Revolt of Asandros and the Relations between the Bosporan Kingdom and Rome under Caesar', in M. Nollé, P.M. Rothenhöfer, G. Schmied-Kowarzik, H. Schwarz & H.C. von Mosch (eds.), *Panegyrikoi Logoi. Festschrift für Johannes Nollé zum 65. Geburtstag*, Bonn, 125–146.

Coşkun, A. 2020: 'The Course of Pharnakes II's Pontic and Bosporan Campaigns in 48/47 BC', *Phoenix* 73.1–2, 2019 (ca. Dec. 2020), 86–113.

Couvenhes, J.-C. 2009: 'L'armée de Mithridate Eupator d'après Plutarque, *Vie de Lucullus* VII, 4–6', in H. Bru, F. Kirbihler & S. Lebreton (eds.), *L'Asie Mineure dans l'Antiquité*, Rennes, 415–438.

Dąbrowa, E. 2018: 'Arsacid Dynastic Marriages', *Electrum* 25, 73–83.

D'Agostini, M. 2016: 'The Multicultural Ties of the Mithridatids: Sources, Tradition and Promotional Image of the Dynasty of Pontus in 4th–3rd Centuries B.C.', *Aevum* 90, 83–96.

De Callataÿ, F. 1997: *L'Histoire des Guerres Mithridatiques vue par les monnaies*, Louvain-la-Neuve.

De Callataÿ, F. & Lorber, C.C. 2011: 'The Pattern of Royal Epithets on Hellenistic Coinages', in P. Iossif, A.S. Chankowski & C. Lorber (eds.), *More than Men, Less than Gods. Studies on Imperial Cult and Royal Worship*, Leuven, 417–455.

De Jong, A.F. 1999: 'Xvarenah', in K. Van den Toorn, B. Becking & P.W. van der Horst (eds.), *Dictionary of Deities and Demons in the Bible*, 2nd ed. Leiden, 481–483.

Desmond, W. 2004: 'Punishment and the Conclusion of Herodotus' *Histories*', *GRBS* 44, 19–40.

Dobesch, G. 1996: 'Caesar und Kleinasien', *Tyche* 11, 51–77.

Engels, D. 2014: '"Je veux être calife à la place du calife"? Überlegungen zur Funktion der Titel "Großkönig"', in V. Cojocaru, A. Coşkun & M. Dana (eds.), *Interconnectivity in the Mediterranean and Pontic World during the Hellenistic and Roman Periods*, Cluj-Napoca, 333–362.

Fisher, N. 2002: 'Popular Morality in Herodotus', in F.J. Bakker, I.J.F. de Jong & H. van Wees (eds.), *Brill's Companion to Herodotus*, Leiden, 199–224.

Fleischer, R. 2009: 'The Rock-Tombs of the Pontic Kings in Amaseia', in Højte 2009a, 109–119.

Fleischer, R. 2017: *Die Felsgräber der Könige von Pontos in Amasya*, Tübingen.

Freber, P.S.G. 1993: *Der hellenistische Osten und das Illyricum unter Caesar*, Stuttgart.

Funck, B. 1998: 'Politische Orientierungen im Bosporanischen Reich im Spiegel der Königstitulatur nach Mithradates VI. Eupator', in C.-F. Collatz (ed.), *Dissertatiunculae Criticae. Festschrift für G.-Ch. Hansen*, Würzburg, 155–170.

Gaertner, J.F. & Hausburg, B. 2013: 'Caesar and the Bellum Alexandrinum: An Analysis of Style, Narrative Technique, and the Reception of Greek Historiography', Göttingen.

Gajducevič, V.F. 1971: *Das Bosporanische Reich*, 2nd ed., Berlin.

Gatzke, A.F. 2019: 'Mithridates VI Eupator and Persian Kingship', *AHB* 33, 60–80.

Gehrke, H.-J. 1982: 'Der siegreiche König. Überlegungen zur hellenistischen Monarchie', *AKG* 64, 247–277.

Gnoli, G. 1999: 'Farr(ah)', *EncIr* 9312–319. URL: http://www.iranicaonline.org/articles/farrah.

Golenko, K.V. & Karyszkowski, P.J. 1972: 'The Gold Coinage of King Pharnaces of the Bosporus', *NC* ser. 7.12, 25–38.

Gourova, N. 2014: 'What Did Ancient Greeks Mean by the "Cimmerian Bosporus"?', *AWE* 13, 29–48.

Griffiths, J.G. 1953: 'βασιλεὺς βασιλέων: Remarks on the History of a Title', *CP* 48, 145–154.

Heinen, H. 1994: 'Mithradates von Pergamon und Caesars bosporanische Pläne. Zur Interpretation von Bellum Alexandrinum 78', in R. Günther & S. Rebenich (eds.), E fontibus haurire. *Beiträge zur römischen Geschichte und zu ihren Hilfswissenschaften*, Paderborn, 63–79.

Heinen, H. 2001: 'Die Mithradatische Tradition der bosporanischen Könige – ein mißverstandener Befund', in K. Geus & K. Zimmermann (eds.), *Punica – Libyca – Ptolemaica. Festschrift für Werner Huß zum 65. Geburtstag*, Leuven, 355–370.

Heinen, H. 2005: 'Mithradates VI. Eupator, Chersonesos und die Skythenkönige. Kontroversen um Appian, Mithr. 12f. und Memnon 22,3f.', in A. Coşkun (ed.), *Roms auswärtige Freunde in der späten Republik und im frühen Prinzipat*, Göttingen, 75–90.

Heinen, H. 2008: 'Romfreunde und Kaiserpriester am Kimmerischen Bosporos. Zu neuen Inschriften aus Phanagoreia', in A. Coşkun (ed.), *Freundschaft und Gefolgschaft in den auswärtigen Beziehungen der Römer (2. Jahrhundert v.Chr. – 1. Jahrhundert n.Chr.)*, Frankfurt, 189–208.

Hardie, P. 2007: 'Images of the Persian Wars in Rome', in E. Brigdes, E. Hall & P.J. Rhodes (eds.), *Cultural Responses to the Persian Wars. Antiquity to the Third Millenium*, Oxford, 127–143.

Hoben, W. 1969: *Untersuchungen zur Stellung kleinasiatischer Dynasten in den Machtkämpfen der ausgehenden römischen Republik*, Diss., Mainz.

Højte, J.M. (ed.) 2009a: *Mithridates VI and the Pontic Kingdom*, Aarhus.

Højte, J.M. 2009b: 'The Death and Burial of Mithridates VI', in Højte 2009a, 121–130.

Hornblower, S. 2003: 'Panionios of Chios and Hermotimos of Pedasa (Hdt. 8.104–6)', in P. Derow & R. Parker (eds.), *Herodotus and His World. Essays in Memory of G. Forrest*, Oxford, 37–57.

Ivantchik, A.I. & Tokhtas'ev, S.R. 2011: 'Queen Dynamis and Tanais', in E. Papuci-Władyka, M. Vickers, J. Bodzek & D. Braund (eds.), *Pontika 2008. Recent Research on the Northern and Eastern Black Sea in Ancient Times. Proceedings of the International Conference, 21st–26th April 2008, Kraków*, Oxford, 163–173.

Kuhrt, A. 2007: *The Persian Empire. A Corpus of Sources from the Achaemenid Period*, London.

Lerouge-Cohen, C. 2014: 'Les amis des Arsacides: pistes de réflexion', *Ktèma* 39, 123–141.

Lerouge-Cohen, C. 2016: 'Prestige et généalogie: le cas des royaumes alliés et amis de Rome', in R. Baudry & F. Hurlet (eds.), *Le Prestige à Rome à la fin de la République et au début du Principat*, Paris, 149–160.

Lerouge-Cohen, C. 2017: 'Persianism in the Kingdom of Pontic Kappadokia. The Genealogical Claims of the Mithridatids', in Strootman & Versluys 2017, 223–234.

Luther, A. 2002: 'Zwietracht am Fluß Tanais: Nachrichten über das Bosporanische Reich bei Horaz', in M. Schuol, U. Hartmann & A. Luther (eds.), *Grenzüberschreitungen. Formen des Kontakts zwischen Orient und Okzident im Altertum*, Stuttgart, 259–277.

Makhlaiuk, A.V. 2015: 'Memory and Images of Achaemenid Persia in the Roman Empire', in J.M. Silverman & C. Waerzeggers (eds.), *Political Memory in and after the Persian Empire*, Atlanta, 299–324.

Marek, C. 1993: 'Stadt, Ära und Territorium in Pontus-Bithynia und Nord-Galatia', Tübingen.

Mastrocinque, A. 1999: *Studi sulle Guerre Mitridatiche*, Stuttgart.

McGing, B.C. 1986a: *The Foreign Policy of Mithridates Eupator, King of Pontus*, Leiden.

McGing, B.C. 1986b: 'The Kings of Pontus: Some Problems of Identity and Date', *RhM* 129, 248–259.

McGing, B.C. 2014: 'Iranian Kings in Greek Dress? Cultural Identity in the Mithradatid Kingdom of Pontus', in T. Bekker-Nielsen (ed.), Space, Place and Identity in Northern Anatolia, Stuttgart 2014, 21–37.

Melikian-Chirvani, A.S. 1993: 'L'emblème de gloire solaire d'un roi iranien du Pont', *Bulletin of the Asia Institute* 7, 21–29.

Michels, C. 2017: The Persian Impact on Bithynia, Commagene, Pontus, and Cappadocia, in S. Müller, T. Howe, H. Bowden & R. Rollinger (eds.), *The History of the Argeads. New Perspectives*, Wiesbaden 2017, 41–56.

Muccioli, F. 2013: *Gli epiteti ufficiali dei re ellenistici*, Stuttgart.

Nefiodkin, A.K. 2004: 'On the Origin of the Scythed Chariots', *Historia* 53, 369–378.

Olbrycht, M.J. 2009: 'Mithridates Eupator and Iran', in Højte 2009a, 163–190.

Olbrycht, M.J. 2019: 'The Memory of the Past. The Achaemenid Legacy in the Arsakid Period', *Studia Litteraria Universitatis Iagellonicae Cracoviensis* 14, 175–186.

Olshausen, E. 1980: 'Pontos und Rom (63 v.Chr. – 64 n.Chr.)', *ANRW* II 7.2, 903–912.

Olshausen, E. & Biller, J. 1984: *Historisch-geographische Aspekte der Geschichte des Pontischen und Armenischen Reiches*. Teil 1: *Untersuchungen zur historischen Geographie von Pontos unter den Mithradatiden*, TAVO Beihefte 29.1, Wiesbaden.

Palazzo, S. 2016: 'Immagini di re e paradigma di regalità. Mitridate *basileus* tra Asia ed Europa', in G. De Sensi Sestito & M. Intrieri (eds.), *Sulle sponde dello Ionio: Grecia Occidentale e greci d'Occidente*, Pisa, 355–370.

Panitschek, P. 1987/8: 'Zu den genealogischen Konstruktionen der Dynastien von Pontos und Kappadokien', *RSA* 17/8, 73–95.

Petković, Z. 2009: 'Mithridates II and Antiochos Hierax', *Klio* 91, 378–383.

Primo, A. 2010: 'The Client Kingdom of Pontus between Mithridatism and Philoromanism', in T. Kaizer & M. Facella (eds.), *Kingdoms and Principalities in the Roman Near East*, Stuttgart, 159–179.

Portanova, J.J. 1988: *The Associates of Mithridates VI of Pontus*, Diss. Columbia University, Ann Arbor.

Reinach, T. 1890: *Mithridate Eupator, roi de Pont*, Paris.

Ritter, H.W. 1965: 'Diadem und Königsherrschaft. Untersuchungen zu Zeremonien und Rechtsgrundlagen des Herrschaftsantritts bei den Persern, bei Alexander dem Grossen und im Hellenismus', Munich.

Rop, J. 2013: 'Reconsidering the Origin of the Scythed Chariot', *Historia* 62, 167–181.

Rostovtzeff, M.I. 1919: 'Queen Dynamis of Bosporus', *JHS* 39, 88–109.

Rostovtzeff, M.I. & Ormerod, H.A. 1933: 'Pontus and Its Neighbours: The First Mithridatic War', *CAH* IX, Cambridge, 211–260.

Russo, F. 2014a: 'Echoes of the Persian Wars in Roman Propaganda', *AW* 44, 160–176.

Russo, F. 2014b: 'Il ricordo delle Guerre Persiane a Roma nello scontro con Filippo V e Antioco III', *Latomus* 73, 303–337.

Sekunda, N. & De Souza, P. 2003: 'Military Forces', in Ph. Sabin, H. van Wees & M. Whitby (eds.), *The Cambridge History of Greek and Roman Warfare*, Cambridge, 325–367.

Schipmann, K. 2014: 'Azerbaijan iii. Pre-Islamic History', *EncIr* 3.2. URL: http://www.iranicaonline.org/articles/azerbaijan-iii.

Sherwin-White, A.N. 1984: *Roman Foreign Policy in the East 168 BC to AD 1*, London.

Sherwin-White, S. & Kuhrt, A. 1993: *From Samarkhand to Sardis: A New Approach to the Seleucid Empire*, Berkeley.

Spawforth, A. 1994: 'Symbol of Unity? The Persian Wars Tradition in Rome', in S. Hornblower (ed.), *Greek Historiography*, Oxford, 232–247.

Strootman, R. 2014: 'Hellenistic Imperialism and the Idea of World Unity', in C. Rapp & H.A. Drake (eds.), *The City in the Classical and Post-Classical World. Changing Context of Power and Identity*, Cambridge, 38–61.

Strootman, R. 2016: '"The Heroic Company of My Forebears": the Ancestor Galleries of Antiochos I of Kommagene at Nemrut Daği and the Role of Royal Women in the Transmission of Hellenistic Kingship', in A. Coşkun & A. McAuley (eds.), *Seleukid Royal Women*, Stuttgart, 209–230.

Strootman, R. 2017: 'From Culture to Concept: the Reception and Appropriation of Persia in Antiquity', in Strootman & Versluys 2017, 9–32.

Strootman, R. 2020: 'Hellenism and Persianism in Iran: Culture and Empire after Alexander the Great', *Dabir* 7, 201–227.

Strootman, R. & Versluys, M. (eds.) 2017: *Persianism in Antiquity*, Stuttgart.

Sullivan, R.D. 1980: 'The Dynasty of Cappadocia', *ANRW* II, 7.2, 1125–1168.

Sullivan, R.D. 1990: *Near Eastern Royalty and Rome, 100–30 B.C.*, Toronto.

Sweeney, J.M. 1978: 'The Career of Cn. Domitius Calvinus', *AncW* 1, 179–185.

Syme, R. 1995: *Anatolica. Studies in Strabo*, Oxford.

Tuplin, C. 2010: 'Revisiting Dareios' Scythian Expedition', in J. Nieling & E. Rehm (eds.), *Achaemenid Impact in the Black Sea. Communication of Powers*, Aarhus, 281–312.

Walbank, F.W. 1984: *A Historical Commentary on Polybius*, vol. 3, Oxford.

Wiesehöfer, J. 1996: '"King of Kings" and "Philhellên". Kingship in Arsacid Iran', in P. Bilde, T. Engberg-Pedersen, L. Hannestad & J. Zahle (eds.), *Aspects of Hellenistic Kingship*, Aarhus, 55–66.

Wilker, J. 2007: *Für Rom und für Jerusalem. Die herodianische Dynastie im 1. Jahrhundert n.Chr.*, Frankfurt.

Yaïlenko, V.P. 1985: 'Novÿe épigraficheskie dannÿe o Mitradate Evpatore i Farnake' (New Epigraphic Evidence on Mithradates Eupator and Pharnakes), in *Prichernomor'e v épokhu éllinizma: Materialÿ 3. Vsesoyuznogo Simpozyuma po drevneï istorii Prichernomor'ya* (Tskhaltubo 1982), Tbilisi, 617–626. (English summary: 727f.).

Yaïlenko, V.P. 2010: *Tÿsyacheletnïï Bosporskiï reikh. Istoria i épigrafika Bospora VI v. do n. é.–V v. n. é.* (One Thousand Years of The Bosporan Kingdom. History and Epigraphy of Bosporos 6th c. BC–5th c. AD), Moscow.

Zograph, A.N. 1977: *Ancient Coinage. Part II. The Ancient Coins of the Northern Black Sea Littoral*, Oxford.

QUEEN DYNAMIS AND PRINCE ASPURGOS IN ROME?

Revisiting the South Frieze of the *Ara Pacis Augustae* (13/9 BC)

Altay Coşkun & Gaius Stern

Abstract: The Senate voted to build the *Ara Pacis* to welcome home Augustus after restoring order in the western provinces, while Agrippa pursued a similar mission in the East. Agrippa had settled the turmoil in the Bosporus by arranging for Queen Dynamis to marry King Polemon of Pontos, thereby uniting the two realms. Brian Rose (1990) explained that two boys on the *Ara Pacis* who do not wear togas are foreign princes in Rome in 13 BC exactly when Augustus and Agrippa returned from their foreign tours. Rose considered the older boy on the south frieze an eastern prince, probably Aspurgos, the future king of the Bosporus. He speculated that Queen Dynamis had come to Rome with Agrippa, and that she is the woman who puts her hand on the boy's head. Rose exposed the frailty of Giuseppe Moretti's theory, who regarded the two boys as Gaius and Lucius Caesar dressed as Trojans. We agree with Ann Kuttner, Gaius Stern, John Pollini, Ilaria Romeo that the boys are barbarians, not Romans, but cannot accept the identifications with Dynamis and Aspurgos, (1) on prosopographical lines, because the placement of Dynamis on the Ara Pacis relies upon identifying her as the mother of Aspurgos, which claim the ancient sources do not support; (2) on practical terms, since Dynamis should have stayed in her kingdom to help Polemon consolidate his new throne (not speculation but positive evidence would be needed to counter this view); (3) iconographically, as the woman on the *Ara Pacis* does not closely resemble the image of Dynamis; (4) because Dynamis was a mature, middle-aged queen by 13 BC, as her portrait on two gold staters indicate, whereas the *Ara Pacis* teenager is far too young. She is actually Agrippa's least famous daughter, wearing not a diadem, but a brill appropriate for a Roman teenage girl close to marrying age. Her hand is resting on the head of a Parthian prince, a 'guest' in Rome, hosted by the family of Agrippa.

Абстракт: Динамия и принц Аспург в Риме? Возвращаясь к южному фризу Алтаря Мира Августа (13/9 г. н.э.): Сенат проголосовал за создание Алтаря Мира, чтобы приветствовать Августа на родине после восстановления порядка в западных провинциях, в то время как Агриппа выполнял аналогичную миссию на Востоке. Агриппа подавил беспорядки на Босфоре, договорившись о том, чтобы царица Динамия вышла замуж за Понтийского царя Полемона, тем самым объединив два царства. Брайан Роуз (1990) объяснил, что два мальчика на Алтаре Мира, которые не носят тоги, являются иностранными принцами, которые пребывали в Риме в 13 г. до н.э. именно тогда, когда Август и Агриппа вернулись из своих зарубежных поездок. Роуз считал старшего мальчика на южном фризе восточным принцем, вероятно Аспургом, будущим царем Боспора. Он предположил, что царица Динамия приехала в Рим с Агриппой и что изображенная на Алтаре женщина, которая кладет руку на голове мальчика, – это именно Динамия. Роуз обнаружил слабость теории Джузеппе Моретти, который считал, что два мальчика, одетые как троянцы – это Гай и Люций Цезарь. Авторы согласны с Энн Каттнер, Гайусом Стерном, Джоном Поллини и Иларией Ромео, что мальчики – это варвары, а не римляне, но авторы не могут согласиться с идентификацией этих персон с Динамией и Аспургом (1) по просопографическим причинам, потому что присутствие Динамии на Алтаре Мира зависит от ее идентификации как матери Аспурга, а древние источники этого не подтверждают; (2) с практической точки зрения, поскольку Динамия должна была остаться в своем царстве, чтобы помочь Полемону укрепить свой новый трон (чтобы противостоять этой точке зрения необходимы будут настоящие доказательства, а не спекуляция); (3) по иконографическим причинам, поскольку женщина на Алтаре Мира не очень похожа на изображения Динамии; (4) потому что Динамия к 13 г. до н.э., как показывает ее портрет на двух золотых статерах, была зрелой женщиной средних лет, в то время как подросток, изображенный на Алтаре Мира, слишком молод, чтобы считать его Аспургом. В действительности женщина на Алтаре – это наименее известная дочь Агриппы. На ее голове нет диадемы, а есть лента которую надевали римские девушки в брачном возрасте. Ее рука лежит на голове парфянского принца, «гостя» Рима, принятого семьей Агриппы.

I. THE *ARA PACIS AUGUSTAE*: NEW EVIDENCE FOR BOSPORAN DYNASTIC HISTORY?

1. Dynamis, the Most Prominent and Controversial Queen of the Bosporus

Queen Dynamis is the most illustrious example of a royal female from both the Bosporan kingdom and the Mithradatid dynasty. The former was a realm that began to develop from the city of Pantikapaion (modern Kerch, located close to the easternmost tip of the Crimea) in the 5th century BC. It controlled substantial portions of the European and Asian Kimmerian Bosporus (the modern Strait of Kerch) until the 4th century AD. The aforementioned dynasty gained fame as rulers of a small principality in northwest Asia Minor under the Achaimenids, before establishing itself as a kingdom in Paphlagonia and northern Kappadokia around Amaseia (Amasya) in the 3rd century. After the conquest of Sinope (Sinop) starting in the early-2nd century BC, its orientation shifted towards the Black Sea coast and its kingdom became known as Pontos. The most famous dynast of this ruling house, Mithradates VI Eupator (123/16–63 BC), gradually extended his territory to include the former Bosporan kingdom in the north (ca. 110 BC) and Kolchis (west Georgia) in the East. He became most notorious as the man who defied the Roman Empire for about half a century. His appetite for expansion brought him in constant conflict with the superpower of the Mediterranean world, which resulted in a series of major ('Mithradatic') wars (89–84, 82–80, 73–63 BC).

Having lost his possessions in Asia Minor and Kolchis, Mithradates tried to renew the war once more from Pantikapaion in 63 BC, but his own son Pharnakes opposed him and forced him to commit suicide, after which he took over the Bosporus as Great King of Kings Pharnakes II (63–47 BC).[1] He was the father of our main subject, Queen Dynamis. After gaining recognition from the Romans and extending the boundaries of his northern dominion as far as Tanaïs on the mouth

1 On the Mithradatids of Pontos up to Pharnakes II, see Hoben 1969; Gajdukevič 1971; Sullivan 1990; Ballesteros Pastor 1996; Højte 2009; Roller 2020; see also Payen, chapter VII on Pharnakes I in this volume.

of the homonymous river (now the Don), to fully encircle the Maiotis (Sea of Azov), he seized the opportunity that the Roman civil war offered. Caesar had defeated Pompey at Pharsalos (9 August 48) and was chasing him down as far as Egypt when Pharnakes invaded his ancestral homeland Pontos (48 BC). But despite Pharnakes' initial success, Caesar defeated him at Zela and expelled him from Asia Minor almost exactly a year later, on 2 August 47 BC.[2]

Shortly before the Battle of Zela, Asandros, one of Pharnakes' leading generals, who had been left in control of the Bosporus, revolted and he established himself as ruler after killing Pharnakes in the same summer. Conducive to the stability of his rule was his marriage with princess Dynamis, who gave him at least the appearance of dynastic legitimacy. His position was corroborated when Mark Antony sold him recognition as a *rex amicus populi Romani* in 44 BC. Aged over 90 years, Asandros finally appointed his wife *basilissa* and co-regent, but not even this gesture held Dynamis back from joining the usurper Scribonius, who took control in 20 or 19 BC. Scribonius' bluff about Roman support was soon exposed and he was killed in due course (ca. 16 BC), while Dynamis stayed in power. The Romans, however, insisted on establishing a king they could trust and dispatched Polemon I, then the ruler of Pontos and Kolchis, to take control. Resistance was fierce, even after Polemon was victorious in a first battle (ca. 15 BC). Only when M. Vipsanius Agrippa, the associate emperor, was gathering a fleet in Sinope for a major naval campaign to the Bosporus did Dynamis give in and accept Polemon into her realm as her king and third husband (14 BC).[3]

Her biography is by no means without difficulties thus far, but the timeline here proposed is quite firm.[4] Much more controversial are the remaining parts of her life. Clear evidence for her abruptly ends with the arrival of Polemon, who immediately became the sole minting authority. Likewise, our literary sources turn silent about her after the royal wedding. At least, Strabo mentions Polemon twice as campaigning on the Asian side of the Bosporus, where he was killed by

2   Besides the previous note, see also chapters VIII (Ballesteros Pastor) and X–XII (Coşkun) in this volume for more on Pharnakes II and Roman imperial politics 63–47 BC.

3   The two most important literary sources, Lukian, *Makrobioi* 17 and Cass. Dio 54.24.4–6, are quoted below, in notes 10 and 18 respectively. Easiest access to the numismatic sources is by MacDonald 2005; cf. Frolova 1997. For a selection of royal inscriptions, see Ivantchik & Tokhtas'ev 2011; Coşkun 2016; for a comprehensive discussion, see Coşkun in preparation. See also next note.

4   Scholars have dated the accession of Asandros between 49 and 42 BC and his death to ca. 20/15 BC, but once ideological distortion is set aside and methodological flaws are overcome, the literary, numismatic and epigraphic evidence combined yields exactly the year dates suggested above, see Coşkun 2019a; also 2016; 2017a; 2017b; 2019b; 2020a; 2020b; in preparation, based on or developing further Heinen 1998; 2006; 2008a; 2008b; 2011; forthcoming a; forthcoming b; arguing with Rostovtzeff 1919; Macurdy 1937, 33–38; Golubtsova 1951; Hanslik & Schmitt 1963; Hoben 1969; Gajdukevič 1971; Sullivan 1980; 1990; Anokhin 1986; Saprykin 1990; 1996; 2002, 97–99; Saprykin 2005, 170f.; Nawotka 1991/2; Leschhorn 1993; Frolova 1997; Kozlóvskaia 2003; 2004; Braund 2004; MacDonald 2005; Ballesteros Pastor 2008a; Saprykin & Fedoseev 2009; Primo 2010; Yaylenko 2010; Ivantchik & Tokhtas'ev 2011; Roller 2018a; Zavoykina, Novichikhin & Konstantinov 2018, 682–686.

the so-called Aspurgians in ca. 9 BC. The same geographer also mentions Pythodoris as his widow,[5] who succeeded him in Pontos and in Kolchis, while
Dynamis and the Bosporus remain unmentioned. The most radical conclusion
would be that Dynamis died shortly after her third wedding and thus gave way to
Polemon's remarriage with Pythodoris. This reconstruction is very implausible,
however, since it requires us to date all inscriptions mentioning the queen prior to
the arrival of Polemon.

Except for one inscription that names her together with Asandros, the others
do not specify a king by her side, but only her father, suggesting that they fall into
a period of sole rule. But the window after the death of Scribonius and before her
marriage with Polemon (15/4 BC) is simply too short to accommodate most of the
royal inscriptions from more than half a century. In addition, most inscriptions
provide her epithet *Philorhomaios* – 'Friend of the Romans', which she certainly
did not bear after betraying the *amicus populi Romani* Asandros and while defy-
ing the Roman candidate Polemon. Inscriptions with this title thus date to her second period of sole rule (after 9 BC) after the death of Polemon. A series of dated
gold staters from the Bosporus that depict Augustus on the obverse and Agrippa
on the reverse imply the same chronology. They do not have a legend to name the
royal minting authority, but use instead a monogram (Fig. 1), which can easily be
deciphered as *Dynamis*. As with the previous and subsequent coins, most specimens are dated according to the dynastic era. The Mithradatic years 289–304
equal the time span from 9/8 BC to AD 7/8. It would be utterly unconvincing to
posit an otherwise unknown king or Roman governor behind this monogram.

Fig. 1: *Monogram of the Gold Staters of Dynamis as Sole Ruler with
Portraits of Augustus and Agrippa, dated 9/8 BC to AD 7/8. Source:
Minns 1913, 595; cf. MacDonald 2005, 68.*

Most scholars thus rightly assume that Dynamis enjoyed an extended period of
sole rule until AD 7/8. Over the next few years, there is a quick succession of
monograms on the coinage before a steady series of coins was minted with a
monogram unanimously attributed to *Basileus Aspurgos* (AD 14/5–37/8). The
frequently changing monograms in the previous decade are normally understood
as reflecting dynastic turmoil. Many scholars interpret the monograms of AD
10/11 and 13/4 as earlier reigns of Aspurgos (perhaps without the royal title).
Since there is a second coin type of AD 13/4 with a monogram similar to that of
Dynamis, there are at least four possible scenarios: Aspurgos may have contended
for the throne and been expelled intermittently or he co-ruled. This second poten-
tate may either have been Queen Dynamis herself or someone else. The full dis-

5   Polemon and Pythodoris: Strab. *Geogr*. 11.2.18 (499C); 12.3.29 (555f.C). Cf. on Pythodoris:
     12.3.31 (557C); 12.3.37 (559C); 14.1.42 (649C). See Olshausen 1980; Braund 2005; Roller
     2018a; 2018b. See also the previous note and further Coşkun, chapter XI in this volume (on
     the territories of Pythodoris).

cussion would require close involvement with the epigraphic and numismatic sources, which would exceed the scope of the present chapter.[6]

What matters for our current investigation is that many (especially Russian) scholars advocated the view that Dynamis and Aspurgos were closely related and thus co-ruled until ca. AD 14. Mikhail Rostovtzeff, for instance, suggested that Aspurgos was the fourth husband of Dynamis, which is quite unlikely in the face of the great age difference between the two. Instead, many more historians believe rather that Aspurgos was the son of Dynamis and Asandros. In favour of this view, they refer to several inscriptions which call Aspurgos the son of king Asandrochos.[7] The same scholars regard Asandrochos as the original Iranian version of the Hellenized form Asandros, and they relate this Asandr(och)os and Aspurgos in some way to the abovementioned Aspurgianoi who revolted against Polemon near Gorgippia (Anapa) on the Taman Peninsula. Dynamis is believed to have cooperated with the insurgents or even led the uprising against Polemon. As the background to these conflicts, Rostovtzeff saw a lasting opposition between the Iranian population on the one hand and the Greek colonists ready to collaborate with the Roman imperial power on the other. Mithradates VI Eupator and his offspring are viewed as fighters against Roman oppressors.[8]

This reconstruction does not stand up to scrutiny. The assumption of ethnically-based hostilities that lasted over centuries is no more than a speculation, and quite implausible at that, in a multi-ethnic kingdom with a dynasty that had been intermarrying in the Hellenistic and Near-Eastern world for centuries. The theory further conflicts with the fact that nearly every ruler of the Bosporus from Mithradates Eupator until the disintegration of the kingdom in the 4th century AD was keen on becoming a friend of the Roman Emperor and people, or at least to be seen as such (as in the case of Scribonius).[9]

Moreover, the identification of Asandros with Asandrochos is fraught with problems of its own. That none of the many inscriptions of Bosporan royals mentions Dynamis and Aspurgos together should give us pause. Of course, the mere possibility of name variants or name switching should not be denied in an inter-ethnic context, but such a claim makes little sense in the case of Asandros. This is how the first husband of Dynamis is persistently called in the literary, numismatic and epigraphic evidence, whereas the inscriptions set up in honour of Aspurgos

6   Frolova 1997, 24–73 proposed Dynamis' death in or around 14 BC; cf. 2009. For documentation of the monogrammatic coinage, see also Minns 1913, 599–604 and MacDonald 2005, 48–68, who ascribe these coins to Dynamis. For discussion and further references, see Cojocaru 2014, 6 and the references above in n. 4.

7   *CIRB* 40 and see Heinen 2008b, 191–201 for a critical discussion of the evidence, including more recent finds as well as the questionable supplement of Asandrochos' name in *CIRB* 39. Add the latest epigraphic discovery from Anapa published by Zavoykina, Novichikhin & Konstantinov 2018, 682–686, who also document the earlier evidence.

8   E.g., Saprykin 2005, 170f. and Saprykin & Fedoseev 2009, a good example of 'creative' epigraphy that produces 'evidence' by supplementing whatever is desired (cf. the criticism by Avram 2011). For further references, see n. 4 above.

9   See Braund 2004; Coşkun 2016, besides further references in n. 4 above.

name his father Asandrochos throughout. They do not, however, ascribe him the epithet *Philorhomaios*, although it is attested for the later Asandros. Aspurgos was not indifferent to such titulature, since he proudly bore the titles *Philokaisar kai Philorhomaios*.[10]

Also, one must mention biological implications that are so significant that even some of the scholars who subscribe to Rostovtzeff's theory of a lasting ethnic divide reject the equation. Literary sources record that Asandros made his wife co-ruling queen when he was 91 or 92 years old, before dying at 92 or 93 during the revolt of Scribonius.[11] Had he had a capable son by either Dynamis or by any other woman, he would certainly have chosen him over his wife. And if Dynamis had had a son of any age with someone else, she would hardly have taken Scribonius' side, but ensured the succession of her own blood. By all means, Asandros' advanced age is incompatible with him having a minor son too young to rule, especially a late-born son. We conclude that the absence of any talk of such a boy under Scribonius or Polemon (who would have wanted to get rid of him), and, further, the lack of any sign of co-rule during the first short and second long, sole reign of Dynamis indicate that no such prince existed.[12]

2. Dynamis and Aspurgos in Rome?

The discussion was reignited by Brian Rose in his investigation of the *Ara Pacis Augustae* in Rome. He suggested that a lady on the south frieze may well be Dynamis, touching the head of her little boy Aspurgos. The relief was sculpted in 13–10 BC, and both figures are grouped with M. Vipsanius Agrippa, so that this son-in-law and deputy of Augustus may have taken the queen and her son to Rome after his eastern campaign. Some scholars happily embraced this reinterpretation and thought it explains why Dynamis is no longer attested in the Bosporan kingdom after Polemon's arrival. They further conclude that it was from Rome that she approached the tribe of the Aspurgianoi to rise against Polemon.[13]

But, once again, there are unsurmountable problems involved with such speculations, besides the lack of any positive evidence. First, we should wonder why the Aspurgians, if understood as loyal to their former ruler Asandr(och)os and his

---

10 See especially Heinen 2008b, 191–201, besides further references in notes 4–8 above.
11 Lukian, *Makrobioi* 17: Ἄσανδρος δὲ ὁ ὑπὸ τοῦ θεοῦ Σεβαστοῦ ἀντὶ ἐθνάρχου βασιλεὺς ἀναγορευθεὶς Βοσπόρου περὶ ἔτη ὢν ἐνενήκοντα ἱππομαχῶν καὶ πεζομαχῶν οὐδενὸς ἥττων ἐφάνη· ὡς δὲ ἑώρα τοὺς ἑαυτοῦ ὑπὸ τὴν μάχην Σκριβωνίῳ προστιθεμένους ἀποσχόμενος σιτίων ἐτελεύτησεν βιοὺς ἔτη τρία καὶ ἐνενήκοντα. 'Asandros, who, after being ethnarch, was proclaimed king of the Bosporus by the divine Augustus, at about 90 (89?) years proved himself a match for anyone in fighting from horseback or on foot; but when he saw his own entourage going over to Scribonius before the battle, he starved himself to death at the age of 93 (92?).' For in-depth discussion, see Coşkun 2019a.
12 Nor does the latest epigraphic find from Anapa compel us to believe this; see Zavoykina, Novichikhin & Konstantinov 2018 and Coşkun in preparation.
13 Rose 1990. And see next note.

assumed son Aspurgos, would have supported Dynamis, who had betrayed this family by allying herself with Scribonius. Secondly, the abovementioned concerns about Asandros' age still stand. But even if we accept a 'Biblical potency' for that man in his early 90s and admit the possibility that he left behind a pregnant wife, who later appeared on the *Ara Pacis* with that hypothetical son, such a child would have been six and a half years in 13 BC, while the boy on the south frieze looks closer to a five-year old.[14] Viktor Parfenov has proposed an alternative, accepting that Asandros was too old to be Asandrochos, the father of Aspurgos. Instead, he regards Scribonius as the father of Dynamis' child.[15]

At least age-wise, there is nothing to object to this speculation, for a queen in her early 40s could in theory still conceive. But, at the time of the *Ara Pacis* procession in 13 BC (on which see more in the next sections), she would have been between 47 and 50 years old. We can infer this from the fact that she should be the daughter whom Pharnakes offered to Caesar in marriage before the Battle of Zela in spring 47 BC, which makes her at least 12, probably 13 to 15. She might even have been older, if she was already married to Asandros at the time of the revolt; in this case, the king would have to look around for a new match for her after putting down the insurgent (which he failed to do).[16] Admittedly, there is the possibility that she had sisters who escape our record. But we happen to have two coin portraits of her, one dated to 21/20 (Fig. 2) and the second to 17/6 BC. They both show her as a mature (and well-nourished) lady barely younger than 40.[17]

Fig. 2: Gold Stater of Dynamis as Sole Ruler, dated to the Mithradatic Era Year 281 (AΠΣ) = 17/16 BC. State Historical Museum, no. A 393. Cf. Rostovtzeff 1916/19; RPC I 1864; Frolova 1997; Heinen 1997; Frolova – Irland 2002, 49; MacDonald 2005, 53f.

14 Pollini, 1987, 23 says the child is approximately 86 cm. But more important is the ratio of child to adult size, which suggests that the boy is about five. See below, with n. 42 for details.
15 Parfenov 1996, followed by Romeo 1998, 130f. Cojocaru 2014, 6f. is positive, but undecided. Ballesteros Pastor 2008b remains uncommitted. Roller 2018a, 96, 145 is somewhat hesitant to regard the boy as a child of Dynamis due to age reasons, but still considers her trip to Rome possible. Saprŷkin 2002, 98f. rejects both Rose's and Parfenov's view, without, however, closely engaging with their arguments.
16 Pharnakes was killed by Asandros some four to five weeks later. See App. *Mith.* 120.590–595, with Coşkun 2019a; 2019b; 2020a.
17 Cf. Coşkun 2019a.

Fig. 3: 'Hermitage Dynamis'. Hermitage PAN.1726. Source: Rostovtzeff 1919.

Many readers will be more familiar with a bust found in Novorossiysk on the south-eastern margins of the former Bosporan kingdom, generally known as the 'Dynamis' of the Hermitage in St. Petersburg, ever since Rostovtzeff proposed this identification in 1916 (Fig. 3). The identification is not based on an inscription or monogram, but mainly due to the fact that Dynamis was the single-most prominent female in the area throughout antiquity.[18] Rostovtzeff connected the star on her head gear with the dynastic iconography. But the Mithradatid dynasty typically used that combination of star and crescent (as did other Near-Eastern potentates), which is not depicted on the Hermitage bust. Moreover, the 'Phrygian cap' of the bust may instead point to an Anatolian context, perhaps even hinting at the Phrygian Kybele, who was indeed venerated as a celestial goddess in many parts of the ancient world.[19] Given the limitation of our visual evidence for the appearance of Dynamis, it would be risky to draw on the bust of the Hermitage.

Besides iconography, aspects of political plausibility must be considered as well. One might speculate about the chances of survival of a very young son of Asandros during the revolt against him, likewise those of an offspring of Scribonius when the subjects rose up against him. Proponents of the identification of Dynamis and a son of hers on the *Ara Pacis* might suggest, on the one hand, that Dynamis sought Roman protection from Polemon; on the other hand, it would hardly have been a wise choice to go against the popular queen (and her son, if she had one), considering the people's affection for her. In fact, Agrippa's insistence on the dynastic marriage carries an implication for Polemon that violence against Dynamis (and therefore her putative son) would have angered the Romans and for Dynamis that her place was at the Bosporus, to allow Polemon to rule and campaign. Dynamis was not in need of Roman protection, once the deal had been accepted. If she had chosen to defy Polemon and Rome, she would have had to withdraw to one of the confederate Scythian tribes rather than to Rome.

This is at least the conclusion we draw from the combined literary, numismatic and epigraphic evidence, most of all Cassius Dio, our main witness for the turbulent dynastic successions in the Bosporan kingdom. About 14 BC, he writes:

---

18 Opinions over the accuracy when matching coin portraiture to a statue can vary greatly, but we do not find the resemblance between the Hermitage bust and the Dynamis stater to be very strong. The facial features of the bust are timeless and ageless, they are more ideal than personal. The coin displays maturity and regal markers.

19 See Rostovtzeff 1919, based on an earlier publication in Russian from 1916. He is still followed, e.g., by Kozlóvskaia 2003 and 2004, 125, 133; Halamus 2017, 162. For a more sceptical view, see, e.g., Roller 2018a, 94. On Mithradatid dynastic symbolism, see also Ballesteros Pastor 2021.

(4) And the revolt among the tribes of the Kimmerian Bosporus was put down. For someone named **Scribonius**, claiming himself to be a grandson of **Mithradates** and claiming to have received the kingdom from Augustus after **Asandros** had died, married Asandros' wife, the one named **Dynamis**, who also had been **entrusted with the regency by her husband**. She was really both the daughter of **Pharnakes** and the granddaughter of Mithradates, and thus Scribonius was holding the Bosporus under his control. (5) And so **Agrippa**, upon learning of these matters, sent against him **Polemon**, the king of that part of Pontos bordering on Kappadokia. Polemon learned Scribonius was no longer alive (for the people of Bosporus, learning of this invasion against them, pre-emptively killed him), but they were still opposing Polemon out of fear that they would be given over to be ruled by him and come under his hand. (6) And even though he defeated them, he was unable to bring them to terms before Agrippa came to Sinope, to campaign against them, also. Then they laid down their arms and gave themselves up to Polemon; and the lady **Dynamis** became his wife, for it was clear that Augustus was approving of it.[20]

Having unfolded the evidence for the dynastic history of the Bosporus, we conclude that a visit of Dynamis (and a potential son of hers) to Rome in 14 or 13 BC is quite unlikely and entirely unsupported by the ancient sources investigated so far. The question is whether the re-interpretation of the *Ara Pacis* yields new evidence urging us to reconsider our previous reconstruction of Bosporan history.[21]

II. THE *ARA PACIS*: A SNAPSHOT OF AUGUSTAN IMPERIAL IDEOLOGY

The Roman Senate voted on 4 July 13 BC to erect the *Ara Pacis Augustae* to honour the successes of Augustus and Agrippa achieved in Germany, Gaul and the East by 13 BC as well as those that were yet to come.[22] Augustus himself reports that the altar was built on the *Campus Martius* on the west side of the *Via Flaminia* (today's Corso), one Roman mile from the *pomerium*, near the present Piazza San Lorenzo in Lucina.[23] Palazzo Fiano-Almagià (Palazzo Ottoboni-Fiano until

---

20 Cass. Dio 54.24.4–6 ed. Boissevain 1898: τά τε ἐν τῷ Βοσπόρῳ τῷ Κιμμερίῳ νεοχμώσαντα κατέστη. Σκριβώνιος γάρ τις τοῦ τε Μιθριδάτου ἔγγονος εἶναι καὶ παρὰ τοῦ Αὐγούστου τὴν βασιλείαν, ἐπειδήπερ ὁ Ἄσανδρος ἐτεθνήκει, εἰληφέναι λέγων, τὴν γυναῖκα αὐτοῦ Δύναμίν τε καλουμένην καὶ τὴν ἀρχὴν παρὰ τοῦ ἀνδρὸς ἐπιτετραμμένην, ἣ τοῦ τε Φαρνάκου θυγάτηρ καὶ τοῦ Μιθριδάτου ἔγγονος ἀληθῶς ἦν, ἠγάγετο, καὶ τὸν Βόσπορον διὰ χειρὸς ἐποιεῖτο. (5) πυθόμενος οὖν ταῦτα ὁ Ἀγρίππας τὸν Πολέμωνα ἐπ' αὐτόν, τὸν τοῦ Πόντου τοῦ πρὸς τῇ Καππαδοκίᾳ ὄντος βασιλεύοντα, ἔπεμψε· καὶ ὃς Σκριβώνιον μὲν οὐκέτι περιόντα κατέλαβε μαθόντες γὰρ οἱ Βοσπόριοι τὴν ἐπιβολὴν αὐτοῦ προαπέκτειναν αὐτόν, ἀντιστάντων δὲ οἱ ἐκείνων δέει τοῦ μὴ βασιλεύεσθαι αὐτῷ δοθῆναι, ἐς χεῖράς σφισιν ἦλθε. (6) καὶ ἐνίκησε μέν, οὐ μὴν καὶ παρεστήσατό σφας πρὶν τὸν Ἀγρίππαν ἐς Σινώπην ἐλθεῖν ὡς καὶ ἐπ' αὐτοὺς στρατεύσοντα. οὕτω δὲ τά τε ὅπλα κατέθεντο καὶ τῷ Πολέμωνι παρεδόθησαν· ἥ τε γυνὴ ἡ Δύναμις συνῴκησεν αὐτῷ τοῦ Αὐγούστου δῆλον ὅτι ταῦτα δικαιώσαντος. Our translation has modified the Loeb edition by Cary & Foster 1914. See § 1 for the consular year date.

21 Cf. Cojocaru 2014, 6f., who is undecided between Rose, Parfenov and Frolova, calling for a thorough scrutiny of the *Ara Pacis*.

22 *CIL* IX 4192 = *EDCS*-55800157: *... quod eo die ara Pacis Augustae in campo Martio constituta est Nerone et Varo cos.*

23 Aug. *RG* 12 = *EDCS*-20200013 = Mitchell & French 2012, no. 1, p. 76, col. 2, ll. 36–41 (though reading *co[nsacrandum* in l. 39): *[Cum] ex [Hisp]ania Gal[liaque, rebus in eis]*

1898) and the San Lorenzo Church today occupy the grounds where the *Ara Pacis Augustae* stood, one kilometre away from the *Ara Pacis* Museum. The *Ara Pacis* was a temple-like sanctuary, but it did not 'house' a divinity or have a roof.[24]

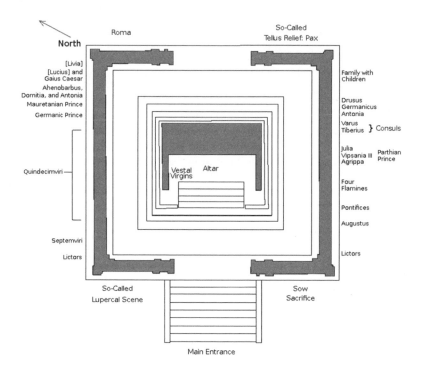

Fig. 4: Iconographic Plan of the Ara Pacis. Drawing by Gaius & Ben Stern, 2020, based on the 'Plan Ara Pacis Augustae' by Wikimedia user 'Augusta 89', updated by Wikimedia user 'Vigneron', translated into English and adapted by Gaius and Ben Stern. Public Domain (CC BY-SA 3.0). URL: https://en.wikipedia.org/wiki/Ara_Pacis#/media/File:Plan_Ara_Pacis_Augustae.svg.

provinciis pros[p]ere [gestis Romam redi] Ti. N[er]one P. Qui[ntilio consulibus] aram [Pacis A]u[g]ust[ae senatus pro] reditu meo co[nsecrandam censuit] ad cam[pum Martium, in qua m]agistratus et sac[erdotes et virgines] V[est]ales [anniversarium sacrific]ium facer[e decrevit].   'When I returned to Rome from Spain and Gaul, since matters were faring prosperously in the provinces in the year when Ti. (Claudius) Nero and P. Quintilius (Varus) held office as consuls, the Senate decided an Altar of August Peace should be consecrated on the Campus Martius, at which the magistrates and priests and Vestal Virgins should make a sacrifice annually.'

24 It measures 11.625 m (east and west) by 10.655 m (north and south) and stands 3.7 meters tall (40 x 37 x 10 feet). On the dimensions, see Pasqui 1903, 568; Rizzo 1919/20, 4; Moretti 1948, 149–151, 1939/75, 8f.

All of the walls of the *Ara Pacis*, inside and out, bear very high-quality sculpture. Its complex imagery (Fig. 4) models and proclaims the unprecedented peace, stability, prosperity, and collaboration of leadership. The Roman World was about to enter this new era of Peace due to the sage guidance of Augustus and his closest allies, who had brought Rome through the terrible era of civil wars and restored the *mos maiorum* ('ways of the ancestors'). The *Ara Pacis* stands out due to the sophistication with which the artists displayed a series of positive ideals to predict and to assure the public that the new Golden Age had just begun and that the wisdom of Roman internal cooperation had even greater benefits than previously expected.[25] However, to understand the more complex messages, one must look at the monument and the imagery as did the Roman audience rather than with modern eyes. It was a monument made by Romans for Romans, so messages that were obvious to them are not always obvious to us and can seem to be coded in a foreign language.

The east and west sides (front and back) of the altar show people and deities in allegorical rather than realistic scenes. Modern drawings fill in the very badly damaged NW and NE panels to recreate the original Lupercal scene (a cave with the legendary she-wolf suckling Romulus and Remus) and a seated Roma, respectively, but neither drawing is at all secure. The better-preserved SW and SE panels show a sacrifice of a sow and a seated goddess in a very tranquil setting. The sow sacrifice is widely interpreted as Aeneas sacrificing either to Juno or to the Penates, but many discrepancies hinder that interpretation, such as the absence of the miraculous thirty piglets, so that we prefer to identify the priest with king Numa. The latter panel was long called the Tellus Relief, but she is surely Pax Augusta herself, for she differs from contemporary Tellus representations, and her presentation advertises the benefits of peace.[26]

The north and south friezes of the *Ara Pacis* display three colleges of priests (the Pontifical College, the *Septemviri Epulones* and the *Quindecimviri Sacris Faciundis*), twelve lictors, four cult attendants, and the imperial family, totalling most probably 96 figures.[27] They wear their finest apparel, marching in a religious ceremony, which bears distinct political overtones, most likely, in a *supplicatio*, a ritual of public thanksgiving to the gods.[28] John Pollini has shown that within each

---

25 The bibliography for the *Ara Pacis* is vast. For post WW II reception, see chapter 2 of the much-cited dissertation by Pollini 1978 as the state of scholarship before Rose 1990. For a post 2000 view, see Stern 2006b. The latest is Cornwell 2017, esp. 155–186.

26 See Rehak 2001 and Stern 2015a. For discussion, see also Cornwell 2017, 163–177.

27 Different scholars suggest between 91 and 100 figures. E.g., the old view repeated in Pollini 1978 placed 46 and 45 figures on the north and south friezes for a total of 91; Koeppel 1987 counted 49 and 48, totalling 97; Rossini 2006 surmises twice 48, likewise Kleiner & Buxton 2008; Stern & Coşkun in preparation.

28 Polacco 1960/1 proposed that the *Ara Pacis* ceremony was a *supplicatio*, followed by Simon 1967, 22, and argued more forcefully by Billows 1993 (cf. Cornwell 2017, 177f.). On *supplicatio* in general, see Wissowa 1931. In the Middle Republic, the Senate would occasionally mark a great military victory by declaring a *supplicatio* for one to three days. The length of these *supplicationes* began to increase along with the inflation of other honours. In 29 BC, the Senate declared several of Augustus' victories were to be repeated as annual *supplicationes*,

religious college an attendant carried an incense box that identifies the college.[29] The uneven distribution of the lictors on the two friezes actually unites the two friezes, for only when taken together do the dozen lictors form a coherent escort for Augustus, who was not a consul but held consular power. Thus, the *Ara Pacis* depicts a single event represented on its two sides.[30]

A lengthy debate divided scholars from 1880 to the 1940s over the date of the procession (13 BC vs. 9 BC) on the friezes, the participants, and the messages those individuals transmitted (the latter two debates continue to this day), since some candidates for inclusion died or were born between the dates in question. Every Roman magnate wanted to be included on the friezes to advertise his own importance and loyalty to the regime in 13 BC, so which participants made the cut helped scholars date the event to 13. Crucial for this was the identification of Augustus (S-13, Fig. 5) and Agrippa (S-25, below, Fig. 8) on the south frieze.

Fig. 5: Left Half of South Frieze, Nearly Beginning with Augustus (S-13).
Photo by Gaius & Ben Stern, 2020.

We assert that the festival on the *Ara Pacis* occurred well after 4 July. The preparations took time to arrange, so weeks or even months passed before the elaborate

including the Battle of Naulochus (3 Sept. 36 BC), the Battle of Actium (2 Sept. 31 BC), his Triple Triumph (13–15 Aug. 29 BC) and his birthday (23 Sept.). The Senate also declared a *supplicatio* should be held every time Augustus returned to Rome, an honour he found obsequious and frequently frustrated by entering the city unannounced at night. See Wissowa 1931, 949f.; Kienast 2004, 61–64, 363–366; Stern 2006b, 168–176. Kleiner & Buxton 2008, 68–71 argue that the replacement of the triumph was not solely due to modesty or jealousy.

29 Pollini 1978, 84f.

30 In reconstructing the *Ara Pacis* in the 1930s, Moretti (1938/75 and 1948) allocated space for twelve lictors at the front of the south frieze beyond the pair he identified on the north frieze. We propose that the correct number for the south frieze is seven lictors, plus five on the north frieze: Stern & Coşkun in preparation.

supplicatio took place. And several important officials, whose participation was essential, were not present in early July. To mark the start of the construction of the *Ara Pacis*, i.e. the digging of the first spade of earth, they probably selected a pre-existing annual holiday, such as the anniversary of the Battle of Actium (2 September) or Augustus' 50th birthday (23 September).[31] Augustus evidently wanted to show his modesty in the Roman spirit of *satis gloriae* ('enough glory' for one person) to avoid adding another holiday celebrating him in swift succession.[32]

Fig. 6: Right End of South Frieze, with Antonia (S-35) and Drusus (S-38).
Photo by Gaius & Ben Stern, 2020.

Several of Augustus' former enemies had eventually joined him and attended the celebration of the successes of the regime.[33] The extended imperial family populates more or less the back third of the *Ara Pacis*. The artists arranged the relatives of Augustus in family groups to make sure the audience would recognize the participants of 13 BC. One of the more important of these family groups consists of the emperor's niece Antonia Minor and his younger stepson Drusus, together with

---

31 Stern 2006b, 175–179. The Romans counted inclusively in many situations, so one was born one year old. Thus, what we call the 50th birthday was 51st in customary Roman reckoning, but this day could well be seen as the conclusion of his 50th year of birth.

32 It is possible that since the *consecratio* took place on his birthday, Livia's 50th birthday (49th for us) was selected for the *inauguratio*.

33 We mention here only the pardoned son of Antony, Iullus Antonius (S-45), who was *praetor urbanus* in 13 BC, and M. Aemilius Lepidus (S-15), the *Pontifex Maximus,* who was obliged to participate, even if unwilling, to fulfil a religious duty. See Cass. Dio 54.15.5f. See Stern & Coşkun in preparation.

their son Germanicus (S-35, S-38, S-36 on Fig. 6), whose exact birthday is known to be 24 May 15 BC.[34] Obviously, his sister Livilla (born 14 BC)[35] was too young to participate, and the later emperor Claudius was not yet born.[36] Drusus had just returned to Rome from Gaul, so he was able to participate.[37]

Fig. 7: Fragment of Left End of North Frieze, with Gaius Caesar (N-45) and German Prince ('Buttock Boy', N-35). Photo by Gaius & Ben Stern, 2020.

The real-life age of Germanicus perfectly matches the toddler who clings to his mother's finger. He serves to verify the date in late Summer 13 BC against the argument for 9 BC when Germanicus turned six. This child is obviously too small to be a six-year old, even though advocates for 9 BC acknowledged him as Germanicus! The conversation between Antonia and Drusus is one of the many human touches we find in the friezes. Another example would be the weary toddler

---

34 *CIL* VI 28028.31; cf. *PIR²* I 221; Mommsen 1878, 245; Kienast 2004, 68; 79–81.

35 Claudia Livia (Livilla) Iulia, daughter of Drusus the Elder and Antonia the Younger, married to Gaius Caesar in ca. Sep. 1 BC. See *PIR²* L 303; Mommsen 1878; Kienast 2004, 69, 74; cf. Bleicken 1998/2010, 641f., who mistakenly assumes that she was about the same age as Gaius, who was born in 20 BC. Since she was born at least ten months after her older brother Germanicus, thus after March 14 BC, she would have been just 14 when she married Gaius. The minimum age for girls to marry was 12, but around 14 was far more common. Given Gaius' assignment to the East, perhaps an early marriage was deliberately selected.

36 Claudius was born 1 Aug. 10 BC, see Suet. *Claud.* 2.1: *Claudius natus est Iullo Antonio Fabio Africano conss. Kal. Aug. Luguduni eo ipso die quo primum ara ibi Augusto dedicata est.* Cf. *PIR²* C 942; Kienast 2004, 90–92.

37 Von Domaszewski 1903, 63 n. 27, 1909, 98 n. 5, argued that the testimony of Cass. Dio 54.25.1 prevents the participation of Drusus in the *Ara Pacis* ceremony on 4 July 13 BC, because Augustus left his stepson in command in Germany when he returned to Rome that summer, followed by Löwy 1926, 57; Toynbee 1961, 155; Pollini 1978, 124; Torelli 1982, 55; Polacco 1960/1, 629; Billows 1993, 91; Holliday 1990, 548. That objection vanishes entirely when the ceremony is moved to 23 September.

on the north frieze (N-35, see Fig. 7), who is too young and too exhausted to understand the ceremony and wants to be carried. We shall argue below that this 'Buttock Boy' was actually a German prince.[38] Regarding chronology, we refrain from adducing further prosopographical evidence, except for the most obvious case: the presence of Agrippa (S-25, Fig. 8), whose death in March 12 BC provides another firm *terminus ante quem*.[39]

*Fig. 8: Centre-Right Part of South Frieze, with Agrippa (S-25), Eastern Prince (S-27)
and Putative Dynamis (S-28). Photo by Gaius & Ben Stern, 2020.*

The south and north friezes of the *Ara Pacis* combined thus presented an idealised and yet naturalistic snapshot of the procession of September 13 BC. It was of course a construction, since it represented a selection from among the thousands of participants and ensured that the most prominent aristocrats be recognisable by their realistic portrait and dignified habitus. Our genealogical table (Fig. 9) tries to visualise in an alternative format the members of the imperial family as we suppose that they participated in the celebration.

38 More on him below, with n. 64.
39 Agrippa died in March 12 BC during the *Quinquatrus*, a five-day festival to Minerva (19–23 March): Cass. Dio 54.28.3; see also Vell. 2.98.1; Liv. *Per.* 138; Plin. *NH* 7.8.45. The presence of Lepidus, whom we identify as S-15, also invalidates a date of 9 BC for the procession: Lepidus died before 6 March 12 BC when Augustus was elected *Pontifex Maximus* to replace him, perhaps due to exhaustion after the *Ara Pacis* ceremony; cf. Aug. *RG* 10; the *Fasti Praenestini*; Ov. *Fasti* 3.415–428; Cass. Dio 54.27.2. For more prosopography, see Stern & Coşkun in preparation.

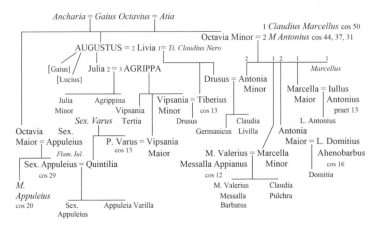

Fig. 9: A Genealogical Table of the Imperial Family in 13 BC.
Draft: Gaius Stern & Altay Coşkun. Drawing: Stone Chen. Berkeley & Waterloo 2020.
Gaius and Lucius Caesar were adopted in 17 BC and appear in square brackets as Augustus'
adoptive sons. Italics indicate 'deceased before 13 BC'. Children are in smaller font than adults.

Agrippa (S-25, Fig. 8 above and Fig. 11 below) plays central roles in many regards, not only for our present argument that is concerned with the individuals
besides him. He also serves as a crucial transition figure at the centre of the south
frieze, commanding the attention of the audience. He was veiled because he was
sharing the duties of officiation with Augustus on behalf of the Pontifical College.
His location, height, and prominence display him as a second emperor, although
the regime would have denied that they were rulers, because it maintained the
trappings of a republic to deflect charges of monarchy and dynasty. Other clues
indicate his status as well. He was a *pontifex,* but unlike the other priests, he has
his wife Julia by his side (S-29) (rather than his step-mother-in-law Livia, who
was somewhere in the now-lost left part of the north frieze, Fig. 7 above). Indeed,
Julia's posture closely mimics Agrippa's, a clear sign they are man and wife.

   A little behind this couple come other families with children, the latter all
dressed in the *toga praetexta* (Fig. 10). Germanicus (S-36), a boy and a girl who
have mistakenly been identified as Domitius and Domitia (S-40, S-42). In between Agrippa and Julia stands a young child (S-27, Fig. 8 above and Fig. 11 below) clinging to Agrippa's toga. Earlier scholars took him for Lucius Caesar and
then later Gaius Caesar, the biological sons of Agrippa and Julia, whom their
grandfather Augustus adopted in 17 BC.[40] In fact, this boy is not their son at all,

---

40   In favour of Lucius Caesar: Petersen 1894, 177–228, 1902, 108; Reisch 1902, 427f.; von
     Domaszewski 1903, 62; Dissel 1907, 17; Strong 1907, 29; Studniczka 1909, 913; Sieveking
     1917, 91; Lugli 1935, 382; Riemann 1942, 2099; Toynbee 1953, 84; Polacco 1960/1, 614,
     Bonanno 1976, 29. The opinion was close to unanimous before 1920. Petersen 1902, 109 n. 1,
     provides a correspondence in which Benndorf argues against Lucius, based on age. Benndorf

but a foreign prince they are fostering for reasons of state, whose presence in Rome counts as a diplomatic success for Augustus (and Agrippa), and whose presence on the *Ara Pacis* brags of Rome's global stature, since foreign royalty come to Rome to learn Roman ways.[41]

Fig. 10: Three Roman Children on Left Side of South Frieze: Germanicus (S-36) and the Putative Domitius and Domitia (S-40 and S-42). Photo by Gaius Stern, 2013.

preferred 13 to 9 BC, but does not name Gaius as such. See also Stuart Jones 1903/4, 256f.: 'Gaius or Lucius'; Strong 1907, 49: 'may be one of his (Agrippa's) sons'; Ducati 1920, 683: 'one of the two sons of Agrippa'; Kleiner 1978, 758: 'Gaius or Lucius'. The consensus shifted to Gaius Caesar, starting with Löwy 1926, 60 n. 28, adopted by Monaco 1934, 32f.; Moretti 1948, 229; 270–273; Kleiner 1978, 758; 1992, 93; Torelli 1982, 48–51, n. 72; Koeppel 1982, 507–535, esp. 527; 1987, 124; Pollini 1978, 105f., initially concurred with Löwy (and others), but changed his mind by 1987, 22–25. However, Löwy's view persists today, even after Rose's decisive 1990 article: e.g., Holliday 1990, 548; Polacco 1991/2, 24, pl. 5.1. The view of Zanker 1987/2009, 218–220 has remained unchanged even in the 5th ed., which still fails to account for Rose 1990 and Kuttner 1995. However, Zanker 1987/2009, 348 does speak out against the identification of the boys as 'Barbarenprinzen' by Simon 1968, surmising that 'die Künstler der *Ara Pacis* auch sonst nicht an Wirklichkeitstreue in der Körperwiedergabe interessiert [sind]'.

41   See below, notes 71–73 for sources on child hostages under Augustus. Compare Rome in 13 BC to Athens in 431 BC from Perikles' 'Funeral Speech', Thuk. 2.41.1: ξυνελών τε λέγω τήν τε πᾶσαν πόλιν τῆς Ἑλλάδος παίδευσιν εἶναι καὶ καθ' ἕκαστον δοκεῖν ... 'In a nutshell, I declare that our entire city seems to be the school of Hellas and in every way ...'

In the background reaching her hand out to touch this boy's head stands a woman (S-28), whom Rose formerly suggested identifying as Queen Dynamis of the Bosporus. Her true identity is also tied to those directly adjacent to her on the south frieze, especially the boy in foreign dress, on whose head her hand is resting (S-27). These two figures form the centre of this study. So, one must understand the implications of the identity of each figure around them to understand why Rose thought the boy might be her son, the future king Aspurgos. Giuseppe Moretti, however, thought the boy was Gaius Caesar in Trojan dress for the Troy Games of 13 BC.[42] We maintain that he is a foreign 'guest', though rather a Parthian prince, while the young lady behind him most likely is a daughter of Agrippa.[43]

III. THE YOUNG LADY AND THE BOY IN THE COMPANY OF AGRIPPA: DYNAMIS AND ASPURGOS?

1. Moretti's Troy-Games Theory

We have so far identified the scene of the south and north frieze as representing the *Ara Pacis* ceremony of September 13 BC, and further specified Augustus and – more importantly for our present concern – Agrippa in the procession of priests on the south frieze. Agrippa appears as one of the last *pontifices* (S-25, cf. Fig. 8 above), veiled to show he shares officiating duties with Augustus.[44] His wife Julia (S-29) follows him. A boy tugs at his toga, and the background woman puts her hand on the child's head in an affectionate gesture (S-27 and S-28).

Looking at the boy's figure, Moretti decided he was Gaius Caesar, despite the fact that he does not wear Roman shoes or the *toga praetexta* – which all the other Roman children wear, even the two girls. Moretti did not even notice that the boy has no *bulla*. Moretti explained the discrepancies by claiming that Gaius was dressed for the Troy Games of 13 BC and thus appears like a Trojan.[45] Before we

---

42 We concur with Pollini 1987, 21–23 (cf. Zanker 2009, 348) that this boy, approximately five years old, is too small to be the seven-year-old Gaius Caesar. A father's elbow is likely to touch the head of a child when about turning five. At least, this was the case with all four of A. Coşkun's children, with a variance of plus minus three months. But his size is merely one of many problems barring an identification as either Gaius or Lucius. If he were Gaius, he would be dressed in a toga and accompany his adoptive parents Augustus and Livia, as well as his brother Lucius. Instead, he is dressed as a foreigner and is too short and appears without a sibling. Nor does he wear the *bulla*, a *sine qua non* for a Roman aristocratic boy.

43 For more on this lady, see below, III.3.

44 For Lepidus and Agrippa as *pontifices*, see Stern 2006a; 2006b, 126–30; 2015b 70–75; Rüpke 2007, 12 and 232. Tiberius was also a *pontifex*, and maybe Varus. But he certainly appears as a consul in his red toga, rather than in a religious role on this occasion.

45 In total, three boys appear in the processional friezes in non-Roman clothing, of whom the most obvious is the one clinging to Agrippa. The other two are the German toddler (N-35, Fig. 7 above) in the foster care of M. Lollius and the boy (N-38) close behind him, whom Pollini 1987, 26f. and Rose 1990, 463f. consider to be Gaius dressed as a *camillus* (an aristocratic boy serving as cult attendant). He could be the grandson of Mark Antony and Kleopatra

explain why Rose regarded them as queen Dynamis and her (assumed) son Aspurgos, we shall enquire more broadly into the foreign nature of the boy. This had
not been a subject for Eugen Petersen or Emanuel Löwy, who just identified him
as Lucius or Gaius Caesar respectively. It is Moretti who first attempted to explain
why this boy is not represented as a young Roman citizen. Moretti read in Dio that
Troy Games were among the festivities of 13 BC.[46] Thinking that Augustus had
returned to Rome on 4 July and accepting that date for the festivities both that
greeted Augustus' return and that included the procession on the *Ara Pacis*, he
added the Troy Games to the mix. In his view, 4–5 July 13 BC witnessed major
celebrations in Rome that comprised the welcome home for Augustus, the *consti-
tutio* of the *Ara Pacis* in the Senate and the Troy Games.

The Romans held Troy Games at irregular intervals to recall their national heritage, to parade the origins of the leading Roman families from Troy, and to emphasize the teamwork that had made Rome successful. They may have included a
lesson lost to us about unity regarding plebeians and patricians or the success of
the unified state against all odds. Two squads of cavalry, composed of boys of
noble families aged eight to thirteen, performed a parade which displayed horse-
manship and teamwork, but not combat. Occasionally someone was hurt, leading
to complaints. Sometime after the Troy Games of 2 BC, in which his grandson
Agrippa Postumus rode, Augustus temporarily halted the games, because P. Nonius Asprenas and M. Claudius Aeserninus both broke a leg, as we learn from
Suetonius.[47] Vergil's fifth book of the *Aeneid* has much detail on athletic contests.
In particular, the following section caught the attention of the Italian scholar:

> *omnibus in morem tonsa coma pressa corona*
> *cornea bina ferunt praefixa hastilia ferro,*
> *pars levis umero pharetras; it pectore summo*
> *flexilis obtorti per collum circulus auri.*

> The hair of all pressed by a twisted crown, according to custom;
> Some carry two cornel lances tipped with iron each,
> Some hang polished quivers from their shoulders, while at their throats
> A pliant circle of twisted gold hangs around their necks.[48]

The most important potential link between the Troy Games and the boy clinging
to Agrippa on the *Ara Pacis* is that the boy wears what appears to be a torque.
Matching Suetonius and Dio with Vergil, Moretti devised a theory that the Troy
Games were part of the welcome festivities for Augustus. This would explain why
Gaius was dressed as a Trojan in the procession seen on the *Ara Pacis*. Moretti's

---

VII, the future king Ptolemy of Mauretania, in Rome to learn Roman ways from his kinsmen.
He is walking before his aunt Antonia Maior (N-40), who may have been hosting him, perhaps together with his own mother, if the woman in the exotic dress preceding him is Kleopatra Selene (N-36) and not Julia, as many others have suggested; see Stern & Coşkun in
preparation for discussion and references.

46   Cass. Dio 54.26.1; Moretti 1948, 116–146; 270f.
47   Suet. *Aug.* 43.2, who leaves it unclear whether the two fractures occurred in the same year.
48   Verg. *Aen.* 5.556–559.

proposal was universally accepted, even though it is extremely tenuously formulated. When Erika Simon first disputed it, she was quite ridiculed by some scholars.[49] It took another 25 years before Brian Rose and Ann Kuttner came forth to refute Moretti, with very strong evidence against him, largely based on iconographic grounds. In his 2006 dissertation, Gaius Stern followed Rose and Kuttner and added historical reasons to refute Moretti as well.[50]

Moretti did not maintain the chronology Dio established. Dio does not link the Troy Games of 13 directly to Augustus' return. He associates them with the inauguration of the Theatre of Marcellus and with Augustus' birthday on 23 September, which he explains as the occasion for the deaths of so many wild beasts.[51] Moretti's Troy Games were not held in conjunction with the return of Augustus in July, at all. According to Moretti's own reconstruction of the events, the Troy Games were still two months away, but he already dressed Gaius as a Trojan (sic) all the same and somehow lost his bulla, in order to ride in Troy Games that were still months away! As a further problem, we are far from convinced that the boy's costume is Trojan. It does not especially resemble the ensemble of Ascanius/Iullus seen in Roman art and most of all lacks a Phrygian cap, which Ascanius/Iullus frequently wears. We note both the anachronism of Moretti's Troy Games theory and the fact that all of the Roman children from the imperial family wear the toga praetexta, even the girls. The real Gaius Caesar wore a toga praetexta in the ceremony; this is not a depiction of the real Gaius or any Gaius at all.

Moretti's theory has additional holes. To start with, Vergil does not use the word torques in association with the equestrian performance of Iullus/Ascanius and company, although Torelli (falsely) attributes it.[52] Nor is there any evidence that participants in historical times ever wore a torque when riding in the Troy Games. Suetonius says only that Augustus gave a torque to one participant, not to the others, and he did so the last time boys rode in the games.[53] If boys always wore a torque in those performances, little Nonius Asprenas would have no need for a new torque, he already would have had one. In addition, what makes Augustus' gift special is clearly its uniqueness for a child, whereas it is otherwise attested only as a rare distinction for bravery in combat.[54] Be this as it may, the torque

49 Simon 1963, 9 theorized that the foreign-dressed boys N-35, S-27 are eastern princes, reworded 1968, 18 to 'barbarians'. Torelli 1982, 60, n. 72, flippantly dismissed Simon's view as 'perfect nonsense'. But he also claims to see 'the two-year-old Lucius Caesar' (p. 50), who was four in 13 BC!
50 Rose 1990; Kuttner 1995; Stern 2006b, 175f.
51 Cass. Dio 54.26.1–3.
52 Torelli 1982, 60, n. 72: 'The boys wore also "torques flexilis obtorti per collum circulus auri" (1.559), a jewel of Trojan, oriental flavour.' Torelli's use of quotation mark seems to be attributing the torque surreptitiously to Vergil, who makes no mention of torques albeit.
53 Suet. Aug. 43.2.
54 The torque was originally perceived as a piece of typically Gallic jewellery and is first mentioned in the report of the Battle of the Anio (361 BC), in which T. Manlius volunteered for a man-to-man combat. He defeated his Gallic opponent, ripped off his torque and put it on himself, as the aetiology for the cognomen in this family goes. Livy (7.10) makes it clear that the torque was not the reward granted by the commander, which was rather a golden crown. In

on the south frieze may not even be authentic. Independently from our current concern, Gerhard Koeppel has suggested that the torque is the result of a misleading modern restoration.[55] If he is correct, Moretti's speculative theory will even lose its material premise.

Moretti's reconstruction has as a further premise that the boy must be Gaius. But we have shown in the previous chapter that neither the age of the child nor his grouping between Agrippa and Julia is compatible with this identification. If, however, Moretti were right, Augustus would have exposed his adoptive sons Gaius (and Lucius) in a way that challenges the subtlety of his PR efforts. We would rather like to endorse the conclusion of Diane Kleiner and Bridget Buxton:

> In 13 B.C.E., Augustus continued to shield his adoptive sons from the kind of public attention that might spur resentment and suspicion. Dressing them as Eastern princes would have been a needlessly provocative gesture ... This situation positioned the boys not as sole heirs to Augustus' unique political authority but as part of a community of Roman and foreign children representing the next generation of aristocratic peers.[56]

2. A Foreign Prince in Augustan Rome

In fact, we assert that three of the boys are not even Roman citizens, given that they neither wear a toga nor a *bulla*. Our argument owes much to the observations of Simon, Rose, and Pollini, who first argued for the representation of foreign princes on the *Ara Pacis*.[57] Other scholars have added more detail, approaching the problem from different angles. Kuttner realized the toddler boy identified as 'Lucius' by Moretti (N-36) bore a striking resemblance to a Germanic or Gallic child on the Boscoreale cup, whose family begs Augustus for clemency. She theorized that the nearly identical boys corresponded to a Roman standard image of Germanic or Gallic children in a pose of submission to the Romans. In no way could a figure in such a pose associated with submissive barbarians be a Roman noble, let alone a member of the imperial family.[58]

Rose provided a number of iconographic reasons why the boy at Agrippa's side (S-27), who had inversely been regarded as 'Gaius' or 'Lucius' respectively,

Roman context, the torque reappears only in the Social War when the consul Pompeius Strabo adorned Iberian horsemen with torques, beside Roman citizenship (*ILS* 8888, line 58). It is uncertain whether the torque became a more regular reward under Augustus or whether some Roman soldiers, as the *centurio* M. Caelius who fell in the Teutoburg Forest (AD 9), simply liked to show off torques as the most distinguished spoil. The tombstone of his cenotaph (now in the Rheinisches Landesmuseum Bonn, with the epitaph *CIL* 8648 = *ILS* 2244 = *EDCS*-11100742) displays two torques hanging over his shoulder. See https://www.livius.org/pictures/germany/xanten-cut/cenotaph-m-caelius/xanten-cenotaph-of-marcus-caelius/.

55   Koeppel 1987, 124, no. 31; 135, no. 35.
56   Kleiner & Buxton 2008, 75f.
57   Kleiner & Buxton 2008, 72f.; Simon 1968, 18; 21; Pollini 1987, 27.
58   Kuttner 1987; 1995, 99–101; 107–111. Kleiner & Buxton 2008, 72–74 agree, but offer an alternative argument for a British prince in n. 68. Cornwell 2017, 179–181 accepts that the children are foreign and the notion is imperial, but remains agnostic as to their identities.

cannot be either, but better fits a prince from the East. Firstly, the boy's face does not at all resemble the portraiture of Gaius or Lucius. We cannot overemphasize this fact. Secondly, Rose believes the boy's distinctive cork-screw curls are a non-Roman hair style for this era, but acceptable for a foreigner. Thirdly, his shoes are un-Roman compared to the *mulleus* or *calceus* of the other participants on the friezes. They do resemble the shoes seen on eastern barbarians in Roman art and also match representations of the Phrygian Attis. The torque (which Rose accepts as authentic) appeared to be another indicator of an eastern context. He was thinking specifically of Galatians, although many cultures actually used this ornament as a prestige object.[59] Identity with Antiochos (III), the son of King Mithradates and Queen Iotape of Kommagene, as suggested by Kuttner, seemed possible to him, though he inclined towards Aspurgos, the later king of the Bosporus, in 1990, before expressing preference for a Parthian prince in 2005.[60]

Ilaria Romeo is one of few Italian scholars to break with Moretti over the identity of the boys on the *Ara Pacis*.[61] She found the size of the foreign boy on the south frieze to be too small (more appropriate for a five year old) for Gaius and the Germanic toddler (under or perhaps around two years old) on the north frieze to be too small for Lucius. She also agreed with Rose and Kuttner that the hairstyle and clothes were too un-Roman for Augustus' sons. She inclined to Rose's identification of Dynamis and Aspurgos, though she considers Aspurgos a son of Scribonius on chronological grounds (as Parfenov). For her the two foreign boys represent the success of Agrippa's and Augustus' work in the East and West in bringing peace to the world.

In principle, we can endorse this general conclusion and even extend it by adding the African Hellenized prince, probably Ptolemy of Mauretania, a grandson of Kleopatra VII and Mark Antony through Kleopatra Selene, who was married to Juba II of Mauretania in 25/20 BC. Their son Ptolemy was born sometime after 20 BC, to become co-ruler with his father in AD 21 (N-38, Fig. 7 above). Hence, we suggest that this boy represented Africa as the third continent.[62] However, we do not agree with Romeo on the identities of Dynamis and Aspurgos, although the political message for the Roman audience amounts to much the same.

Kleiner and Buxton largely followed in Kuttner's, Rose's, Romeo's and Stern's paths, claiming exotically un-Roman looks for three children altogether: 'each shows off his atypical attire with an audacious impropriety. The sleeveless tunic of N-34 *[N-35 in our count]* flips up to reveal his chubby buttocks, the neckline of N-37 *[N-38]*'s belted tunic slides off his right shoulder, and S-30 *[S-27]*

59 Rose 1990, 456. Note that Latène torques are attested nearly throughout the Celtic-speaking world and beyond, though not among the earliest Galatian settlers in central Anatolia. They are first attested in Asia Minor in the context of later eastern-European mercenaries; see Coşkun 2014, 143–149.
60 Rose 1990, 458f. and 2005, 38–42 with n. 105; Kuttner 1995, 104. Uzzi 2005, 149 joined in the rejection of the boys' identification with Gaius and Lucius, but hesitated to accept Bosporan royals on the frieze.
61 Romeo 1998, 130–132.
62 See over-next note.

has a similar cascading left collar'. They even ascribe a second torque to the German toddler (N-35).[63] They, too, considered all three propositions for the eastern prince beside Agrippa feasible, though favoured Kommagenian origin. For the toddler of the north frieze, they assumed German or Gallic roots. They share our view that the older 'Hellenistic' boy (N-38) was a grandson of Kleopatra VII and Mark Antony through Kleopatra Selene, though settle on an otherwise unknown brother of the abovementioned Ptolemy, the later king of Mauretania.

This way, the *Ara Pacis* frieze includes princes from the three different continents which accepted Roman hegemony under the principate of Augustus – a harmonious and powerful expression of Augustan imperial ideology or *pax*.[64]

3. A Hellenistic Queen Grouped with Agrippa and Julia?

Simon, Pollini, Kuttner, Rose, Romeo, Stern, Kleiner and Buxton extended their argument also to the woman (S-28, Fig. 8 above) standing behind the eastern-looking boy (S-27) and gently holding her hand on his head. Those who thought he was 'Gaius' or 'Lucius' often avoided committing to an identification, but some of them called the woman 'Julia' (S-28),[65] if the lady (S-29) following the boy was 'Livia', and not Julia, as we suggest, while others simply thought of a nurse.[66] In contrast, the new trend takes her (S-28) for his mother. Her headband is identified as a diadem, which no Roman would ever wear publicly in the very city of Rome. This seemed to be giving strong support for an oriental interpretation also of the boy and significantly limited the options for identifications.

Those who identify the boy as Antiochos (III) of Kommagene address the lady behind him as Jotape, a princess of Media Atropatene and wife of King Mithradates III of Kommagene. In support of this identification, Kuttner says the headband closely resembles that of her granddaughter Jotape, wife of Antiochos IV, on a coin, just as the boy's shoes resemble those on two Republican decorative bronze lamp holders (ca. 40–20 BC) with boys from that kingdom. If this

---

63   Kleiner & Buxton 2008, 72 '... N-34 *[N-35]* and S-30 *[S-27]* wear foreign torques and are groomed in a decidedly non-Roman way, with full curls rather than the customary Augustan comma-shaped locks'.

64   Kleiner & Buxton 2008, 68–87. The idea that non-Roman children represented the three known continents was first expressed by Stern 2006b, 401–405: chapter 8.3 'The Boy with a Chlamys (N-38)', where preference is given to Ptolemy, the later king, and the idea of an undocumented brother is rejected. Stern had shared these ideas with Buxton and although they ultimately settled on different princes, he would have appreciated to be credited for his ideas.

65   Ducati 1920, 683 invented the idea of placing 'Julia' in the background, followed by E. Strong 1923, 31, Lugli 1935, 382; Riemann 1942, 2099; Ryberg 1949, 83–85; Toynbee 1953, 85; Kähler 1954, 76; Moretti 1957, 10f. (a posthumously revised version reversing earlier opinions); Polacco 1960/1, 616, tav. 5.1; Bonanno 1976, 185 n. 156.

66   Monaco 1934, 33, Moretti 1948, 272; Toynbee 1953, 84f.; Kleiner 1978, 758; Pollini 1978, 118f. One can ignore eccentric suggestions, such as that of Kebric 2014, who misidentified the lady (S-28) as the deceased first son-in-law of Augustus, Claudius Marcellus, claiming an allegedly perfect match of the earlobes with another Marcellus statue.

were right, we doubt that any average member of the Roman elite might have been able to draw the connection. Although we agree with Kuttner in so far as the affiliation with Agrippa would certainly be a more efficient indicator for a contemporary, we do not see any evidence for the claim that the son-in-law of Augustus had personally interfered in Kommagene.[67] A further weakness of Kuttner's interpretation is that we have no hint of dynastic turmoil which may have resulted in the queen's travel to Rome.

Therefore, the pair Jotape / Antiochos appear as an even less satisfactory proposition to us than Dynamis plus any Bosporan prince. Despite all the reasons for our concern with Rose's interpretation, we can at least concede that Agrippa's personal involvement is sufficiently attested by Dio, as is the potential for tensions within the Bosporan royal family. We acknowledge Rose's more recent work on this matter and his change of mind. Surprisingly, his original view continues to draw supporters. At the 2011 CAMWS conference, a paper was presented that sought to corroborate Rose's identification of Dynamis and incidentally Aspurgos on the *Ara Pacis*, based on matching the earrings of the background lady on the south frieze (S-28) with those on the assumed bust of Dynamis in the

Hermitage (Fig. 3 above). As we have argued in the introduction, however, nothing compels us to accept the speculative interpretation of Rostovtzeff, which is based mostly on wishful thinking rather than positive evidence.[68] Furthermore, using a simple, removable form of jewellery to identify a historical character is especially shaky, unless it is very distinct. Altogether, nothing on the *Ara Pacis* friezes points to the Bosporus or Mithradatid dynasty in any meaningful way, nor do we have independent sources to even support the likelihood of Dynamis visiting Rome for whatever reason.

Fig. 11: Verona Girl with Brill, 1ˢᵗ Century AD. Verona Museo Lapidario Maffeiano. Photo by Gaius Stern, 2014.

We actually doubt that the mysterious lady represents a queen. The presence of female royals in Rome was quite unusual and less likely to escape our (admittedly lacunose) sources.[69] At any rate, the headband of figure S-28 does not seem to be a diadem, which was usually worn higher on the head and not over the forehead. More likely, it is a brill, the typical adornment for teenage girls in Roman times. We found a good example (Fig. 11) in the

---

67   Kuttner 1995, 104; cf. Kleiner & Buxton 2008, 74. But note that Cass. Dio 54.9.3 mentions Augustus' interference in the dynastic succession of Kommagene without reference to Agrippa's eastern mission (17–13 BC, see Kienast 2004, 72). See, in a different context, Facella 2006, 314, who rightly questions that the *Tabula Peutingeriana* presents valid evidence for any kind of Agrippa's involvement with Kommagene.

68   Gorham 2011.

69   Our sources take special interest in the visits of foreign royalty to Rome. E.g., Cato the Elder scorned the visiting ally Eumenes II by saying 'yes he is a great friend of Rome, but a king is a creature that feeds on human flesh' (Plut. *Cat. Mai.* 8.8). More examples could be added.

Museo Lapidario Maffeiano in Verona: the head of a Roman teenage girl wearing such a brill over her forehead, although it is erroneously labelled as 'diadem' on the exhibition plate.

The clothing of the lady on the *Ara Pacis* is neither that of a queen or princess, but fully in line with the dressing style of her Roman peers. If anything, a queen might wear considerably more jewellery, such as necklaces and rings, neither of which are visible. While her neck is in very low relief, both of her hands are in plain view and lack signet rings or any other ring. And her affectionate relation with the boy does not necessarily make her his mother. Given her placement so close to Agrippa, she is most likely one of his many daughters from his previous marriages, perhaps Vipsania III. Why not assume that she had taken particular care of her royal foster brother?

IV. CONCLUSIONS

The south and north friezes of the *Ara Pacis* present the *supplicatio* procession of September 13 BC, which shows Augustus, his extended family and his friends acting in harmony with the Roman senatorial elite. While the *Ara Pacis* scene avoids making an impression of monarchical rule by Augustus and the exposure of a potential successor designate, the dignity and vitality of the imperial family is certainly implied amidst the expressions of Roman prosperity. More subtly formulated is the claim of Rome's global hegemony through the presence of one prince each from the three continents onto which the Empire's influence extended. At least two of them had been given into the foster care of most distinguished Roman noblemen who were also strong supporters of Augustus: the German boy, cared for by Augustus' friend M. Lollius (N-35) and the oriental prince warmly received into the family of M. Agrippa (and Julia, the emperor's daughter). We are less certain about the son of the Mauretanian rulers, who is most likely in the company of his aunt Antonia (N-40), daughter of Augustus' sister Octavia the Younger and the *triumvir* Mark Antony, because he may just be visiting Rome on the occasion of Augustus' birthday together with his mother Kleopatra Selene (N-36), if our identification is acceptable: the caution against queens sojourning in Rome that we have expressed above does not have to apply to a royal woman who was at the same time a member of the imperial family.[70] The Roman audience could not easily recognize the different costumes of these foreign nations, even if the sculptors used great accuracy in their work, but they were certainly in a position to recognize their alien nature, which they were likely to map out geographically thanks to the affiliation with their host families.

Regarding the foster child of Agrippa, we would finally like to align ourselves with Rose's more recent interpretation of a Parthian prince. The Romans were

---

70   The figures that we regard as Antonia and Kleopatra have previously been identified as Julia (whom we place beside Agrippa) and Octavia Minor (who most likely walked at the now-lost end of the south frieze); see n. 45 above.

always far more fascinated by and preoccupied with the bordering super-power than with the client kingdoms of Armenia, Bosporus or Kommagene. These kingdoms may have had short moments of crisis, but Parthia remained a constant concern, owing to its considerable power. For that reason, contemporary Romans and Romans of future generations could be expected to recognize a Parthian prince and recall with pride why he would be represented on the *Ara Pacis*, where his presence demonstrates the successes of Augustan foreign policy and verifies Rome's status as a centre of global attention. Rose has shown that the presence of Parthians in person and in imagery was already ubiquitous in Rome by 13 BC. In 30 BC, Tiridates I fled to Syria from Parthia, taking with him one of the sons of Phraates IV, whom he held as a hostage. Augustus gave Tiridates asylum in Syria, but confiscated the hostage, allegedly to ensure his safety. Subsequently, Augustus sent the (unnamed) son back to Parthia, in return for which he demanded the standards and maybe the captives from Crassus' ill-fated invasion of 53 BC, the campaign of Decidius Saxa in 40 BC and the attack of Mark Antony in 36 BC. The kidnapped son was likely the crown-prince, given how badly Phraates IV wanted him back.[71]

Later, Phraates IV sent four of his sons, Seraspadanes, Rhodaspes, Phraates and Vonones, as well as four grandsons to Augustus to keep them safe from his new wife and new son, who were suspected of trying to murder their rivals. Eventually, prince Phraatakes (later Phraates V) did murder Phraates IV and took the throne in 2 BC, through both patricide and fratricide, following the 'good' example of Phraates IV in 37 BC. Although this overthrow was much later than the *Ara Pacis* procession, it is not at all surprising that Phraates IV would send his sons abroad to Augustus for safe-keeping, given the perils at the Parthian court. Also, it indicates the triumph of his favoured wife, Musa, whose aforementioned son had become the favourite to succeed as Shah.[72] Augustus welcomed the Parthian princes to display to the Roman people as proof that Rome was now the centre of the world and the city with which foreign kings wanted to be 'friends and allies' (*amici et socii populi Romani*) and to which they sent their sons for an education in civilization and lessons on how to stay on the Romans' good side. Besides providing shelter to dethroned kings or unsuccessful rivals, the city on the Tiber had thus become the 'school' of princes of the Mediterranean world and beyond, or even a 'princely kindergarten', to use an expression coined for Augustan Rome by Olivier Hekster.[73]

71 On Tiridates I and Phraates, see, e.g., Aug. *RG* 32; Just. 42.5.6–9; Cass. Dio 51.18.2f.; cf. Luther 2008/19; Bräckel 2019, 128–130. For more on Parthian princes in Augustan Rome, see Strab. *Geogr.* 6.4.2 (288C); 16.1.28 (748f.C); Tac. *Ann.* 2.1f.; Suet. *Aug.* 43.4; 48 and Cass. Dio 53.33.2; cf. Timpe 1975; Sullivan 1990, besides the next two references.
72 On Vonones and his siblings, see Aug. *RG* 32f.; Just. 42.5.12; Strab. *Geogr.* 16.1.28 (748f.C); cf. Alidoust 2017/9; Bräckel 2019, 131–133.
73 Hekster 2010, 54. On hostage-taking in general, see Aug. *RG* 26, 27, 29, 32, 33; Suet. *Aug.* 21.2f. On hostages from among the Herodians, see Jos. *AJ* 18.6.1 (143); 18.6.4 (165); 18.6.6 (191); 20.2.3 (37). For German princes, see Tac. *Ann.* 1.58; 11.16. See also Coşkun 2008;

To return to our initial concern with Bosporan dynastic history, the *Ara Pacis* clearly does not provide us with the firm evidence we would require to reconsider our previous conclusions: first, that Dynamis was childless, at least by the time Asandros appointed her *basilissa* around 22 BC, secondly, that Agrippa needed her to be in the Bosporan kingdom in and after 14 BC, to support Polemon, and, thirdly, that she had fulfilled Augustus' expectation of being a loyal wife to his friend. As a result, she could be trusted with sole rule after the death of her third husband.

Acknowledgments

Gaius Stern's interest in the family of Augustus goes back to childhood and in the Ara Pacis *to 1997, resulting in his PhD (Berkeley 2006). Altay Coşkun began investigating Bosporan dynastic history in 2014. The two of us began exchanging our views and sharing our largely unpublished materials at the 113th CAMWS conference (Kitchener-Waterloo, April 2017) and designed the plan of this joint study for a Black Sea studies workshop at Waterloo (November 2018). Our dedication to the topic and our friendship combined were stronger than the numerous personal and global challenges that delayed, but never crippled our cooperation, of which we present the first part here. We expect that another major study on the chronology and prosopography of the* Ara Pacis *will follow in due course (Stern & Coşkun in preparation). We would like to thank Gaius' brother Ben Stern for his generous help with editing the images of the* Ara Pacis *and our friend Lola Petrova for much support 'along the way', especially translating various texts from Russian. We are further grateful to Elena Caliri and Alfio Mazza for sending us photographs of Moretti's book during the COVID-19 library closure, and to Stone Chen for bringing into shape our draft of the Augustan stemma. We would like to add special thanks to Joanna Porucznik and Germain Payen for their feedback on an earlier version of this paper. Last but not least, the Social Sciences and Humanities Council of Canada is to be thanked for funding the project 'Ethnic Identities and Diplomatic Affiliations in the Bosporan Kingdom', which Altay conducts at the University of Waterloo (2017–2022).*

Bibliography – Ancient Literary Sources

Translations are ours unless stated otherwise.

Boissevain, U.P. 1895/8: *Cassii Dionis Cocceiani Historiarum Romanarum quae supersunt*, 2 vols., Berlin.
Cary, E. & Foster, H.B. 1914: *Cassius Dio*, London. (Loeb ed., drawn from the Perseus Collection)
Mitchell, S. & French, D. 2012: *The Greek and Latin Inscriptions of Ankara*, vol. I, Munich.
Mynors, R.A.B. 1994: *P. Vergilii Maronis Opera*, 11th ed., Oxford.

2017c and Baltrusch & Wilker 2015 on Roman *amicitia*; Allen 2006 and Bräckel 2019, 37–47 on hostages in Rome; also Bräckel 2019 *passim* on refugees in Rome.

Bibliography – Modern Scholarship

Allen, J. 2006: *Hostages and Hostage-Taking in the Roman Empire*, New York.

Alidoust, F. 2017/9: 'Vonones I., König des Partherreiches', *APR, s.v.*

Anokhin, V.A. 1986: *Monetnoe delo Bospora (Coinage from the Bosporus)*, Kiev.

Avram, A. 2011: Notes épigraphiques (I), *Pontica* 44, 137–140.

Ballesteros Pastor, L. 1996: *Mithrídates Eupátor, rey del Ponto*, Granada.

Ballesteros Pastor, L. 2008a: 'Asandros, King of the Bosporos', *APR, s.v.*

Ballesteros Pastor, L. 2008b: 'Dynamis', *APR, s.v.*

Ballesteros Pastor, L. 2021: 'The Land of the Sun and the Moon: An Interpretation of the Emblem on Royal Pontic Coins', forthcoming in J. Boardman, A. Avram, J. Hargrave and A. Podossinov (eds.), *Connecting East and West. Studies Presented to Prof. Gocha R. Tsetskhladze*, Leuven 2021, chapter 5. (*non iam vidimus*)

Baltrusch, E. & Wilker, J. (eds.) 2015: *Amici – Socii – Clientes. Abhängige Herrschaft im Imperium Romanum*, Berlin.

Billows, R. 1993: 'The Religious Procession of the Ara Pacis Augustae: Augustus' *supplicatio* in 13 B.C.', *JRA* 6, 80–92.

Bleicken, J. 1998/2010: *Augustus. Eine Biographie*, Berlin 1998, repr. Reinbek bei Hamburg 2010.

Bonanno, A. 1976: *Portraits and Other Heads on Roman Historical Relief up to the Age of Septimius Severus*, Oxford.

Bräckel, O. 2019: *"Ad me supplices confugerunt." Die Flucht auswärtiger Eliten ins Römische Reich. 2. Jh. v.Chr. – 2. Jh. n.Chr.*, (unpublished) PhD, Leipzig.

Braund, D. 2004: 'King Scribonius', in V. Chržanovskij et al. (eds.), *Bosporskij Fenomen (The Bosporan Phenomenon)*, Part 1, Saint Petersburg, 81–87.

Braund, D. 2005: 'Polemo, Pythodoris and Strabo. Friends of Rome in the Black Sea Region', in Coşkun 2005, 253–270.

Cojocaru, V. 2014: 'Once More about Antonia Tryphaina', *Journal of Ancient History and Archaeology* 1.2, 1–18.

Cornwell, H. 2017: *Pax and the Politics of Peace*, Oxford.

Coşkun, A. (ed.) 2005: *Roms auswärtige Freunde in der späten Republik und im frühen Prinzipat*, Göttingen.

Coşkun, A. (ed.) 2008: *Freundschaft und Gefolgschaft in den auswärtigen Beziehungen der Römer (2. Jh. v.Chr. – 1. Jh. n.Chr.)*, Frankfurt.

Coşkun, A. 2014: 'Latène-Artefakte im hellenistischen Kleinasien: ein problematisches Kriterium für die Bestimmung der ethnischen Identität(en) der Galater ', *IstMitt* 64, 129–162.

Coşkun, A. 2016: *Heinz Heinen und die Bosporanischen Könige – Eine Projektbeschreibung*, in V. Cojocaru & A. Rubel (eds.), *Mobility in Research on the Black Sea. (Iaşi, July 5–10, 2015)*, Cluj-Napoca, 51–71.

Coşkun, A. 2017a: 'Addendum to Asandros', *APR, s.v.*

Coşkun, A. 2017b: 'Addendum to Dynamis', *APR, s.v.*

Coşkun, A. 2017c: *Amicitia, fides* und Imperium der Römer aus konstruktivistischer Perspektive. Überlegungen zu Paul Burton's *Friendship and Empire* (2011), *Latomus* 76, 910–924.

Coşkun, A. 2019a: 'The Date of the Revolt of Asandros and the Relations between the Bosporan Kingdom and Rome under Caesar', in M. Nollé, P.M. Rothenhöfer, G. Schmied-Kowarzik, H. Schwarz & H.Ch. von Mosch (eds.), Panegyrikoi Logoi. *Festschrift für Johannes Nollé zum 65. Geburtstag*, Bonn, 125–146.

Coşkun, A. 2019b: 'Chersonesos Taurike, Asandros and Rome – A New Interpretation of the Embassy of C. Julius Satyrus to Rome, 46 BC (*IOSPE* I² 691)', in A. Bencivenni, A. Cristofori, F. Muccioli & C. Salvaterra (eds.), *Philobiblos – Scritti in onore di Giovanni Geraci*, Milan, 281–306.

Coşkun, A. 2020a: 'The Course of Pharnakes' Pontic and Bosporan Campaigns in 48/47 BC', *Phoenix* 73.1–2, 2019 (ca. Dec. 2020), 86–113.

Coşkun, A., 2020b: 'The Bosporan Kings in-between the Mithridatic Tradition and Friendship with Rome: The Usurpation of Asandros Revisited', forthcoming in D. Braund, A. Chaniotis & E. Petropoulos (eds.), *Roman Pontos*, Athens ca. 2020.

Coşkun, A. (ed.), in preparation: *Studien zur Herrschaft über das Bosporanischen Reich. Mit bisher unveröffentlichten Kapiteln von H. Heinen.* (preliminary title)

Dissel, K. 1907: *Der Opferzug der Ara Pacis Augustae,* Hamburg.

Ducati, P. 1920: *L'Arte Classica,* Turin.

Facella, M. 2006: *La dinastia degli Orontidi nella Commagene ellenistico-romana,* Pisa.

Frolova, N.A. 1997: *Monetnoe delo Bospora (seredina I v. do n.é. – seredina IV v. n.é.) (Coinage from the Bosporus, Mid-1st Century BC to Mid-4th Century AD),* vol. 1 *(49/48 BC to AD 210/11),* Moscow.

Frolova, N.A. 2009: 'Eshche raz o vremeni pravleniya Dinamii tsaritsÿ Bospora (21/20–17/16 gg. do n. é.)' (Once again on the Times of the Rule of Dynamis, Queen of Bosporus [21/20–17/16 BCE]), in I.V. Shiryakov (ed.), *Numizmaticheskie chteniya 2009 goda: k 80-letiyu A.S. Mel'nikovoï i 90-letiyu V.V. Uzdenikova. Moskva, 19–20 noyabrya 2009 g. Tezisÿ dokladov i soobshcheniï*, Moscow, 8–10.

Gajdukevič, V.F. 1971: *Das Bosporanische Reich*, 2nd ed., Berlin.

Golubtsova, E.S. 1951: *Severnoe Prichernomor'e i Rim na rubezhe nasheï érÿ (Rome and the Northern Black Sea Area around the Beginning of the Common Era),* Moscow.

Gorham, R.B. 2011: 'Dynamis on the Ara Pacis: The Importance of Identity', (unpublished) paper given at the annual meeting of CAMWS, 6–9 April 2011, Grand Rapids, MI.

Halamus, M. 2017: 'Barbarization of the State? The Sarmatian Influence in the Bosporan Kingdom', *VDI* 77, 160–167.

Hanslik, R. & Schmitt, H.H. 1963: 'Pythodoris', *RE* 24, 581–586.

Heinen, H. 1998: 'Fehldeutungen der ἀνάβασις und der Politik des bosporanischen Königs Aspurgos', *Hyperboreus* 4, 340–361.

Heinen, H. 2006: *Antike am Rande der Steppe. Der nördliche Schwarzmeerraum als Forschungsaufgabe,* Stuttgart.

Heinen, H. 2008a: 'La tradition mithridatique des rois du Bosphore, de Rostovtzeff à l'historiographie soviétique', in J. Andreau & W. Berelowitch (eds.), *Michel Ivanovitch Rostovtzeff*, Bari, 137–152.

Heinen, H. 2008b: 'Romfreunde und Kaiserpriester am Kimmerischen Bosporos. Zu neuen Inschriften aus Phanagoreia', in Coşkun 2008, 189–208.

Heinen, H. 2011: 'Kaisareia und Agrippeia: das Tor zur Maiotis als augusteisches Monument', in N. Povalahev & V. Kuznetsov (eds.), *Phanagoreia und seine historische Umwelt. Von den Anfängen der griechischen Kolonisation (8. Jh. v.Chr.) bis zum Chasarenreich (10. Jh. n.Chr.)*, Göttingen, 225–240.

Heinen, H. forthcoming a: 'Vom Tode des Mithradates des Großen bis zum Tode des Augustus (63 v.Chr.–14 n.Chr.) (1997)', forthcoming in Coşkun (ed.), in preparation.

Heinen, H. forthcoming b: 'Die Münzprägung des Aspurgos und seiner Vorgänger (8 v.Chr.?–37/38 n.Chr.) (1997)' forthcoming in Coşkun (ed.), in preparation.

Hekster, O. 2010: 'Trophy Kings and Roman Power: Roman Perspective on Client Kingdoms', in M. Facella & T. Kaizer (eds.), *Kingdoms and Principalities in the Roman Near East*, Stuttgart, 45–56.

Hoben, W. 1969: *Untersuchungen zur Stellung kleinasiatischer Dynasten in den Machtkämpfen der ausgehenden römischen Republik*, Diss. Mainz.

Holliday, P. 1990: 'Time, History, and Ritual on the Ara Pacis Augustae', *Art Bulletin* 72.4, 542–557.

Højte, J.M. (ed.) 2009: *Mithridates VI and the Pontic Kingdom*, Aarhus.

Ivantchik, A.I. & Tokhtas'ev, S.R. 2011: 'Queen Dynamis and Tanais', in E. Papuci-Władyka et al. (eds.), *Recent Research on the Northern and Eastern Black Sea in Ancient Times. Proceedings of the International Conference, 21ˢᵗ–26ᵗʰ April 2008*, Kraków, 163–173.

Kebric, R. 2014: 'Identifying Augustus' Deceased Nephew and Heir Marcellus on the Ara Pacis Augustae', (unpublished) paper given at the at the XIV A.D. Saeculum Augustum Conference, Lisbon, 24–26 Sept. 2014.

Kienast, D. 2004: *Römische Kaisertabelle. Grundzüge einer römischen Kaiserchronologie*, 3ʳᵈ ed., Darmstadt.

Kleiner, D. 1978: 'The Greek Friezes of the Ara Pacis Augustae. Greek Sources, Roman Derivatives, and Augustan Social Policy', *MEFRA* 90, 753–785.

Kleiner, Diana. 1992: *Roman Sculpture*, New Haven.

Kleiner, D. & Buxton, B. 2008: 'Pledges of Empire: The Ara Pacis and the Donations of Rome', *AJA* 112, 57–89.

Koeppel, G. 1982: 'Official State Reliefs of the City of Rome (in the Imperial Age)', *ANRW* II.12, 477–506.

Koeppel, G. 1987: 'Die Historischen Reliefs der römischen Kaiserzeit V: Ara Pacis Augustae', Teil 1, *BJbb* 187, 101–157.

Kozlóvskaia, V.I. 2003: 'Mujer Póntica en la Antigüedad, en Representación, Construcción e Interpretación de la Imagen Visual de La Mujer', in *X Coloquio Internacional de la AEIHM*, Madrid, 15–34.

Kozlóvskaia, V.I. 2004: 'La mujer griega del Bósforo: su imagen visual y papel social a la luz de la epigrafía antigua', *Gerión. Revista de Historia Antigua* 22, 121–134.

Kuttner, A. 1987: 'Lost Episodes in Augustan History', Conference abstract for the 1986/7 AIA-APA, *AJA* 91, 297f.

Kuttner, A. 1995: *Dynasty and Empire in the Age of Augustus. The Case of the Boscoreale Cups*, Berkeley.

Leschhorn, W. 1993: *Antike Ären. Zeitrechnung, Politik und Geschichte im Schwarzmeerraum und in Kleinasien nördlich des Tauros*, Stuttgart.

Löwy, E. 1926: 'Bemerkungen zur Ara Pacis', *Jahrbuch des Österreichischen Archäologischen Instituts* 23, 53–61.

Lugli, G. 1935: 'In attesa dello scavo dell' Ara Pacis Augustae', *Capitolium* 11, 365–383.

Luther, A. 2008/19: 'Tiridates I., Usurpator im Partherreich', *APR, s.v.*

MacDonald, D. 2005: *An Introduction to the History and Coinage of the Kingdom of the Bosporus. Including the Coinage of Panticapaeum (with "Apollonia" and "Myrmecium"), Phanagoria, Gorgippia, Sindicus Limen or the Sindoi, Nymphaeum, Theodosia, and the Kings of the Cimmerian Bosporus*, Lancaster.

Macurdy, G.H. 1937: *Vassal-Queens and Some Contemporary Women in the Roman Empire*, Baltimore.

Minns, E.H. 1913: *Scythians and Greeks. A Survey of Ancient History and Archaeology on the North Coast of the Euxine from the Danube to the Caucasus*, Cambridge.

Mommsen, Th. 1878: 'Die Familie des Germanicus', *Hermes* 13, 245–265.

Monaco, G. 1934: 'L' iconografia imperiale nell'Ara Pacis Augustae', *Bolletino della Commissione Archeologica communale di Roma* 62, 17–40.

Moretti, G. 1948: *Ara Pacis Augustae*, Rome.

Moretti, G. 1938/75: *The Ara Pacis Augustae*, based on the Italian original of 1938, transl. by Veronica Priestley.

Moretti, G. 1957: *The Ara Pacis Augustae*, 3ʳᵈ ed. Rome 1957.

Nawotka, K. 1991/2: 'Asander of the Bosporus: His Coinage and Chronology', *AJN* 3–4, 21–48, pls. 3–4.

Olshausen, E. 1980: 'Pontos und Rom (63 v.Chr.–64 n.Chr.)', *ANRW* II 7.2, 903–912.

Parfenov, V.N. 1996: 'Dynamis, Agrippa und der Friedensaltar. Zur militärischen und politischen Geschichte des Bosporanischen Reiches nach Asandros', *Historia* 45, 95–103. Cf. *Dinamiya, tsaritsa Bospora. Neskol'ko shtrikhov k politicheskomu portretu* – (Dynamis, Queen of Bosporus. Several Strokes of a Political Portrait), in *Bospor i antichnyï mir (Bosporus and the Ancient World)*, Nizhniy Novgorod 1997, 126–136.

Pasqui, A. 1903: 'Scavi dell'Ara Pacis Augustae', *Notizie degli Scavi*, 549–574.

Petersen, E. 1894: 'L'Ara Pacis Augustae', *Röm. Mitt.* 9, 171–228.

Petersen, E. 1902: *Ara Pacis Augustae*, Vienna.

Polacco, L. 1960/1: 'La festa della Pace nell'Ara Pacis', *Venezia: Istituto Veneto di Scienze, Lettere ed Arte* 119, 605–642.

Pollini, J. 1978: *Studies in Augustan 'Historical' Reliefs*, (unpublished) PhD at the University of California at Berkeley.

Pollini, J. 1987: *The Portraiture of Gaius and Lucius Caesar*, New York.

Primo, A. 2010: 'The Client Kingdom of Pontus between Mithridatism and Philoromanism', in T. Kaizer & M. Facella (eds.), *Kingdoms and Principalities in the Roman Near East*, Stuttgart, 159–179.

Rehak 2001: 'Aeneas or Numa? Rethinking the Meaning of the Ara Pacis Augustae', *The Art Bulletin* 83.2, 190–208.

Reisch, E. 1902: 'Zur Ara Pacis Augustae', *Wiener Studien* 24, 425–436.

Riemann, H. 1942: '*Pacis Ara Augustae*', *RE* 18.2, 2082–2107.

Rizzo, G.E. 1919/20: 'Pro Ara Pacis Augustae', *Atti del reale accademia di Napoli di archeologia. lettere e belle arti* 7, 1–22.

Roller, D.W. 2018a: *Cleopatra's Daughter and Other Royal Women of the Augustan Era*, Oxford.

Roller, D.W. 2018b: *A Historical and Topographical Guide to the Geography of Strabo*, Cambridge.

Roller, D.W. 2020: *Empire of the Black Sea: the Rise and Fall of the Mithridatic World,* Oxford.

Romeo, I. 1998: *Ingenuus leo: l'immagine di Agrippa*, Rome.

Rose, Ch.B. 1990: '"Princes" and Barbarians on the Ara Pacis', *AJA* 94, 453–467.

Rose, Ch.B. 2005: 'The Parthians in Augustan Rome', *AJA* 109, 21–75.

Rossini, O. 2006: *Ara Pacis*, Rome.

Rostovtzeff, M. 1919: 'Queen Dynamis of Bosporus', *JHS* 39, 88–109.

Rüpke, J. 2007: *Römische Priester in der Antike*, Stuttgart.

Ryberg, I.S. 1949: 'The Procession of the Ara Pacis', *Memoirs of the American Academy in Rome* 19, 77–101.

Saprykin, S.Y. 1990: 'Unikal'nyï stater bosporskoï tsaritsÿ Dinamii' (A Unique Stater of the Bosporan Queen Dynamis), in *Sovetskaya Arkheologiya* 1990.3, 204–214.

Saprykin, S.Y. 1996: *Pontiïskoe tsarstvo. Gosudarstvo grekov i varvarov v Prichernomor'e* (The Pontic Kingdom. A State of the Greeks and Barbarians on the Black Sea), Moscow.

Saprykin, S.Y. 2002: *Bosporskoe tsarstvo na rubezhe dvukh épokh* (The Kingdom of Bosporus on the Verge of Two Epochs), Moscow.

Saprykin, S.J. 2005: Thrace and the Bosporus under the Early Roman Emperors, in D. Braund (ed.), *Scythians and Greeks. Cultural Interactions in Scythia, Athens and Early Roman Empire (Sixth Century BC – First Century AD)*, Exeter, 167–175 & 216f. (notes).

Saprykin, S.Y. & Fedoseev, N.F. 2009: 'Epigraphica Pontica II: New Inscription of Pythodoris from Panticapaeum', *VDI* 270, 2009.3, 138–147.

Sieveking, J. 1907: 'Zur Ara Pacis Augustae', *Jahreshefte des Österreichischen Archäologischen Institutes in Wien* 10, 175–190.

Sieveking, J. 1917: 'Die Kaiserliche Familie auf der Ara Pacis', *Röm. Mitt.* 32, 90–93.

Simon, E. 1963: 'Das neugefundene Bildnis des Gaius Caesar in Mainz', *Mainzer Zeitschrift* 58, 1–18.

Simon, E. 1968: *Ara Pacis Augustae*, New York.

Stern, G. 2006a: 'M. Aemilius Lepidus and the Four *Flamines* on the Ara Pacis Augustae', in C. Mattusch, A.A. Donohue & A. Brauer (eds.), *Common Ground: Acta of the XVI International Congress of Classical Archeology*, 2003, 293–297.

Stern, G. 2006b: *Women, Children and Senators on the Ara Pacis Augustae: A Study of Augustus' Vision of a New World Order in 13 BC*, (unpublished) PhD at the University of California at Berkeley.

Stern, G. 2015a: 'The New Cult of *Pax Augusta* 13 BC – AD 14', *Acta Antiqua* 55, 1–16.

Stern, G. 2015b: 'Augustus, Agrippa, the Ara Pacis, and the Coinage of 13 BC', *Acta Antiqua* 55, 61–78.

Stern, G. & Coşkun, A. in preparation: 'Chronology and Prosopography of the *Ara Pacis Augustae*'.

Strong, E.S. 1907/23: *Roman Sculpture from Augustus to Constantine*, London 1907, revised 1923.

Stuart Jones, H. 1903/4: 'The Ara Pacis Augustae', *Journal of British and American Archaeological Society of Rome* 3, 251–258.

Studniczka, F. 1909: 'Zur Ara Pacis', *Abhandlungen der philologische Klasse der königl.-sächsischen Gesellschaft der Wissenschaften* 27 (26), 899–944.

Sullivan, R.D. 1980: 'Dynasts in Pontus', *ANRW* II.7, 913–930.

Sullivan, R.D. 1990: *Near Eastern Royalty and Rome, 100–30 BC*, Toronto.

Timpe, D. 1975: 'Zur augusteischen Partherpolitik zwischen 30 und 20 v.Chr.', *Würzburger Jahrbücher für die Altertumswissenschaft* 1, 155–169.

Torelli, M. 1982: *Typology and Structure of Roman Historical Reliefs*, Ann Arbor.

Toynbee, Jocelyn.M.C. 1953: 'The *Ara Pacis* Reconsidered and Historical Art in Roman Italy', *PBA* 39, 67–95.

Toynbee, J.M.C. 1961: 'The "Ara Pacis Augustae"', *JRS* 51, 153–156.

Uzzi, J.D. 2005: *Children in the Visual Arts of Ancient Rome*, Cambridge.

von Domaszewski, A. 1903: 'Die Familie des Augustus auf der Ara Pacis', *Jahreshefte des Österreichischen Archäologischen Instituts in Wien* 6, 57–66.

von Domaszewski, A. 1909: *Abhandlungen zur römischen Religion,* Leipzig.

Wissowa, G. 1931: 'Supplicatio', *RE* 4A.1, 942–951.

Yaylenko, V.P. 2010: *Tysyacheletniĭ bosporskiĭ reikh. Istoriya i épigrafika Bospora VI v. do n. é. – V v. n. é.* (The Thousand-Year Bosporan Kingdom. History and Epigraphy of the Bosporus from the 6th Cent. BC to the 5th AD), Moscow.

Zanker, P. 1987/2009: *Augustus und die Macht der Bilder*, Munich ¹1987, ⁵2009. Cf. the English translation (based on the 1st ed.) by Alan Shapiro: *The Power of Images in the Age of Augustus*, Ann Arbor 1988.

Zavoykina, N.V., Novichikhin, A.M. & Konstantinov, V.A. 2018: 'Novaya posvyatitel'naya nadpis' Aspurga iz Gorgippii' (New Dedicatory Inscription of King Aspurgos from Gorgippia), *VDI* 78, 2018.3, 680–692.

C. Studies in the Historical Geography
of Pontos and Kolchis

DEIOTAROS PHILORHOMAIOS, PONTOS UND KOLCHIS

Altay Coşkun

Abstract: Deiotaros Philorhomaios, Pontos and Kolchis: Towards the end of the Third Mithradatic War (64 BC), Pompey promoted the Tolistobogian tetrarch Deiotaros to become the most powerful king of Asia Minor. Strabo describes his new territories as follows (*Geogr.* 12.3.13 [547C]): 'the other (part of the Gadilonitis) Pompey gave to Deiotaros, such as the areas around Pharnakeia and Trapezus, up to Kolchis and Armenia Minor (μέχρι Κολχίδος καὶ τῆς μικρᾶς Ἀρμενίας)'. One can precisely specify these territories. A first argument addresses the *chora* of the exclave Amisos, which was likely limited by the Iris River before Actium. A second argument suggests that there was a land bridge between Galatia and the Gadilonitis along the Halys, as well as an inland connection between the latter and the major parts of the Pontic realm. Scholars are divided regarding the meaning of μέχρι: most consider it exclusive, assuming that Kolchis never belonged to Deiotaros' kingdom (where Pompey appointed a certain Aristarchos), whereas Armenia Minor was supposedly given to him only later in 59 BC. Others try to overcome the difficulties by emending the text to μέχρι Κολχίδος καὶ τ<ὴν> μικρ<ὰν> Ἀρμενία<ν>, thus accepting Armenia Minor as granted by Pompey, while denying Kolchis. However, neither solution is convincing, because a comparison with the description of the territories conquered by Mithradates VI Eupator (*Geogr.* 12.3.1 [541C]) or granted to Polemon I and Pythodoris (*Geogr.* 11.2.18 [499C]) firmly proves that Strabo considered Kolchis a part of Deiotaros' assignment. The Galatian king hence appears to have held at least a supervisory function over Kolchis, possibly resulting in some tension with Aristarchos. The latter's role as a vassal was probably like that of the Tektosagian Kastor Tarkondarios, the Trokmian Brogitaros and the Paphlagonian dynasts Pylaimenes and Attalos.

Абстракт: Дейотар Филоромайос, Понт и Колхида: К концу Третьей Митрадатской войны (64 г. до н.э.) Помпей сделал тетрарха толистобогов Дейотара самым могущественным царём Малой Азии. Страбон описывает его новые территории следующим образом (Геогр. 12.3.13 [547C]): «другую же Помпей отдал Дейотару, так же как и области около Фарнакии и Трапезусии, вплоть до Колхиды и Малой Армении» (μέχρι Κολχίδος καὶ τῆς μικρᾶς Ἀρμενίας). В настоящей статье автор пытается начертать эти территории. Первый аргумент касается хоры эксклавного Амисоса, которая, вероятно, была ограничена рекой Ирис (до сражения при Акциуме). Далее автор предполагает, что между Галатией и Гадилонитис вдоль Галиса находился сухопутный мост, а также другие внутренние маршруты между Гадилонитис и основными частями Понтийского царства. Исследователи расходятся во мнении о значении слова μέχρι: большинство считают его отделенным, полагая, что Колхида никогда не принадлежала царству Дейотара (поскольку Помпей назначил там некоего Аристарха), а Малая Армения предположительно была передана ему позднее, в 59 году до н. э. Другие пытаются выйти из ситуации, исправляя текст на μέχρι Κολχίδος καὶ τ<ὴν> μικρ<ὰν> Ἀρμενία<ν>, тем самым принимая, что Малая Армения была подарена Помпеем, и отрицая, что Дейотар получил Колхиду. Однако ни одно из решений не является убедительным, поскольку сравнение с описанием территорий, завоеванных Митридатом VI Евпатором (Геогр. 12.3.1 [541C]) или переданных Полемону I и Пифодориде (Геогр. 11.2.18 [499C]), ясно доказывает, что Страбон считал Колхиду частью территорий, которую получил Дейотар. Следовательно, получается, что галатский царь выполнял надзорную функцию над Колхидой, что, возможно, приводило к некоторой напряженности с Аристархом. Роль последнего в качестве вассала, вероятно, была аналогична роли тектосага Кастора Таркондара, трокма Брогитара и пафлагонских династов, Пилаймена и Атталоса.

I. DIE VERTEILUNG DES MITHRADATESREICHES DURCH POMPEIUS

Der Dritte Mithradatische Krieg (73–63 v.Chr.) forderte Opfer auf vielen Seiten, aber – zumindest in machtpolitischer Perspektive kannte er auch einige Gewinner. Schon vor dem Tod des Mithradates VI. Eupator begannen zuerst L. Licinius Lucullus und dann – mit viel nachhaltigeren Ergebnissen – Cn. Pompeius Magnus, das Pontische Reich aufzuteilen. Die meisten Entscheidungen wurden von 65 bis 63 v.Chr. getroffen, bevor die *lex Vatinia* des Jahres 59 v.Chr. einen rechtlichen Schlussstrich unter die Neuordnung des Ostens durch den Proconsul zog. Der hier vorgelegte Beitrag ist weniger an der Chronologie oder den politischen Prozessen in Kleinasien und Rom interessiert als vielmehr an der Klärung strittiger Detailfragen zur historischen Geographie, die ihrerseits Aspekte der Herrschaftspolitik im Südosten des Schwarzmeerraumes berühren.[1]

Von Norden nach Süden lassen sich die Änderungen auf der politischen Landkarte wenigstens grob wie folgt zusammenfassen: Der Mithradates-Sohn Pharnakes II stand, obwohl er mit dem Bosporanischen Reich einen erheblichen Anteil erhielt, als Verlierer da, weil er nicht, wie erhofft, die pontischen Kernländer zugesprochen bekam. Und selbst der Bosporos war weit von seiner größten Ausdehnung entfernt, da mehrere Untertanen oder Bündner abgefallen waren. Am besten dokumentiert ist dies für Phanagoreia, das vorübergehend unter der Führung des Kastor von Rhodos unabhängig wurde.[2] Für Kolchis ist ein Regent na-

1　Hier und im Folgenden vgl. die Darstellungen zu Kleinasien von Magie 1950, der die Neuordnung während Pompeius' Winteraufenthalte in Amisos 65/4 und 63/2 sowie Antiocheia 64/3 datiert (S. 368; vgl. Marek 1993, 26); Hoben 1969; Stein-Kramer 1988, bes. 53–58; 89–96; Sullivan 1990; Mitchell 1993, Bd. 1; Syme 1995, 87–176; Payen 2020, 381–394; zusammenfassend Olshausen 1980, 906–910; Marek 2010, 354–368. Daneben s. die Pompeius-Biographien z.B. von van Ooteghem 1954; Seager 2002; Christ 2004; Dingmann 2007; zusammenfassend Will 2001. Zum politischen Gerangel unter den Römern und zur Feinchronologie s. Murphy 1993; Kallet-Marx 1995, 311–335; Tröster 2005; Dingmann 2007, 294–333; Coşkun 2007, 123–140. Zur Topographie von Pontos s. Wagner 1983 (*TAVO*-Karte); Olshausen 1980, 905 (Karte); 2014 (mit Karte S. 42); Olshausen & Biller 1984 (mit Kartenbeilage); Braund & Sinclair 1997/2000 (*BA* Karte 87); Marek 1993, 7–25 (mit Kartenbeilage); Mladiov ca. 1999 (online Karte); Wittke et al. 2010, bes. 158–171 (mit Karten, für die E. Olshausen die Hauptverantwortung trägt, s. S. V). Zum archäologischen Befund s. Olshausen & Biller 1984; Burcu Erciyas 2007 und 2009; zu den fassbaren Veränderungen infolge der römischen Machtübernahme auch Højte 2006. Einen beeindruckenden Bildband hat Marek 2003 vorgelegt. Weitere Hinweise finden sich in den folgenden Anmerkungen.

2　Pharnakes II musste große Teile der Taman-Halbinsel in den kommenden Jahren zurückerobern; s. von Bredow 2000; Ballesteros Pastor 2008/19a; Coşkun ca. 2020d sowie Ballesteros

mens Aristarchos belegt, mit dem wir uns weiter unten näher beschäftigen werden.[3] Die geringeren Landschenkungen in Kleinasien sind kaum besser bezeugt. Wohl um Gangra (Çankırı) herum entstanden zwei paphlagonische Fürstentümer, die den Nachkommen des pylaimenidischen Königsgeschlecht unterstellt wurden, Pylaimenes und Attalos. Zudem wurde die Tetrarchie des Trokmers Brogitaros nach Osten ausgedehnt und schloss nunmehr auch die Festung Mithridation ein, es sei denn, dass diese einem anderen Galater mit ähnlichem Namen übergeben wurde.[4] In manchen Heiligtümern des pontischen Kernlandes setzte Pompeius ihm loyal ergebene Priesterfürsten ein. Derjenige des Ma-Enyo-Tempels von Komana Pontike (nahe Tokat) ist uns mit Archelaos namentlich bekannt, während die Nachfolge in der Kultstätte der Anaïtis von Zela (Zile) anonym bleibt.[5]

Die Hauptmasse der kleinasiatischen Länder wurde in zwei große Blöcke geteilt. Die westlichen Territorien der paphlagonischen und pontischen Reichsteile wurden an der Küste von Amastris (Amasra) bis Amisos (Samsun), im Hinterland weit bis ins Tal des Lykos (Kelkit Çayı) hinein, dem römischen Statthalter von *Pontus et Bithynia* unterstellt.[6] Insgesamt wurde das Gebiet auf elf lokal autonome Städte verteilt, von denen Pompeius bis zu acht neugegründet hatte.[7]

Der größte Einzelgewinner war der Tetrarch der tolistobogischen Galater Deiotaros I. Philorhomaios,[8] der im Nordosten den pontischen Küstenstreifen mit-

---

Pastor und Coşkun, Kapitel VIII und XII in diesem Band. Phanagoreia: App. *Mith.* 120.590 mit Coşkun 2014a und ca. 2020d.

3 S. Coşkun 2007/19a. Zu weiteren Details, s. unten III.3.

4 Zusammenfassend: Strab. *Geogr.* 12.3.1 (541C); vgl. Mitchell 1993, Bd. 1; Coşkun 2008. Im Einzelnen s. Coşkun 2007/19b (Attalos); 2010/19a (Bogodiataros: entweder ein weiterer galatischer Dynast oder verschrieben für Brogitaros); 2010/19b (Brogitaros, vgl. Biffi 2010, 129f.); 2007/19d (Pylaimenes). Syme 1995, 113 vermutet, dass Pompeius nur einen Dynasten in Paphlagonien eingesetzt habe; vgl. Ballesteros Pastor 1996, 283, den man so verstehen könnte, dass ursprünglich nur von Attalos, einem Nachfahren des homerischen Pylaimenes, die Rede gewesen sei.

5 Komana: Boffo 2007, 117; Burcu Erciyas 2009. Zela: Strab. *Geogr.* 11.8.4 (512C); 12.3.37 (559f.C). Neben dem Heiligtum von Zela gründete Pompeius auf dem großen Territorium auch eine Stadt, s. unten Anm. 7. Zu beiden Heiligtümern s. auch unten Anm. 83.

6 Entgegen Wesch-Klein 2001, die den Namen Bithynia für die frühe Zeit vorschlägt, halte ich an der traditionellen Terminologie von Beginn an fest. S. Kapitel XI in diesem Band, Anm. 1.

7 App. *Mith.* 117.576 zählt acht Gründungen ‚in Kappadokien'; vgl. Cass. Dio 37.20.2. Strab. *Geogr.* 12.3.1 (541C) spricht von elf Städten, die zu Bithynien ergänzt wurden. Die frühere Forschung (z.B. Magie 1950, Bd. 2, 1232–1234) identifizierte sie mit Amastris, Sinope und Amisos an der paphlagonischen Küste, Pompeiopolis und Neapolis im paphlagonischen Hinterland sowie Zela, Amaseia, Diospolis (ehemals Kabeira), Magnopolis (mit Eupatoria), Nikopolis und Megalopolis im pontischen und kleinarmenischen Hinterland. Abweichend schlug Marek 1993, 33–46 Amaseia zu Zela und ergänzte Abonuteichos. Demgegenüber argumentiere ich anderswo (Kapitel XI in diesem Band), dass Herakleia und Tieion hinzuzuzählen sind, während Pompeius Amaseia und Zela gewiss Deiotaros unterstellte.

8 Zu Person und Politik des Deiotaros s. bes. Cicero, *Pro rege Deiotaro*, mit Coşkun 2005 sowie zusammenfassend Coşkun 2019d und 2020a; ferner die Detailstudien Coşkun 2007 und 2008 (Genealogie); 2011 und 2018 (Konflikte mit benachbarten Tetrarchen); 2013a (Städtebau); 2014b (Hof und Diplomatie); 2015 (tetrarchische Ordnung). Frühere Darstellungen sind häufig wegen der unkritischen Übernahme antigalatischer Stereotype verzerrt (Beispiele sind

samt den ‚Pontischen Alpen', also den Gebirgsketten Paryadres and Skydises, erhielt. Hinzu kam noch das Gebiet nördlich und westlich des Oberlaufs des Euphrats, das später unter dem Namen Kleinarmenien geführt wurde. Allerdings ergibt eine systematische Untersuchung der Quellen, dass *Armenia Minor* bereits die offizielle Bezeichnung des gesamten hinzugewonnenen Reiches war. Mithin wurde Deiotaros gen Süden Nachbar des Ariobarzanes II., des Königs von Kappadokien, und gen Osten des Tigranes II., der nur noch König eines reduzierten armenischen Territoriums sein durfte.[9] Strabon von Amaseia beschreibt den Zugewinn des Galaterkönigs mit diesen Worten:

> ‚Von diesem Land' – d.h. der Gadilonitis, dem Schwemmland östlich der Mündung des Halys ins Schwarze Meer und westlich von Amisos –[10] ‚besitzen einen Teil die Amisener, den anderen gab Pompeius dem Deiotaros, wie auch die Gegend um Pharnakeia und das Gebiet um Trapezus, bis Kolchis und Kleinarmenien (μέχρι Κολχίδος καὶ τῆς μικρᾶς Ἀρμενίας). Und er ernannte ihn zum König dieser Länder, außer welchen er noch die ererbte Tetrarchie der Galater, die Tolistobogier, besaß. Nach seinem Tod aber sind viele Erbschaften[11] aus seinen Besitzungen hervorgegangen.'[12]

Diese Zusammenfassung ist nur auf den ersten Blick präzise. Strabons Beschreibung wirft bei näherer Betrachtung zahlreiche Fragen auf. In der Forschung wird beispielsweise recht unterschiedlich diskutiert, ob seine Angaben vollständig sind oder die Deiotaros zugeschriebenen Landstriche miteinander verbunden waren. Ziel des vorliegenden Beitrags ist eine möglichst genaue geographische Beschreibung der östlichen Territorien dieses Galaters, die ich auf einer nunmehr revidierten Karte festgehalten habe (Map 2 am Ende dieses Bandes). Weiterhin anzusprechen sind die politischen Implikationen der Raumordnung.[13]

unten in Anm. 51 und 53 zitiert); s. hierzu die grundlegenden Studien von Strobel 1994a und Kistler 2009, zudem Coşkun 2013b zur Rezeptionsgeschichte des Deiotaros und 2014c zu antiken und modernen Konstrukten galatischer Ethnizität. Daneben sind auch faktische Fehlannahmen sehr zahlreich. Verdienstvoll bleiben aber weiterhin die Darstellungen von Hoben 1969, 64–126; 169–173; Mitchell 1993, Bd. 1; Syme 1995, 127–143; Dingmann 2007, 315–319. Einen umfassenden Forschungsbericht bietet Coşkun ca. 2021a.

9   Zur späthellenistischen Ordnung Kleinasiens s. die Verweise oben in Anm. 1. Zu *Armenia Minor* s. den Anhang.

10 Vgl. Roller 2018a, 701f., allerdings ohne weiteres Interesse am Territorium des Deiotaros.

11 Die von Strabon angedeuteten komplizierten Nachfolgeregelungen nach dem Tod des Deiotaros 41/40 v.Chr. habe ich an anderer Stelle untersucht (Coşkun 2007; 2008 mit Karte 4).

12 Strab. *Geogr.* 12.3.13 (547C), ed. Radt 2008, 438: ταύτης δὲ τῆς χώρας τὴν μὲν ἔχουσιν Ἀμισηνοί· τὴν δ᾽ ἔδωκε Δηϊοτάρῳ Πομπήϊος, καθάπερ καὶ τὰ περὶ Φαρνάκειαν καὶ τὴν Τραπεζουσίαν μέχρι Κολχίδος καὶ τῆς μικρᾶς Ἀρμενίας· καὶ τούτων ἀπέδειξεν αὐτὸν βασιλέα, ἔχοντα καὶ τὴν πατρῴαν τετραρχίαν τῶν Γαλατῶν, τοὺς Τολιστοβωγίους. ἀποθανόντος δ᾽ ἐκείνου πολλαὶ διαδοχαὶ τῶν ἐκείνου γεγόνασι. Die früheren Ausgaben von Meineke 1877, vol. 2, 768 und Lassère 1981, 75 variieren geringfügig betreffs der Orthographie und Interpunktion. Neben den Übersetzungen von Radt und Lassère habe ich auch diejenigen von Hamilton & Falconer 1903 und Jones 1924 (durch die Perseus Collection) sowie Roller 2014, 524 eingesehen.

13 Erstmals habe ich mich dem Problem in meiner Habilitationsschrift zum antiken Galatien zugewandt (Coşkun 2007, 125–131, mit Karte 1 auf S. 505f., leicht verbessert in Coşkun 2008, Karte 3 im Anhang).

II. DIE PONTISCHEN TERRITORIEN DES DEIOTAROS PHILORHOMAIOS

1. Von der Gadilonitis über Amisos bis Themiskyra und Sidene (Polemonion)

Eine sichere Ausgangsbasis für eine Rekonstruktion der Territorien des Deiotaros ist der Halys (Kızılırmak). Östlich seiner Mündung befand sich die fruchtbare Gadilonitis (Gazelonitis), das dem heutigen Bafra vorgelagerte Schwemmland.[14] Später ist sie vollständig im Besitz von Amisos bezeugt, während sie laut Strabon von Pompeius geteilt und zur Hälfte dem Deiotaros zugesprochen wurde. Das Motiv für diese Abtrennung ist nicht sicher zu erkennen. Vermutlich hatte erst Lucullus die Gadilonitis zu Amisos geschlagen. Hierauf deutet auch der Hinweis Plutarchs, der von einer Erweiterung des Stadtterritoriums um 120 Stadien berichtet, allerdings ohne Lokalisierung. Die Umrechnung dieser Maßeinheit ist zwar höchst unzuverlässig, aber selbst, wenn wir den (oft angenommenen, meist aber nicht erreichten) Standardabstand vom achten Teil einer römischen Meile (185 m) ansetzen,[15] ergeben sich gerade einmal 22.2 km. Obwohl sich heute ungefähr 36 km entlang des Strandes messen lassen, sind rund 20 km dennoch eine plausible Schätzung für den Küstenverlauf von vor 2000 Jahren. Der Unterlauf des Halys war ja weder begradigt noch kanalisiert und das Schwemmland noch nicht soweit ins Meer vorgeschoben.[16]

Pompeius hätte sich dann bei der Reduktion des Amisener Territoriums womöglich von persönlichen Rivalitäten leiten lassen, wie auch in vielen anderen Fällen.[17] Allerdings sind durchaus strategische Vorteile in dieser Neuordnung zu erkennen. Abgesehen vom hohen landwirtschaftlichen Ertrag, der einen großen

---

14 Nach Strab. *Geogr.* 12.3.13 (546f.C) grenzte die Gadilonitis an die ansonsten unbekannte ‚Saramene und Amisos'. Es ist unklar, ob die Saramene zwischen der Gadilonitis und Amisene lag oder aber Amisos und seine Chora einschloss, wie Olshausen & Biller 1984, 162 und Roller 2018a, 702 meinen; ähnlich Lassère 1981, 239, der sogar den Anschein erweckt, dass die Saramene noch über den Iris hinausging. Gadilon wird regelmäßig mit dem heutigen Bafra identifiziert; vgl. Olshausen & Biller 1984, 130; auch Burcu Erciyas 2006, 63–65, welche die Gadilonitis etwas widersprüchlich zuerst zum früheren Amisos und sodann erst zum römischen Amisos zählt. Nach Lassère 1981, 210 erstreckt sich die Gadilonitis 25 km nördlich von den letzten Gebirgsausläufern, was ein Blick auf das Satellitenfoto bei *Google Maps* grob bestätigt (s. auch im Folgenden zu meinen Berechnungen). Dies sagt aber noch wenig darüber aus, wie weit das Deiotaros zugewiesene Herrschaftsgebiet ins Hinterland ausgedehnt war. Zu verschiedenen Namensformen s. Radt 2008, vol. 7, 363.

15 S. Geus & Guckelsberger 2017, 167–170 und 173; vgl. Coşkun 2020c.

16 Die direkte Linie vom Halysufer im Westen Bafras gen Osten, wo sich die Gebirgsausläufer, die Schwemmlandebene und das Meer bei Engin berühren, sind 23 km (Berechnungen auf der Grundlage von *Google Maps*). Die Alternative, dass die Landerweiterung in Richtung Osten stattfand, ist nicht plausibel, da schon das Land um Themiskyra rund doppelt so lang gewesen wäre und bei Hinzunahme der Sidene einen etwa fünf- bis sechsmal so langen Küstenabschnitt gebildet hätte.

17 Plut. *Luc.* 19.5f. Großzügige Behandlung bzw. Neugründung von Amisos: Strab. *Geogr.* 12.3.14 (547C); App. *Mith.* 83.373–375; Memn. *FGrH* 434 F 30.4. Zu den Rivalitäten s. Tröster 2005, der allerdings Amisos (S. 98) nicht unter die strittigen Punkte zählt.

Teil des stehenden Galaterheeres ernähren konnte,[18] bot der Unterlauf des Halys einen geeigneten Stützpunkt, um einen mittleren Küstenabschnitt Nordanatoliens zu kontrollieren. Es wäre vor allem dann ein günstiges Aufmarschgebiet, wenn die ehemalige Königsresidenz Sinope oder auch das mächtige Amisos Gefahr liefen, zurück in die Hände der Mithradatiden zu fallen, die ja weiterhin den Kimmerischen Bosporos regierten.[19]

Im Osten reicht das Territorium von Amisos mindestens bis an den Unterlauf des Iris (Yeşilırmak). Im Anschluss an seinen historischen Überblick zu Amisos stellt Strabon indes fest, dass auch das östlich jenes Flusses gelegene ‚Themiskyra, die Heimat der Amazonen, und die Sidene' im Besitz jener Stadt seien. Derselbe Geograph erwähnt allerdings auch, dass Caesar, M. Antonius und Octavian einschneidend in die Verfassung von Amisos eingegriffen hätten. Wie tiefgreifend die Neukonstitution der Polis unter dem letztgenannten Herrscher war, belegt nicht zuletzt die Verwendung der aktischen Ära, also der Beginn einer neuen Zeitrechnung mit Octavians Sieg bei Actium am 2. September 31 v.Chr.[20] Dass die Chora jener Stadt in mithradatischer Zeit über den Iris hinausgereicht hätte, ist sehr unwahrscheinlich. Mit der oben genannten Streckenangabe für die Landschenkung des Lucullus ist sie jedenfalls nicht vereinbar.[21] Vielmehr dürfte Themiskyra den Küstenstreifen östlich der Mündung beansprucht haben, sofern nicht zusätzlich mit der Möglichkeit königlicher Güter entlang des Schwarzmeerufers zu rechnen ist. Die Erwähnung von Themiskyra nutzt der Geograph vor allem zur Beschreibung des pontischen Hinterlandes. Namentlich nennt er den Fluss Thermodon und führt seinen Bericht bis in die Phanaroia, in der Iris und Lykos zusammenfließen.[22] Man gewinnt den Eindruck, dass Themiskyra zu Strabons Lebzeiten keine Stadt mehr war. Zuletzt bezeugt ist sie während der Belagerung durch Lucullus 71 v.Chr. Vielleicht wurde sie damals vollständig zerstört.[23]

Jedenfalls zwingt nichts zu der Annahme, dass das Gebiet um Themiskyra vor der Ankunft des Pompeius zu Amisos gehört hätte.[24] Der Widerstand des Lucullus gegen die Neuordnung des Ostens betraf in Pontos am ehesten die Teilung der

---

18 Vgl. Niese 1883, 579, der hierin eine Stärkung des Deiotaros zu dem Zwecke sah, die wilden Bergvölker in Zaum zu halten.

19 Sinope erschien auch Appian (*Mith.* 120.591) als das Einfallstor des Pharnakes II im Jahr 48, obwohl eine Untersuchung des Feldzuges zu dem eindeutigen Ergebnis kommt, dass er den Landweg durch die Kolchis beschritt; s. Coşkun ca. 2020d, mit Cass. Dio 42.45.3. Allgemein zur Topographie und Geschichte von Sinope s. Dan 2009.

20 Amisos: Strab. *Geogr.* 12.3.14 (547C); die Zugehörigkeit der Sidene wird in 12.3.16 (548C) wiederholt. Ära: Marek 1993, 50 und 53; Leschhorn 1993, 106–115, mit Modifikation: 32/1 v.Chr., vielleicht auch mit Blick auf die Beseitigung des Tyrannen Straton. Zum ideologischen Kontext neuer Ären im ehemaligen Mithradates-Reich s. Højte 2006, 20–23.

21 S. oben mit Anm. 15–17.

22 Strab. *Geogr.* 12.3.15 (547f.C). Vgl. Radt 2008, Bd. 7, 365f. und Roller 2018a, 702f., jeweils ohne Problematisierung. Zum Thermodon vgl. Dan 2015.

23 So auch Olshausen 2002, 302 mit App. *Mith.* 78.345f. Weniger klar ist Burcu Erciyas 2006, 63: ‚The Iris delta was called Themiscyra, and both the Iris and Thermodon rivers watered it.'

24 So aber die allgemeine Sicht, vgl. z.B. Burcu Erciyas 2006, 63f.; E. Olshausen in Wittke et al. 2010, 161. Ohne diese Annahme ist noch die Darstellung der Quellen von Hirschfeld 1894.

Gadilonitis, die vielleicht er selbst erstmals den Amisenern zugesprochen hatte. Die massive Erweiterung des Territoriums gen Osten fiel mithin wohl erst in die Zeit Octavians, nachdem Amisos vorübergehend wieder Teil des Pontischen Reiches unter Dareios und sodann unabhängig unter dem ‚Tyrannen' Straton gewesen war. Der Sieger des Bürgerkrieges steigerte so langfristig die Einnahmen des römischen Staates, während Polemon I. als treuer Gefolgsmann des M. Antonius zwar in seiner pontischen Herrschaft belassen, nicht aber mit der Ausweitung seines pontischen Territoriums zu Lasten Roms belohnt wurde.[25]

Über die östlich benachbarte Sidene weiß Strabon wenig zu sagen, außer dass es sich um eine relativ fruchtbare Ebene handelt, die zu seiner Zeit ebenfalls im Besitz von Amisos war. Er benennt dort die drei Orte Side, Chabaka und Phauda, bei denen es sich angesichts fehlender Attribute damals wohl um Dörfer handelte.[26] Side selbst wurde später zur königlichen Residenzstadt Polemonion ausgebaut, ein Name, der heute noch im türkischen Ortsnamen Bolaman nachhallt. Dass Polemon I. der Erbauer dieser Stadt sei, wird zwar heute weitgehend angenommen,[27] ist aber nicht mit dem Zeugnis des Geographen vereinbar. Gewiss ist bei der oft zufälligen Auswahl seiner Informationen Vorsicht geboten und ein *argumentum e silentio* bleibt riskant; allerdings wäre es angesichts seines großen Inte-

25 Die meisten Verfassungswechsel von Lucullus bis zur erneuten ‚Befreiung' unter ‚Caesar Augustus' nach Actium bezeugt Strab. *Geogr.* 12.3.14 (547), ohne freilich Pompeius zu nennen; auch Steuerprivilegien lässt er unerwähnt, was die Möglichkeit innerer Autonomie mit Steuerpflicht zulässt. Zwar berichtet Plinius der Jüngere, dass die Amisener ihre Freiheit Trajan verdankt hätten, doch könnte dies – mit Millar 1999, 111f. – so zu verstehen sein, dass die Freiheit von jedem Kaiser erneut bestätigt worden sei. Der fiskalische Status von Amisos ist also unklar. Man könnte sich gut vorstellen, dass Caesar den Amisenern nach den Misshandlungen durch Pharnakes Freiheit mit Immunität gewährt hatte, doch sicher bezeugt ist dies nicht. Aber auch im positiven Fall muss eine solche Abgabefreiheit nicht von Octavian erneuert worden sein. Kaum wird sie sich auf die der Stadt neu unterstellten Gebiete bezogen haben (vgl. Marek 1993, 44 mit Anm. 324). Jedenfalls schloss der römische Freiheitsbegriff im Fall von Provinzstädten eine Leistungspflicht nicht grundsätzlich aus (s. Bernhardt 1999, bes. 67f.). Zu Lucullus und Pompeius s. die Verweise oben in Anm. 1 und 17; zu Pharnakes Ballesteros Pastor, Kapitel VIII in diesem Band; zu Dareios s. unten Anm. 65; allgemein zu Polemon s. auch Coşkun & Stern, Kapitel IX in diesem Band; vgl. Sørensen 2016, 124–137.
26 Strab. *Geogr.* 12.3.16 (548C).
27 Jones 1937/71, 170; 428 erklärt dies u.a. damit, dass Magnopolis gescheitert und ihre Einwohner von Polemon I. umgesiedelt worden seien; die Gadilonitis, die für das 2. Jh. n.Chr. wieder ganz im Besitz von Amisos nachgewiesen ist, sei am ehesten unter dem Tyrannen Straton gegen die Sidene getauscht worden. Während Marek 1993, 52; 62 das erste Argument widerlegt (Magnopolis sei zwar als Polis aufgelöst worden, habe als urbane Siedlung aber zumindest bis in die Zeit des Plinius fortexistiert), hält auch er es für plausibel, dass Dareios mit Straton die (halbe) Gadilonitis gegen die Sidene getauscht haben könnte. Freilich erwägt Marek alternativ, dass der Galaterkönig Kastor die Gadilonitis geerbt habe, was aber durchaus nicht durch Cass. Dio 48.33.5 gedeckt ist. Jedenfalls habe Polemon I. die Sidene übernommen und dort Polemonion gegründet (vgl. Marek 2010, 403). Olshausen 1974 (gefolgt von Michels 2009, 306; Sørensen 2016, 136: ‚in order to rival the adjacent coastal city of Pharnakeia') wiederum stützt sich für dieselbe Zuschreibung auf das mittelalterliche *Etymologicum Genuinum Gudianum Magnum s.v.* Πολεμώνιος. Für Polemon II. spricht sich nun aber Vitale 2012, 151–154 aus.

resses an diesem König und vielleicht noch mehr an dessen Witwe Pythodoris
sonderbar, dass ihn selbst Jahre nach dem Tod des Königs noch keine Kunde von
diesem wichtigen Bauprojekt erreicht hätte.[28] Es kommt wohl nicht von ungefähr,
dass unser frühestes Zeugnis für Polemonion von Plinius stammt, was mit großer
Sicherheit auf Polemon II. (38–64 n.Chr.), den Enkel von Polemon I. und Pythodoris sowie Jugendfreund des Caligula, hinweist.[29]

Da weiter im Osten Pharnakeia (mit Kotyora) und Trapezus folgten, können
wir also festhalten, dass, von der Exklave Amisos abgesehen, der ganze ostpontische Küstenstreifen vom Halys bis Kolchis der Herrschaft des Deiotaros unterstellt war. Damit dürfte das Territorium etwa zwei- bis dreimal so umfangreich
gewesen sein, wie oft angenommen wird, wenn die Schenkung allein auf die nähere Umgebung von Trapezus und Pharnakeia beschränkt wird (s. unten).[30]

2. Die Bergstämme der Pontischen Alpen

Das Hinterland der pontischen Küste ist in der oben zitierten Strabon-Stelle zum
Territorium des Deiotaros nicht ausdrücklich genannt. Allerdings hatte er im vo-
rangehenden Buch das spätere Pontische Reich der Pythodoris mit den Worten
‚Trapezus und Pharnakeia und die oberhalb (d.h. im gebirgigen Hinterland) woh-
nenden Barbaren' beschrieben.[31] Und in seiner Beschreibung der pontischen
Landschaft vor ihrer Eroberung durch Mithradates Eupator sagt Strabon ferner:

> ‚Oberhalb der Gegenden um Pharnakeia und Trapezus leben die Tibarener und Chaldaier bis
> nach Kleinarmenien. Es ist dies ein hinreichend wohlhabendes Land. Aber Dynasten hielten
> es immer im Besitz, so wie die Sophene, wobei sie bald mit den anderen Armeniern befreun-
> det waren und bald ihre eigene Politik betrieben. Sie hielten aber die Chaldäer und Tibarener
> untertan, so dass sich ihre Herrschaft wohl bis Trapezus und Pharnakeia erstreckte.'[32]

---

28 Der Verbindung zwischen Strabon und Pythodoris, die zuletzt 17/8 n.Chr. in Rom war, sind
    gewiss einige der jüngsten Nachrichten der *Geographie* zu verdanken, während der Hauptteil
    zu Kleinasien wohl im letzten Jahrzehnt vor der Zeitenwende geschrieben wurde. S. Syme
    1995, 357f.; vgl. Braund 2005, 254–260; 264–269; auch Radt 2008, 388; Roller 2014, 13–16;
    2018a, 709–716; Sørensen 2016, 138–153, bes. 146.
29 Plin. *NH* 6.4.11. Allgemein zu Polemon II. s. Sullivan 1980, 925–930; von Bredow 2001;
    Sørensen 2016, 162–164. Die Schenkung erfolgte entweder durch Caligula oder Claudius.
30 Die kleinste Version, d.h. weitgehend eine Beschränkung auf Pharnakeia und Trapezus mit
    ihren Choren, vermuten z.B. Judeich 1885, 154; Sherwin White 1984, 228; Syme 1995, 111;
    132; 135; Mladiov ca. 1999; Dingmann 2007, 302. Aber meist wird der Küstenabschnitt jen-
    seits der Trapezusia bis zur Akampsis-Mündung ergänzt: z.B. Wellesley 1953, 304–307; Mit-
    chell 1993, Bd. 1, 33; Ballesteros Pastor 1996, 283; Payen 2020, 422 Fig. 22. Selten geht man
    über jenen Fluss hinaus, so ca. 50 km in der Karte von E. Olshausen in Wittke u.a. 2010, 161;
    Stein-Kramer 1988, 90 zieht die Grenze erst mit dem Phasis. Zur Geschichte und Topogra-
    phie von Pharnakeia und Trapezus s. Janssens 1967; Burcu Erciyas 2007.
31 Strab. *Geogr.* 11.2.18 (499C), zitiert unten mit Anm. 61.
32 Strab. *Geogr.* 12.3.28 (555C): ὑπὲρ μὲν δὴ τῶν περὶ Φαρνάκειαν καὶ Τραπεζοῦντα τόπων οἱ
    Τιβαρηνοὶ καὶ Χαλδαῖοι μέχρι τῆς μικρᾶς Ἀρμενίας εἰσίν. αὕτη δ᾽ ἐστιν εὐδαίμων ἱκανῶς
    χώρα· δυνάσται δ᾽ αὐτὴν κατεῖχον ἀεί, καθάπερ καὶ τὴν Σωφηνήν, τοτὲ μὲν φίλοι τοῖς ἄλλοις

Vor diesem Hintergrund kann kaum bezweifelt werden, dass Strabon diese Berglandschaft als Teil des neuen pontisch-galatischen Reiches des Deiotaros betrachtet. Er bezeichnet die Stämme an anderer Stelle als Chald(ä)er, Tibarener, Sanner, Appaïten und Heptakometen.[33] Zwar lässt sein Bericht nur eine annähernde Lokalisierung zu, doch erstreckten sich die Siedlungsgebiete der Chald(ä)er (so sie ganz oder teilweise mit den früheren Chalybern gleichgesetzt werden dürfen) bis in die westlichen Ausläufer des Paryadres und mithin bis ins Iristal.[34] Da nun kein weiterer ‚Klientelfürst' des Pompeius für diesen Streifen bekannt ist, steht nichts der Annahme im Weg, dass jene Gebiete geschlossen dem Deiotaros unterstellt wurden, so wie auch die südliche Berglandschaft Kleinarmeniens.[35]

Im Süden wird die pontische Gebirgskette durch das Tal des Lykos begrenzt. Bis zu seiner Mündung in den Iris reihten sich die Choren dreier neugegründeter Poleis aneinander, welche die römische Provinz nach Osten verlängerten:

Ἀρμενίοις ὄντες, τοτὲ δὲ ἰδιοπραγοῦντες· ὑπηκόους δ᾽ εἶχον καὶ τοὺς Χαλδαίους καὶ Τιβαρηνούς, ὥστε μέχρι Τραπεζοῦντος καὶ Φαρνακείας διατείνειν τὴν ἀρχὴν αὐτῶν. S. auch den Anhang zu Kleinarmenien.

33 Vgl. Strab. *Geogr.* 12.3.18f. (548f.C) und 28 (555C). Viele moderne Darstellungen (z.B. Wellesley 1953; Hoben 1969; Sullivan 1990; Mitchell 1993) übergehen diese Stämme in ihrer Darstellung völlig. Ähnliches gilt auch für die Karten von Wellesley 1953, 307; Wagner 1983 (*TAVO* B V 7; vgl. Wirth 1983, 32a); Mitchell 1993, Bd. 1, 40a; Will 2001, 101, welche aber die Siedlungsgebiete der Bergvölker, wenn auch ohne ihre namentliche Nennung, dem Territorium des Deiotaros zuweisen. Demgegenüber schließt Marek 1993, 38 die Tibarener und Chald(ä)er ausdrücklich als Untergebene des Deiotaros aus, wobei er freilich vom Texteingriff Ungers abhängt, s. unten Anm. 58. Liebmann-Frankfort 1969, 285 schlägt die Gebirgskette des Paryadres der römischen Provinz zu, lässt aber die Verhältnisse im Skydises unerwähnt; im Gegensatz dazu vermutet sie aber S. 286, dass Pompeius die Barbaren des Paryadres dem Deiotaros unterstellt habe. Magie 1950, Bd. 1, 374 spricht wiederum von einem zusammenhängenden Territorium vom Schwarzen Meer bis Kappadokien; ähnlich Olshausen 1980, 907: ‚die Küste von Pharnakeia über Trapezus bis an die Grenze nach Kolchis; dazu kam das sich südlich anschließende Hinterland bis nach Kleinarmenien'. Explizit sind die Tibarener und Chald(ä)er eingeschlossen bei Jones 1971, 157 (zudem die Sanner, Appaïten und Heptakometen); Sherwin-White 1984, 228 Anm. 109; 1994, 265. – Eine sichere Lokalisierung der Bergvölker ist wegen der lückenhaften, teils anachronistischen, teils ausdrücklich aus ferner Vergangenheit stammenden Informationen unmöglich. Trotzdem scheint Strabon aber wegen seines Versuchs, die Überlieferung zu aktualisieren, der bessere Gewährsmann für die Zeit des Deiotaros zu sein als Plin. *NH* 6.3.10–6.4.12. Vgl. Arr. *PPE* 7–16: § 11, der die Sanner als östliche, kriegerische Nachbarn der Trapezunter auch für das 2. Jh. n.Chr. belegt; ähnlich Plin. *NH* 6.12. Es ist unklar, warum Olshausen 1980, 904a in seiner Karte für die Zeitenwende die Appaïten zwischen Pharnakeia und Trapezus sowie die Sanner in deren Hinterland lokalisiert; ebensowenig entspricht es Strabons Beschreibung, die Heptakometen in das nahe Hinterland von Side und Kotyora südlich davon zu verorten.

34 Vgl. Strab. *Geogr.* 12.3.18–28 (548–555C), bes. § 19: Zuordnung der Chald(ä)er bzw. Chalyber zu Pharnakeia. Xen. *An.* 5.5.1 lokalisiert die Chalyber westlich von Pharnakeia; dazu Lendle 1995, 266–268. Zu den Chalybern, mit denen die Griechen recht unterschiedliche (meist Eisen verarbeitende) Bergvölker Nordostkleinasiens bezeichneten, s. Olshausen 2012.

35 Diese Annahme behielte selbst dann ihre Plausibilität, wenn die Ebenen von Themiskyra und Side entgegen der oben geäußerten Vermutung doch nicht zum Reich des Deiotaros gehört hätten. Vgl. in anderem Kontext Braund & Sinclair 1997/2000, 1226: ‚The Roman empire made no effort to control the interior of Chaldia until the reign of Justinian.'

(ehemals Eupatoria) am Zusammenfluss, Diospolis (ehemals Kabeira) und
Nikopolis bildeten eine Exklave, welche tief in das Reich des Deiotaros ein-
schnitt.[36] Selbst wenn die städtischen Territorien – wie im Fall von Diospolis – bis
an die Südhänge des Paryadres gereicht haben mögen, so deutet nichts darauf hin,
dass sie um weiträumige, von kriegerischen Stämmen bewohnte Bergregionen
ergänzt worden wären. Pompeius stattete sie vielmehr mit fruchtbaren Teilen der
Phanaroia bzw. des Lykostales aus.

3. Verbindungen zwischen den alten und neuen Reichsteilen des Deiotaros

Die bisherigen Ausführungen zum pontischen Gebiet des Galaterkönigs mögen
den Eindruck erwecken, dass die Gadilonitis und der mit dem Iris beginnende
Hauptteil unverbunden gewesen seien. Dies war auch der Hauptkritikpunkt von
Kenneth Wellesley, welcher den früheren, auf Arnold Jones und David Magie fu-
ßenden Rekonstruktionen einen kühnen Gegenentwurf entgegenstellte (Figs. 1–2).

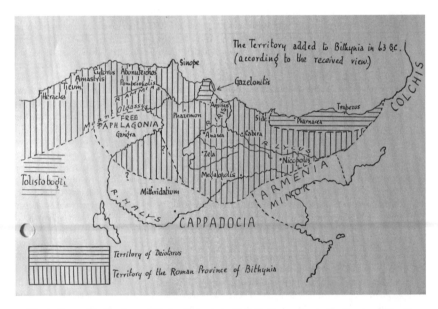

Fig. 1: Karte ‚The Territory Added to Bithynia in 63 B.C. (According to the Received View)‘.
Author: K. Wellesley, in Wellesley 1953, 306.

---

36 So die gewiss zutreffende Mehrheitsmeinung. Zur Begründung s. Coşkun, Kapitel XI in die-
    sem Band. Zur Lage von Magnopolis, Kabeira und Nikopolis s. Olshausen & Biller 1984, 27–
    54 und 151; zu Magnopolis s. jetzt die Modifikation von Sørensen 2016, 153–162.

Dieser war von der Vorstellung geleitet, dass Deiotaros als Nachfolger in der Herrschaft von Pontos kaum mit zwei kleinen und unzugänglichen pontischen Restterritorien abgespeist worden wäre. Zudem versteht er Strabon als Zeugen dafür, dass der Galater auch die Kontrolle über die Bergvölker erhalten habe. Weiterhin nahm er Anstoß daran, dass nach bisheriger Auffassung manche Städte (wie Amaseia und Zela) später wieder aus der Provinz herausgelöst und Königen unterstellt worden seien; stattdessen seien sie vorerst noch im pontisch-galatischen Reich verblieben.[37]

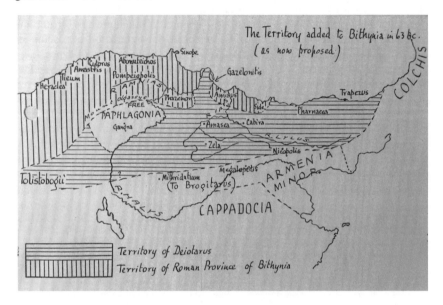

Fig. 2: Karte ‚The Territory Added to Bithynia in 63 B.C. (As Now Proposed)‘.
Author: K. Wellesley, in Wellesley 1953, 307.

Wellesleys teils spekulativer Vorschlag blieb freilich hinsichtlich der Benennung der elf pontischen Städte aporetisch und ist zu Recht auf massive Kritik gestoßen.[38] So auch auf jene von Luis Ballesteros Pastor, der zudem ein mit der Küste

37 Wellesley 1953 versus Magie 1950, Bd. 2, 1232–1234 und Jones 1937. Näher an der kompakten Lösung Wellesleys ist z.B. Judeich 1885, 154; positiv urteilt Olshausen 1980, 906 Anm. 11; unentschieden bleibt Murphy 1993, 138. Mitchell 1993 übergeht ihn stillschweigend. Guinea Díaz 1999, 320 schlägt eine ähnliche Aufteilung vor, ohne Wellesley zu zitieren oder Argumente vorzubringen.
38 Zu den elf Städten s. Coşkun, Kapitel XI in diesem Band. Die Sonderstellung der Priesterstaaten deutet Wellesley 1953 weder in seinem Text noch auf seinen Karten an. Obwohl seine Karte (S. 307) Nikopolis zum Reichsteil des Deiotaros schlägt, erweckt er S. 313 den Eindruck, dass es eine Exklave der Provinz sei. Die Schenkung von Megalopolis und Südkleinarmeniens an Brogitaros ist weder bezeugt noch wahrscheinlich. Unhaltbar ist die Ausdeh-

verbundenes Kontinuum der ostpontischen Provinzstädte fordert; dem Deiotaros gesteht er ab dem Iristal ein geschlossenes Territorium bis zum Akampsis zu.[39] Allerdings scheint mir Strabon von einer kohärenten Landmasse des Königreiches beginnend mit der Halysmündung auszugehen, abgesehen von Amisos.[40] Denn dies dürfte doch der Hauptgrund dafür sein, dass die östlichen Territorien des Deiotaros von eben diesem Fluss an aufgelistet werden. Andernfalls müsste man μέχρι Κολχίδος καὶ τῆς μικρᾶς ᾿Αρμενίας auf τὴν Τραπεζουσίαν beziehen, was das Umland jener Stadt bei weitem übertreiben würde.[41] Sollte dies zutreffen, dann war Amisos wohl eine vollständige Enklave, und die kurzen Land-verbindungen zu Neapolis und Magnopolis führten durch königliches Territorium. Allerdings sollte dies mit Blick auf die bekannte Schutzfunktion des Deiotaros sowie die maritime Ausrichtung von Amisos kein Problem dargestellt haben.

Ausführlicher ist der Widerspruch zu Wellesley bei Marek, der in vielen Fäl-len eine präzisere Deutung des Strabon-Textes vorlegt. Jedoch schüttet er das Kind mit dem Bade aus, wenn er die praktische Frage danach, wie der Galater die Verbindung zu seinen neuen Reichsteilen herstellen konnte, um etwa Truppen oder Proviant zu transportieren, als anachronistisch abtut.[42] Gewiss berücksichtig-te Pompeius bei seiner Planung die effektiven Bedürfnisse sowohl der Städte als auch der Dynasten, da ihm ja an einer Stabilität seiner Neuschöpfung gelegen war. Dabei musste es sowohl um topographische Zugänglichkeit als auch um Sicher-heitserwägungen gegangen sein.

Eine Revision der ‚elf‘ Provinzstädte lässt nun tatsächlich ein viel kohärente-res pontisch-galatisches Reich erkennen, als dies aufgrund weit verbreiteter Miss-

---

nung des Tolistobogierterritoriums über ganz Nordgalatien bis weit in den Halysbogen hin-ein. Zu den Tolistobogiern s. jetzt Coşkun 2019e; vgl. Strobel 1994b und 1999.

39   Ballesteros Pastor 1996, 283–286, unter anderem mit Verweis auf Liv. *Per.* 102: *Cn. Pom-peius in provinciae formam Pontum redegit.* Allerdings bleiben viele Fragen offen: The-miskyra und Side bleiben ungenannt, und die elf Städte werden nicht definiert.

40   So könnte man auch Roller verstehen, dessen Übersetzung (Roller 2014, 524: ‚but the rest Pompeius gave to Deiotaros, as well as that around Pharnakeia and Trapezusia as far as Kol-chis and Lesser Armenia‘) und Kommentar (2018, 701f.) aber ohne Problembewusstsein bleiben. Dennoch nützlich sind die Hinweise zu den Städten und Stämmen im pontischen Reichsteil (2018, 702–709).

41   So aber angenommen z.B. von Marek 1993, 38. Allerdings sprechen unsere Quellen klar dagegen, dass die Trapezusia bis zur Mündung des Akampsis gereicht hätte. Nach Xen. *An.* 5.2 befanden sich die feindlichen Drillai ein bis zwei Tagesmärsche östlich von Trapezus. Strabon ist uneindeutig: Zwar nennt er in *Geogr.* 12.3.17 (548C) Trapezus als letzten heraus-ragenden Ort vor Kolchis, doch erweckt er in *Geogr.* 12.3.29 (555C) den Eindruck, dass die Stammesgebiete der Tibarener und Chaldäer zwischen Trapezus und Kolchis lagen, wohl un-ter Einschluss des schmalen Küstenstreifens (der aber nicht ausdrücklich erwähnt wird). Am detailliertesten ist Arr. *PPE* 3–7, der nach Trapezus die Orte Hyssu Limen (bzw. Hyssos), Ophis (mit der Thiannike), Athenai, Prytanis und Apsaros vor der Mündung des Akampsis nennt, ohne irgendwie den Eindruck einer Abhängigkeit von Trapezus zu erwecken. Athenai war gewiss ein unabhängiger Ort, vielleicht bekannt unter dem Namen Limen in spätarchai-scher Zeit und neugegründet als Athenai unter Perikles, während Amisos damals in Peiraieus umbenannt wurde; s. Coşkun 2019b.

42   Marek 1993, 29–33.

verständnisse bisher zu sehen war. Wir brauchen nun nicht mehr über schmale Landbrücken zwischen den Tolistobogiern und den drei neuen Landesteilen zu spekulieren, die einerseits die territoriale Integrität anderer Herrschaften verletzt hätten und andererseits ohne Zugriff auf das jeweilige Hinterland nicht zu schützen gewesen wären. Für jeden Streckenabschnitt muss es einen klaren Plan gegeben haben. In Richtung Westen ist davon auszugehen, dass der tolistobogische Tetrarch als einziger galatischer König (und zudem Schwiegervater des Tektosagen Kastor Tarkondarios und Trokmer Brogitaros) eine Art Oberaufsicht über deren Tetrarchien ausübte und somit keine Hindernisse bei der Durchquerung ihrer Territorien zu erwarten hatte.[43]

Eine Schlüsselfunktion kam den riesigen Territorien von Amaseia und Zela zu. Wie an anderer Stelle gezeigt, wurden diese Territorien ‚Königen‘ unterstellt, und zwar entgegen weit verbreiteten Ansichten nicht erst durch M. Antonius.[44] Von dort aus waren die Gadilonitis (vielleicht entlang des Halys, westlich von Neapolis) und Kleinarmenien (über Megalopolis, das zu jenem Landesteil gehörte) erreichbar. Kontakt zum pontischen Küstenstreifen mit seiner Gebirgskette bestand zum einen mit dem Schiff von Gadilon aus, zum anderen am Oberlauf des Lykos und entlang des Akampsistals, welche vollständig in der Hand des Deiotaros waren. Offen bleibt indes, ob noch eine weitere, zwischen Neapolis, Amisos und Magnopolis verlaufende Landbücke von Amaseia ins Iristal bestand oder nicht. Beides ist denkbar. Handelsbeziehungen bestanden gewiss über die politischen Grenzen hinaus, und zudem ist von einer engen Kooperation mit dem römischen Statthalter auszugehen, um den freien Verkehr und die Sicherheit der Einwohner in jener Gegend zu gewährleisten. Es war ja die enge Zusammenarbeit des Deiotaros mit den römischen Autoritäten während der drei Mithradatischen Kriege, welche ihn zum Empfänger jener Territorien bestimmt hatten. Ein entsprechendes Dienstverständnis blieb bis zu seinem Lebensende charakteristisch für ihn.[45] Zudem ist Deiotaros auch als Städte- und Festungsbauer in den östlichen Territorien bezeugt, was als klares Zeichen dafür zu verstehen ist, dass er seine Ordnungsfunktion bewusst und energisch wahrnahm.[46]

III. DEIOTAROS PHILORHOMAIOS UND DIE KOLCHIS

1. ‚Bis nach Kolchis und Kleinarmenien‘?

Die nächste Frage betrifft die östliche Ausdehnung des Deiotaros-Reiches. Hier stehen wir vor mehreren Problemen. Zuerst einmal gilt es festzustellen, dass der

---

43 S. auch unten Abschnitt IV. Eine solche Kontrollfunktion setzt z.B. auch Stein-Kramer 1988, 92–94 voraus, aber lediglich mit Blick auf die Befolgung römischer Interessen; die Frage nach Transport und Kommunikation zwischen den disparaten Reichsteilen stellt sie nicht.

44 Vgl. Strab. *Geogr.* 12.3.39 (561C): ἐδόθη δὲ καὶ ἡ Ἀμάσεια βασιλεῦσι, νῦν δ᾽ ἐπαρχία ἐστί, mit Coşkun, Kapitel XI in diesem Band.

45 S. bes. Cic. *Deiot.* 2, 8–14, dazu Coşkun 2005 und die weiteren Hinweise oben in Anm. 8.

46 S. Coşkun 2013a.

Landschaftsname Kolchis keineswegs klar definiert war. Eine extreme Position ist beispielsweise durch Xenophon bezeugt, der selbst noch Trapezus als Teil der Kolchis betrachtet (um 400 v.Chr.), obwohl andere Quellen das Territorum ein bis zwei Tagesmärsche östlich enden lassen. Demgegenüber verortet Prokop das eigentliche Kolchis erst nördlich des Flusses Phasis (6. Jh. n.Chr.).[47] Plausibler ist es indes, die Andeutungen bei Plinius und Appian dahingehend zu verstehen, dass der Akampsis (heute Çoruh Nehri / Tchoroki) bzw. sein nördlicher Seitenarm, der Apsaros (heute Acharistskali), wenigstens in geographischer Sicht die südliche Grenze bildete.[48] Von einer stabilen Grenze in Nordwesten ist kaum auszugehen, sie lag irgendwo nordwestlich von Dioskurias.[49]

Für den Kontext der Politik des Pompeius ist aber eine ganz andere Schwierigkeit viel bedeutsamer: Setzt Strabon in der eingangs zitierten Beschreibung der Deiotaros-Territorien einen inkludierenden oder exkludierenden Charakter für die Präposition μέχρι an? Mit anderen Worten: Betrachtet er die Kolchis (und Kleinarmenien) als hinzugewonnene Länder oder vielmehr als Nachbargebiete des pontisch-galatischen Reiches? Für beide Territorien gibt es zwar unabhängige, aber widersprüchlich erscheinende Quellen. Denn sie bezeugen, dass Pompeius den oben genannten Aristarchos in Kolchis einsetzte, wohingegen Deiotaros mehrfach als König von (Pontos und) Kleinarmenien belegt ist.[50] In der Forschung wurden bislang fünf unterschiedliche Wege beschritten, um diesen Befund zu erklären.

Zahlreiche Interpreten glaubten die Lösung darin zu erkennen, dass beide Gebiete *nicht* oder zumindest *nicht offiziell* der Schenkung durch Pompeius während seines Aufenthaltes im Osten zuzurechnen seien. Gerade für Kleinarmenien sei

---

47 Xen. *An.* 4.8.8 und 22; Prokop. 8.2.4.29, nach dem es südlich des Phasis keine Siedlungen gab mit Ausnahme der spätrömischen Stadt Petra; vgl. Arr. *PPE* 7 und 11; *Tab. Peut.* Segment 10.2.1. Das Problem ist komplex und bisher kaum angesprochen. Ohne Problembewusstsein ist Janssens 1967 (trotz des Titels *Trébizonde en Colchide*). Zur Ausdehnung der Trapezusia s. oben Anm. 41. Zur angeblichen Lage von Trapezus in oder neben der Kolchis s. Podossinov 2012; Coşkun 2019c, 100f. Anm. 77; ca. 2021b; in Vorbereitung. Ballesteros Pastor & Álvarez Ossorio 2001, 4 (ohne Berücksichtigung des hier vorliegenden Problems) erkennen ,que el termine „Cólquide" fue durante mucho tiempo un topónimo referido a una región de límites difusos, ... que servía de manera general para designar a todos los territorios ignotos que se hallaban bordeando el Mar Negro al este del Halis, un rio que habría representado por mucho tiempo la frontera del mundo conocido'.

48 Zu diesen Flüssen s. Plin. *NH* 6.4.12; 6.10.29 und App. *Mith.* 101.465, mit Coşkun 2019c. Wagner 1983 (*TAVO* B V 7) geht sogar noch rund 50 km über den Akampsis hinaus, während Wellesley 1953, 307 das Territorium ca. 50 km weiter südwestlich enden lässt. Man beachte, dass Strabon, bes. *Geogr.* 11.2.17 (498C), den Akampsis gar nicht zu kennen scheint. Auch hierfür kann man auf eine breitere literarische Tradition verweisen; s. Coşkun ca. 2021b.

49 Sie lag jenseits von Dioskurias (wohl Ochamchire, nicht Sukhumi, wie weithin angenommen wird) und schloss wohl auch Pityus (Bizunta ist eine spätere Neugründung) noch ein; vielleicht ging sie bis Herakleion (Adler). S. Strab. *Geogr.* 11.2.12–17 (496–499C); Arr. *PPE* 18; Coşkun 2020b und 2020c.

50 Zu Aristarchos s. App. *Mith.* 114.560; Eutr. 6.14.1; Coşkun 2007/19a; vgl. Magie 1950, Bd. 2, 1238 Anm. 41; Hoben 1969, 70 Anm. 80 mit Verweis auf numismatische Quellen. App. *Mith.* 114.558–561 verschweigt Deiotaros' Gebietszuwächse gänzlich und spricht lediglich von seiner galatischen Tetrarchie.

nämlich in der zeitgenössischen Literatur die Rede davon, dass der Senat Deiotaros dort zum König eingesetzt habe. (1) Der größere Teil der Forschung sieht die Schenkung deswegen als Teil der Senatsverhandlungen über die *acta Pompei* im Jahr 59 v.Chr.,[51] während andere (2) den Anlass erst mit der seit 53 drohenden Gefahr einer Invasion der Parther vermuten.[52] (3) Eine dritte Variante versucht Strabons Aussage dadurch zu harmonisieren, dass Deiotaros allein der Königstitel erst nach 60 v.Chr. verliehen worden sei.[53] Alle diese Ansätze gehen jeweils davon aus, dass das Zeugnis der Geographie so bei einem exklusiven Verständnis der Präposition wenigstens für die 60er Jahre korrekt sei.

Allerdings kann dieser Ansatz nicht wirklich überzeugen, da doch Strabon ausdrücklich Pompeius und nicht den Senat zum Urheber der Landzuteilung und Königswürde macht.[54] Davon abgesehen schreiben etwa auch Eutrop (vermutlich

---

51 Den Senat nennen *BAlex.* 67.1 sowie Cic. *Div.* 2.79; ähnlich *Phil.* 2.94, besprochen von Coşkun ca. 2020e. Die Unterstellung Kleinarmeniens wird überwiegend auf 59 v.Chr. datiert: Judeich 1885, 154 (Volksbeschluss); Zwintscher 1892, 12f. (Senatsbeschluss); Stähelin 1907, 8f. mit Anm. 7 (Deiotaros habe Kleinarmenien nach 62 usurpiert und 59 v.Chr. die erzwungene Einwilligung des Senats erhalten); Drumann & Groebe 1908/64, Bd. 4, 477f. (Senatsbeschluss); Adcock 1937, 12f. (Strabons Quelle sei das 63 v.Chr. entstandene Geschichtswerk des Poseidonios); Hoben 1969, 69–73 (Caesar habe Deiotaros 47 v.Chr. die Pompeius zu verdankenden pontischen Gebiete belassen, aber die ihm selbst geschuldeten Gaben entzogen); Lassère 1981, 161; Stein-Kramer 1988, 91 (inkonsequent ist aber die Zuweisung von Kolchis bis zum Phasis auf S. 90). Ähnliches scheinen vorauszusetzen: Buchheim 1960, 57 mit Anm. 130; Lob 1968, 88f. (die Ernennung zum *rex Armeniae Minoris* sei ein über den Volksbeschluss hinausgehender Gunsterweis des Senats); Jones 1937/71, 157 (,shortly afterwards added to Deiotarus' kingdom by a decree of the senate, Pompey's nominee having presumably died, probably, to judge by Deiotarus' exploits in Galatia, not by a natural death'); Wirth 1983, 36 Anm. 102 (,läßt vermuten, daß man in diesem Gebiet ein Vakuum gelassen hatte ..., dies vorerst möglicherweise unter römischen Kommandanten zur Sicherung der Verbindung nach Kolchis'); Sartre 1995, 132 (,et, un peu plus tard, l'Arménie Mineure'); Sherwin-White 1994, 265 (,possibly by stages'); 269 (,now or a few years later the principality of Lesser Armenia'); Kallet-Marx 1995, 325f. (,perhaps').
52 Adcock 1937, 15–17 vermutet, dass Pompeius Kleinarmenien an Brogitaros übergeben und Deiotaros es erst ca. 52 v.Chr. nach dessen Tod erhalten habe; ebenso z.B. Wellesley 1953, 298; 307; van Ooteghem 1954, 250 (vgl. S. 224); Syme 1995, 132; 137–143; Christ 2004, 81; ähnlich Wirth 1983, 34 mit Anm. 100.
53 So z.B. Freber 1993, 86; Strobel 1999, 399. Sullivan 1990, 165 mit Anm. 63f. erklärt die Rangerhebung als Stärkung seiner Position angesichts einer innergalatischen Opposition, für welche er lediglich auf Ramsay 1941, 44f. verweist; dieser wiederum behauptet ohne jede Quellengrundlage: ,Deiotarus made himself king and killed all tetrarchs whom he could get hold of.' Nach Sherwin-White 1984, 228 mit Anm. 109 gab Pompeius das Territorium, der Senat ,the much sought title of *rex*'. Auch Gasti 1997, 183 datiert die Verleihung der Königswürde nach der Anerkennung der Territorialreform durch den Senat.
54 Vgl. die Umstände der Thronfolge des Ariobarzanes II., der sich zunächst sträubte, der Aufforderung seines Vaters zur Sukzession Folge zu leisten; s. Val. Max. 5.7, ext. 2 (ed. Kempf 1966): *... nec ullum finem tam egregium certamen habuisset, nisi patriae voluntati auctoritas Pompei adfuisset: filium enim et regem appellavit et diadema sumere iussit et in curuli sella considere coegit.* Freilich ist das kappadokische Ereignis nur bedingt mit Galatien vergleichbar, weil für Deiotaros kein väterliches Diadem existierte. Dennoch belegt es die effektive Vollmacht des Pompeius auf der Grundlage der *lex Manilia*.

unter Rückgriff auf Livius) und die Suda die Schenkung Kleinarmeniens Pompeius zu.[55] Außerdem hat es den Anschein, dass in der vorliegenden Passage die Reichsteile des Deiotaros *umfassend* aufgezählt werden sollen, worauf auch die Erwähnung der ererbten Tetrarchie über die Tolistobogier sowie die Andeutung über das Schicksal seines Reiches nach seinem Ableben hinweisen.[56]

(4) Andere Forscher haben deswegen nach einer sprachlichen Lösung gesucht. Benedikt Niese schlug eine scheinbar einfache Harmonisierung vor und deutet die Ausdehnung des Territoriums als ‚bis (ausschließlich) Kolchis und (einschließlich) Kleinarmenien'.[57] Das ist freilich grammatikalisch bedenklich und hat zu Recht wenige Sympathien gefunden. Im Gegensatz hierzu hat der Vorschlag Herbert Ungers viel größere Zustimmung gefunden. Er hielt einen Texteingriff für unerlässlich. Die Schwierigkeiten würden tatsächlich ganz entfallen, wenn ein Kopist das ursprünglich letzte Akkusativobjekt der Schenkung (τὴν μικρὰν Ἀρμενίαν) durch den Einfluss der voranstehenden Worte (μέχρι Κολχίδος καὶ) in den Genitiv gesetzt hätte.[58] Indes hat dieser Ansatz paläographisch wenig für sich, da die Kasusendung gleich dreimal geändert worden wäre. So etwas lässt sich nicht einfach als Flüchtigkeit abtun.

2. Tatsächlich ‚bis (einschließlich) nach Kolchis und Kleinarmenien'!

Unser vermeintlich besseres Wissen von den ‚tatsächlichen' Zuständen sollte uns nicht dazu verleiten, die eigentlich klare Aussage unseres Gewährsmannes leichtfertig zu verwerfen und so die Gelegenheit zu verpassen, aus seinem Verständnis etwas Neues zu lernen. Für die vorliegende Streitfrage ist vor allem die Ausdrucksweise bedeutsam, derer sich Strabon am Anfang desselben Kapitels über Pontos bedient hat. Er beginnt seine Skizze der Gebietserweiterungen durch Mithradates Eupator mit den westlich an Pontos angrenzenden Territorien, um dann

---

55 Vgl. Eutr. 6.14; Suda *s.v.* Πομπήιος. Vgl. Lenaghan 1969, 137, der auch auf die unspezifischen Zeugnisse App. *Mith.* 114.560 und *Syr.* 50.254 verweist.

56 Freilich bleibt der spätere Erwerb des Trokmer- (ca. 53 v.Chr.) und des Tektosagenlandes (ca. 42/1 v.Chr.) hier unerwähnt, obwohl beides wenig später von Strab. *Geogr.* 12.5.1, 3 (567f. C) angedeutet wird. Allerdings liegt der Akzent in *Geogr.* 12.3.13 (547C) auf den Schenkungen des Pompeius, wobei Deiotaros nur als Tetrarch der Tolistobogier identifiziert wird.

57 Niese 1883, 579; dementsprechend bleibt die 1901, 2401 gebotene Chronologie auch ohne Problematisierung; *contra* z.B. Hoben 1969, 70 mit Anm. 81.

58 So Unger 1896, 249f., der μέχρι Κολχίδος καὶ τ<ὴν> μικρ<ὰν> Ἀρμενία<ν> lesen will und eine Bestätigung in Eutr. 6.14.1 (ed. Müller 1995) erkennt: *Pompeius ... Armeniam minorem Deiotaro Galatiae regi donavit.* Ihm schließen sich Magie 1950, Bd. 2, 1237f. und Liebmann-Frankfort 1969, 280f. an; zudem Marek 1993, 37f.; auch 33f. mit Anm. 244 (zur Ambivalenz der Präposition bei Strabon). Unentschieden bleiben etwa Lenaghan 1969, 137; Olshausen 1980, 907; Mitchell 1993, Bd. 1, 33; Seager 2002, 211 Anm. 94; Radt 2008, 364 (‚vielleicht'). Ähnlich, aber ohne Quellen und geographisch unklar ist Mitford 2013, 2566: ‚Deiotaros in Galatia, augmented with eastern Armenia Minor and the Pontic coast as far as Trapezos, to control important roads leading from Satala into northern Armenia'. Yoshimura 1961, 481 Anm. 25 spricht von einem ‚Irrtum'.

wie folgt fortzufahren: „... auf der Gegenseite aber bis Kolchis und Kleinarmenien, was er bekanntlich auch zu Pontos dazuschlug'.[59] Hier werden exakt dieselben Wörter miteinander verbunden wie in § 13, ohne dass je an der Eroberung Kleinarmeniens oder der Kolchis durch Mithradates gezweifelt worden wäre. Und über die Behandlung eben dieser sich östlich bzw. südlich an Pontos anschließenden Gegenden wollte Strabon auch in § 13 eine Aussage treffen.[60]

Darüber hinaus hat der Schriftsteller noch im vorangehenden Buch davon berichtet, dass sein Großonkel Moaphernes von Mithradates als Satrap über die Kolchis eingesetzt worden sei. In diesem Kontext heißt es weiter:

> ,Nachdem aber Mithradates vernichtet worden war, wurde gleichzeitig auch alles von ihm Beherrschte aufgelöst und unter viele verteilt. Zuletzt aber besaß Polemon die Kolchis, und nach seinem Tod herrscht (jetzt) seine Frau Pythodoris, welche auch Königin über die Kolcher und Trapezus und Pharnakeia und die oberhalb (d.h. im gebirgigen Hinterland) wohnenden Barbaren ist'.[61]

Wiederum sind hier Trapezus und Pharnakeia metonymisch für die ostpontische Küstenlinie genannt, welche wie schon unter Mithradates und Deiotaros, so auch unter Polemon und Pythodoris mit Kolchis verbunden war. Strabon macht auch hier eine Aussage über die Ausdehnung des Territoriums ins Hinterland, allerdings ohne Kleinarmenien zu nennen, das schon von Caesar zu Kappadokien geschlagen worden war.[62]

Betrachtet man die drei Textstellen zusammen, so entsteht der Eindruck, dass Strabon von einer kontinuierlichen Zugehörigkeit der Kolchis sowie auch Kleinarmeniens zum Pontischen Reich ausging: ersteres Gebiet von Mithradates Eupator bis Pythodoris und letzteres zumindest bis in die Zeit des römischen Bürgerkrieges. Abweichende Vorannahmen zum Deiotaros-Reich oder zur Kolchis haben bislang verhindert, Strabons Zeugnis in seiner Klarheit zu erfassen. Vielleicht

---

59 Strab. *Geogr.* 12.3.1 (541C): ... ἐπὶ δὲ τάναντία μέχρι Κολχίδος καὶ τῆς μικρᾶς Ἀρμενίας, ἃ δὴ καὶ προσέθηκε τῷ Πόντῳ.

60 Ähnliches hat Strab. *Geogr.* 12.3.1 (541C) auch im unmittelbaren Anschluss an den soeben zitierten Satz getan: καὶ δὴ καὶ Πομπήϊος καταλύσας ἐκεῖνον (sc. *Pontum*) ἐν τούτοις τοῖς ὅροις οὖσαν τὴν χώραν ταύτην παρέλαβε· τὰ μὲν πρὸς Ἀρμενίαν καὶ τὰ περὶ τὴν Κολχίδα τοῖς συναγωνισαμένοις δυνάσταις κατένειμε, τὰ δὲ λοιπὰ εἰς ἕνδεκα πολιτείας διεῖλε καὶ τῇ Βιθυνίᾳ προσέθηκεν ... Dass hier – ähnlich wie auch in *Geogr.* 11.2.18 (499C) – von Dynasten im Plural die Rede ist, steht der in § 13 (547C) geäußerten Ansicht nicht zwingend entgegen, da auch nach der Auffassung Strabons noch genügend Raum für kleinere Herrschaften verblieben sein konnte. An dieser Differenzierung ist dem Geographen aber nicht gelegen; er stellt hier lediglich die Bildung von Klientelfürstentümern und die Provinzialisierung gegenüber. Nur nebenbei sei erwähnt, dass die Verfügung über Gebiete πρὸς Ἀρμενίαν (also bis an die Grenze des von Tigranes beherrschten Großarmeniens) sowie περὶ τὴν Κολχίδα (also die Kolchis mit Umland) den einschließenden Charakter des vorherigen μέχρι nochmals eindeutig unter Beweis stellen.

61 Strab. *Geogr.* 11.2.18 (499C): καταλυθέντος δὲ Μιθριδάτου συγκατελύθη καὶ ἡ ὑπ' αὐτῷ πᾶσα καὶ διενεμήθη πολλοῖς. ὕστατα δὲ Πολέμων ἔσχε τὴν Κολχίδα, κἀκείνου τελευτήσαντος ἡ γυνὴ Πυθοδωρὶς κρατεῖ, βασιλεύουσα καὶ Κόλχων καὶ Τραπεζοῦντος καὶ Φαρνακείας καὶ τῶν ὑπερκειμένων βαρβάρων. Vgl. Tröster 2006/19; Kuin 2017.

62 S. den Anhang zum weiteren Verbleib Kleinarmeniens.

hat hierzu ferner beigetragen, dass der Geograph viele Fragen offenlässt: Den oben genannten Dynasten Aristarchos scheint er gar nicht zu kennen;[63] die konkreten Auswirkungen der Feldzüge des Pharnakes II. und Mithradates von Pergamon beschränkt er weitgehend auf die Plünderung des Leukothea-Heiligtums (gewiss gingen seine *Historien* näher auf diese Kampagnen der Jahre 48–46 v.Chr. im Südosten der Schwarzmeerregion ein);[64] ebenfalls unerwähnt bleibt bei Strabon der wenig bekannte Dareios, ein Sohn des Pharnakes II., den M. Antonius nach Philippi in Pontos einsetzte.[65] All diese Unsicherheiten ändern freilich nichts daran, dass Strabon das Bild einer dauerhaften Verbindung von Pontos und Kolchis seit der Zeit Eupators zeichnet.

3. KOLCHIS IN DER NACHMITHRADATISCHEN TERRITORIALORDNUNG DES POMPEIUS

Es wäre verlockend, die fragmentarischen Quellen oder kontroversen Ansichten zur politischen Ordnung der Kolchis im 1. Jahrtausend v.Chr. hier *revue* passieren zu lassen, doch würde dies zu weit führen. An dieser Stelle genügt der Hinweis darauf, dass die einheimischen Stämme traditionell ein hohes Maß an Autonomie genossen und stark in Richtung Unabhängigkeit drängten, sobald die ihnen übergeordnete Instanz – ob auswärtiger Herrscher oder landeseigener König oder Gouverneur – Schwäche zeigte.[66] Dies war sicher der Fall nach den Rückschlägen, die Eupator im Ersten Mithradatischen Krieg erlitten hatte (85/3 v.Chr.),[67] und man sollte ähnliches für den Verlauf des Dritten Mithradatischen Krieges annehmen, spätestens nachdem der König Pontos und den Bosporos verloren hatte (71/70). Jedenfalls scheint Mithradates bei seiner Flucht vor Pompeius 66/5 v.Chr. erst bei seiner Ankunft in der nördlich gelegenen griechischen Stadt Dioskurias in Sicherheit gewesen zu sein.[68]

Allerdings stieß auch sein Verfolger Pompeius in der Kolchis auf Widerstand. Zerstörungsschichten in Surion / Vani wurden in der Literatur wenig überzeugend Pharnakes II. (48/7 v.Chr.) und Mithradates von Pergamon (46 v.Chr.) zugeschrieben, doch sind diese wohl niemals so tief ins Landesinnere eingedrungen.

---

63 Leider können wir sein verlorenes Geschichtswerk nicht mehr daraufhin überprüfen.
64 Zum Leukotheion s. Strab. *Geogr.* 11.2.17 (498C), mit Coşkun, Kapitel XII in diesem Band.
65 S. App. *BCiv.* 5.75.319; vgl. die Andeutung in Strab. *Geogr.* 12.3.14 (547) sowie Primo 2010, 162; Sørensen 2016, 121–123 (trotz fragwürdiger Prämissen zum Zustand von Pontos) und 125f.; auch Ballesteros Pastor, Kapitel VIII in diesem Band.
66 S. vor allem Strab. *Geogr.* 11.2.13 (496C) und 18 (498C) zu den *skeptuchoi*; vgl. Braund 1994; Lordkipanidze 1996; Tsetskhladze 1998; Sens 2009; Roller 2018a, 638f.; 642.
67 Zur Revolte sowie zum Vizekönig Mithradates, dem Sohn Eupators, s. Memn. *FGrH* 434 F 24 [34].4 = Phot. 231b und App. *Mith.* 64.265f.; dazu Braund 1994, 157–161; Ballesteros Pastor 1996, 191f.; Coşkun 2018/19.
68 Zur Flucht Eupators s. Strab. *Geogr.* 11.2.13 (496C), der den Widerstand der *skeptuchoi* erst unter den Heniochern (also jenseits von Dioskurias) erwähnt; App. *Mith.* 101.467, nach dem Mithradates sein Winterlager in Dioskurias einrichtete.

Pompeius musste hingegen ein Interesse daran haben, die bestgesicherte Stadt unter seine Kontrolle zu bringen und sie anschließend einem ihm gegenüber loyalen Dynasten zu übergeben. Damit wurde die Kolchis ein Bollwerk gegen eventuelle Angriffe vom Bosporos, dessen Schicksal damals ungewiss war.[69]

Appian nennt diesen Vasallen Aristarchos, ohne seinen Rang zu präzisieren. Dass ihm Eutrop den Königstitel zulegt, ist unglaubwürdig, denn auf Münzlegenden begegnet er als Ἀριστάρχο(υ) τοῦ ἐπὶ Κολχίδο(ς).[70] Der Verzicht auf das Diadem könnte implizieren, dass die Neuordnung der Region im Jahr 65 v.Chr. noch zu ungewiss war. Etwa ein Jahr später, als Pompeius Deiotaros zum hauptsächlichen Nachfolger der pontischen Herrschaft sowie als Garanten einer stabilen romfreundlichen Ordnung im Osten etablierte, mochte die Vorenthaltung der Königswürde eine Art lose Unterstellung des Aristarchos unter den Galaterkönig oder wenigstens eine übergeordnete Verantwortung des Tolistobogiers bedeuten.[71] Dazu würde passen, dass Deiotaros auch die Hauptlast der römischen Verteidigung gegen die Parther nach dem katastrophalen Untergang des Crassus bei Karrhai (53) sowie in der Abwehr des Pharnakes II. unter dem Kommando des Cn. Domitius Calvinus (48 v.Chr.) und Caesars (47 v.Chr.) trug.[72]

Ob Aristarchos noch lebte, um die kolchische Flotte für Pompeius während des Bürgerkrieges 49/8 v.Chr. zu führen, ist ungewiss. Dass Kolchis eigene Schiffe stellte,[73] zeigt jedenfalls einerseits, dass es dem Herrschaftsbereich des Deiotaros nicht oder nicht direkt einverleibt worden war, und andererseits, dass Pharnakes seinen Eroberungszug südlich des Großen Kaukasus erst nach der Niederlage des Pompeius bei Pharsalos eröffnete.

---

69   S. Diod. 40.4; Plut. *Pomp.* 34; App. *Mith.* 103.477–104.484; vgl. Braund 1994, 161–169. Zu Vani s. Dundua in Dundua & Lordkipanidze 1979 (mit numismatischen Quellen); Coşkun, Kapitel XII in diesem Band. Anders Lordkipanidze 1996, 262–264; Tsetskhladze 1998, 115.

70   App. *Mith.* 114.560; Eutr. 6.14.1 (vgl. Fest. *Brev.* 16). Münzlegende: Head 1967, 496; vgl. Hoben 1969, 70 Anm. 80. Braund 1994, 168f. hält ihn für einen *skeptouchos* und erwägt, dass die Münze das Porträt des Pompeius zeige. Als König sieht ihn hingegen Ballesteros Pastor 1996, 283. Zur Ikonographie der Münzen s. Bergmann 1998, 85–88, der das Porträt (mit langen Locken und Strahlenkranz) auf der Vorderseite Pompeius zuweist. Der Vergleich mit den bekanntesten Pompeius-Porträts (Bergmann, Taf. 18) zeigt aber wenig Ähnlichkeit. Zwar mögen die Sonnengottsymbolik des Strahlenkranzes mit der griechisch-kolchischen Argonautentradition und die lanegn Haare mit einer Alexander-Imitation oder Apollo-Angleichung erklärt werden, aber die Nase passt nicht zur Stupsnase des Pompeius. Demgegenüber erinnert sie an die vorragende Nase des Crassus, auf den auch die Datierung 54/3 v.Chr. hindeutet; vgl. Coşkun 2010/19a zur auch im Vorfeld des Partherfeldzugs geschlagenen Münzen des Brogitaros. Vermutlich handelt es sich auf der kolchischen Münze aber um den Sonnengott selbst, dessen Porträt lediglich an das des verbündeten römischen Feldherrn angeglichen ist.

71   In anderem Kontxt stellt z.B. auch Wirth 1983, 36 Anm. 102 fest: ‚Deiotaros muß zugleich auch die Verbindung mit Aristarchos in Kolchis übernommen haben'. Ich vermute, dass Wirth damit ausdrücken will, dass der Kontakt zu Rom über Deiotaros hätte verlaufen sollen.

72   So Coşkun 2007/19a; vgl. Dreher 1996, 205f. Anders Hoben 1969, 70; vgl. Ballesteros Pastor 1996, 283. Zu Calvinus als Freund des Deiotaros s. Coşkun 2005; zu seiner Karriere und seinem Rang als *legatus pro praetore* s. Habicht, Corey Brennan & Blümel 2009.

73   Cic. *Att.* 9.9.2 = 176.2 ed. Shackleton Bailey 1965-1970; angedeutet in App. *BCiv.* 2.211. Dies scheint Dingmann 2007, 313 zu übersehen, wenn er Kolchis eine Beteiligung abspricht.

Wenig Aufmerksamkeit hat bisher die Tatsache gefunden, dass die erhaltenen Münzen des Aristarchos in sein 12. Regentschaftsjahr datieren. Kaum signifikant wäre das Datum, wenn es mit Lordkipanidze in 52/1 v.Chr. aufgelöst würde.[74] Allerdings ergibt sich wohl 54 oder 54/3 v.Chr., da Aristarchos etwa im Sommer 65 v.Chr. in sein Amt eingesetzt worden war und wir die in der Antike typische Inklusivzählung ansetzen müssen. Damit könnte die Münzprägung in einem Zusammenhang mit der Rekrutierung des Crassus für den Partherkrieg gestanden haben, wie dies auch für die einzige belegte Münzemission des Trokmers Brogitaros anzunehmen ist. Deiotaros verweigerte sich bekanntlich dem Werben des Proconsuls, den er als Rivalen des Pompeius betrachtete. Brogitaros und Aristarchos sahen indes in Crassus einen potenziellen Freund, mit dessen Unterstützung sie sich künftig aus der Kontrolle des Deiotaros befreien könnten. Da sie aber nach der Schlacht von Karrhai nicht mehr bezeugt sind, dürften sie zusammen mit Crassus untergegangen sein.[75]

Deiotaros füllte sogleich das bei den Trokmern entstandene Vakuum, was der Senat angesichts der Parthergefahr ohne Zögern billigte. Demgegenüber gibt es keine Hinweise auf ein analoges Vorgehen in der Kolchis. Dass Aristarchos einen Stellvertreter (oder vielleicht sogar einen Sohn) zurückgelassen hatte, ist eine wahrscheinliche Annahme. Dass sich derselbe nach der Niederlage 53 v.Chr. hätte halten können, ist weniger sicher, aber doch möglich. Immerhin belegt die Unterstützung des Pompeius mit Schiffen, dass eine hinreichend starke Fraktion Pompeius die Treue hielt.[76] Nach dessen Niederlage am 9. August 48 (bzw. 7. Juni gemäß dem Julianischen Kalender) dürfte die prorömische Fraktion schwerlich ihre Position behauptet haben.

4. Kolchis von Caesar bis Augustus

Die galatische Schutzmacht erlitt ebenfalls erhebliche Verluste und musste außerdem noch Caesar neue Truppen für den römischen Bürgerkrieg stellen. Zudem überschlugen sich mit der Invasion des Pharnakes bis Herbst 48 die Ereignisse: Kolchis, Pontos, Kleinarmenien und Nordkappadokien wurden überrannt und die Allianz aus römischen, galatischen und kappadokischen Truppen wurde bei Nikopolis zurückgeschlagen (Herbst 48). Zwar siegte Caesar dann blitzartig bei Zela am 2. August 47 (bzw. 21. Mai nach dem julianischen Kalender),[77] aber er zog

---

74 Lordkipanidze 1996, 293 n. 487. Nach Braund 1994, 169 ist 54 das frühestmögliche Datum, und 52/1 vielleicht zutreffend. Als Grund gibt er Unruhe an, für welche die Ausraubung des Tempels der Leukothea (Strab. *Geogr.* 11.2.17f. [498f.C]) symptomatisch sei, obwohl dies erst 47 v.Chr. eintrat; s. Coşkun, Kapitel XII in diesem Band.

75 S. Coşkun 2007/19a zu Aristarchos und 2010/19b zu Brogitaros.

76 Die Republikaner erwarteten noch am 17. März 49 v.Chr., dass auch ein Kontingent von Kolchis kommen werde (Cic. *Att.* 9.9.2 = 176 ed. Shackleton Bailey 1965–1970); vgl. App. *BCiv.* 2.51.211, wo Unterstützung von allenthalben aus dem Schwarzen Meer bezeugt ist.

77 Zu Pharnakes II. s. Coşkun 2019a und ca. 2020d sowie Kapitel VIII (Ballesteros Pastor) und XII (Coşkun) in diesem Band.

sogleich nach Westen ab und schwächte Deiotaros aus Misstrauen erneut, indem er ihm Kleinarmenien (im engeren Sinn) und das Trokmerland entzog.[78] Damit waren keine Signale nach Kolchis gesendet, dass sich Caesar oder Deiotaros in jener Gegend weiter engagieren würden. Mithradates von Pergamon wird den Kolchern vor seinem Raubzug im Jahr 46 noch unbekannt gewesen sein, aber da sein Hauptgegner Asandros, der Usurpator am Bosporos, war, wird er damals kaum die Besetzung der Kolchis angestrebt haben. Dies hätte wohl seine gesamten Kräfte absorbiert. Ohnedies überlebte er das Jahr 46 v.Chr. nicht, womit selbst die letzten verbliebenen romtreuen Dynasten ihren Hoffnungsträger verloren.[79]

Deiotaros konnte seine Herrschaft über Galatien nach den Niederlagen von Pharsalos und Nikopolis konsolidieren und nach Caesars Tod sogar das Trokmerland zurückgewinnen (44 v.Chr.), aber der Fortgang des römischen Bürgerkrieges und die andauernde Gefahr einer parthischen Invasion verhinderten jegliches Engagement in Richtung Kolchis. Nach Philippi (42 v.Chr.) ordnete M. Antonius Kleinasien grundlegend neu. Es entstand nun erstmals ein vereinigtes galatisches Reich, dass neben den Tetrarchien auch Paphlagonien einschloss, während ein weiter geschrumpftes Pontos Dareios, einem Sohn des Pharnakes II., übergeben wurde (42/1–37 v.Chr.). Dessen Herrschaft war schon zu schwach, um Pontos in den Griff zu bekommen, so dass an die Wiedergewinnung der Kolchis nicht zu denken war.[80]

Anders verhielt es sich mit seinem Nachfolger Polemon I. Eusebes, den M. Antonius als Nachfolger des Dareios im verminderten Pontischen Reich einsetzte (37/6 v.Chr.). Er vermochte es mit dem Segen Roms, seine Herrschaft allmählich auch wieder über die Kolchis auszudehnen. Seine Energie, sein Territorium sowie seine Treue – zuerst zu M. Antonius, zuletzt zu Augustus – empfahlen ihn später dem ‚Vizekaiser' M. Agrippa (15/4 v.Chr.) auch als Nachfolger im Bosporanischen Reich, so dass unter ihm erstmals wieder die gesamte östliche Hälfte der Schwarzmeerküste vereint war. Nach seinem Tod (ca. 9 v.Chr.) wurde die Herrschaft unter seinen Witwen aufgeteilt: Dynamis erhielt die Alleinherrschaft über den Bosporus zurück, während Pythodoris Pontos und Kolchis erbte.[81]

IV. SCHLUSSFOLGERUNGEN UND AUSBLICK

Die Beschreibung der Landschenkung des Pompeius an Deiotaros bei Strabon (*Geogr.* 12.3.13 [547C]) wird erst dann voll verständlich, wenn man sie im Zusammenhang mit den Skizzen der Pontischen Reiche des Mithradates Eupator und der Pythodoris liest. Dann ergibt sich, dass der Tolistobogierfürst neben seiner

78 S. den Anhang zu Kleinarmenien.
79 Zu Mithradates von Pergamon s. Coşkun ca. 2020d sowie Kapitel XII in diese Band. Vgl. Heinen 1994; Primo 2010, 160f.; Ballesteros Pastor 2008/19b.
80 Zu Dareios s. oben Anm. 65.
81 Zu Polemon, Dynamis und Pythodoris s. Olshausen 1980, 910f.; Sullivan 1980, 913–922; 1990, 151–163; 323f.; Marek 1993, 49–58; Roller 2018a, 710; 2018b; Coşkun & Stern, Kapitel IX in diesem Band.

ererbten galatischen Tetrarchie auch die Herrschaft über Ostpontos von der Gadilonitis an der Halys-Mündung bis zum Akampsis (allerdings ohne Amisos) sowie
im Hinterland über die Gebirgsstämme des Paryadres und Skydises hinaus bis
nach Kleinarmenien im engeren Sinne erhielt. Es ist recht wahrscheinlich, dass
die verschiedenen Reichsteile des Deiotaros miteinander verbunden waren. Seine
Hegemonie über die übrigen galatischen Stämme sowie die Neubewertung des
Schicksals von Amaseia und Zela zeigen, wo diese Landbrücken verliefen. Des
Weiteren erstreckte sich die Macht des Königs bis in die Kolchis hinein. Dass dies
zumindest die Ansicht Strabons war, sollte nicht mehr in Frage gestellt werden.

Weniger sicher ist, wie die dortige Stellung des Deiotaros zu verstehen ist, da
andere Quellen (vor allem Appian und Münzlegenden) den Regenten Aristarchos
bezeugen. So bleibt offen, ob der Galater lediglich eine Schutzmacht im Hintergrund war, direkt eingreifen konnte oder vielleicht sogar Einnahmen von dort bezog. Mit Blick auf Deiotaros' Schwiegersöhne Kastor Tarkondarios und Brogitaros ist jedenfalls von einer Art Unterordnung bzw. Aufsichtsfunktion über benachbarte Dynasten ohne Königswürde auszugehen. Die politische Kontrolle und
finanziellen Vorteile des Kybele-Heiligtums von Pessinus (worüber es zu einem
gewaltsamen Konflikt mit Brogitaros kam) erwuchsen Deiotaros jedenfalls nicht
aus seiner Rolle als tolistobogischer Tetrarch, sondern aus seiner Position als einzig verbleibender König in Zentralanatolien.[82] Wahrscheinlich übte er als ponti-
scher König ebenso eine entsprechende Rolle in Komana (Ma-Enyo), Zela
(Anaïtis) und Ameria (Men Pharnaku) bei Diospolis (dem ehemaligen Kabeira)
aus.[83] Eine Vorrangstellung ist zudem gegenüber den Paphlagonierfürsten Pylaimenes und Attalos zu erwarten, die wie auch die galatischen Tetrarchen in eine
Heiratsallianz mit der Familie des Tolistobogiers getreten zu sein scheinen.[84]
Selbst die Enkelin des ehemaligen Großkönigs der Armenier wurde in den 50er
Jahren mit seinem Sohn Deiotaros II. Philopator verlobt.[85]

Das Geflecht direkter und indirekter Herrschaften, das Deiotaros Philorhomaios infolge der Neuordnung des Ostens durch Pompeius entfaltete, ist

---

82   S. Cic. *Harusp. Resp.* 27–29, mit Coşkun 2018; vgl. Lenaghan 1969; Boffo 2007, 117f. Dass
     P. Clodius Pulcher dem Tetrarchen Brogitaros 58 v.Chr. auch zum Königstitel verhalf, ist
     wohl in erster Linie als Versuch zu verstehen, ihn der Aufsicht durch Deiotaros zu entziehen.
83   Komana: s. Strab. *Geogr.* 12.3.32–36 (557–559C); vgl. Christmann 2004/19. Allerdings
     scheint *BAlex.* 66.3f. Komana 47 v.Chr. Kappadokien zuzurechnen; Caesar unterstellte es
     ‚dem Bithynier Lykomedes ... aus königlichem Geschlecht der Kappadoker'. Entweder folgte
     er hierin einer Entscheidung des Pompeius, der Archelaos ebenso Ariobarzanes II. unterstellt
     haben könnte, oder es handelt sich hier um eine weitere indirekte Schwächung des Deiotaros,
     wie ich eher annehme; s. den Anhang zur Teilung Kleinarmeniens. Zela: s. oben mit Anm. 5.
     Ameria: s. Strab. *Geogr.* 12.3.31 (556f.C); vgl. Olshausen & Biller 1984, 113. Zu den klein-
     asiatischen Priesterfürstentümern, aber ohne die Annahme der Kontrolle durch Deiotaros, s.
     Boffo 1985; 2007; speziell zu Pontos: Olshausen 1987, bes. 188f. allgemein zur Aufsichts-
     funktion des Königs, sowie Dalaison 2014, 145–149 zu den Kulten.
84   So lassen genealogische Beobachtungen hinsichtlich der Nachkommen des Amyntas sowie
     die spätere Herrschaft seines Enkels Kastor und Urenkels Deiotaros Philadelphos vermuten.
     S. Coşkun 2008 (Zusammenfassung) sowie 2007/19c; 2007/19e; 2008/19b; 2008/19c.
85   S. Cic. *Att.* 5.21.2 = 114 ed. Shackleton Bailey 1965–1970, mit Coşkun 2008/19a.

vielleicht am besten mit der gewaltigen Expansion Pergamons unter Eumenes II. zu vergleichen, der Rom im Kampf gegen Antiochos III. den Großen unschätzbare Dienste geleistet hatte.[86] Die historischen Verdienste des Deiotaros um Rom waren sogar noch umfangreicher, bedenkt man seine Rolle in den drei (bzw. vier) Mithradatischen Kriegen und überdies in seinen entscheidenden Beiträgen zur Abwehr der Parther nach 53 v.Chr.

Allerdings verhinderte der Ausbruch des römischen Bürgerkrieges (49 v.Chr.) und Deiotaros' rückhaltlose Parteinahme für den Verlierer Pompeius bei Pharsalos (48 v.Chr.), dass es zu einer ebenso großen und dauerhaften Glanzentfaltung der tolistobogischen Dynastie kam. Auch unter Caesar und den Triumvirn leisteten Deiotaros, seine Nachkommen und sein ehemaliger Sekretär Amyntas erhebliches für die römischen Machtkämpfe nach innen und außen, aber letztlich zu dem Preis, dass ihre Ressourcen und Unabhängigkeit erodierten. Der ermordete König Amyntas wurde 26/5 v.Chr. ohne jedes Aufsehen durch einen römischen Provinzstatthalter ersetzt, und kleinere galatische Herrschaften in Nordanatolien folgten diesem Beispiel in den kommenden Jahrzehnten.[87]

Anhang: Kleinarmenien, Armenien oder Pontos?

Die Gebiete Nordostanatoliens wurden besonders im 1. Jh. v.Chr. recht unterschiedlich als Pontos, Kappadokien am Pontos, Kappadokien, Armenien, Kleinarmenien und sogar Assyrien bezeichnet. Diese terminologische Vielfalt hat zu mancherlei Verwirrung geführt, weniger betreffs der Territorien, welche Pompeius dem Deiotaros unterstellte, als vielmehr betreffs derjenigen, welche Caesar ihm nach der Schlacht von Zela 47 v.Chr. beließ. Das Problem ist dadurch erschwert, dass die detaillierteste zeitgenössische Quelle ([Caes.] BAlex. 66.4) an entscheidender Stelle korrupt ist. In der Forschung gingen die Meinungen weit auseinander. Carl Nipperdey vermutete, dass Deiotaros Kleinarmenien mit dem kappadokischen Prinzen Ariarathes teilte. Benedikt Niese war der Auffassung, dass dem König gar nichts entzogen worden sei, sondern dass er sein östliches Reich gegen die Tetrarchie der Tektosagen, welche direkt an sein ererbtes Tolistobogierland angrenzte, eingetauscht habe. Walther Judeich vertrat hingegen eine Teilung des gesamten Ostreiches mit Ariobarzanes III. von Kappadokien etwa entlang des Lykostales.[88] Die größte Zustimmung findet heute die von Hans-Werner Ritter vertretene Auffassung, dass nur jener Teil südlich des Lykos

---

86 Weitere Parallelen bilden die Spannungen mit anderen kleinasiatischen Dynasten, Roms inkonsequente Haltung in solchen Konflikten, Sanktionen wegen als zu gering wahrgenommener Loyalität sowie die Resilienz des Eumenes und Deiotaros. Zu Eumenes II. s. nun Payen 2020 mit weiterer Literatur.

87 S. Coşkun 2007/19e; 2008; ca. 2021a; vgl. Mitchell 1993, Bd. 1; 1994; Olshausen 1996.

88 Nipperdey 1847, 200f.; Judeich 1885, 149–155; Niese 1883, 579; 588; 594f.; vgl. 1901, 2402. Vgl. Zwintscher 1892, 21–23; Stähelin 1907, 92f.; Magie 1950, Bd. 1, 413; Bd. 2, 1267, Anm. 31; Hoben 1969, 72; 92f.; 171f.; Stein-Kramer 1988, 155–157; Speidel 2019, 126; Michels 2013, 687.

zwischen Deiotaros und Ariobarzanes geteilt worden sei.[89] Eine angemessene Untersuchung würde den Rahmen dieses Beitrags sprengen und soll daher an anderer Stelle erfolgen. Hier seien lediglich die Ergebnisse zusammengefasst.[90]

In Anknüpfung an frühere Namenstraditionen nannte Mithradates seine Territorien westlich des Halys oder vielleicht erst westlich des Iris ‚Pontos‛ und östlich davon (neben Kappadokien am Pontos auch) oftmals ‚Armenien‛.[91] Im Wesentlichen behielt Pompeius diese Unterscheidung bei, indem er die überwiegend westlichen Gebiete zum Provinzteil *Pontus* formte,[92] während die weitgehend östlich angrenzenden Gebiete des Deiotaros offiziell *Armenia Minor* genannt wurden ([Caes.] *BAlex.* 34.1 and 67.1; cf. Eutr. 6.14). Oftmals wurden sie kurz *Armenia* genannt (Cic. *Div.* 1.27; 2.79; *Phil.* 2.94; [Caes.] *BAlex.* 34–36; Plut. *Caes.* 50.1; Cass. Dio 42.45.3; 42.47.1; 42.48.4), doch in all unseren Quellen ist der Kontext so eindeutig, dass eine Verwechslung mit dem armenischen Reich der Orontiden ausgeschlossen werden kann.

Das entscheidende Zeugnis [Caes.] *BAlex.* 66.4 muss – bei sorgfältiger Berücksichtigung von Kontext und Parallelüberlieferung – etwa wie folgt gelautet haben: *fratri autem Ariobarzanis Ariarathi, cum bene meritus uterque eorum de re publica, ne aut regni hereditas Ariarathen sollicitaret aut heres regni terreret,* [*partem Armeniae Minoris, Cappadociam*] *Ariobarzani attribuit, qui sub eius imperio ac dicione esset.*[93] Das Kleinarmenien dieser Quelle ist identisch mit dem

89 Ritter 1970, mit Buchheim 1960, 57 mit Anm. 133. Ihm folgen z.B. Olshausen 1980, 909 mit Anm. 20; Sullivan 1990, 166 mit Anm. 69 (mit weiterer Differenzierung); Mitchell 1993, Bd. 1, 36; Schottky 1996, 1083. Auch Freber 1993, 85 und 89: ‚Während die pontischen Besitzungen ihm offenbar weitgehend erhalten blieben, mußte er einen Teil Kleinarmeniens an das kappadokische Königshaus abtreten‛; auf S. 104f. folgt er indes Hoben (und offensichtlich Nipperdey) in der Ansicht, dass ein ‚Teil Kleinarmeniens‛ zunächst an Ariarathes gefallen, aber bis 45 v.Chr. von Ariobarzanes eingezogen worden sei. Ohne Hinweis auf Ritter zieht Marek 2010, 377 eine ähnliche Schlussfolgerung (vgl. S. 366 zu seiner Terminologie).

90 Coşkun ca. 2020e.

91 Zu Armenien s. Ballesteros Pastor 2007; 2013, 183–185; 189; 2016, 274–276; 280; darauf aufbauend schlägt Coşkun 2020d und ca. 2020e vor, dass die Scheidelinie allmählich vom Halys zum Iris verschoben wurde und dass Armenia Minor eine Prägung ist, die der Neuordnung des Pompeius entstammte. Demgegenüber wird das Mithradatidenreich traditionell pauschal als Pontos bezeichnet, wobei territorial gegebenenfalls zwischen Paphlagonien und dem durch den Halys getrennten pontischen Kappadokien unterschieden wird; so z.B. Olshausen 2014, 40f.; 47 und weitere Literatur oben in Anm. 1. Neuerdings gehen indes Mitchell 2002 und Dan 2014, bes. 46–50 davon aus, dass sich unter den Untertanen der Mithradatiden eine Art historisch-kulturelle Identität herausgebildet habe, welche Mitte des 1. Jhs. nachträglich zur ‚Erfindung‛ von Pontos als ehemaliges Königsterritorium und später auch als Provinzname geführt habe. Demgegenüber vermute ich mit Ballesteros Pastor, dass der seit jeher frei mit der Region des Schwarzen Meeres verbundene Name schon früh mit den Mithradatiden und ihrer Herrschaft verbunden war und deswegen dem Kerngebiet ihrer Herrschaft um Kimitata und Amaseia seinen Namen gab; s. Coşkun ca. 2020e.

92 S. Kapitel XI in diesem Band, bes. Anhang I zum Namen der neuen Provinz.

93 Vgl. dagegen die weithin akzeptierte Textergänzung von Carl Nippedey: [*Ariobarzanen, partem Armeniae minoris concessit eumque*], dem die meisten Herausgeber direkt (z.B. die Loeb-Ausgabe, Way 1955, 118) gefolgt sind; anders etwa der Herausgeber der Edition Budé (Andrieu 1954, 65), der aber dennoch den historischen Implikationen zustimmt.

Armenien, das Deiotaros sich nach Cassius Dio (41.63.3) mit Ariobarzanes III. teilen musste. Nimmt man die Aussage Ciceros hinzu (*Att.* 13.2a = 301 ed. Shackleton Bailey 1965–1970), dann wird klar, dass Ariarathes dort wenig später seinem älteren Bruder hatte weichen müssen.

Strabon ist sich bewusst, dass die gesamte Landschaft nördlich des Euphrat-Oberlaufs (*Geogr.* 11.12.3 [521C]; 12.3.28 [555]; cf. 11.14.2 [527]) bis zur Schwarzmeerküste historisch eine Einheit bildete. So zählt er drei aufeinanderfolgende Herrscher auf (*Geogr.* 12.3.28 [555C]): Antipatros von Sisis (oder Antipatros Sisines), Mithradates Eupator (vgl. 12.3.1 [540f.C]) und Deiotaros (vgl. 12.3.13 [547C]). Zur Abfassungszeit seines Werkes unter Augustus schreibt er Kleinarmenien indes Archelaos, dem damaligen König Kappadokiens, zu, während dessen Gattin Königin Pythodoris als Witwe Polemons I. ihrerseits die Gebiete der ,Tibarener und Chaldäer bis Kolchis sowie Pharnakeia und Trapezus' beherrschte (12.3.29 [555f.C]). Das eigentliche Kleinarmenien befand sich für ihn also südlich der Pontischen Alpen (11.14.1 [527]), so wie es etwa auch in seiner oben zitierten Beschreibung der Territorien des Mithradates (12.3.1 [540f.C])[94] und Deiotaros (12.3.13 [547C]) impliziert ist.

Die engere und heute geläufigere Bezeichnung Armenia Minor für das Gebiet zwischen dem Lykos und dem Oberlauf des Euphrats geht also auf die strategische Entscheidung Caesars im Jahr 47 zurück.[95]

Danksagung

Für Ermutigung, kritisches Feedback oder Hilfe bei der Literaturbeschaffung danke ich Germain Payen und Vera Sauer. Für die kartographische Umsetzung gilt mein Dank Chen Stone. Der erste Entwurf entstand im Rahmen meiner von Heinz Heinen betreuten Trierer Habilitationsschrift zu den Galatern (Coşkun 2007). Wesentliche Verbesserungen erlaubten meine Untersuchungen zur historischen Geographie im Schwarzmeerraum, speziell im Rahmen des Projekts ,Ethnic Identities and Diplomatic Affiliations in the Bosporan Kingdom', das ich an der University of Waterloo verfolge und das dankenswerter Weise vom Social Sciences and Humanities Council of Canada gefördert wird (2017–2022).

Bibliographie – Editionen, Übersetzungen und Kommentare zu Strabon

Hamilton, H.C. & Falconer, W. 1903/6: *Strabo*, 3 Bde., 1st ed. by H. Bohn, 1854–1857; 2nd ed. London 1903–1906.

Jones, H.L. 1924: *The Geography of Strabo*, Bd. 5, Cambridge, MA.

---

94 Problematisch ist Strabons Behauptung, dass Armenia Minor einschließlich der Küste erst von Mithradates VI. erobert worden sei, zumindest wenn man davon ausgeht, dass Pharnakes I. Pharnakeia gegründet habe. Vgl. z.B. Payen, Kapitel VII in diesem Band, mit Anm. 60 und Karte 2. Die Karte von E. Olshausen (in Wittke et al. 2010, 159) setzt voraus, dass Eupator das Küstengebiet bis auf wenige Kilometer vor Trapezus geerbt habe.

95 Es besteht kein Grund zu vermuten, dass die Grenze unter Dareios (s. oben Anm. 25 und 65), den Strabon nicht nennt, anders verlaufen wäre.

Lassère, F. 1981: *Strabon, Géographie*. Bd. 9: *Livre XII*, Paris.
Meineke, A. 1877: *Strabonis Geographica*, Bd. 2, Leipzig.
Radt, S. 2002–2011: *Strabons Geographika*, 10 Bde., Göttingen (vol. 3, 2004; vol. 7, 2008).
Roller, D.W. 2014: *The* Geography *of Strabo*. Translated by D.W. Roller, Cambridge.
Roller, D.W. 2018a: *A Historical and Topographical Guide to the Geography of Strabo*, Cambridge.

Bibliographie – Weitere antike Quellen

Andrieu, J. 1954: *César, Guerre d'Alexandrie*, Paris.
Gasti, F. 1997: *Marco Tullio Cicerone, Orazioni Cesariane: Pro Marcello, Pro Ligario, Pro Rege Deiotaro*, Mailand.
Kempf, C. 1966: *Valerii Maximi Factorum et dictorum memorabilium libri novem. Cum Iulii Pariidis et Ianuarii Nepotiani Epitomis*, Nachdr. der 2. Aufl. von 1888, Stuttgart.
Lenaghan, J.O. 1969: A Commentary on Cicero's Oration *De Haruspicum Responso*, den Haag.
Lendle, O. 1995: *Kommentar zu Xenophon's Anabasis*, Darmstadt.
Lob, M. 1968: *Cicéron, Discours*, Bd. 18, Paris. (inkl. *Deiot.*)
Müller, F.L. 1995: *Eutropii Breviarium ab urbe condita*, Stuttgart.
Nipperdey, C. 1847: *C. Iulii Caesaris Commentarii, cum supplementis A. Hirtii et aliorum*, Leipzig.
Shackleton Bailey, D.R. 1965–1970: *Cicero's Letters to Atticus*, 7 Bde., Cambridge.
Way, A.G. 1955: *Caesar, Alexandrian, African and Spanish Wars. With an English Translation*, London.

Bibliographie – Weitere Literatur

Adcock, F.E. 1937: ‚Lesser Armenia and Galatia after Pompey's Settlement of the East', *JRS* 27, 1937, 12–17.
Ballesteros Pastor, L. 1996: *Mithrídates Eupátor, rey del Ponto*, Granada.
Ballesteros Pastor, L. 2007: ‚Del reino Mitridátida al reino del Ponto: orígenes de un término geográfico y un concepto politico', *OrbTerr* 9, 3–10.
Ballesteros Pastor, L. 2008/19a: ‚Pharnakes II, King of Pontos', *APR, s.v.*
Ballesteros Pastor, L. 2008/19b: ‚Mithradates (VII) of Pergamon, King of the Trokmoi, King Designate of Kolchis and the Bosporos', *APR, s.v.*
Ballesteros Pastor, L. 2013: ‚*Nullis umquam nisi domesticis regibus*. Cappadocia, Pontus and the Resistance to the Diadochi in Asia Minor', in V. Alonso Troncoso & E.M. Anson (eds.), *After Alexander: The Time of the Diadochi (323–281 BC)*, Oxford, 183–198.
Ballesteros Pastor, L. 2016: ‚The Satrapy of Western Armenia in the Mithridatid Kingdom', in V. Cojocaru & A. Rubel (eds.), *Mobility in Research on the Black Sea Region*, Cluj-Napoca, 273–287.
Ballesteros Pastor, L. & Álvarez-Ossorio, A. 2001: ‚Las Fronteras de la Cólquide: Espacio mítico y realidad geográfica en el sur del Ponto Euxino', *OrbTerr* 7, 3–11.
Bergmann, M. 1998: *Die Strahlen der Herrscher. Theomorphes Herrscherbild und politische Symbolik im Hellenismus und in der römischen Kaiserzeit*, Mainz.
Bernhardt, R. 1999: ‚Enstehung, *immunitas* und *munera* der Freistädte. Ein kritischer Überblick', *MedAnt* 2.1, 49–68.
Biffi, N. 2010: *Scampoli dei* Mithridatika *nella Geografia di Strabone*, Bari.
Boffo, L. 1985: *I re ellenistici e i centri religiosi dell'Asia Minore*, Florenz.

Boffo, L. 2007: ‚I centri religiosi d'Asia Minore all'epoca della conquista romana‘, in G. Urso (Hg.), *Tra Oriente e Occidente. Indigeni, Greci e Romani in Asia minore. Atti del convegno internazionale, Cividale del Friuli, 28–30 settembre 2006*, Rom 2007, 105–128.

Braund, D. 1994: *Georgia in Antiquity: a History of Colchis and Transcaucasian Iberia, 550BC–AD 562*, Oxford.

Braund, D. 2005: ‚Polemo, Pythodoris and Strabo. Friends of Rome in the Black Sea Region‘, in A. Coskun (Hg.), *Roms auswärtige Freunde in der späten Republik und im frühen Prinzipat*, Göttingen, 253–270.

Braund, D. & Sinclair, T. 1997/2000: 'Map 87 Pontus-Phasis' (1997), *BA Directory* 2000, 1226–1242.

Buchheim, H. 1960: *Die Orientpolitik des Triumvirn Marcus Antonius*, Heidelberg.

Burcu Erciyas, D. 2006: *Wealth, Aristocracy and Royal Propaganda under the Hellenistic Kingdom of the Mithradatids in the Central Black Sea Region of Turkey*, Leiden.

Burcu Erciyas, D. 2007: ‚Cotyora, Kerasus and Trapezus: The Three Colonies of Sinope‘, in D.V. Grammenos & E.K. Petropoulos (Hgg.), *Ancient Greek Colonies in the Black Sea 2*, Bd. 2, Oxford, 1195–1206.

Burcu Erciyas, D. 2009: ‚Komana Pontike: A City or a Sanctuary?‘, in J.M. Højte (Hg.), *Mithridates and the Pontic Kingdom*, Aarhus, 285–308.

Christ, K. 2004: *Pompeius. Der Feldherr Roms. Eine Biographie*, München.

Christmann, K. 2004/19: ‚Archelaos, Priester von Komana Pontike und König von Ägypten‘, *APR*, *s.v.*

Coşkun, A. 2005: ‚*Amicitiae* und politische Ambitionen im Kontext der *causa Deiotariana*‘, in idem (Hg.), *Roms auswärtige Freunde in der späten Republik und im frühen Prinzipat*, Göttingen, 127–154.

Coşkun, A. 2007: *Von der ‚Geißel Asiens‘ zu ‚kaiserfrommen Reichsbewohnern‘. Sudien zur Geschichte der Galater unter besonderer Berücksichtigung der* amicitia populi Romani *und der göttlichen Verehrung des Augustus (3. Jh. v. – 2. Jh. n.Chr.)*, unveröffentlichte Habilitationsschrift, Trier.

Coşkun, A. 2007/19a: ‚Aristarchos, Dynast von Kolchis‘, *APR*, *s.v.*

Coşkun, A. 2007/19b: ‚Attalos, Dynast von Paphlagonien‘, *APR*, *s.v.*

Coşkun, A. 2007/19c: ‚Kastor (III.), König von Paphlagonien‘, *APR*, *s.v.*

Coşkun, A. 2007/19d: ‚Pylaimenes, Dynast von Paphlagonien‘, *APR*, *s.v.*

Coşkun, A. 2007/19e: ‚Amyntas, König und Tetrarch von Galatien‘, *APR*, *s.v.*

Coşkun, A. 2008: ‚Das Ende der „romfreundlichen Herrschaft" in Galatien und das Beispiel einer „sanften Provinzialisierung" in Zentralanatolien‘, in idem (Hg.), *Freundschaft und Gefolgschaft in den auswärtigen Beziehungen der Römer (2. Jh. v.Chr. – 1. Jh. n.Chr.)*, Frankfurt, 133–164.

Coşkun, A. 2008/19a: ‚Deiotaros II. Philopator (I.), König von Galatien, Pontos und Kleinarmenien‘, *APR*, *s.v.*

Coşkun, A. 2008/19b: ‚Deiotaros (III.) Philadelphos, König von Paphlagonien‘, *APR*, *s.v.*

Coşkun, A. 2008/19c: ‚Deiotaros (IV.) Philpator (II.), König von Paphlagonien‘, *APR*, *s.v.*

Coşkun, A. 2010/19a: ‚Bogodiataros, trokmischer Dynast von Mithridation (?)‘, *APR*, *s.v.*

Coşkun, A. 2010/19b: ‚Brogitaros Philorhomaios, König und Tetrarch der galatischen Trokmer, *APR*, *s.v.*

Coşkun, A. 2011: ‚Annäherungen an die lokalen Eliten der Galater in hellenistischer Zeit‘, in B. Dreyer & P.F. Mittag (Hgg.), *Lokale Eliten und hellenistische Könige. Zwischen Kooperation und Konfrontation*, Berlin, 80–104.

Coşkun, A. 2013a: ‚War der Galaterkönig Deiotaros ein Städtegründer? Neue Vorschläge zu einigen kleinasiatischen Toponymen auf *Sin-/Syn-*‘, *Gephyra* 10, 152–162.

Coşkun, A. 2013b: ‚Belonging and Isolation in Central Anatolia: The Galatians in the Graeco-Roman World‘, in S.L. Ager & R.A. Faber (eds.), *Belonging and Isolation in the Hellenistic World*, Toronto, 73–95.

Coşkun, A. 2014a: ‚Kastor von Phanagoreia, Präfekt des Mithradates und Freund der Römer‘, in N. Povalahev (Hg.), *Phanagoreia und darüber hinaus … – Festschrift für Vladimir Kuznetsov*, Göttingen, 131–138.
Coşkun, A. 2014b: ‚Vier Gesandte des Königs Deiotaros in Rom (45 v.Chr.). Einblicke in den galatischen Hof der späthellenistischen Zeit auf onomastischer Grundlage‘, *Philia* 1, 1–13.
Coşkun, A. 2014c: ‚Latène-Artefakte im hellenistischen Kleinasien: ein problematisches Kriterium für die Bestimmung der ethnischen Identität(en) der Galater‘, *IstMitt* 64, 129–162.
Coşkun, A. 2015: ‚Die Tetrarchie als hellenistisch-römisches Herrschaftsinstrument. Mit einer Untersuchung der Titulatur der Dynasten von Ituräa‘, in E. Baltrusch & J. Wilker (Hgg.), *Amici – Socii – Clientes. Abhängige Herrschaft im Imperium Romanum*, Berlin, 161–197.
Coşkun, A. 2018: ‚Brogitaros and the Pessinus-Affair – Some Considerations on the Galatian Background of Cicero’s Lampoon against Clodius in 56 BC (Harusp. resp. 27–29)‘, *Gephyra* 15, 117–131.
Coşkun, A. 2018/9: ‚Mithradates, King of Kolchis‘, *APR*, *s.v.*
Coşkun, A. 2019a: ‚The Date of the Revolt of Asandros and the Relations between the Bosporan Kingdom and Rome under Caesar‘, in M. Nollé, P.M. Rothenhöfer, G. Schmied-Kowarzik, H. Schwarz & H.C. von Mosch (Hgg.), *Panegyrikoi Logoi. Festschrift für Johannes Nollé zum 65. Geburtstag*, Bonn, 125–146.
Coşkun, A. 2019b: ‚Pontic Athens – An Athenian Emporion in Its Geo-Historical Context‘, *Gephyra* 18, 11–31. URL: https://dergipark.org.tr/tr/pub/gephyra/issue/49781.
Coşkun, A. 2019c: ‚Phasian Confusion: Notes on Kolchian, Armenian and Pontic River Names in Myth, History and Geography‘, *Phasis* 21–22, 73–118. URL: http://phasis.tsu.ge/index.php/PJ/issue/view/569.
Coşkun, A. 2019d: ‚The Galatian Kingdoms‘, in Tekin 2019, 146–163.
Coşkun, A. 2019e: ‚The “Temple State” of Phrygian Pessinus in the Context of Seleucid, Attalid, Galatian and Roman Hegemonial Politics (3rd–1st Centuries BC), in G.R. Tsetskhladze (Hg.), *Phrygia in Antiquity: From the Bronze Age to the Byzantine Period (Proceedings of the International Conference ‘The Phrygian Lands over Time: From Prehistory to the Middle of the 1st Millennium AD’, held at Anadolu University, Eskişehir, Turkey, 2nd–8th November, 2015)*, 607–648.
Coşkun, A. 2020a: ‚Deiotaros of Galatia‘, in *EAH*, 2. Aufl. (online)
Coşkun, A. 2020b: ‚(Re-) Locating Greek & Roman Cities along the Northern Coast of Kolchis. Part I: Identifying Dioskourias in the Recess of the Black Sea‘, *VDI* 80.2, 354–376.
Coşkun, A. 2020c: ‚(Re-) Locating Greek & Roman Cities along the Northern Coast of Kolchis. Part II: Following Arrian’s *Periplous* from Phasis to Sebastopolis‘, *VDI* 80.3, 654–674.
Coşkun, A. ca. 2020d: ‚The Course of Pharnakes’ Pontic and Bosporan Campaigns in 48/47 BC‘, *Phoenix* 73.1–2, 2019 (ca. Dec. 2020), 86–113.
Coşkun, A. ca. 2020e: ‚Die Teilung „Armeniens“ durch Caesar und die Entstehung „Kleinarmeniens“‘, demnächst in *OrbTerr* 18.
Coşkun, A. ca. 2021a: ‚A Survey of Recent Research on Ancient Galatia (1993–2019)‘, demnächst in idem (Hg.), *Galatian Victories and Other Studies into the Agency and Identity of the Galatians in the Hellenistic and Early-Roman Periods* (Colloquia Antiqua 33), Leuven.
Coşkun, A. ca. 2021b: ‚Akampsis, Boas, Apsarus, Petra, Sebastopolis: Rivers and Forts on the Southern Littoral of Colchis‘, demnächst in J. Boardman, A. Avram, J. Hargrave and A. Podossinov (eds.), *Connecting East and West. Studies Presented to Prof. Gocha R. Tsetskhladze*, Leuven, chapter 15.
Coşkun, A. in Vorbereitung: ‚Trapezus in Kolchis‘.
Dalaison, J. 2014: ‚Civic Pride and Local Identities: the Pontic Cities and Their Coinage in the Roman Period‘ in T. Bekker-Nielsen (ed.), *Rome and the Black Sea Region: Domination, Romanisation, Resistance*, Aarhus, 125–155.

Dan, A. 2009: ‚Sinope, „capitale" pontique, dans la géographie antique', in H. Bru, F. Kirbihler & S.Lebreton, S. (Hgg.), L'Asie Mineure dans l'Antiquité: Échanges, population et territoires. Regards actuels sur une péninsule, Rennes, 67–131.

Dan, A. 2014: ‚Pontische Mehrdeutigkeiten', *eTopoi, Journal for Ancient Studies* 3, 43–66.

Dan, A. 2015: ‚Le Thermodon, fleuve des Amazones, du Pont-Euxin et de la Béotie: un cas d'homonymie géographique qui fait histoire', in V. Naas & M. Mahé-Simon (Hgg.), *De Samos à Rome: personnalité et influence de Douris*, Paris, 157–193.

Dingmann, M. 2007: *Pompeius Magnus. Machtgrundlagen eines spätrepublikanischen Politikers*, Rahden, Westf.

Dreher, M. 1996: ‚Pompeius und die kaukasischen Völker: Kolcher, Iberer, Albaner', *Historia* 45, 188–207.

Drumann, W. & Groebe, P. 1908/64: Geschichte Roms in seinem Übergange von der republikanischen zur monarchischen Verfassung, oder: Pompeius, Caesar, Cicero und ihre Zeitgenossen nach Geschlechtern und mit genealogischen Tabellen, 6 Bde., Königsberg 1834–44; 2. Aufl. hg. von P. Groebe, Berlin 1899–1929 (Bd. 4, 1908), Nachdr. Hildesheim 1964.

Dundua, G.F. & Lordkipanidze, G.A. 1979: 'Hellenistic Coins from the Site of Vani, in Colchis (Western Georgia)', *NC* 7.19 (139), 1–5.

Freber, P.-S.G. 1993: *Der hellenistische Osten und das Illyricum unter Caesar*, Stuttgart.

Geus, K. & Guckelsberger, K. 2017: ‚Measurement data in Strabo's Geography', in D. Dueck (Hg.), *The Routledge Companion to Strabo*, Abingdon, 165–177.

Gnoli, T. 2000: ‚Il Ponto e la Bitinia (Strabone XII 3)', in A.M. Biraschi & G. Salmeri (Hgg.), *Strabone e l'Asia Minore, Atti del Convegno Perugia, La Colombella, 25–28 maggio 1997*, Neapel, 543–564.

Guinea Díaz, P.M. 1999: ‚Notas sobre la organización pompeyana de la provincia de Bitinia y Ponto', *Gerión* 17, 317–329.

Habicht, Ch., Corey Brennan, T. & Blümel, W. 2009: ‚Ehren für Cn. Domitius Calvinus in Nysa', *ZPE* 169, 157–161.

Head, B.V. 1967: *Historia Numorum. A Manual of Greek Numismatics*, New and Enlarged Edition, Chicago.

Heinen, H. 1994: ‚Mithradates von Pergamon und Caesars bosporanische Pläne. Zur Interpretation von Bellum Alexandrinum 78', in R. Günther & S. Rebenich (Hgg.), E fontibus haurire. Beiträge zur römischen Geschichte und zu ihren Hilfswissenschaften (FS H. Chantraine), Paderborn, 63–79.

Hirschfeld, O. 1894: ‚Amisos', *RE* 1.2, 1839f.

Hoben, W. 1969: *Untersuchungen zur Stellung kleinasiatischer Dynasten in den Machtkämpfen der ausgehenden römischen Republik*, Diss. Mainz.

Højte, J.M. 2006: ‚From Kingdom to Province: Reshaping Pontos after the Fall of Mithridates VI', in T. Bekker-Nielsen (Hg.), *Rome and the Black Sea Region*. Aarhus, 15–30.

Janssens, E. 1967: *Trébizonde en Colchide*, Brüssel.

Jones, A.H.M. 1937/71: *The Cities of the Eastern Roman Provinces*, Oxford, 1. Aufl. 1937, 2. Aufl. 1971.

Judeich, W. 1885: *Caesar im Orient. Kritische Übersicht der Ereignisse vom 9. August 48 bis October 47*, Leipzig.

Kallet-Marx, R.M. 1995: *Hegemony to Empire. The Development of the Roman Imperium in the East from 148 to 62 B.C.*, Berkeley.

Kistler, E. 2009: *Funktionalisierte Keltenbilder. Die Indienstnahme der Kelten zur Vermittlung von Normen und Werten in der hellenistischen Welt*, Frankfurt.

Kuin, I.N.I. 2017: ‚Rewriting Family History: Strabo and the Mithridatic Wars', *Phoenix* 71.1–2, 102–118.

Leschhorn, W. 1993: *Antike Ären. Zeitrechnung, Politik und Geschichte im Schwarzmeerraum und in Kleinasien nördlich des Tauros*, Stuttgart.

Liebmann-Frankfort, Th. 1969: *La frontière orientale dans la politique extérieure de la république romaine depuis le traité d'Apamée jusqu'à la fin des conquêtes asiatiques de Pompée (189/88–63)*, Brüssel.

Lordkipanidze, O. 1996: *Das alte Georgien (Kolchis und Iberien) in Strabons Geographie. Neue Scholien*, deutsch von Nino Begiaschwili, Amsterdam.

Magie, D. 1950: *Roman Rule in Asia Minor to the End of the Third Century after Christ*, 2 Bde., Princeton.

Marek, C. 1993: *Stadt, Ära und Territorium in Pontus-Bithynia und Nord-Galatia*, Tübingen.

Marek, C. 2003: *Pontus et Bithynia. Die römischen Provinzen im Norden Kleinasiens*, Mainz.

Marek, C. 2010: *Geschichte Kleinasiens in der Antike*, München.

Marek, C. 2019: ‚Anadolu'daki Roma Eyaletlerine Genel Bakış. An Overview to the Roman Provinces in Anatolia', in Tekin 2019, 262–273.

Marshall, A.J. 1968: ‚Pompey's Organization of Bithynia-Pontus: Two Neglected Texts', *JRS* 58, 103–109.

Michels, Ch. 2009: *Kulturtransfer und monarchischer 'Philhellenismus'. Bithynien, Pontos und Kappadokien in hellenistischer Zeit*, Göttingen.

Michels, Ch. 2013: ‚Ariobarzanid Dynasty', *EAH*, 1. Aufl., 686f. Vgl. *EAH Online* 2012.

Millar, F. 1999: ‚*Civitates liberae, coloniae* and Provincial Governors under the Empire', *MedAnt* 2.1, 95–113.

Mitchell, S. 1993: *Anatolia. Land, Men, and Gods in Asia Minor*. Bd. 1: *The Celts in Anatolia and the Impact of Roman Rule*; Bd. 2: *The Rise of the Church*, Oxford.

Mitchell, S. 1994: ‚Termessos, King Amyntas, and the War with the Sandaliôtai. A New Inscription from Pisidia', in D. French (Hg.), *Studies in the History and Topography of Lycia and Pisidia. In Memoriam A.S. Hall*, Ankara, 95–105 und Taf. 6.1–2.

Mitford, T.B. 2013: ‚Euphrates Frontier (Roman)', *EAH*, 1. Aufl., 2566–2570.

Mladiov, I. ca. 1999: (Map of) ‚Asia Minor, c. 63 BC'. *Ian Mladiov's Resources*, University of Michigan, Ann Arbor (ohne Jahr). URL: https://sites.google.com/a/umich.edu/imladjov/maps.

Murphy, J. 1993: ‚Pompey's Eastern *Acta*', *AHB* 7, 1993, 136–142.

Niese, B. 1883: ‚Straboniana', *RhM* 38, 567–602.

Niese, B. 1901: ‚Deiotarus [2–3]', *RE* 4.2, 2401–2404.

Olshausen, E. 1974: ‚Polemonion', *RE Suppl.* 14, 427f.

Olshausen, E. 1980: ‚Pontos und Rom (63 v.Chr.–64 n.Chr.)', *ANRW* II 7.2, 903–912.

Olshausen, E. 1987: ‚Der König und die Priester. Die Mithradatiden im Kampf um die Anerkennung ihrer Herrschaft in Pontos', in E. Olshausen & H. Sonnabend (Hgg.), *Stuttgarter Kolloquium zur Historischen Geographie des Altertums I, 1980*, Bonn, 187–212.

Olshausen, E. 1996: ‚Amyntas [9]', *DNP* 1, 637.

Olshausen, E. 2002: ‚Themiskyra', *DNP* 12.1, 302.

Olshausen, E. 2012: ‚Chalyben – Autonym oder Xenonym?', in E. Olshausen & V. Sauer (eds.), *Die Schätze der Erde – Natürliche Ressourcen in der antiken Welt*, 2012, 338–344.

Olshausen, E. 2014: ‚Pontos: Profile of a Landscape', in T. Bekker-Nielsen (ed.), *Space, Place and Identity in Northern Anatolia*, Stuttgart, 39–48.

Olshausen, E. & Biller, J. 1984: *Historisch-geographische Aspekte der Geschichte des Pontischen und Armenischen Reiches. Teil I: TAVO B 29.1*, Wiesbaden.

Payen, G. 2020: *Dans l'ombre des empires. Les suites géopolitiques du traité d'Apamée en Anatolie*, Québec.

Podossinov, A.V. 2012: ‚Bithynia, Paphlagonia and Pontus on the *Tabula Peutingeriana*', in G.R. Tsetskhladze (Hg.), *The Black Sea, Paphlagonia, Pontus and Phrygia in Antiquity. Aspects of Archaeology and Ancient History*, Oxford, 203–206.

Primo, A. 2010: ‚The Client Kingdom of Pontus between Mithridatism and Philoromanism', in T. Kaizer & M. Facella (eds.), *Kingdoms and Principalities in the Roman Near East*, Stuttgart, 159–179.

Ramsay, W.M. 1941: ,Careers of New Cives'; ,Specimen List of Eastern Cives', in *idem*, *The Social Basis of Roman Power in Asia Minor*, hg. von J.G.C. Anderson, Aberdeen, 11–47.

Ritter, H.W. 1970: ,Caesars Verfügungen über Kleinarmenien im Jahre 47', *Historia* 19, 124–128.

Roller, D.W. 2018b: *Cleopatra's Daughter and Other Royal Women of the Augustan Era*, Oxford.

Sartre, M. 1995: *L'Asie Mineure et l'Anatolie d'Alexandre à Dioclétien, IVᵉ siècle av. J.-C. – IIIᵉ siècle ap. J.-C.*, Paris.

Schottky, M. 1996: ,Ariobarzanes [5] III. Eusebes Philorhomaios', *DNP* 1, 1083.

Seager, R. 2002: *Pompey the Great. A Political Biography*, Oxford.

Sens, U. 2009: *Kulturkontakt an der östlichen Schwarzmeerküste*, Langenweißbach.

Sherwin-White, A.N. 1984: *Roman Foreign Policy in the East (168 B.C. to A.D. 1)*, London.

Sherwin-White, A.N. 1994: ,Lucullus, Pompey and the East', in *CAH*² IX, 229–273.

Sørensen, S.L. 2016: *Between Kingdom and* koinon. *Neapolis/Neoklaudiopolis and the Pontic Cities*, Stuttgart.

Speidel, M.A. 2019: ,The Hellenistic Kingdom of Cappadocia', in Tekin 2019, 118–131.

Stähelin, F. 1907: *Geschichte der kleinasiatischen Galater*, 2. Aufl., Leipzig.

Stein-Kramer, M. 1988: *Die Klientelkönigreiche Kleinasiens in der Außenpolitik der späten Republik und des Augustus*, Berlin.

Strobel, K. 1994a: ,Galatien und seine Grenzregionen', in E. Schwertheim (Hg.), *Forschungen in Galatien*, Bonn, 29–65.

Strobel, K. 1994b: ,„Keltensieg und Galatersieger". Die Funktionalisierung eines historischen Phänomens als politischer Mythos der hellenistischen Welt', in E. Schwertheim (Hg.), *Forschungen in Galatien*, Bonn, 67–96.

Strobel, K. 1999: ,Kelten [III.]: Kelten im Osten', *DNP* 6, 1999, 393–400.

Sullivan, R.D. 1980: ,Dynasts in Pontus', *ANRW* II 7.2, Berlin, 913–930.

Sullivan, R.D. 1990: *Near Eastern Royalty and Rome, 100–30 BC*, Toronto.

Syme, R. 1995: *Anatolica. Studies in Strabo*, hg. von A. Birley, Oxford.

Tekin, O. (Hg.): *Hellenistik ve Roma İmparatorluğu dönemlerinde Anadolu – Anatolia in the Hellenistic and Roman Imperial Periods (English-Turkish)*, Istanbul.

Tröster, M. 2005: ,Lucullus, His Foreign *Amici*, and the Shadow of Pompey', in A. Coşkun (Hg.), *Roms auswärtige Freunde in der späten Republik und im frühen Prinzipat*, Göttingen, 91–111.

Tröster, M. 2006/19: ,Moaphernes', *APR*, *s.v.*

Tsetskhladze, G.R. 1998: *Die Griechen in der Kolchis*, Amsterdam.

Unger, G.F. 1896: ,Umfang und Anordnung der Geschichte des Poseidonios, V. Zeit der Reise an den Ocean', *Philologus* 55, 245–256.

van Ooteghem, J. 1954: *Pompée le Grand, Bâtisseur d'empire*, Brüssel.

Vitale, M. 2012: *Eparchie und Koinon in Kleinasien von der ausgehenden Republik bis ins 3. Jh. n.Chr.*, Bonn.

von Bredow, I. 2000: ,Pharnakes [2]', *DNP* 9, 752f.

von Bredow, I. 2001: ,Polemon [5]', *DNP* 10, 8.

Wagner, J. 1983: ,Die Neuordnung des Orients von Pompeius bis Augustus (67 v.Chr.–14 n.Chr.)', *TAVO* B V 7, Wiesbaden.

Wellesley, K. 1953: ,The Extent of the Territory Added to Bithynia by Pompey', *RhM* 96, 292–318.

Wesch-Klein, G. 2001: ,Bithynia, Pontus et Bithynia, Bithynia et Pontus', *ZPE* 136, 251–256.

Will, W. 2001: ,Pompeius [I 3]', *DNP* 10, 99–107.

Wirth, G. 1983: ,Pompeius, Armenien, Parther: Mutmaßungen zur Bewältigung einer Krisensituation', *BJb* 183, 1–60.

Wittke, A.-M., Olshausen, E., Szydlak, R. & Salazar, Ch.F. (eds.) 2010: *Historical Atlas of the Ancient World* (Brill's New Pauly Suppl. 3), Leiden.

Yoshimura, T. 1961: ,Die Auxiliartruppen und die Provinzialklientel in der Römischen Republik', *Historia* 10, 473–495.

Zwintscher, A. 1892: *De Galatarum tetrarchis et Amynta rege quaestiones*, Leipzig.

POMPEIUS UND DIE ‚ELF STÄDTE‘ DER PROVINZ PONTUS

Altay Coşkun

Abstract: Pompey and the 'Eleven Cities' of the Province of Pontus: The line of events from the death of Nikomedes IV through the Third Mithradatic War (73–63 BC) to the ratification of Pompey's Eastern acts in Rome in 59 BC is well documented in our sources and well-studied in modern scholarship. One can say the same of the main outcomes of the war, the Roman province of *Pontus et Bithynia*: there is no other province of the Republican period whose original constitution and circumscription we can delineate with so much precision thanks to Strabo's *Geography*. This said, the devil is in the detail, and a consensus on the cities that Pompey founded, refounded or included into this province is not yet in reach. The main reason for this seems to be that scholars tend to quarry, correct or harmonize pieces of information spread throughout the *Geography* to make them fit their own reconstruction. This way, sight may be lost of the specific view that Strabo held on the Pontic province as a whole. The present approach tries to take his text more seriously and develop the geographical design that he had before his eyes. Strabo (*Geogr.* 12.3.1 [541C])) attests that there were 'eleven cities' located in the previous realm of Mithradates VI Eupator and added to the former kingdom of Bithynia to yield the new province. These eleven consisted of the five pre-existing cities Herakleia, Tieion, Amastris, Sinope, Amisos plus his six new foundations Abonuteichos, Pompeiopolis, Neapolis, Magnopolis, Diospolis, Nikopolis. If we add Zela and Megalopolis, which Pompey also organized into *poleis*, the number of his settlements in the area goes up to eight, as Appian (*Mith.* 117.576) and Cassius Dio (37.20.2) attest. The latter two cities were given to King Deiotaros (or possibly other Galatian dynasts), just as was the former royal residence of Amaseia.

Абстракт: Помпей и одиннадцать городов провинции Понт: Линия событий от смерти Никомеда IV, Третьей Митридатовой войны (73–63 г. до н.э.) до ратификации восточных актов Помпея в Риме в 59 г. до н.э. хорошо документирована в письменных источниках и хорошо изучена современной наукой. То же самое можно сказать и об одном из главных результатов войны – римской провинции Понт и Вифиния: нет другой провинции республиканского периода, первоначальную конституцию и очертание которой мы могли бы описать с такой точностью, благодаря «Географии» Страбона. Тем не менее, дьявол кроется в деталях и консенсус относительно городов, которые были основаны, перестроены или включены в эту провинцию Помпеем, является недостижимым. Основная причина этого, по-видимому, заключается в том, что ученые стремятся найти гармонию, добывать, исправлять фрагменты информации, распространяемой по всей «Географии», чтобы они соответствовали их собственной реконструкции. Таким образом, мы можем упустить из виду особый взгляд, который Страбон имел на целую Понтийскую провинцию. Автор по-другому подходит к этому вопросу, пытается отнестись к нему более серьезно и развить ту географическую картину, которая развивалась перед глазами Страбона. Страбон (Геогр. 12.3.1 [541C])) подтверждает, что в предыдущем царстве Митридата VI Евпатора находились «одиннадцать городов», которые были присоединены к бывшему царству Вифинии для создания новой провинции. Эти одиннадцать городов состояли из пяти раньше существовавших: Гераклея, Тиейон, Амастрис, Синопа, Амисос, а также шести новых:

Абонутейхос, Помпейополис, Неаполис, Магнополис, Диосполис, Никополис. Если мы добавим Зелу и Мегаполис, которые Помпей тоже превратил в *полис*, число его поселений в этом районе возрастет до восьми, что подтверждают Аппиан (Митр. 117,576) и Кассий Дион (37.20.2). Последние два города были отданы царю Дейотару (или, возможно, другим галатским династам), как бывшая царская резиденция Амасия.

I. FORSCHUNGSÜBERBLICK ÜBER DIE EINRICHTUNG DER PROVINZ *PONTUS (ET BITHYNIA)* UNTER POMPEIUS

Eng verbunden mit dem Thema des vorangehenden Aufsatzes zum Pontisch-Galatischen, oder, wie wir soeben gesehen haben, ‚Kleinarmenischen' Reich des Deiotaros ist die Frage nach der Beschaffenheit der römischen Provinz, welche der siegreiche Feldherr Cn. Pompeius Magnus aus den zentralen und westlichen Teilen des untergegangenen Mithradatidenreich formte (65–63 v.Chr.). Bereits der Name jener neuen mit Bithynien verbundenen Verwaltungseinheit ist umstritten. Sie wird bald *Bithynia et Pontus*, bald *Pontus et Bithynia*, bald sogar nur *Bithynia* genannt, jedoch scheint die Quellenlage insofern eindeutig, als der aus der Konkursmasse des Mithradates VI. Eupator hervorgegangene Sprengel seinen Namen *Pontus* als Unterprovinz oder Eparchie beibehielt. Dessen ungeachtet sollte der für ihn zuständige *proconsul* seinen Hauptsitz in Nikomedeia in der westlichen Unterprovinz haben. Dementsprechend wurde sein Amtssprengel rund ein Jahrhundert lang meistens kurz *Bithynia* genannt. Erst seit der Zeit Neros dominiert der Name *Pontus et Bithynia* für die Doppelprovinz.[1]

Im Mittelpunkt des vorliegenden Aufsatzes steht also der Versuch, die exakte geographische Organisation jener Provinz Pontus zu bestimmen. Zusammen mit weiteren (hier ausgeklammerten) konstitutionellen und fiskalischen Fragen gehen diese weit über ein rein antiquarisches Interesse hinaus, sind sie doch eng mit dem Erfassen der Zielsetzungen und Mechanismen römischer Herrschaft sowie einer historischen Bewertung des ‚großen' Pompeius verbunden. Dies hat man lange erkannt, und besonders Christian Marek hat dem einen trefflichen Ausdruck verliehen, indem er das Verständnis jener Details an das ‚Urteil über die staatsmännische und organisatorische Leistung des Pompeius [knüpft] und und ihn je nachdem als überehrgeizigen „empirebuilder" oder maßvollen Realisten [einstuft]' und das Thema ferner mit Fragen des römischen Imperialismus verbindet.[2]

Hauptanliegen des vorliegenden Beitrags ist die Identifizierung der Stadtstaaten jenes pontischen Provinzsprengels. Die Forschungsdiskussion wurde intensiv

---

1   Zum Namen der Provinz s. Anhang I.
2   Marek 1993, 26. Vgl. Olshausen 1980, 905f., der betont, dass ein abschließendes Urteil noch ausstehe; Marshall 1968, 103: ‚In this inquiry whether Pompey was the tool of commercial interests, the saviour of eastern Hellenism, or a vainglorious imperialist, a variety of subjective evaluations of his work have been offered, ranging from lavish praise to heavy censure.' Positive Urteile haben in jüngerer Zeit z.B. Guinea Díaz 1999, 322–329 und Fernoux 2004, 132f. gefällt. Madsen 2014 beleuchtet die Frage der zivilisatorischen Leistung des Pompeius aus der Sicht Strabons; vgl. Madsen 2020 (*non iam vidi*).

bis in die 1990er Jahre hineingeführt, flaute dann aber ab, da man die Umrisse der neuen Provinz bis auf wenige Nuancen konsensfähig rekonstruieren zu können glaubte. Diese allgemeine Auffassung möchte ich hinterfragen und einen mit den Hauptquellen besser vereinbaren Gegenentwurf vorschlagen.

Ausgangspunkt aller Untersuchungen ist das Zeugnis Strabons, obwohl dieser die Leistung des Pompeius nur kursorisch und summarisch erwähnt sowie in den Abschnitten zu den jeweiligen Städten oft ganz verschweigt. Dennoch ist von höchster Bedeutung, dass der Geograph von Amaseia die Zahl der in die Provinz *Pontus* eingegangenen Städte auf elf beziffert. Seine Aussage werden wir uns bald genauer anschauen.[3] Erwähnt sei aber auch die Zahl derjenigen Städte, die Pompeius ‚in Kappadokien' neugegründet habe: Appian spricht von acht; auch diese beiläufige Information wird sich gegen Ende der vorliegenden Diskussion als relevant erweisen.[4]

Die Identifizierung jener Poleis hat die Forschung lange Zeit beschäftigt. Nachdem Benedikt Niese (1883) die wissenschaftliche Diskussion eröffnet hatte, basierten die Synthesen von Arnold Jones (1937) und David Magie (1950) bereits auf relativ soliden Grundlagen. Ihre Auffassungen wurden in den meisten Punkten durch die erneuten Untersuchungen im Kontext des *TAVO* (1982) bestätigt und sind für einen Teil der Forschung bis heute maßgeblich. Gemäß dieser Tradition besteht die Liste aus folgenden elf Städten: Amastris, Sinope und Amisos an der paphlagonischen Küste, Pompeiopolis und Neapolis (ehemals Phazimon) im paphlagonischen Hinterland sowie Zela, Amaseia, Diospolis (ehemals Kabeira), Magnopolis (als Neugründung von Eupatoria), Nikopolis und Megalopolis im pontischen Hinterland.[5]

Weniger erfolgreich war demgegenüber der radikale Ansatz von Kenneth Wellesley, welcher das Provinzterritorium im Wesentlichen durch den Halys begrenzt sah und nur noch Amisos und Neapolis hinzurechnete.[6] Unter den weiteren

3   Strab. *Geogr.* 12.3.1 (541C), zitiert unten mit Anm. 9.
4   App. *Mith.* 117.576. Vgl. Cass. Dio 37.20.2, der von acht Gründungen (ohne Lokalisierung) spricht, sowie Plut. *Pomp.* 45.2f., bei dem von insgesamt 39 Gründungen die Rede ist. Die Zahlen sind unten mit Anm. 29 diskutiert, die Lage ‚in Kappadokien' in Anhang II, wo auch die Quellen zitiert werden.
5   Jones 1937, 159f.; Magie 1950, Bd. 2, 1232–1234; Wagner 1983 (*TAVO* B V 7), die stark von Niese 1883 abweichen; zu diesem s. unten mit Anm. 11. Vgl. Lassère 1981, 62 Anm. 2; Sherwin-White 1984, 229f. und 257f. mit Anm. 46; Stein-Kramer 1988, 90 (Karte); Mitchell 1993, Bd. 1, 31f. (mit Karte 3 auf S. 40f.); Kallet-Marx 1995, 327–329; Mladiov ca. 1999; Will 2001; Seager 2002, 60; Payen 2020, fig. 22 (Karte). Die Zugehörigkeit des östlich des Halys gelegenen Phazimonitis zu Paphlagonien bestreitet neuerdings Bekker-Nielsen 2014b, 64–70; s. auch unten Anm. 17 zu Sørensen 2016, 136–153. Zahlreiche Kommentierungen zu den Städten bietet Roller 2018, 694–701; 709–717, aber er hat ebensowenig ein Interesse an der Politik des Pompeius wie Radt 2008, 347. Für ältere Literatur s. Marshall 1968, 103 Anm. 2f.
6   Wellesley 1953; seine Karten sind als Fig. 2–3 in Coşkun, Kapitel X in diesem Band wieder abgedruckt. Positiv urteilt vor allem Olshausen 1980, 906 mit Anm. 11, so dass er die politische Zugehörigkeit der meisten Städte offenlässt. Die wichtige historisch-geographische Studie Olshausen & Biller 1984 klammert diese Frage systematisch aus, da ihr Untersu-

weniger einflussreichen Varianten hebe ich noch die posthum erschienene Rekon-
struktion von Ronald Syme hervor, welcher Abonuteichos einbezog, allerdings
Nikopolis wegen der Lage in Kleinarmenien trotz der Unterstellung unter den
Statthalter von *Pontus et Bithynia* nicht mitrechnete.[7]

Demgegenüber gilt heute zumeist Mareks erste Monografie zu *Pontus et
Bithynia* (1993) als richtungsweisend. Er identifiziert die elf Städte mit Amastris,
wahrscheinlich Abonuteichos, Sinope und Amisos an der paphlagonischen Küste
sowie Pompeiopolis, Neapolis, Zela (vermutlich mit Amaseia) und Diospolis,
Magnopolis, Nikopolis und Megalopolis im paphlagonisch-pontischen Hinterland.
Herakleia und Tieion, die ebenfalls in die Provinz eingegliedert wurden, zählt
Marek wie auch schon die meisten Forscher nach Niese nicht zu den elf, da sie
historisch und geographisch eher zu Bithynien gehörte hätten. Mareks wichtigste
Neuerungen liegen in der Erwägung, dass Amaseia in die Zelitis eingegliedert
worden sei sowie Abonuteichos, welches Strabon indes nur flüchtig als *polichnion*
zwischen Amastris und Sinope auflistet, zu ergänzen sei. Trotz bedeutender Änderungen in der Argumentation und inneren Struktur blieben die rekonstruierten
Umrisse der Provinz aber fast unverändert.[8]

---

chungsfeld im Wesentlichen mit Eupator endet; vgl. bes. 15–20 zu Neapolis; 27–44 zu Magnopolis; 44–54 zu Kabeira / Diospolis / Neokaisareia. S. auch Olshausen 1991, an den anschließend auch Kleu 2013, 5421 die Liste der Städte offenlässt.

7   Syme 1995, 112–115, z.B. 115: ‚the eleven *politiae* were all contiguous‘. Bereits an anderer
Stelle (Kapitel X in diesem Band) habe ich die Position von Wellesley 1953 besprochen, der
das Reich des Deiotaros an der Küste mit dem Halys beginnen ließ und nur die Exklave Amisos sowie im Hinterland Phazemon / Neapolis zugestand und ein ‚Herumraten‘ zur Identifika-
tion der elf Städte als nutzlos verwarf (S. 296f.); *contra* auch Marek 1993, 29–33 und Ballesteros Pastor 1996, 283–285. S. unten Abschnitt 3 zu Nikopolis. Die Variante von Strobel
1997 behandle ich ebenfalls weiter unten. Vgl. auch Chaniotis 2018, 220f., dessen Resümee
der Neuordnungen des Pompeius von Marek beeinflusst ist (denn er erkennt Abonuteichos als
dessen Neugründung an), jedoch liegt seine Darstellung gegenüber den traditionellen Diskus-
sionen quer, da er bei Amastris, Sinope und Amisos von einer Wiederherstellung des alten
Polisstatus spricht, obwohl dieser ja erst durch die jüngsten Kriegseinwirkungen infrage gestellt war, aber womöglich nach der Neukonstituierung durch Lucullus ca. 70 bestehen geblieben waren; s. Marek 1985, 144–152 und 1993, 26f.; 32f. Des Weiteren setzt Chaniotis
Neukonstituierungen für Zela und Kabeira / Diospolis an, während er im Fall von Nikopolis,
Megalopolis, Magnopolis, Pompeiopolis und Neapolis von Neugründungen spricht. Dies
ergibt eine Summe von zehn relativ neuen Provinzstädten, ohne den Versuch zu machen, die
von Strabon und Appian bezeugten Zahlen zu erklären. Zur Annahme von zehn pontischen
Städten infolge der falschen Lesung einer Inschrift von Herakleia s. Marek 1993, 73.

8   Marek 1993 basiert auf seiner Marburger Habilitationsschrift 1988; s. bes. 33–46 mit Karte in
Beilage 1; vgl. auch Marek 2003, 36–40 (ohne Details, aber mit vielen Fotos); 2010, 366f.
(‚...wahrscheinlich auch Abonuteichos ... Zela ... vermutlich einschließlich Amaseias ...‘);
2019, 303. Auf Marek berufen sich unter anderem Guinea Díaz 1999, 321f. (obwohl seine eigene unvollständige Liste nicht mit derjenigen Mareks vereinbar ist); Gnoli 2000, 545–550,
der betreffs Amaseia und Abonuteichos unentschieden ist; Coşkun 2007, 125–131 (allerdings
mit Einschränkungen betreffs des Territoriums von Amisos); Michels 2009, 308 Anm. 1601;
Primo 2010, 159 mit Anm. 3 (allerdings begrenzt er die Provinz irrtümlich durch den Halys);
Vitale 2012, 164f. (bes. zu Amaseia); Olshausen 2014, 47; ähnlich E. Olshausen in Wittke
u.a. 2010, 160f.: ‚Amastris, Amisus, Eupatoria ..., Cabira ..., Phazemon ..., Sinope, Zela, ...

II. EIN NEUER LÖSUNGSWEG

1. Herakleia Pontika und Tieion

In nahezu allen Fällen zeigt sich eine ähnlich – problematische – Vorgehensweise wie schon im vorangehenden Kapitel zu Strabon's Beschreibung der Territorien des Deiotaros beobachtet wurde. Informationen werden zu schnell aus dem Text herausgelesen, ohne den näheren und weiteren Zusammenhang hinreichend zu beachten. Denn liest man den einführenden Abschnitt zum Mithradates-Reich an einem Stück, so sollte klar sein, dass der Geograph hier Herakleia (und somit implizit auch Tieion) zu den Eroberungen des Königs zählt und daher auch in die von Pompeius umgeformte ‚Konkursmasse‘ einrechnet:

> Über Pontos etablierte sich Mithradates Eupator als König. Er hatte aber das Land jenseits des Halys bis zu den Tibarenern und Armeniern sowie diesseits des Halys bis Amastris und einige Landstriche Paphlagoniens. Indes **eroberte dieser zusätzlich die Küste bis Herakleia gen Westen**, die Heimat des Platonikers Herakleides, und zudem in entgegengesetzter Richtung bis zur Kolchis und Kleinarmenien, **welche er also zu Pontos hinzufügte**. Und so übernahm denn auch Pompeius, nachdem er jenen vernichtet hatte, dieses Land innerhalb dieser Grenzen: zum einen teilte er also das Land in Richtung Armenien und in der Gegend von Kolchis den Dynasten zu, die mit ihm zusammen gekämpft hatten; zum anderen **teilte er den Rest in elf Stadtstaaten (*politeiai*) auf** und **schloss es an Bithynien an, so dass aus zwei Provinzen eine würde**. Unterdessen gab er aber Teile des Binnenlandes der Paphlagonen den Nachfahren des Pylaimenes als ihre Herrschaft, so wie auch die Galater den angestammten Tetrarchen.[9]

Ganz offensichtlich vertritt Strabon die Ansicht, dass die Eroberungen Eupators Herakleia einschlossen, so wie er es noch mehrfach in nachfolgenden Abschnitten tut. Einmal situiert er diese Stadt sogar explizit ‚in der Provinz Pontos, die mit Bithynien verbunden war‘. Damit sollte klar sein, dass sowohl Herakleia als auch Tieion zu den elf Stadtstaaten der Provinzerweiterung gezählt werden müssen.

Megalopolis, Nicopolis, Pompeiopolis and one other city, possible candidates for which are Abonuteichus, Amasia, Heraclea Pontica and Tium‘. Vgl. Eck 2007, 190f.; Edelmann-Singer 2015, 71f. Indes vermeidet Sørensen 2016, 107–120 in seiner Besprechung der Provinzialisierung von Pontus das Thema in auffälliger Weise; die entscheidende Strabon-Stelle (s. nächste Anm.) zitiert er S. 122 im Kapitel zu M. Antonius.

9   Strab. *Geogr.* 12.3.1 (541C) ed. Radt 2004, Bd. 3, 422 (vgl. Lassère 1981 und Roller 2014): τοῦ δὲ Πόντου καθίστατο μὲν Μιθριδάτης ὁ Εὐπάτωρ βασιλεύς, εἶχε δὲ τὴν ἀφοριζομένην τῷ Ἅλυϊ μέχρι Τιβαρανῶν καὶ Ἀρμενίων καὶ τῆς ἐντὸς Ἅλυος τὰ μέχρι Ἀμάστρεως καί τινων τῆς Παφλαγονίας μερῶν. προσεκτήσατο δ᾽ οὗτος καὶ τὴν μέχρι Ἡρακλείας παραλίαν ἐπὶ τὰ δυσμικὰ μέρη, τῆς Ἡρακλείδου τοῦ Πλατωνικοῦ πατρίδος, ἐπὶ δὲ τἀναντία μέχρι Κολχίδος καὶ τῆς μικρᾶς Ἀρμενίας· ἃ δὴ καὶ προσέθηκε τῷ Πόντῳ. καὶ δὴ καὶ Πομπήϊος καταλύσας ἐκεῖνον ἐν τούτοις τοῖς ὅροις οὖσαν τὴν χώραν ταύτην παρέλαβε. τὰ μὲν <δὴ> πρὸς Ἀρμενίαν καὶ τὰ περὶ τὴν Κολχίδα τοῖς συναγωνισαμένοις δυνάσταις κατένειμε, τὰ δὲ λοιπὰ εἰς ἕνδεκα πολιτείας διεῖλε καὶ τῇ Βιθυνίᾳ / προσέθηκεν ὥστ᾽ ἐξ ἀμφοῖν ἐπαρχίαν γενέσθαι μίαν· μεταξὺ <δὲ> τῶν Παφλαγόνων τῶν μεσογαίων τινὰς βασιλεύεσθαι παρέδωκε τοῖς ἀπὸ Πυλαιμένους, καθάπερ καὶ τοὺς Γαλάτας τοῖς ἀπὸ γένους τετράρχαις. Mit Blick auf die Erstellung einer Liste ist es wichtig, Strabons Aussage nicht mit Syme 1995, 114 misszuverstehen: ‚Strabo states that Pompeius created eleven "politiae".‘

Denn es spielt ja keine Rolle, ob man – wie etwa Marek – der Auffassung ist, dass die Eroberung Herakleias im Jahr 72 v.Chr. keine wirkliche Zugehörigkeit zum Pontischen Reich des Mithradates begründe; entscheidend ist doch, dass Strabon es ausdrücklich so sieht und damit wohl den römischen Standpunkt wiedergibt, womöglich wegen einer engagierten Parteinahme für den König seitens der Stadt.[10] Ähnlich dachten vielleicht schon Benedikt Niese, Bernhard Rémy und Karl Strobel, freilich ohne Problembewusstsein zu zeigen oder Erklärungen anzubieten. Vielleicht reicht auch lediglich der seit der Kaiserzeit übliche Name Herakleia Pontika, um die Plausibilität von Strabons Aussage weiter zu untermauern.[11] Vor diesem Hintergrund ist die Liste der elf Provinzstädte neu zu verhandeln, da mindestens zwei der oben genannten Kandidaten auszuscheiden haben, um Herakleia und Tieion zu weichen.

2. Amaseia und Kabeira-Diospolis

Strobel spricht der pompejanischen Provinz von den oben genannten Städten Amaseia, Abonuteichos und Nikopolis ab.[12] Begründungen liefert er nicht, aber

---

10   Strab. *Geogr.* 12.3.6 (543C): ἡ δὲ πόλις ἐστὶ τῆς Ποντικῆς ἐπαρχίας τῆς συντεταγμένης τῇ Βιθυνίᾳ. Vgl. 12.3.2 (541C) und 12.3.9 (544C), obwohl der letzte Satz – in Verbindung mit dem nächsten Abschnitt 12.3.10 (544C) – den Eindruck erweckt, dass das eigentliche Pontos erst mit dem Fluss Parthenios beginne; vgl. Syme 1995, 113: ‚the Parthenius is a boundary after all' (aber s. auch die nächste Anm.). Zuvor hat er in 12.3.5 (542C) diesen – bei Tieion ins Schwarze Meer mündenden Fluss – als Grenze der Mariandyner bezeichnet. Deren Territorium war wiederum nach 12.3.4f. (542C) zwischen Bithynien und Paphlagonien geteilt; vgl. Radt 2008, Bd. 7, 347. Kaum überzeugend argumentiert Marek 1993, 35f.; 39–41 gegen den Einschluss von Herakleia und Tieion. Nur scheinbar logisch ist sein ergänzender Beweis, dass Tieion als ehemaliger Besitz bithynischer Könige sicher bithynisch und damit nicht pontisch sei; dies verkennt aber die Tatsache, dass sich historisch-geographische Bezeichnungen überlagern und ablösen konnten, was nicht nur hier bei Strabon offensichtlich ist (und damit auch der Beweiskraft des Parthenios als Grenzfluss den Boden entzieht), sondern auch durch die unscharf getrennten Begriffe Pontos, Kappadokien und Armenien (s. Anhang II; zudem den Anhang in Coşkun, Kapitel X in diesem Band) oder die Diskussion um den Namen der römischen Provinz (s. Anhang I) dokumentiert wird. Vitale 2014, 50f. verstrickt sich in Widersprüche: Er folgt angeblich Marek, zählt aber Herakleia zu den elf Mitgliedstaaten von ‚this province and league of cities', präsentiert dann aber eine Karte mit 13 *poleis*. Vgl. auch in anderem Kontext Madsen 2014, 77: ‚Strabo opens his survey of Pontos in Herakleia Pontike.'
11   Strab. *Geogr.* 12.3.10 (545C) sowie Niese 1883, 577 und 581; Rémy 1986, 20f., der nur sechs der elf Städte nennt (darunter Abonuteichos), während er ‚le reste du royaume' als an Dynasten übergeben betrachtet. Strobel 1997 verweist für die Begründung seiner Liste auf den bislang nicht erschienenen Bd. 2 von *Die Galater*. Vgl. auch Syme 1995, 112f., der betont, dass es sich hier um die Sicht Strabons handle, wenngleich diese manche historische Schwierigkeit impliziere; allerdings rechnet er S. 114 Herakleia und Tieion **nicht** zu den elf *poleis*. Ähnlich Kallet-Marx 1995, 327, der von einer Bestrafung Herakleias wegen ‚adherence to the cause of Mithridates' spricht, ohne Konsequenzen für die Zusammensetzung der Provinz zu ziehen. Zu Herakleia Pontika (allerdings in anderem Kontext) s. z.B. Mitchell 2002, 40.
12   Strobel 1997. Unklar ist, warum Strobel Amastris zwar (richtig) unter die elf Provinzstädte zählt, es ferner auch dem pontischen Provinziallandtag (nach 40) zuordnet, aber zugleich von

für die erste Stadt kann mit Strabon darauf verwiesen werden, dass es wohl im Anschluss an den Dritten Mithradatischen Krieg ‚an Könige gegeben wurde, jetzt aber eine (Unter-) Provinz ist'.[13] Der Kontext spricht dagegen, die Vergabe an einen oder mehrere Monarchen dem Triumvirn M. Antonius zuzuschreiben, wie allgemein geschieht,[14] da der Geograph sonst doch wohl eher einen anderen zeitlichen Bezugspunkt gewählt hätte. Das Schicksal der Städte in der Phanaroia wich von dieser Ereignisfolge ab, denn für sie hat Strabon noch kurz zuvor gesagt, dass ‚Pythodoris das ganze den Barbaren benachbarte Gebiet in ihrem Besitz hält'. Konkret nennt er die ehemalige mithradatische Residenzstadt Kabeira, die Pompeius unter dem neuen Namen Diospolis der Provinz zugeschlagen habe, später aber von Pythodoris zu ihrer Hauptresidenz Sebaste ausgebaut worden sei. Daneben sind auch die Zelitis und Megalopolitis in ihrem Besitz genannt, nicht aber Amaseia selbst, das Zentrum des früheren Mithradatidenreiches, dessen Territorium sich tatsächlich weit gen Westen erstreckte und sogar an das Trokmergebiet angrenzte.[15]

Man gewinnt also den Eindruck, dass Pompeius Amaseia noch im Pontischen Reichsverband beließ, so dass der neue König Deiotaros eine effektive Landbrücke zwischen Galatien und seinen ‚kleinarmenischen' Territorien nutzen konnte. Wahrscheinlich verblieb Amaseia auch unter M. Antonius in der pontischen Restmasse, welche er dem Dareios übergab. Der Mithradates-Enkel wird besonde-

einer Befreiung der Stadt (vor Caesar?) ausgeht. Er verweist hierfür auf einen Aufsatz von B. McGing, der aber Amastris mit keiner Silbe erwähnt. Das Ärenjahr 71/70 (s. Leschhorn 1993, 162–166) fällt jedenfalls in die Zeit des Lucullus. Niese 1883, 580f. zählt wiederum Zela zur Provinz, nicht aber Amaseia.

13 Strab. *Geogr.* 12.3.39 (561C): ἐδόθη δὲ καὶ ἡ Ἀμάσεια βασιλεῦσι, νῦν δ' ἐπαρχία ἐστί. Damit ist wohl auf die Verwaltungseinheit *Pontus Galaticus* in der Provinz Galatien bezeichnet; s. hierzu unten mit Anm. 17.

14 So explizit z.B. Marek 1993, 49–51 (‚Der eigentliche Zerstörer der pompejanischen Ordnung') und Leschhorn 1993, 122f. je mit weiterer Literatur; vgl. Lassère 1981, 107 Anm. 3; Ballesteros Pastor 1996, 286. Anzunehmen ist das auch für die anderen Forscher, die Amaseia zunächst der Provinz des Pompeius zuschreiben; s. oben in Anm. 5. *Contra* indes Wellesley 1953, 295f. S. auch Sørensen 2016, 117, der die Zuteilung an Dynasten ebenfalls auf Pompeius bezieht.

15 Strab. *Geogr.* 12.3.30 (556C) nennt Amaseia in seinem Exkurs zum Iristal und der Phanaroia, aber lokalisiert sie außerhalb jener fruchtbaren Ebene; vgl. Olshausen 2014, 44. Die Ausdehnung der Chora von Amaseia behandelt er in 12.3.39 (561C). Strab. *Geogr.* 12.3.31 (556C) betrifft Kabeira bis unter Mithradates Eupator. In § 31 (557C) fährt Strabon fort: ταύτην δὴ τὴν χώραν ἔχει πᾶσαν ἡ Πυθοδωρὶς προσεχῆ οὖσαν τῇ βαρβάρῳ τῇ ὑπ' αὐτῆς κατεχομένῃ, καὶ τὴν Ζηλῖτιν καὶ Μεγαλοπολῖτιν. τὰ δὲ Κάβειρα Πομπηΐου σκευάσαντος εἰς πόλιν καὶ καλέσαντος Διὸς πόλιν, ἐκείνη προσκατεσκεύασε καὶ Σεβαστὴν μετωνόμασε, βασιλείῳ τε τῇ πόλει χρῆται. Die fünf kursiven Worte tilgt Radt Bd. 7, 2008, 391 (vgl. Bd. 3, 2004, 466, kommentarlos gefolgt von Roller 2014, 533) als eine ‚offenkundige von 559, 21 (sc. 12.3.37 [559C]) inspirierte Interpolation'. Dies ist zwar möglich, aber nicht zwingend und ändert nichts an der geohistorischen Interpretation, da Strab. *Geogr.* 12.3.37 (559C) die Besitzverhältnisse unter Pythodoris klärt. Die Neubenennung von Diospolis in Sebaste scheint zu bestätigen, dass der Ausbau nicht schon durch Polemon erfolgte, jedenfalls nicht vor Actium, und ich sehe keinen Grund, daran zu zweifeln, dass die alleinige Urheberschaft der Pythodoris auf einen Zeitpunkt nach 9 v.Chr. verweist, wie weiter unten im Text ausgeführt ist.

ren ideologischen Wert auf die Stadt seiner Ahnen gelegt haben. Mit dem Herr-
schaftswechsel zu Polemon 37/6 bot sich theoretisch die Möglichkeit recht freizü-
giger Umverteilungen an. Strabon lässt uns hier leider im Dunkeln.[16] Man könnte
erwägen, dass Amaseia damals in den Besitz des neu eingesetzten Galaterkönigs
Amyntas gelangte, da die Stadt später als Teil des *Pontus Galaticus* in die Provinz
Galatien eingegliedert wurde. Allerdings steht dem nicht nur die südliche Ausdehnung des Amyntasreiches unter M. Antonius und Augustus entgegen, sondern
auch der in der Kaiserzeit gut dokumentierte Ärenbeginn von 3/2 oder 2/1 v.Chr.
Eher ist deswegen daran zu denken, dass Augustus Pontos erst nach dem Tod Polemons 9 v.Chr. verringerte und Amaseia dem sich als loyal erwiesenen galatischen Dynasten Ateporix unterstellte, welcher damals in der Karanitis (im Süden
der ehemaligen Zelitis) herrschte und der wohl etwa 3 v.Chr. starb.[17] Die Loslö-
sung Amaseias aus dem Pontischen Reich bei dessen Übergabe an Pythodoris
führte dazu, dass sie einer neuen Residenz bedurfte. Wie soeben gesehen, bezeugt
Strabon das ehemalige Diospolis als Sebaste und noch nicht das von ihrem Enkel-
sohn erbaute Polemonion in dieser neuen Funktion.[18]

16 Seltsamerweise lässt Strabon das Schicksal seiner Heimatstadt Amaseia in seinen Details
    offen; zur dynastischen Geschichte (ohne meinen Vorschlag) s. Roller 2018, 716 und Bal-
    lesteros Pastor, Kapitel VII in diesem Band. Zu Dareios s. Sørensen 2016, 121–125 und
    Coşkun, Kapitel X in diesem Band, bes. Anm. 65.
17 S. Leschhorn 1993, 115–119 für die Dokumentation zu Amaseia (vgl. Dalaison 2014, 125)
    und 119–122 zu Karana-Sebastopolis-Herakleiopolis. Mitchell 1993, Bd. 1, 93; 107f. schlägt
    vor, die Herrschaft eines nur durch ein Inschriftenfragment bezeugten Galaterkönigs Brigatos
    anzusetzen, doch betrachte ich jenen als den direkten Nachfolger des Deiotaros I. und Vor-
    gänger des Amyntas in Galatien; s. Coşkun 2007/19c; 2007/19e. Wenig überzeugend argu-
    mentiert Sørensen 2016, 136–153, dass alle pontischen Ortschaften östlich des Halys (nach
    dem Tod des Deiotaros Philadelphos von Paphlagonien auch die Phazemonitis mit Neapolis,
    s. dazu auch oben Anm. 5) der Pythodoris zugesprochen und erst unter Nero Teil des *Pontus
    Galaticus* geworden seien. Seine Vorstellungen zu den jeweiligen städtischen Ären oder seine
    Deutung des Eides von Gangra (und Neapolis) sind aber unhaltbar. – Gegen König Amyntas
    spricht die Tatsache, dass sein Territorium nach Süden hin ausgerichtet wurde und er 26/5
    v.Chr. starb; ich vermute, dass ihm nach seinem Tod ein gleichnamiger Sohn im Tetrarchen-
    rang im Trokmerland nachfolgte, welches ca. 20 v.Chr. zur Provinz kam; s. Coşkun 2007/19e
    und 2007/19f. Verlockend wäre die Annahme einer Verbindung mit Paphlagonien unter Kas-
    tor (III.) und Deiotaros (III.) Philadelphos, jedoch starb letzterer ca. 6/5 v.Chr.; s. Coşkun
    2007/19g; 2008/19b. Ebensowenig können wir den Herrscher von Amaseia mit Dyteutos
    identifizieren, der als Hohepriester von Komana Pontike nachgewiesen ist; s. Coşkun 2007/19i.
    Damit verbleibt als wahrscheinlichster Kandidat Ateporix, der bald nach 29 v.Chr. das Dorf
    Karana zur Residenzstadt Sebastopolis ausbaute, die für jene Stadt später ein Ärenbeginn von
    3/2 v.Chr. galt; s. Coşkun 2007/19h (aber noch ohne die Vermutung betreffs Amaseia). In
    diesem Jahr wurden Amaseia, Zela und die Karanitis als *Pontus Galaticus* an die Provinz Ga-
    latien angeschlossen; s. Ptol. *Geogr.* 5.6.9; vgl. Leschhorn 1993, 123. Zusammenfassend zu
    den spätgalatischen Dynasten Coşkun 2008.
18 Anders schreibt die große Mehrheit der Forscher Polemon I. die Neugründung von Side als
    Polemonion zu, aber s. die Diskussion bei Coşkun, Kapitel X in diesem Band, Anm. 27, wo
    mit Vitale 2012, 151–154 für Polemon II. argumentiert wird.

3. Nikopolis (und Magnopolis)

Diospolis und Nikopolis konnte M. Antonius offenbar leicht aus der Provinz aus-
gliedern und organisch mit dem Pontischen Reich verbinden, wozu gehörig die
Städte auch bei Strabon bzw. Appian erscheinen.[19] Problematisch wäre dies indes
für den Fall, dass Pompeius Nikopolis zu einer römischen Veteranenkolonie gemacht hätte. Dies wird vielfach angenommen, da Cassius Dio von der Ansiedlung
verletzter Soldaten neben anderen Einheimischen spricht.[20] Doch scheint gerade
das spätere Schicksal der Stadt gegen diese Behauptung zu sprechen, es sei denn,
der Feldherr beschränkte die Kolonie auf nichtrömische Veteranen seiner Bündner. Strabon bezeugt die Niederlassung von Veteranen allein für das wesentlich
günstiger gelegene Magnopolis in der Phanaroia, das auf dem Territorium der
unvollendeten Stadt Eupatoria errichtet wurde, und Appian nennt einzig Nikopolis
und Magnopolis als die Neugründungen des Pompeius auf dem Boden des ehemaligen Mithradatischen Reiches. Vielleicht hat Dio also schlicht verschiedene Berichte kontaminiert.[21] Verkehrstechnisch blieben die Nikopolitaner und Diospoli-

19 Diospolis: Strab. *Geogr.* 12.3.30 (556C), s. oben Anm. 15. Nikopolis erscheint bei App. *Mith.*
    115.561 als Teil Kleinarmeniens, wie schon im *BAlex.* im Kontext des Berichts zur Schlacht
    von Nikopolis im Jahr 48 (36.3); allerdings wird hier wiederholt der ganze Reichsteil des
    Deiotaros als Armenia Minor bezeichnet (34–36), während später (allerdings auch mit Blick
    auf das weiter westlich gelegene Zela) durchweg von Pontos gesprochen wird (67–77). Wei-
    terhin bezeugt Cass. Dio 49.33.2 und 49.44.2, dass M. Antonius Armenia Minor 34/3 v.Chr.
    an Polemon übergab, bevor es – vielleicht nach seinem Tod 9 v.Chr. – wieder mit Kappado-
    kien verbunden war, wie Strab. *Geogr.* 12.3.29 (555C). S. auch den Anhang in Kapitel X in
    diesem Band.
20 Cass. Dio 36.50.3; vgl. Oros. 6.4.7 (mit falscher Lokalisierung zwischen Euphrat und Ara-
    xes). Sturm 1936, 536, setzt die Ansiedlung römischer Veteranen voraus, geht aber ohne Dis-
    kussion – wie Niese 1883, 580 und Strobel 1997 – von der sofortigen Eingliederung der Stadt
    in Kleinarmenien aus. Unklar betreffs der Zugehörigkeit bleiben Wellesley 1953, 307 (exklu-
    diert) und 313 (inkludiert) sowie Sherwin-White 1984, 258 Anm. 46. Entschieden für römi-
    sche Veteranen spricht sich Magie 1950, Bd. 2, 1233 aus, aber sein ergänzter epigraphischer
    Beleg hat wenig Aussagekraft. Dreizehnter 1975, 235–237 setzt Veteranenansiedlung und
    Provinzzugehörigkeit voraus und diskutiert die Möglichkeit einer Alexander-*imitatio* (mit
    wenig überzeugendem negativem Ergebnis). Syme 1995, 112 betont, dass der Bürgerstatus
    eine Anbindung an die Provinz notwendig gemacht habe, auch wenn die Kolonie territorial
    nicht damit verbunden gewesen sei. Kallet-Marx 1995, 329 spricht von ,first foundation of
    Roman citizens in the East'. ,Roman military settlers' setzt etwa auch Chaniotis 2018, 221
    voraus; ebenso Sørensen 2016, 120, der aber zugleich von erheblichen Verlusten unter diesen
    im Krieg gegen Pharnakes II spricht. Bei Marek 1993, 37–39 spielt die Kolonie wiederum
    keine Rolle, und er stellt auch zutreffend fest, dass weder Strabon noch Appian eine Aussage
    über die politische Zugehörigkeit der Stadt machen. Die Aufnahme in die Provinz ergibt sich
    für Marek vor allem aus der numerischen Notwendigkeit. Vgl. in anderem Kontext Sartre
    2001, 112: ,... Pompée en (sc. des vétérans) avait installé à Nicopolis du Lycos, en Arménie
    Mineure, colons qui auraient revendiqué plus tard le statut de colonie romaine bien que toute
    la tradition considère la ville comme une banale polis grecque'.
21 Zu Magnopolis s. Strab. *Geogr.* 12.3.30 (556C), der die Herkunft der Siedler (wie Dio) offen-
    lässt: συμβάλλουσι δ' ἀμφότεροι κατὰ μέσον που τὸν αὐλῶνα, ἐπὶ τῇ συμβολῇ δ' ἵδρυται
    πόλις, ἣν ὁ μὲν πρῶτος ὑποβεβλημένος Εὐπατορίαν ἀφ' αὑτοῦ προσηγόρευσε, Πομπήϊος δ'

ten jedenfalls durch das Lykostal weiterhin mit Magnopolis und der übrigen pontischen Provinz verbunden, aber die ökonomische Ausrichtung auf Kleinarmenien
war offenbar stärker. Hier hatte sich die Aufteilung des Pompeius als nicht allzu
nachhaltig erwiesen.[22]

4. Zela und Megalopolis

Spätestens die Neubewertung des Schicksals von Amaseia erfordert auch eine
Revision der Geschichte des Priesterstaates Zela, auf dem Pompeius eine Polis
gründete. Ebenfalls zu überdenken ist die Behandlung der an Kleinarmenien angrenzenden Landschaften Kulupene, Kamisene und Laviansene, welche derselbe
Proconsul der Neugründung Megalopolis zuwies. Beide zählen weithin zu den elf
Provinzstädten, was aber bei genauerer Betrachtung fragwürdig ist. Strabon
spricht jedenfalls weder von der dortigen Ansiedlung römischer Veteranen noch
von der Zugehörigkeit der Städte zur Provinz. Zudem begegnen Zela und Megalopolis als Teile des Reiches der Pythodoris wieder, wobei Zela nunmehr um drei
Landstriche verringert wurde: Das Heiligtum von Zela wurde von einem (namentlich unbekannten) Hohepriester regiert; Augustus hatte den Galater Dyteutos zum
Priester von Komana eingesetzt und ihm einen kleinen Teil der Zelitis dazuge-
schenkt; zuvor hatte dessen Bruder Ateporix von M. Antonius die Karanitis (das
spätere Sebastopolis), einen anderen Distrikt der Zelitis, erhalten, welcher unter
Augustus zur Provinz kam. Hier entsteht doch der Eindruck, dass Pompeius weder
das Priestertum von Zela aufgehoben noch die beiden am weitesten gen Süden
bzw. Südosten gelegenen Territorien in die Provinz einbezogen hatte.[23] Womöglich war die Stadt Zela mitsamt seiner neuen Chora an Deiotaros gegangen, der

ἡμιτελῆ καταλαβών, προσθεὶς χώραν καὶ οἰκήτορας Μαγνόπολιν προσεῖπεν. Sowie App.
    Mith. 115.561.

22   S. oben Anm. 15 und 19 zu Diospolis. Zu Nikopolis s. Strab. *Geogr.* 12.3.28 (555C) und App.
     Mith. 105.494; im späten 1. oder frühen 2. Jh. n.Chr. wurde es metropolis des armenischen
     koinon; s. Vitale 2014, 56. Zur Lage und Geschichte von Magnopolis, das zuletzt bei Çevresu
     2 km westlich von Kızılçubuk in Tokat Province lokalisiert wurde s. Sørensen 2016, 153–
     162.

23   Megalopolis und Zela, mit Pythodoris, Ateporix and Dyteutos: Strab. *Geogr.* 12.3.37
     (559f.C). Zu Zela und Pompeius s. auch *Geogr.* 11.8.4 (512C). Marek 2010, 402f. geht davon
     aus, dass Pompeius neben der Provinzstadt Zela einen verringerten Tempelstaat fortbestehen
     ließ, den erst Octavian auflöste, um Polemon zumindest teilweise für den Verlust von Arme-
     nia Minor zu entschädigen. Die Karten von E. Olshausen in Wittke et al. 2010, 160f.; 168–
     171; 180f. vermitteln den Eindruck, dass die ostpontischen Provinzgrenzen von Pompeius bis
     Augustus stabil geblieben seien. Nach Sørensen 2016, 122f.; 126 hob Pompeius den Tempel-
     staat, um die Zelitis dem Deiotaros zu übergeben, bevor M. Antonius das Priestertum unter
     Abschaffung der Polis wiedereingeführt habe, welches Polemon unterstellt worden sei. Roller
     2018, 714 bleibt betreffs Zela unklar: ‚Pompeius secularized it somewhat ...'. Zu Dyteutos s.
     auch *Geogr.* 12.3.35 (558f.C), mit Coşkun 2007/19i; ferner 2007/19h zu Ateporix. Zu den
     Territorien der Zelitis s. Olshausen 1987, 190–193. Zum Fortleben des Heiligtums in römi-
     scher Zeit Dalaison 2014, 145–147.

wohl indirekt auch das (mit verringertem Territorium) fortbestehende Heiligtum kontrollierte.[24]

5. Abonuteichos oder Mazaka?

Ein weiterer Problemfall ist Abonuteichos (das spätere Ionopolis), das für den Geographen noch ein obskures *polichnion* war. Es ist nur ganz am Rande erwähnt, ohne jeden Zusammenhang mit Pompeius.[25] Doch das muss kein Hindernis sein, denn es impliziert die Möglichkeit, dass für den Küstenabschnitt ab Amastris eine Quelle verwendet wurde, welche den Zustand einer früheren Zeit widerspiegelt. Gleiches mag selbst noch für den Bericht zu Sinope gelten, der mit der Tätigkeit des Lucullus endet. Ein Blick auf die Landkarte lässt jedenfalls die von Marek vorgeschlagene Statuserhöhung plausibel erscheinen, vergleicht man die Abstände zwischen Herakleia, Amastris, Sinope und Amisos. Ein interessanter Streitpunkt betrifft indes die Frage, ob die Neugründung von Abonuteichos als Polis vielleicht sogar schon auf Lucullus zurückgeht, der bekanntlich die Küstenstädte Amastris, Sinope und Amisos neu konstituiert hatte. Der einzige bekannte Beleg für ein Ärenjahr gibt das Jahr 274 irgendwann unter Septimius Severus (193–211 n.Chr.), was den Beginn grob in den Zeitraum 80/62 v.Chr. verweist. Für Magie implizierte das noch einen Bezug zu Pompeius, während in der neueren Forschung seit Marek – mit Blick auf die Nachbarstädte – Lucullus der Vorzug gegeben wird.[26]

Tatsächlich ist die Entscheidung dieser Frage nur von begrenzter Relevanz, da wir in beiden Fällen dazu genötigt wären, Abonuteichos als eine der elf Städte des pontischen Provinzteils zu zählen. Wichtiger ist unsere Antwort vielmehr mit

---

24   Im procaesarischen Bericht (*BAlex.* bes. 67–77, bes. 72.1) wird Zela ein *oppidum* genannt und in Pontos lokalisiert, wobei weder die politische Zugehörigkeit (zu Deiotaros oder zur Provinz) noch das sicher nicht allzu weit entfernte Heiligtum erwähnt werden. Dass Cn. Domitius Calvinus in Zela epigraphisch als Gott bezeichnet (und vielleicht sogar verehrt) wurde (s. Marek 2010, 377), könnte auf seine Verdienste bei der Schlacht von Zela oder vielleicht auch der vorangehenden Verteidigung der Stadt nach der Niederlage von Nikopolis hinweisen.

25   Strab. *Geogr.* 12.3.10 (545C). Strobel 1997 berücksichtigt Abonuteichos erst für die nach 40 oder nach 30 v.Chr. erfolgte Gründung des pontischen Provinziallandtages.

26   Magie 1950, Bd. 2, 1233 *versus* Marek 1985, 144–152 und 1993, 26f. zu Amastris sowie 1993, 32f. zu Amisos und 36f. zu Abonuteichos; vgl. Leschhorn 1993, 169; Højte 2006, 22. Ohne tiefgehendes Problembewusstsein bleibt die Diskussion von Dalaison, Delrieux & Ferriès 2015, 96–99, da sie behaupten, Strabon nenne 14 Städte der Provinz Pontus, ohne im Einzelfall zwischen der Terminologie des Geographen oder auch seiner territorialen Zuordnung zu unterscheiden, geschweige denn, den Versuch einer Auswahl von elf Städten zu unternehmen; sie neigen vorsichtig zur Annahme, Abonuteichos sei eine *polis* unter Pompeius gewesen. Interessant ist die Tatsache, dass der Ort bereits in der Mitte des 2. Jhs. v.Chr. ein respektables Zeus-Heiligtum hatte (S. 91–96). Zur Diskussion vgl. auch Madsen 2014, 79: ‚Strabo does not offer any clear definition of what constituted a Greek city and he is, except in the case of Sinope, not particularly informative about what kind of institutions he thought characterised a Greek community. … This focus on origin Strabo's definition of Greekness was more dependent on heritage than cultural adaption ... ‘; weiteres hierzu unten in III. mit Anm. 31.

Blick auf die ‚acht Stadtgründungen in Kappadokien'. Abonuteichos dieser Gruppe zuzuweisen erscheint mir recht verlockend, da wir somit einen problembehafteten Kandidaten aus der Liste streichen könnten: Mazaka, das allgemein als achte Neugründung (im Sinne Appians) gezählt wird.[27] Zwar wurde dies während des Dritten Mithradatischen Krieges sicher neu gegründet, nachdem Tigranes die Einwohner jener Stadt nach Tigranokerta entführt und Lucullus sie im Jahr 69 wieder heimgeschickt hatte. Doch besteht kein hinreichender Grund zu der Annahme, dass er oder Pompeius in jenem autonomen und verbündeten Königreich als ‚Städtegründer' im Sinne eines Verfassungsgebers aufgetreten war. Appian bezeugt lediglich den Einsatz des Pompeius für die Hilfe beim Wiederaufbau der Hauptstadt des befreundeten Reiches. Dabei unterscheidet der Geschichtsschreiber diese Tätigkeit aber relativ deutlich von anderswo erfolgten Neugründungen mit konstitutivem Charakter.[28]

Zu dieser Sichtweise passt ferner, dass Appian seine Information auf ein Plakat des Triumphzuges zurückführt, auf dem mit acht Gründungen in Kappadokien und zwanzig in Kilikien und Koilesyrien geprahlt wurde. Damit scheint Kappadokien hier zumindest auf den ersten Blick synonym für die kleinasiatischen Reichsteile des Mithradates verwendet zu sein, für das antike Schriftsteller eine Vielzahl sich teils überlagernder Namen kannten. Bei genauerer Untersuchung zeigt sich indes, dass hier im Zuge der Überlieferung stümperhaft gekürzt wurde und im Original von Pontos, (Klein-) Armenien und Paphlagonien die Rede war.[29]

III. Ergebnisse: Strabons elf und Appians acht Städte

Folglich setzten sich Strabons elf ‚pontische' Städte sicher aus den fünf alten (Herakleia, Tieion, Amastris, Sinope, Amisos) und sechs neuen (Abonuteichos,

27 So bereits Niese 1883, 582; vgl. Magie 1950, Bd. 2, 1232; Dreizehnter 1975, 236.
28 S. Strab. *Geogr.* 12.2.9 (539C) und Plut. *Luc.* 29.4 zu Lucullus sowie App. *Mith.* 115.562 zu Mazaka; vgl. Coloru 2013, 723. Eine ähnliche Unterscheidung trifft auch Sherwin-White 1984, 257 Anm. 46, allerdings ohne Bezug auf das numerische Problem. Dass Lucullus und nicht Pompeius als *ktistes* von Mazaka gegolten habe, betont auch Dreizehnter 1975, 238f., jedoch ohne die Zählung (S. 236) zu ändern: ‚Offenbar handelt es sich bei dieser Nachricht wieder um die Übertragung von Verdiensten, die anderen zukommen, auf Pompeius, wie es in seiner Biographie nicht selten ist.' Zur hellenistischen Geschichte von Mazaka bis Tigranes s. Panichi 2018, bes. 217.
29 S. App. *Mith.* 117.576 zum Triumphzug. Dreizehnter 1975, 220–222 bezweifelt die Zahlen und deutet sie zudem durch einen Texteingriff (ἐκτήσαν statt ἐκτίσθησαν) auf Eroberungen statt Gründungen um. Während ihm Engster 2011, 198 (s. dieselbe auch zum ideologischen Kontext von Pompeius' Triumphzug) und Kleu 2013, 5421 folgen, weist Goukowsky 2017, 252 zu Recht darauf hin, dass eine solche Korrektur mit Blick auf Plut. *Pomp.* 45.2 (der von insgesamt 39 κατοικίαι πόλεων im Osten spricht) und vor allem Cass. Dio 37.20.2 unstatthaft ist; allerdings behauptet Goukowsky irrtümlicher Weise mit Verweis auf Strab. *Geogr.* 12.3.1 (541C): ‚Pompée fonda onze πολιτεῖαι' (der gleiche Fehler ist Michels 2009, 308 Anm. 1601 unterlaufen). Freilich betrachtet Dreizehnter 1975, 216–220 auch die Plutarch-Stelle als verderbt. S. die Zitate und Diskussion in Anhang II.

Pompeiopolis, Neapolis, Magnopolis, Diospolis, Nikopolis) Poleis zusammen.
Dabei wird die von Appian und Cassius Dio genannte Zahl von acht Neugründun-
gen ,in Kappadokien', worunter wir hier speziell das ehemalige Reich des Mith-
radates Eupator verstehen sollten, durch die Hinzunahme von Megalopolis und
Zela erreicht, welche vorerst in der Hand der Monarchen von Pontos verblieben.[30]
Diese Ergebnisse sind auf der Karte 3 am Ende des Bandes festgehalten.

Stadtgründungen durch den Proconsul außerhalb des neuen Provinzterritori-
ums mögen auf den ersten Blick überraschen, sind indes leicht zu erklären, wenn
man die Entscheidungen zeitnah zur militärischen Vertreibung des Mithradates
Ende 66 oder Anfang 65 v.Chr. ansetzt und als Alexander-*imitatio* begreift. Damit
müssen sie noch nicht Teil eines ausgereiften Plans zur politisch-kulturellen Neuordnung gewesen sein, und die spätere gleichförmige Verfassungsgebung durch
die *lex Pompeia* hatte keine Auswirkungen auf die Gründungen außerhalb der
Provinz. Anfangs wird es ohnehin vielmehr um die Rekrutierung von Bewohnern
und die Bereitstellung von Ressourcen gegangen sein, weniger um die Ausarbei-
tung einer Verfassung, welche Pompeius nach der Konkretisierung seiner Pläne
sicherlich Deiotaros für Megalopolis und Zela überließ. Sollten sich römische
Veteranen in einer dieser Neugründungen niedergelassen haben, dann könnte diesen bei der Festlegung der Provinzgrenzen im Jahr 64 v.Chr. ein Umzug etwa
nach Magnopolis angeboten worden sein. Unüberwindbare Hindernisse gab es für
eine ,Refeudalisierung' jedenfalls nicht, wie die spätere Neuordnung des Ostens
unter M. Antonius zeigte, da er nicht nur in Nordanatolien ehemalige Provinzstädte an befreundete Dynasten und Könige übergab.

Insgesamt scheinen mir die hier vorgeschlagenen Revisionen betreffs der
Ordnung Kleinasiens durch Pompeius erhebliche historische Neubewertungen
erforderlich zu machen. Sie bestätigen zum Beispiel ein Ergebnis, das Jesper
Madsen in seiner Untersuchung der kulturellen Auswirkungen von Pompeius'
Städtegründungen gemacht hat, dass nämlich die Klassifikation als *polis* bei
Strabon praktisch gar nichts über die Organisationsform aussagt.[31] Konsequenzen
gehen aber weit über die Zeit jenes großen Feldherren hinausgehen. Mit den Hinweisen auf die Maßnahmen des M. Antonius und Octavians oder die dynastische
Politik von Polemon I. und Pythodoris sind hier einige geopolitische Andeutungen
gemacht worden, während das vorangehende Kapitel zu den Territorien des Deiotaros Philorhomaios einen solchen Faden in seiner größeren Komplexität ver-
folgt.

---

30 So bereits in Ansätzen auf meiner Karte in Coşkun 2007, 505f., Karte 1 = 2008, Karte 3.
    Allerdings ist Abonuteichos dort noch nicht berücksichtigt und Amaseia liegt dort außerhalb
    des Deiotaros-Reiches.
31 Madsen 2014, 81, bes. mit Blick auf die *poleis* in den ehemaligen oder auch zeitgleichen
    Tempelstaaten Komana und Zela: ,More often than not, the term *polis* is used in the sense
    simply of an urban centre, with no specific political, constitutional or cultural connotations at-
    tached to its meaning.' S. auch oben Anm. 26.

ANHANG I: ZUM NAMEN DER PROVINZ *PONTUS (ET BITHYNIA)*

Von Nero bis in die severische Zeit ist der Provinzname *Pontus et Bithynia* regelmäßig in den Cursus- und Ehreninschriften der römischen Stadthalter nachgewiesen. Traditionell gilt dieser Doppelname auch als der offizielle Titel der Provinz seit ihrer Einrichtung durch Pompeius ca. 64 v.Chr., wenngleich die Dominanz des alleinigen Namens *Bithynia* sowie der reguläre Statthaltersitz im bithynischen Nikomedia (İzmit) nicht selten zu einer Umkehrung der Bestandteile geführt haben. Am auffälligsten ist eine diesbezügliche Unsicherheit in der wichtigen Studie zur römischen Raumerfassung in Kleinasien von Bernhard Rémy (1986): Er spricht zwar in seinen Texten von ‚Pont-Bithynie‘, das entsprechende Territorium auf seinen Karten nennt er aber durchweg ‚Bithynie-Pont‘. Konsequenter heißt es bei Marek regelmäßig *Pontus et Bithynia* oder ähnlich.[32]

Demgegenüber forderte bereits Kenneth Wellesley (1953), dass der Amtssprengel der Provinzstatthalter vom ersten Nachfolger des Pompeius an, dem Proconsul C. Papirius Carbo (62–59 v.Chr.), bis in die neronische Zeit schlicht *Bithynia* gelautet habe. Immerhin verweist er hierfür auf neun literarische Zeugnisse. Vor nicht allzu langer Zeit hat Gabriele Wesch-Klein Wellesleys wenig beachtetes Plädoyer wiederaufgegriffen und durch vier weitere epigraphische Belege, die vielleicht alle in die Zeit des Augustus fallen, zu bekräftigen versucht. Eine geringfügige Nuance ist dabei, dass die früheste Inschrift mit der Nennung der Doppelprovinz Pontus-Bithynia wohl schon unter Claudius datieren könnte.[33]

Freilich variiert die Qualität der einzelnen angeführten Belege zur Stützung von *Bithynia*. So ist nur mit Vorsicht von Cassius Dios Aussage, dass Carbo ‚Bithynien regiert‘ habe (36.40.4 ed. Boissevain 1895; vgl. Cary & Foster 1914: τῆς τε Βιθυνίας καὶ αὐτὸς ὕστερον ἄρξας), auf dessen offiziellen Titel der Provinz zurückzuschließen. Auch hat nicht jede von Wesch-Klein herangezogene Inschrift Beweiskraft für unsere Fragestellung, so *CIL* IX 2335 = *ILS* 961, in der *[Bithynia]* ergänzt ist. In ihrer Gesamtheit zeigen diese Quellen aber dennoch ganz deutlich, dass der etwas stärker urbanisierte westliche Teil der Doppelprovinz intensiver in der allgemeinen Wahrnehmung in Rom verankert war. Dies mag sogar noch bis ins späte 1. Jh. n.Chr. der Fall gewesen sein, wie die Formulierungen *praetor Bithyniae* bzw. *proconsul Bithyniae* bei Tacitus zeigen (*Ann.* 1.74; 16.18).

---

32 So z.B. Rémy 1986, bes. 19–21 (auch 1989, bes. 16) gegenüber 1986, 22 (Karte für 25 v.Chr.); vgl. 26 (14 n.Chr.); 31 (18 n.Chr.); 36 (43 n.Chr.); 38 (55 n.Chr.); 42 (65 n.Chr.); … 97 (unter Septimius Severus). S. nur die Titel der Arbeiten von Marek 1993 und 2003. Weiterhin, allerdings ohne Auseinandersetzung mit Wesch-Klein 2001, auch Eck 2007, 190f.; Marek 2010, 367; Vitale 2012, 178; Olshausen 2014, 47.

33 Wellesley 1953, 294 Anm. 1; Wesch-Klein 2001, gefolgt z.B. von Kleu 2013, 5421; Ballesteros Pastor 2013, 284: ‚aunque extraoficialmente pudo llamarse ya entonces *Pontus et Bithynia* (Cic.Sen.38)‘; Dan 2014, 49 Anm. 17; Edelmann-Singer 2015, 71 mit Anm. 241, obwohl sie die Provinz abweichend *Bithynia-Pontus* nennt; Sørensen 2016, 54 Anm. 70, obgleich er im Text durchgängig von *Pontus et Bithynia* spricht. Vgl. auch ‚Bithynie-Pont‘ auf der Karte von Payen 2020, fig. 22.

Allerdings stellt sich die Frage, was den offiziellen Namenswechsel in der Jahrhundertmitte herbeigeführt haben soll, wo doch Nikomedeia die Hauptresidenz des Statthalters blieb. Zudem wäre bei einer später erfolgten Änderung wohl eine Ergänzung von *Pontus* hinter *Bithynia* plausibler als seine Voranstellung. Weiterhin wäre es eine intuitive Annahme, dass Pompeius als der Besieger des Mithradates und Neuorganisator vor allem der pontischen Reichsteile das Schwergewicht auf jenen östlichen Provinzteil gelegt hätte, zumal dieser mehr als doppelt so groß war wie der westliche.

In dieselbe Richtung weist auch eine beiläufige Charakterisierung des C. Calpurnius Piso Frugi durch seinen Schwiegervater Cicero: Im Jahr 58 v.Chr. habe sich dieser seinen Pflichten in *Pontus et Bithynia* entzogen (*Red. Sen.* 38: *qui Pontum et Bithyniam quaestor prae mea salute neglexit*).[34] Wellesley hat dieses Zeugnis als ‚inconclusive‘ beiseitegeschoben. Gleiches tat er mit der Aussage des Livius, nach der Pompeius die Provinz *Pontus* gegründet habe (*Per.* 102): *Cn. Pompeius in provinciae formam Pontum redegit*.[35] Hier lässt sich die Quellenlage noch deutlich vermehren. So habe ich im vorangehenden Kapitel darauf hingewiesen, dass Pompeius mit der Unterscheidung zwischen *provincia Ponti* und *regnum Armeniae Minoris* bewusst an die Terminologie des besiegten Königs anschloss. Diese Folgerung basiert vor allem auf der procaesarischen Schrift des *Bellum Alexandrinum* aus den 40er Jahren v.Chr. Dort ist durchweg von *Pontus* die Rede, wobei drei Stellen ganz spezifisch die römische Provinz bezeichnen (34.5; 35.3 und 67.1). Strabon liefert weitere Hinweise. Die oben zitierte Kurzversion der pontischen Geschichte bestätigt die Einteilung in *Pontos* und *Armenia Minor* unter Eupator, während der Hinweis, dass aus ‚dem Rest‘ (also Pontos) und Bithynien ‚eine einzige Provinz‘ geworden sei, noch ambivalent ist. Denn der neue Name bleibt in dieser Textstelle ja offen.[36] Aber nur wenige Paragraphen später lokalisiert Strabon Herakleia explizit ‚in der Provinz Pontos, die mit Bithynien verbunden war‘.[37]

Wörtlich spricht der Geograph von einer *eparchia*, ein Ausdruck, der in der Forschungsliteratur gern im Sinn von ‚Unterprovinz‘ oder ‚Distrikt auf der Ebene unterhalb einer Provinz‘ verwendet wird. So auch im Titel des nunmehrigen Standardwerks zum Thema von Marco Vitale, *Eparchie und Koinon in Kleinasien*. Allerdings darf nicht vergessen werden, dass *eparchia* das exakte Pendant zum römischen Ausdruck *provincia* ist, welcher ebenfalls nicht selten für Unterprovinzen verwendet wurde. Eben diesen Binnengrenzen folgten häufig die politisch-kultischen Verbände der provinzialen Eliten (*koina*), die auch unter dem Namen *conventus* (‚Provinzlandtage‘) bekannt sind. Dass Provinz- und Koinon-Grenze deckungsgleich waren wie in Asia, war keineswegs eine feste Regel, wie noch Jürgen Deininger annahm und neuerdings wieder von Stephen Mitchell und Julie

---

34 Vgl. Broughton 1952, 197.
35 Wellesley 1953, 294 Anm. 1, der für Pontus auch noch auf Vell. 2.38.6 verweist.
36 Strab. *Geogr.* 12.3.1 (541C), zitiert o. Anm. 9.
37 Strab. *Geogr.* 12.3.6 (543C), zitiert o. Anm. 10.

Dalaison für die vortrajanische Provinz *Pontus et Bithynia* vertreten wird.[38] Die neuesten analytischen Ansätze zu jener nordanatolischen Provinz (einschließlich eines weiteren Koinons für Paphlagonien) vertreten Vitale und Søren Sørensen im Anschluss an Marek. Dass es sowohl ein pontisches als auch ein bithynisches *koinon* gab, ist mittlerweile weithin akzeptiert. Überdies erkennt man nun auch an, dass die jeweils ersten Belege bereits in die republikanische Zeit fallen. Dass damit zugleich auch Pompeius den Anstoß zu ihren Gründungen gegeben habe,[39] ist damit nicht sicher entschieden, denn wir sollten weder die Gleichförmigkeit republikanischer Provinzen über- oder die Möglichkeit provinzieller Initiativen bei der Organisation eines *koinon* unterschätzen.[40]

Jedenfalls legt der Befund nahe, dass Pompeius seine neue Provinz entsprechend ihrer historisch weitgehend eigenständiger Teile *Pontus et Bithynia* nannte. Zugleich sollten wir akzeptieren, dass die offizielle Terminologie schon in den ersten Generationen schwankend war und subjektive Assoziationen den Ausschlag für die häufige Unterdrückung einer der beiden Eparchienamen gaben.

ANHANG II: ACHT STÄDTEGRÜNDUNGEN DURCH POMPEIUS ‚IN KAPPADOKIEN‘?

Appian berichtet, dass eine der Schautafeln auf dem Triumphzug des Pompeius die Städtegründungen während seines Proconsulats wie folgt beziffert (*Mith.* 117.576 ed. Goukowsky 2003; vgl. McGing 2019): ‚Er gründete acht Städte in Kappadokien, bei den Kilikern aber und in Koilesyrien zwanzig, und in Palästina […], die jetzige (Stadt) Seleukis‘ (πόλεις ἐκτίσθησαν Καππαδοκῶν ὀκτώ, Κιλίκων δὲ καὶ κοίλης Συρίας εἴκοσι, Παλαιστίνης δὲ […] ἡ νῦν Σελευκίς). Cassius Dio (37.20.2) oder wenigstens seiner Zwischenquelle scheint ein ähnlicher Bericht vorgelegen zu haben, da er in der Zusammenfassung von Pompeius' Leistungen die ‚Besiedlung von acht Städten‘ (πόλεις τε ὀκτὼ ἀπῴκισε) anführt, freilich ohne diese näher zu lokalisieren. Sollte bei Appian also ‚Kappadokien‘ pauschal für die anatolischen Territorien des Mithradates Eupator verwendet worden sein?

---

38   Vitale 2012. Sowie Deininger 1965; Mitchell 2002; Dalaison 2014, 126f.

39   So Marek 1993, 73–82, bes. 75–77 zur Auseinandersetzung mit dem prominentesten ‚Unitarier‘ Deininger 1965. Sowie Vitale 2012 (vgl. auch 2014, 54 mit Karte; zudem 52f. gegen die problematische Annahme eines pontischen Bürgerrechts) und Sørensen 2016, 75–89. Vgl. Edelmann-Singer 2015, 71f.

40   Ähnlich vorsichtig ist auch Sørensen 2016, 84 betreffs Pontus et Bithynia. Bei anderen Voraussetzungen geht auch Dalaison 2014, 127 von der Koinon-Gründung in Pontos nicht vor dem späten 1. Jh. n.Chr. aus. Vgl. meine Ausführungen zu Galatien in Coşkun ca. 2021, Kapitel 11, wo es mindestens drei voneinander unabhängige *koina* gegeben hat. Anders Edelmann-Singer 2015, 95–98; eine Zwischenposition vertritt Vitale 2012, 122–131 (Galatien) und 205–229 (Paphlagonien).

Dies wird allgemein angenommen,[41] allerdings ohne eine stichhaltige Begründung. Ein solcher Sprachgebrauch wäre jedenfalls recht ungewöhnlich, da ,Kappadokien' normalerweise nur die Gebiete östlich oder teils auch südlich des Halys bzw. nördlich des Tauros-Gebirges bezeichnet und im Westen an Paphlagonien, Galatien, Phrygien und Lykaonien angrenzte, wie wir etwa von Strabon (*Geogr.* 12.3.8f. [543f.C]) wissen. Persischem Sprachgebrauch folgend unterscheidet derselbe (*Geogr.* 12.1.4 [534C]) ,Kappadokien am Tauros' im Süden und ,Kappadokien am Pontos' (also ,am Schwarzen Meer') im Norden (Strab. *Geogr.* 12.3.2 [541C]: τὴν πρὸς τῷ Πόντῳ Καππαδοκίαν). Der erstere Kontext nennt den gerade gestorbenen König Archelaos, ohne Eupator zu erwähnen, während der letztere ,Kappadokien am Pontos' als einen von Eupators Herrschaftsbereichen neben Paphlagonien und Kolchis nennt, und zwar aus der Perspektive eines Seefahrers, der (vom Thrakischen Bosporos kommend) die ,rechte Seite' des Schwarzen Meeres betrachtet.[42] Strabon hätte Pompeiopolis im paphlagonischen Hinterland also gewiss nicht unter Kappadokien gefasst. Müssen wir es also deswegen von der Zählung Appians ausnehmen, einen weiteren Landschaftsbegriff ansetzen oder nach einer anderen Lösung suchen?

Man beachte, dass Plutarch zumindest indirekt auf dieselbe Informationsquelle zurückgriff: die Triumphaltafeln des Pompeius. Es besteht keine Notwendigkeit, sich dem negativen Urteil von Alois Dreizehnter anzuschließen. Dieser Forscher hat (neben Appians Zeugnis) auch die Plutarch-Stelle für verdorben erklärt, 1. da ein römischer Triumph niemals auf Städtegründungen verweise; 2. da die lateinische Sprachpraxis eher an die Deduktion römischer Veteranen denken lasse; 3. da die Syntax des übergeordneten Satzes eigentlich eine Eroberung jener 39 κατοικίαι πόλεων voraussetzte. Drei schwache (und teils zirkuläre) Argumente ergeben aber kein starkes, und der unabhängige Befund der drei griechisch schreibenden Autoren verdient, ernst genommen zu werden.[43]

Plutarch bietet eine viel längere Liste von Landschaften, über die Pompeius gesiegt und in denen er Städte gebaut habe (*Pomp.* 45.2f. ed. Perrin 1917): ,Pontos, Armenien, Paphlagonien, Kappadokien, Medien, Kolchis, Iberer, Albaner, Syrien, Kilikien Mesopotamien, das Gebiet um Phönikien und Palästina, Arabien, das ganze Piratentum, das zu Wasser und zu Lande besiegt worden ist. Bei diesen

---

41 So ohne Diskussion zuletzt Goukowsky 2017, 252: ,entendons dans le Pont'; ähnlich bereits Niese 1883, 581, dessen Beispiele aber auf das Gebiet östlich des Halys beschränkt bleiben.

42 Vgl. allgemein und ohne Berücksichtigung unseres Problems z.B. Dan 2009, 72f.; Roller 2018, 685–687; Speidel 2019, 118f.; auch Mitchell 2002; Ballesteros Pastor 2007. Eine Ausnahme stellt indes Dan 2014, 47 mit Anm. 10 dar; sie stellt Zeugnisse zusammen, welche Mithradates als ,König von Kappadokien' erscheinen lassen könnten; jedoch sind die Ergänzungen oder Interpretationen der Texte unsicher und werden auch von Dan nicht forciert.

43 Dreizehnter 1975, 216–220 zur Plutarch-Stelle; S. 225–233 stellt er die weiteren Zahlen der Triumphaltafeln in Frage und fordert unter anderem eine Korrektur der 39 in 29; ähnlich kritisch ist er S. 220–222 zu Appian (s. oben Anm. 29). Er zieht S. 240 (vgl. 235–237) den fraglichen Schluss, dass Pompeius eigentlich nur Nikopolis gegründet habe, ohne dabei aber die antike Vorstellung von konstitutiven Neugründungen hinreichend zu würdigen. Mit gebotener Vorsicht bleiben lediglich seine gelehrigen Einzelbeobachtungen für die Frage der Alexander-*imitatio* des Pompeius von Interesse.

hat er nicht weniger als 1000 Festungen eingenommen und nicht viel weniger als 900 Städte, 800 Piratenschiffe; die Gründungen von Städten (κατοικίαι πόλεων) zählen 39'.

Da hier der den gesamten Mittelmeerraum umspannende Piratenkrieg einbezogen ist, sehe ich entgegen Dreizehnter keine Notwendigkeit, Plutarchs Summe zu korrigieren und an die insgesamt 29 bei Appian anzugleichen. Es scheint mir offensichtlich, dass die Informationen unterschiedlich redigiert wurden, aber sich doch gegenseitig in ihrer Glaubwürdigkeit stützen. Die Vermutung liegt nahe, dass Appian oder eine Zwischenquelle den Text sachfremd gekürzt hat und wir die acht Gründungen – mit Blick auf Plutarchs ausführlichere Liste – auf ‚Pontos, Armenien, Paphlagonien (und) Kappadokien' beziehen sollten. Da es sich um Eroberungen handelt, ist hier kaum das verbündete kappadokische Reich des Ariobarzanes I. gemeint, sondern – in der Sprache Strabons – ‚Kappadokien am Pontos'. Obwohl diese Landschaft oft auch ‚Armenien' oder genauer ‚Kleinarmenien' genannt wurde,[44] ist hier natürlich nicht das Königreich des Tigranes II. gemeint. Da Pompeius dessen Unterwerfung annahm, ohne sein Land im Kampf erobern zu müssen, begnadigte er ihn und beließ ihm sein ererbtes Territorium.[45] Zu den acht Städtegründungen kam es damit allein auf dem Gebiet des ehemaligen Mithradates-Reiches.

Danksagung

Für Ermutigung, kritisches Feedback oder Hilfe bei der Literaturbeschaffung danke ich Germain Payen und Vera Sauer. Für die kartographische Umsetzung (Map 3 am Ende des Bandes) gilt mein Dank Chen Stone. Meine Forschungen zur historischen Geographie im Schwarzmeerraum gehören in den Rahmen des Projekts ‚Ethnic Identities and Diplomatic Affiliations in the Bosporan Kingdom', das ich an der University of Waterloo mit Unterstützung des Social Sciences and Humanities Council of Canada (2017–2022) verfolge.

Bibliographie – Editionen und Übersetzungen antiker Quellen

(Alle Übersetzungen stammen von mir)

Boissevain, U.P. 1895/8: *Cassii Dionis Cocceiani Historiarum Romanarum quae supersunt*, 2 vols., Berlin.
Cary, E. & Foster, H.B. 1914: *Cassius Dio*, London.
Goukowsky, P. 2003: *Appien, Histoire Romaine*. Bd. 7: *Livre XII: La Guerre de Mithridate*, 2nd ed., Paris.
Lassère, F. 1981: *Strabon, Géographie*. Bd. 9: *Livre XII*, Paris.
McGing, B. 2019–2020: *Appian, Roman History*, 6 vols., Cambridge, MA.
Perrin, B. 1917: *Plutarch's Lives*. Bd. 5: *Agesilaus and Pompey. Pelopidas and Marcellus*, London.

---

44   S. den Anhang in Coşkun, Kapitel X in diesem Band.
45   S. z.B. Vell. 2.37.

Radt, S. 2002–2011: *Strabons Geographika*, 10 Bde., Göttingen (vol. 3, 2004; vol. 7, 2008).
Roller, D.W. 2014: *The Geography of Strabo*. Translated by D.W. Roller, Cambridge.

Bibliographie – Weitere Literatur

Ballesteros Pastor, L. 1996: *Mithrídates Eupátor, rey del Ponto*, Granada.
Ballesteros Pastor, L. 2007: ‚Del reino Mitridátida al reino del Ponto: orígenes de un término geográfico y un concepto politico‘, *Orbis Terrarum* 9, 3–10.
Ballesteros Pastor, L. 2013: *Pompeyo Trogo, Justino y Mitrídates. Comentario al Epítome de las Historias Filípicas (37,1,6–38,8,1)*, Göttingen.
Bekker-Nielsen, T. (ed.) 2014a: *Space, Place and Identity in Northern Anatolia*, Stuttgart.
Bekker-Nielsen, T. 2014b: ‘To be or not to Be Paphlagonian? A Question of Identity’, in Bekker-Nielsen 2014a, 63–74.
Broughton, T.R.S. 1952: *The Magistrates of the Roman Republic*, Bd. 2, Atlanta.
Chaniotis, A. 2018: *Age of Conquests. The Greek World from Alexander to Hadrian*, Cambridge, MA.
Coloru, O. 2013: ‚Armenia‘, *EAH*, 1. Auf., 722–725. Vgl. *EAH Online* 2012.
Coşkun, A. 2007: *Von der ‚Geißel Asiens‘ zu ‚kaiserfrommen Reichsbewohnern‘. Sudien zur Geschichte der Galater unter besonderer Berücksichtigung der* amicitia populi Romani *und der göttlichen Verehrung des Augustus (3. Jh. v. – 2. Jh. n.Chr.)*, unveröffentlichte Habilitationsschrift, Trier.
Coşkun, A. 2007/19c: ‚Brigatos, König von Galatien‘, *APR, s.v.*
Coşkun, A. 2007/19e: ‚Amyntas, König und Tetrarch von Galatien‘, *APR, s.v.*
Coşkun, A. 2007/19f: ‚Amyntas (II.), galatischer Tetrarch‘, *APR, s.v.*
Coşkun, A. 2007/19g: ‚Kastor (III.), König von Paphlagonien‘, *APR, s.v.*
Coşkun, A. 2007/19h: ‚Ateporix, Dynast der Karanitis‘, *APR, s.v.*
Coşkun, A. 2007/19i: ‚Dyteutos, Hohepriester von Komana Pontike‘, *APR, s.v.*
Coşkun, A. 2008/19b: ‚Deiotaros (III.) Philadelphos, König von Paphlagonien‘, *APR, s.v.*
Coşkun, A. 2008: ‚Das Ende der „romfreundlichen Herrschaft“ in Galatien und das Beispiel einer „sanften Provinzialisierung“ in Zentralanatolien‘, in idem (ed.), *Freundschaft und Gefolgschaft in den auswärtigen Beziehungen der Römer (2. Jh. v.Chr. – 1. Jh. n.Chr.)*, Frankfurt, 133–164.
Coşkun, A. ca. 2021: ‚A Survey of Recent Research on Ancient Galatia (1993–2019)‘, demnächst in idem (ed.), *Galatian Victories and Other Studies into the Agency and Identity of the Galatians in the Hellenistic and Early-Roman Periods* (Colloquia Antiqua 33), Leuven.
Dalaison, J. 2014: ‚Civic Pride and Local Identities: the Pontic Cities and Their Coinage in the Roman Period‘, in Bekker-Nielsen 2014a, 125–155.
Dalaison, J., Delrieux, F. & Ferriès, M.-C. 2015: ‚Abônoteichos-Ionopolis et son atelier monétaire‘, in C. Courrier, L. Passet, Ch. Clément (eds.), *Arcana Imperii. Mélanges d'histoire économique, sociale et politique offerts au Professeur Yves Roman*, vol. 1, 91–205.
Dan, A. 2009: ‚Sinope, „capitale“ pontique, dans la géographie antique‘, in H. Bru, F. Kirbihler & S. Lebreton (eds.), *L'Asie Mineure dans l'Antiquité: Échanges, population et territoires. Regards actuels sur une péninsule*, Rennes, 67–131.
Dan, A. 2014: ‚Pontische Mehrdeutigkeiten‘, *eTopoi, Journal for Ancient Studies* 3, 43–66.
Deininger, J. 1965: *Die Provinziallandtage der römischen Kaiserzeit von Augustus bis zum Ende des dritten Jahrhunderts*, München.
Dreizehnter, A. 1975: ‚Pompeius als Städtegründer‘, *Chiron* 5, 213–245.
Edelmann-Singer, B. 2015: *Koina und Concilia. Genese, Organisation und sozioökonomische Funktion der Provinziallandtage im römischen Reich*, Stuttgart.
Eck, W. 2007: ‘Die politisch-administrative Struktur der kleinasiatischen Provinzen während der Hohen Kaiserzeit‘, in G. Urso (ed.), *Tra Oriente e Occidente. Indigeni, Greci e Romani in*

Asia minore. Atti del convegno internazionale, Cividale del Friuli, 28–30 settembre 2006, Rom, 189–207.

Engster, D. 2011: ‚Der Triumph des Pompeius über Mithridates VI.‘, in N. Povalahev & V. Kuznetsov (eds.), *Phanagoreia und seine historische Umwelt. Von den Anfängen der griechischen Kolonisation (8. Jh. v.Chr.) bis zum Chasarenreich (10. Jh. n.Chr.)*, Göttingen, 197–224.

Fernoux, H.L. 2004: *Notables et élites des cités de Bithynie aux époques hellénistique et romaine (IIIe siècle av. J.-C. – IIIe siècle ap. J.-C.)*, Lyon.

Gnoli, T. 2000: ‚Il Ponto e la Bitinia (Strabone XII 3)‘, in A.M. Biraschi & G. Salmeri (eds.), *Strabone e l'Asia Minore, Atti del Convegno Perugia, La Colombella, 25–28 maggio 1997*, Neapel, 543–564.

Guinea Díaz, P.M. 1999: ‚Notas sobre la organización pompeyana de la provincia de Bitinia y Ponto‘, *Gerión* 17, 317–329.

Højte, J.M. 2006: ‚From Kingdom to Province: Reshaping Pontos after the Fall of Mithridates VI‘, in T. Bekker-Nielsen, *Rome and the Black Sea Region: Domination, Romanisation, Resistance*, Aarhus, 15–30.

Jones, A.H.M. 1937/71: *The Cities of the Eastern Roman Provinces*, Oxford, 1. Aufl. 1937, 2. Aufl. 1971.

Kallet-Marx, R.M. 1995: *Hegemony to Empire. The Development of the Roman Imperium in the East from 148 to 62 B.C.*, Berkeley.

Kleu, M. 2013: ‚Pontos‘, *EAH*, 1. Auf., 5419–5422. Vgl. *EAH Online* 2012.

Leschhorn, W. 1993: *Antike Ären. Zeitrechnung, Politik und Geschichte im Schwarzmeerraum und in Kleinasien nördlich des Tauros*, Stuttgart.

Madsen, J.M. 2014: 'An Insider's View: Strabo of Amaseia on Pompey's Pontic Cities', in Bekker-Nielsen 2014a, 75–86.

Madsen, J.M. 2020: *From Trophy Towns to City-States. Urban Civilization and Cultural Identities in Roman Pontus*, Philadelphia.

Magie, D. 1950: *Roman Rule in Asia Minor to the End of the Third Century after Christ*, 2 Bde., Princeton.

Marek, C. 1985: ‚Katalog der Inschriften im Museum von Amasra. Mit Anhang: Die Inschriften von Amastris und die angebliche Pompeianische Ära der Stadt‘, *EA* 6, 133–156.

Marek, C. 1993: *Stadt, Ära und Territorium in Pontus-Bithynia und Nord-Galatia*, Tübingen.

Marek, C. 2003: *Pontus et Bithynia. Die römischen Provinzen im Norden Kleinasiens*, Mainz.

Marek, C. 2010: *Geschichte Kleinasiens in der Antike*, München.

Marek, C. 2019: ‚Anadolu'daki Roma Eyaletlerine Genel Bakış. An Overview to the Roman Provinces in Anatolia‘, in O. Tekin (ed.), *Hellenistik ve Roma İmparatorluğu dönemlerinde Anadolu – Anatolia in the Hellenistic and Roman Imperial Periods (English-Turkish)*, Istanbul, 262–273.

Marshall, A.J. 1968: ‚Pompey's Organization of Bithynia-Pontus: Two Neglected Texts‘, *JRS* 58, 103–109.

Michels, Ch. 2009: *Kulturtransfer und monarchischer 'Philhellenismus'. Bithynien, Pontos und Kappadokien in hellenistischer Zeit*, Göttingen.

Mitchell, S. 1993: *Anatolia. Land, Men, and Gods in Asia Minor. Bd. 1: The Celts in Anatolia and the Impact of Roman Rule*; Bd. 2: *The Rise of the Church*, Oxford.

Mitchell, S. 2002: ‚In Search of the Pontic Community in Antiquity‘, in A. Bowman et al. (eds.), *Representations of Empire: Rome and the Mediterranean World*, Oxford, 35–64.

Mitford, T.B. 2013: ‚Euphrates Frontier (Roman)‘, *EAH*, 1. Aufl., 2566–2570. Vgl. *EAH Online* 2012.

Mladiov, I. ca. 1999: (Map of) ‚Asia Minor, c. 63 BC‘. Ian Mladiov's Resources, University of Michigan, Ann Arbor (ohne Jahr). URL: https://sites.google.com/a/umich.edu/imladjov/maps.

Niese, B. 1883: ‚Straboniana‘, *RhM* 38, 567–602.

Olshausen, E. 1980: ‚Pontos und Rom (63 v.Chr.–64 n.Chr.)‘, *ANRW* II 7.2, 903–912.

Olshausen, E. 1987: ‚Der König und die Priester. Die Mithradatiden im Kampf um die Anerkennung ihrer Herrschaft in Pontos‘, in E. Olshausen & H. Sonnabend (eds.), *Stuttgarter Kolloquium zur Historischen Geographie des Altertums I, 1980*, Bonn, 187–212.

Olshausen, E. 1991: Zum Organisationskonzept des Pompeius in Pontos: ein historisch-geographisches Argument, in idem & H. Sonnabend (eds.), *Stuttgarter Kolloquium zur historischen Geographie des Altertums 2, 1984 und 3, 1987*, Bonn, 443–455.

Olshausen, E. 2014: ‚Pontos: Profile of a Landscape‘, in Bekker-Nielsen 2014a, 39–48.

Olshausen, E. & Biller, J. 1984: *Historisch-geographische Aspekte der Geschichte des Pontischen und Armenischen Reiches. Teil I: TAVO B 29.1*, Wiesbaden.

Panichi, S. 2018: ‚Mazaca tra Strabone e Polibio: "il luogo più centrale della Cappadocia"‘, *OrbTerr* 16, 209–220.

Payen, G. 2020: *Dans l'ombre des empires. Les suites géopolitiques du traité d'Apamée en Anatolie*, Québec.

Primo, A. 2010: ‚The Client Kingdom of Pontus between Mithridatism and Philoromanism‘, in T. Kaizer & M. Facella (eds.), *Kingdoms and Principalities in the Roman Near East*, Stuttgart, 159–179.

Rémy, B. 1986: *L'évolution administrative de l'Anatolie aux trois premiers siècles de notre ère*, Paris.

Rémy, B. 1989: *Les carrières sénatoriales dans les provinces romaines d'Anatolie au Haut-Empire, 31 av. J.-C. – 284 ap. J.-C: Pont-Bithynie, Galatie, Cappadoce, Lycie-Pamphylie et Cilicie*, Istanbul.

Roller, D.W. 2018: *A Historical and Topographical Guide to the Geography of Strabo*, Cambridge.

Sartre, M. 2001: ‚Les colonies romaines dans le monde grec. Essai de synthèse‘, *Electrum* 5, 111–152.

Seager, R. 2002: *Pompey the Great. A Political Biography*, Oxford.

Sherwin-White, A.N. 1984: *Roman Foreign Policy in the East (168 B.C. to A.D. 1)*, London.

Sørensen, S.L. 2016: *Between Kingdom and koinon. Neapolis / Neoklaudiopolis and the Pontic Cities*, Stuttgart.

Speidel, M.A. 2019d: ‚The Hellenistic Kingdom of Cappadocia‘, in O. Tekin (ed.), *Hellenistik ve Roma İmparatorluğu dönemlerinde Anadolu – Anatolia in the Hellenistic and Roman Imperial Periods (English-Turkish)*, Istanbul, 118–131.

Stein-Kramer, M. 1988: *Die Klientelkönigreiche Kleinasiens in der Außenpolitik der späten Republik und des Augustus*, Berlin.

Strobel, K. 1997: ‚Bithynia et Pontus‘, *DNP* 2, 700–702.

Sturm, J. 1936: ‚Nikopolis [8]‘, *RE* 17.1, 536–538.

Syme, R. 1995: *Anatolica. Studies in Strabo*, hg. von A. Birley, Oxford.

van Ooteghem, J. 1954: *Pompée le Grand, Bâtisseur d'empire*, Brüssel.

Vitale, M. 2012: *Eparchie und Koinon in Kleinasien von der ausgehenden Republik bis ins 3. Jh. n.Chr.*, Bonn.

Vitale, M. 2014: ‚"Pontic" Communities under Roman Rule: *polis* Self-Representation, Provincialisation and the *koina* "of Pontus"‘, in Bekker-Nielsen 2014a, 49–62.

Wagner, J. 1983: ‚Die Neuordnung des Orients von Pompeius bis Augustus (67 v.Chr.–14 n.Chr.)‘, *TAVO* B V 7, Wiesbaden.

Wellesley, K. 1953: ‚The Extent of the Territory Added to Bithynia by Pompey‘, *RhM* 96, 292–318.

Wesch-Klein, G. 2001: ‚Bithynia, Pontus et Bithynia, Bithynia et Pontus‘, *ZPE* 136, 251–256.

Will, W. 2001: ‚Pompeius [I 3]‘, *DNP* 10, 99–107.

Wittke, A.-M., Olshausen, E., Szydlak, R. & Salazar, Ch.F. (eds.) 2010: *Historical Atlas of the Ancient World* (Brill's New Pauly Suppl. 3), Leiden.

SEARCHING FOR THE SANCTUARY OF LEUKOTHEA
IN KOLCHIS

Altay Coşkun

To Mackenzie Lewis, who left us too early

Abstract: Strabo mentions a sanctuary of Leukothea, together with an Oracle of Phrixos, in the *Moschike* somewhere in Kolchis (11.2.17f. 498f.C). O. Lordkipanidze (1972) suggested a location in modern Vani at the confluence of the Sulori and Rioni (Phasis) Rivers. In contrast, D. Braund (1994) proposed an area farther to the east in the Lesser Caucasus (Moschian Mountains), southwest of Borjomi, in the valley of the upper Mtkvari River (Kyros). Both identifications are difficult to accept. First, Ino, the wife of the Theban king Athamas and stepmother of Phrixos, called Leukothea after her *apotheosis*, was a sea goddess. As such, her cult was widespread along the northern coast of the Mediterranean. Its only attested branch in the Black Sea region should therefore not be sought in the hinterland or far-away mountains. Second, Strabo's indications do not point to a location east of the mouth of the Phasis, but rather south of it, where the westernmost foothills of the Lesser Caucasus reach the sea. Third, we can now contextualize Strabo's historical references in detail: the sack of the sanctuary by Pharnakes II occurred after his defeat at Zela in Pontos by Caesar and before his final battle against Asandros near Pantikapaion. Since both battles occurred within no more than a month, Pharnakes had no time to march through the Kolchian hinterland, let alone to lay siege to its fortifications, when sailing back to Pantikapaion in August 47 BC. As a result, the Leukotheion most likely stood out as a landmark for sailors on their way from Trapezus to Phasis. The Mtsvane Kontskhi ('Green Cape'), which is now covered by the Batumi Botanical Garden, might have been an ideal location, and the hills of Tsikhisdziri would offer a feasible alternative.

Абстракт: В поисках святилища Левкофеи в Колхиде: Страбон упоминает святилище Левкофеи вместе с Оракулом Фрикса в «Москике» где-то в Колхиде (11.2.17f. 498f.C). О. Лордкипанидзе (1972) предположил, что это место находилось в современном Вани у слияния рек Сулори и Риони (Фазис). В отличие от него, Д. Браунд (1994) предположил, что оно находилось в регионе намного дальше на восток, в Малом Кавказе (горы Мошиан), к юго-западу от Боржоми, в долине верховья реки Мтквари (Кирос). Обе идентификации трудно принять. Во-первых, Ино, жена фиванского царя Атамаса и мачеха Фрикса, называемая Левкофея после ее апофеоза, была морской богиней. Таким образом, ее культ был широко распространен вдоль северного побережья Средиземного моря, поэтому его единственную аттестованную ветвь в Черноморском регионе не следует искать во внутренних районах или в далеких горах. Во-вторых, показания Страбона указывают не на местоположение к востоку от устья Фазиса, а на юг от него, где самые западные предгорья Малого Кавказа достигают открытого моря. В-третьих, теперь мы можем подробно описать упоминания Страбона об исторических событиях: разграбление храма Фарнаком II произошло после его поражения от Цезария в Зеле в Понте и перед его последней битвой против Асандра под Пантикапеем. Поскольку оба сражения произошли в течение месяца, у Фарнака, при возвращении в Пантикапей в августе 47 г. до н.э, не было времени пройти через внутреннюю часть Колхиды, не говоря уже о том, чтобы осадить ее укрепления. В результате Левкофейон, скорее всего, выделялся как ориентир для моряков на пути из Трапезунда в Фазис. Мцване-Концхи («Зеленый мыс»), который в настоящее время находится на территории Батумского ботанического сада, возможно, был идеальным местом для святилища, а холмы Цихисдзири можно считать возможной альтернативой для его локализации.

I. INTRODUCTION:
THE LEUKOTHEION IN THE CONTEXT OF STRABO'S *GEOGRAPHY*

The sanctuary of Leukothea is among the few *realia* of ancient Kolchis mentioned in the preserved literary tradition. Strabo of Amaseia has encapsulated two references to it in his account of the landmasses between the Black and Caspian Seas. Beginning with the Tanaïs / Don in the north, the geographer introduces his treatment of Asia with a brief outline of the largest mountain ranges, waters and peoples of the region, before going into more detail. The second chapter starts with a description of the Tanaïs and gradually introduces the (Asian parts of the) Bosporan Kingdom as far as the foothills of the northern Caucasus.[1] The flight of Mithradates VI Eupator from Pontos to the Bosporos in 66 BC provides an elegant transition to the exposition of Kolchis.[2] Strabo starts by surveying the coastline of the eastern Black Sea from the Bosporos to Sinope, before presenting yet another overview of the area's mountain ranges. Next comes a section on Dioskurias to the south of the northernmost outliers of the Caucasus, which, at the same time, forms the northern part of Kolchis.[3] Then he describes the Kolchian coast, centering on the Phasis River, i.e. the modern Rioni (though only as of Rhodopolis / Geguti) and its homonymous city at its mouth.[4] This section ends with a few lines on the Leukotheion, which I here present in an English translation adapted from the Loeb edition:

> Above the aforesaid rivers in the Moschian country lies the temple of Leukothea, founded by Phrixos, and the Oracle of Phrixos, where a ram is never sacrificed; it was once rich, but it was robbed in our time by Pharnakes, and a little later by Mithradates of Pergamon.[5]

---

1   Strab. *Geogr.* 11.1 (490–492C) and 11.2.1–12 (492–496C). For a general survey of Strabo's *Geography*, see now Roller 2018, esp. 629–684 for book 11.

2   Strab. *Geogr.* 11.2.13 (496). On Eupator's flight in 66 BC, also see App. *Mith.* 101.463–102.477; Ballesteros Pastor 1996, 269; Biffi 2010, 119–124; Roller 2018, 638f.

3   Strab. *Geogr.* 11.2.14f. (496f.C) and 11.2.16 (497f.C).

4   Strab. *Geogr.* 11.2.17 (498C). The identity of the Kolchian Phasis with the Rioni is widely accepted, see, e.g., Braund & Sinclair 1997/2000, *BA* 87; Dan 2016. But Lordkipanidze 1996, 101–105, 247 (cf. 38–41) and 2000, 9–36 points to a tradition represented by Eratosthenes and Strabo (*Geogr.* 11.2.17f. [498f.C]) that the Phasis was navigable until Sarapana / Shoropani and originated in Armenia. Lordkipanidze thus identifies the Kvirila River as the middle course of the Phasis between Shoropani and Geguti, opting for the Dzirula River (coming from the north-east) as the upper course of the Phasis. However, a broad ancient tradition claims an Armenian source, which recommends the Barimela River. This has its spring in the Lesser Caucasus and unites with the Dzirula into the Kvirila at Shoropani. See Coşkun 2019c.

5   Strab. *Geogr.* 11.2.17 (498C), translation adapted from Jones 1924. For the Greek text, see below, n. 47. I render Strabo's *hieron* with the more neutral term 'sanctuary' (with Roller 2014,

The first of the two plunderers is king Pharnakes II, son and successor of the aforementioned Mithradates Eupator, whom he dethroned in 63 BC at the end of the Third Mithradatic War. The Roman supreme commander Pompey had acknowledged his rule only in the Bosporan Kingdom. Pharnakes therefore waited for an opportunity to re-incorporate Kolchis and Pontos into his realm. He took his chances when Rome was plunged into a civil war, but Caesar defeated him at Zela on 2 August 47 BC and compelled him to retreat. His situation was exacerbated by the revolt of Asandros in the Bosporan Kingdom, and he fell while fighting against the insurgent. Caesar refused to accept the succession of Asandros, and saw the turmoil in the North as an opportunity to promote his friend Mithradates, the son of a priest from Pergamon and a princess from the Trokmian Galatians. He made the Pergamene Mithradates tetrarch of the Trokmoi in central Anatolia and further appointed him king of the Bosporos. The pillage of the sanctuary in Kolchis is the last we hear of this Mithradates. He must have died in the course of 46 BC, either in combat with Asandros or on his march up north.[6]

After these historical notes on the Leukotheion, Strabo touches in passing on the famous myths of Jason and the Argonauts as well as Phrixos. The account is unusually short, perhaps because he had dealt with them in more depth before.[7] He shows a bit more interest in the historical kings of Kolchis, most prominently the aforesaid Mithradates Eupator, in which context Strabo pays homage to his mother's uncle Moaphernes, who had served as the king's governor in the area. The short historical outline ends with Queen Pythodoris, who ruled during the author's time. An additional clarification of geopolitics mentions the Leukotheion again:

> Now the Moschian country, in which is situated the sanctuary, is divided into three parts: one part is held by the Kolchians, another by the Iberians, and another by the Armenians. There is also a small city in Iberia, the City of Phrixos, the present Ideëssa, well fortified, on the confines of Kolchis.[8]

Strabo somewhat misleadingly speaks of a 'Moschian country'. In the Augustan period, *Moschike* no longer referred to a territory inhabited by a Moschian population, since it had ceased to exist by the Hellenistic period. The term rather denoted the mountain range now usually called *Lesser Caucasus* or *Meskheti* according to a Georgian local tradition.[9] The second chapter of Strabo's eleventh

---

482 and Radt 2004, 307) instead of Jones' 'temple' (thus also Hamilton & Falconer 1903/6) or even 'Tempelstadt' (Lordkipanidze 1996, 251). We do not know how the sanctuary looked.

6   See Coşkun 2019a and forthcoming a on the chronology. For further details esp. on Pharnakes and Asandros, see below, Argument 3. For general information on the kings, see, e.g., Heinen 1994; Ballesteros Pastor 2008/19a; 2008/19b; 2017.

7   See esp. Strab. *Geogr.* 1.2.37 (46C), on which see below, ns. 34 and 51.

8   Strab. *Geogr.* 11.2.18 (499C), quoted below, n. 47; translation adapted from Jones 1924.

9   See Strab. *Geogr.* 11.2.1 (492C); 11.2.15 (497C); Plin. *NH* 6.10.28: *per convalles autem proximi Armeniae sunt Menobardi et Moscheni*; 6.10.29: *Colchicae solitudines, quarum a latere ad Ceraunios verso Armenochalybes habitant et Moschorum tractus ad Hiberum amnem in Cyrum defluentem et infra eos Sacasani et deinde Macerones ad flumen Absarrum*. Cf. Herrmann 1933, 351; Roller 2018, 639; 642. According to Lordkipanidze 1996, 151, they seem to have originated in the eastern parts of the Lesser Caucasus, before expanding westwards around

book continues with a doggerel flashback to Dioskurias, which provides further detail on the Caucasian peoples. He thus gradually transitions to the chapters on the Iberians, Albanians and the legendary Amazons, before dealing with the Caspian Sea.[10]

Well into the 20th century, the location of the Leukotheion was accepted as either being unknown or the object of wild speculation.[11] After all, neither the topographical details nor the description of the site itself are overly specific. We do not even know what type of sanctuary or monument to look for.[12] Moreover, although archaeological and historical work in Georgia has increased over the past few generations, the available data do not yet compare with equally important sites elsewhere in the Mediterranean and Black Sea region. To a significant extent, however, the difficulties of writing a History of Ancient Kolchis is due to the area's lack of stone inscriptions, which is further aggravated by the fact that ancient coins are rare to find especially along the coast and the Graeco-Roman literary tradition rarely went beyond celebrating the myths of Phrixos, Jason and Medeia.[13]

Scholars have, of course, made efforts to find the sanctuary. Given the lack of supportive epigraphic and numismatic sources, suggestions have largely been based on the wealth and prominence of a given location, combined with some material evidence pointing to a major cult site. One of the most authoritative Georgian archaeologists of the past generation, Otar Lordkipanidze, has repeatedly argued for Akhvlediani's Hill in Vani, situated on the Sulori River, about 2 km south of where the latter merges into the Rioni. Some 70 km east from Poti Seaport near the ancient city of Phasis,[14] Vani covers a slope from the northernmost ridge of the

300 BC. Also see pp. 256–259, although his effort to explain 'warum Strabon einen Teil der Rioni-Niederung als "das Land der Mos'cher" bezeichnet' (p. 258) is superfluous, since his preferred location of Vani (see below) would still justify the notion of Moschian Mountains. Note, however, that Prokop. *Goth.* 8.2.4.24–26 speaks of the *Meschoi* (*sic*) as subject to the Iberians, being settled in the mountains in-between the Iberians and the Lazoi. I suppose that the name of the mountains had been imposed on its inhabitants by the time.

10   Strab. *Geogr.* 11.2.19 (499C) and 11.3–6 (499–508C).
11   E.g., Eitrem 1925, 2296f. only mentions its location in Kolchis; Jones 1924, vol. 5, 213–215 avoids any specification. For a discussion of the evidence, a historical synthesis and a scholarly survey, see Lordkipanidze 1996, 141–153 and 252f.
12   See above, n. 5.
13   Cf. Braund 1994 and Tsetskhladze 1998. To some degree, Vani is an exception, see Dundua & Lordkipanidze 1979 on the coins found there, although their historical interpretation remains problematic; see below, n. 18.
14   The location of Phasis City in the Poti area is widely accepted, although identifying the site remains a challenge due to the frequent changes of the river bed and the rise of the sea water level; see, e.g., Silberman 1995, 30; Lordkipanidze 1996, 228–232; 2000, 47–53; Tsetskhladze 1998, 7–11; Nawotka 2005, 235. Braund & Sinclair 1997/2000, *BA* 87 and *Directory*, p. 1227 recommend the results of underwater archaeology by Gamkrelidze 1992 for pointing to the Paleostomi Lake; cf. Belfiore 2009, 175 n. 86 with further references. But dislocated evidence mainly from the Byzantine period is insufficient; see Tsetskhladze 2018, 477. The most probable location is slightly north-east of the Paleostomi Lake, see Coşkun 2020b, 658f. with n. 9. The effective distance from Vani may have been closer to a hundred km, depending on road

Lesser Caucasus extending into the Kolchian plain. Now it is only a small town of less than 5,000 inhabitants, but its ancient settlement was extraordinarily affluent, based on easy access to gold washed out of the mountain rivers, fertile farmland and a combination of land and water routes fostering trade. The impressive monuments on Akhvlediani's Hill thus seemed to encourage us to look for the Leukotheion among them.[15]

Lordkipanidze also points to a twofold Hellenistic destruction layer. He ascribes this evidence to conquests in 49 and 47 BC respectively, as he dates the abovementioned campaigns of Pharnakes II and Mithradates VII (I would rather propose August 47 and spring 46 BC).[16] Lordkipanidze's view seems to be accepted, with some hesitation albeit, by Gocha Tsetskhladze, who further emphasizes the long-term historical implications:

> Wenn die Annahme zutreffend ist, daß in Strabos Schilderung des Heiligtums der Leukotheia von der alten Stadt die Rede ist, die in Vani ausgegraben wird, dann ereignete sich das bei den Einfällen in die Kolchis zuerst des bosporanischen Königs Pharnakes (ungefähr 49 v.Chr.) und etwas spatter des Mithridates von Pergamon (47 v.Chr.), von denen Strabo (XI, 2, 17) und Cassius Dio (42, 45) berichte<n>. Der Fall von Souris (Vani) war der Schlußakkord des Zusammenbruchs der alten Kolchis. Der Schwarzmeerküstenteil gerät in Abhängigkeit vom Imperium Romanum, der Ostteil fällt an Iberien.[17]

---

conditions. *Google Maps* now suggests a route via the E60 as the quickest connection by car, which is 88 km.

15 See Lordkipanidze 1996, 251–269, esp. 258–268; cf. 1991, 194; 2000, 99 n. 658. For Vani as the location of the Leukotheion, also see Blázquez 2005, 235; Licheli 2007, 1122; Radt 2008, 255; Belfiore 2009, 173 n. 82 (quoting Lordkipanidze 1972, but also considering Samtredia, as below, n. 21); also Sens 2009, 166f. (despite admitting several difficulties). Tsetskhladze 1998, 114–164 offers a critical survey of the archaeological material of Vani, esp. 119–126 on the gold industry; he also describes the hill as located about 170 m above the sea level (114; 133–143); he identifies it as the 'Akropolis' with the palatial residence of the *skeptouchos* (138f.); he further addresses the traces of rituals (115) and sanctuaries (144–150), esp. on the Mother Goddesses Demeter and Aphrodite (150). Only in a final note (164 n. 92), he admits Leukothea as the central divinity of the 'Tempelstadt ... als eine der möglichen Hypothesen', though he gives more credit to this possibility at the end of his historical account (p. 186, quoted below). For an updated summary of the material evidence and history of Vani, see Tsetskhladze 2018, 485–500, though without mention of the Leukotheion; cf. Balandier 2005; Blázquez 2005; Sens 2009, 172–186. This is convincingly accepted by Tsetskhladze 1998, 132 (*Souris*) and 2018, 485–500. The settlement at the confluence of the Sulori and the Phasis near Vani has been identified with *Surium* by Pliny (*NH* 6.4.13). This is convincingly accepted by Tsetskhladze 1998, 132 (*Souris*) and 2018b, 490f. (cf. 498–500 on a less likely alternative: Zeda Tsikhesulori slightly north of Vani) as well as by Licheli 2007, 1091, further considered closely by Lordkipanidze 1996, 269 (with n. 437), but doubted by Braund 1994, 148. Sens 2009, 167 regards the identification of Vani with *Surium* or the *Leukotheion* as mutually exclusive.

16 See previous note, esp. Lordkipanidze 1991, 194: 'Vani was destroyed in the middle of the first century. If, as I have conjectured, Vani is the sanctuary of Leucothea in Strabo, then the destruction must be credited to two invasions of Colchis – the first ca 49 by the Bosporan king Pharnaces and a second ca 47 by Mithridates VII. Excavations have clearly revealed traces of two destructions within a short period.' On the numismatic evidence, see n. 18 below.

17 Tsetskhladze 1998, 186. Cf. Braund 1994, 147, who considers the chronological interpretation of the destruction possible regardless of rejecting the identification with Vani; also see p. 149:

But it is unconvincing to claim two materially distinct strata for late-Hellenistic Vani and to assign them to such a narrow timeframe without conclusive numismatic evidence.[18] What is more, Strabo does not even mention destruction, but rather speaks of pillage and impoverishment. Vague indications of destruction in the archaeological data are thus not helpful for the identification of the sanctuary's site.

Irrespective of this difficulty, David Braund has put forward the strongest objections to Lordkipanidze's interpretation. While he accepts that the site of Vani formed a large city in the Hellenistic period, he does not see sufficient evidence for conceiving of it as a 'temple city'. Most of the evidence adduced to prove its pre-eminence as a cult site appears ambiguous to him.[19] I would add that the term 'temple state' has been defined as a sanctuary of regional pre-eminence centred around one specific cult and ruled largely autonomously by a high priest, who is, however, appointed by and ultimately a vassal of a king: Lordkipanidze has not even tried to argue that any of these criteria have been met.[20] More importantly, Braund understands Strabo as implying a location in the triple border zone of the Kolchians, Iberians and Armenians. He thus proposes the Akhaltsikhe area or Atsquri, southwest of Borjomi in the valley cut into the mountains by the upper Mtkvari / Kyros.[21]

'Rather, destruction at Vani and at the temple of Leucothea are better seen as symptomatic of the extensive destruction that Strabo mentions.' Lordkipanize's interpretation is further quoted in the German version of *Wikipedia*: https://de.wikipedia.org/wiki/Wani. Without discussing Lordkipanidze's or Tsetskhladze's views or even addressing the Leukotheion, Licheli 2007, 1096 points to Parthian influence around the time.

18 Lordkipanidze bases his argument on the discussion of Dundua & Lordkipanidze 1979, while Dundua actually relates the destruction to the Kolchian revolt around 83 and Pompey's campaign in 65 BC. For the latter, one may add reference to Diod. 40.4; Plut. *Pomp.* 34; App. *Mith.* 103.477–104.484; cf. Braund 1994, 161–169. There is no clearly attributable destruction layer after this, and an isolated coin from Pergamon dated to 50/49 BC does not justify the postulate of one, let alone, two destruction layers in the early 40s. Lordkipanidze 1996, 262–264 adds further detail, but the chronological implications remain circular. The evidence has been addressed only in passing by other scholars (e.g., Braund 1994, 159–161; Tsetskhladze 1998, 115; Coşkun 2018b) and passed over in silence by de Callataÿ 1997.
19 Braund 1994, 146f., only excepting a female terracotta statue possibly representing Hekate. He is followed by Tsetskhladze 1995, 298, although Tsetskhladze 1998 fully endorses the view of Lorkipanidze, speaking of 'ein Heiligtum oder eine Tempelstadt' (133) and 'Kultzentrum' (158); also see pp. 164 and 184f. (comparison with Anatolian 'temple states' and explanation with Greek or Anatolian immigration); Tsetskhladze 2018, 496f. is more cautious and concludes 'that Vani was both the administrative (sceptuchal) and religious 'capital' of central Colchis'. For Vani as a 'temple city' also see Blázquez 2005; cf. Henkel 2007, 1: 'eine Tempelstadt des 7. bis 1. Jahrhunderts v. Chr.', and is considered *communis opinio* by Lordkipanidze 1996, 251 (with further references); cf. 262.
20 Foundational is Boffo 1985. Also see Coşkun 2018a with further bibliography.
21 Braund 1994, 146–151, speaking vaguely of 'a body of water or a water source' in the 'general region of modern Akhaltsikhe' (followed by Roller 2018, 641). He mentions Atsquri as a site on the upper Mtkvari, where 5th-century Athenian ware has been found, but not as the sanctuary's site (p. 185). Braund 1997/2000, 1260 specifies Atsquri as the location of the Leukotheion, tentatively followed by Belfiore 2009, 173 n. 82, who, however, also considers near Samtredia (following Lordkipanidze 1972). For another description of Atsquri, without reference to the Leukotheion, see Licheli 2007, 1124f.

But a site in the Borjomi area is at least as problematic to accept as in Vani, for mainly three reasons, as I shall unfold in this paper. First, such inland locations would be quite unusual for the sea goddess Leukothea; second, they do not seem to be fully in line with Strabo's topographical description; and third, they appear to be incompatible with the implications of the military campaigns which induced the downfall of the Leukotheion in the times of the Roman Civil War. My argument will conclude with a tentative suggestion of two sites in the sea-facing foothills of the Lesser Caucasus that would ideally fit Strabo's description. A new map of ancient Kolchis drawn by my student Stone Chen will help the reader navigate through this investigation (Map 4 at the end of the volume).

II. ARGUMENT 1:
MYTHICAL IMPLICATIONS OF THE LEUKOTHEA CULT

1. Ino-Leukothea and Phrixos in the Graeco-Roman Tradition

Throughout the ancient world, Leukothea was identified with Ino, one of the daughters of king Kadmos, the mythical founder of Thebes. She was also married to his successor Athamas, son of the god Aiolos. One branch of the broad tradition emphasizes the couple's role in the upbringing of Dionysos: Zeus had engendered him with Ino's sister Nephele, Athamas' first wife, whom Hera killed out of jealousy. The goddess also took revenge on Dionysos' foster parents, striking them with madness, so that Ino would live on as a Maenad in the wilderness. Her sons Learchos and Melikertes were also persecuted, if not slaughtered, by Athamas or his next wife Themisto, though some accounts even name Athamas' oldest son Phrixos as the evildoer, if not Ino herself. The older versions of the myth had Learchos die first at the hands of Athamas, whereas Ino escaped, together with Melikertes. In despair, they leaped into the sea, imploring Dionysos for help. According to some, they were saved by a dolphin, while others have them stranded or even drowned, but they were eventually transformed into divinities. Ino turned into Leukothea, named after the white foam of the sea, or perhaps after the shimmering reflection of the sunlight. Her son became Palaimon. Henceforth, they were called upon by sailors in distress. As such, Ino-Leokothea was already known to Homer by the end of the 8th century BC, who had her save Odysseus when erring all over the sea on a raft.[22]

---

22 Hom. *Od.* 5.333–353, 461–463; cf. Hyginus, *Fab. 'Odyssea': Ibi cum fluctibus iactaretur, Leu-cothoe, quam nos Matrem Matutam dicimus, quae in mari exigit aevum, balteum ei dedit, quo sibi pectus suum vinciret, ne pessum abiret.* And see Eitrem 1925, 2293 (on possible etymological explanations); 2297–2302 (on the myths involving her); 2304 (Leukotheai as an epithet of the Nereids according to Hesychios *s.v.*); Krauskopf 1981 (discussing her cults throughout the Mediterranean); Nercessian 1990 (cataloguing her ancient depictions); *theoi.com s.v.* 'Leu-kothea' (with many literary sources quoted in English translation); Gantz 1993, 176–180; 183f.; 473–478. For further references on Ino, see below, ns. 31–33.

While this is the basis of my assumption that Leukothea should be regarded as a maritime divinity, an important variation of the legend may have the potential to alter our view. According to some ancient sources, Ino was not so much the protectress of Dionysos than an 'evil queen'. She is said to have plotted against Athamas' son by Nephele, her nephew and stepson Phrixos. There is repeated talk of a vicious design: she had the seeds in her land secretly roasted, in order to prevent them from sprouting, so that the ensuing famine would cause demand for the most costly sacrifice: Phrixos, the oldest son of Athamas. Just before his father slaughtered him on the altar, Hermes or Nephele intervened by sending a miraculous golden ram for his and his sister Helle's escape. Riding on the ram's back, she made it only to the Dardanelles, duly called *Hellespont* after she drowned there. Her brother reached Aia, the exotic realm of Aiëtes, which the later tradition identified with Kolchis (by ca. 500 BC). There, he became the son-in-law of the king.[23]

Ino's perilous role is now considered the standard version of the myth.[24] According to Paul Dräger, this was already the case in the plot that he reconstructs for the pre-Homeric period.[25] If this were true, it would raise serious question about her 'qualification' as a helper in need. What is more, Phrixos would seem to be a very unlikely candidate for 'founding' a sanctuary of his evil stepmother, as the local aetiology seems to have claimed. Despite these problems, however, the connection of Leukothea with Phrixos has largely been accepted without further comment so far.[26] There are only few exceptions to such views. On the one hand, Duane Roller thinks that 'a sanctuary to his evil stepmother might be a reasonable precaution for Phrixos to have taken' – as if she continued to pose a risk to his safety even in far-away Kolchis.[27] On the other hand, Otar Lordkipanidze finds the combination so hopelessly unlikely that he simply rejects it wholesale, together with all the clues to the sanctuary's location. But his radical approach fails to explain why Strabo or anyone before him would have wanted to connect the two cults in the first place.[28]

A diachronic reading of our sources reveals, however, that Ino's role as the blameful stepmother is an Athenian innovation which can be dated to the 430s, if not early 420s. The myth seems to originate in rites, probably involving human sacrifice, that were meant to avert the failure of the crops. Its earliest narratives elaborations focused on how king Athamas navigated between the ritual demands

---

23 See previous n. for references to scholarship and add Keyßner 1941; Bruneau 1994. The learned but speculative musings by Graves 1952/71 need not distract us: he rashly equates Leukothea with Nephele (as the mother of the Centaurs) (p. 63), further links her indirectly with Io (p. 62) and a pristine lunar goddess, the 'White Lady' (p. 24).

24 E.g., Keyßner 1941; Nercessian 1990; Bruneau 1994; Antoni 2006/18; Radt 2008, 255; Węcowski 2009, on *BNJ* 6 F 11; Roller 2018, 641. It is called 'the common story' in *theoi.com*.

25 Dräger 2001, 8.

26 There is no reference to the connection in Keyßner 1941; Nercessian 1990; Bruneau 1994. The link is mentioned but not explained by Eitrem 1925, 2296f.; Krauskopf 1981, 145 ('seltsamerweise'); Braund 1994, 149 (he merely comments on the higher popularity of Phrixos compared to Jason on p. 32); Martin 2001, 172; Radt 2008, 255; *theoi.com*.

27 Roller 2018, 641.

28 Lordkipanidze 2000, 99 n. 658.

of the gods and the physical needs of his subjects. We find the first clear attestation for a misogynistic reinterpretation in Pindar's *Fourth Pythian Ode* (462 BC), which, however, leaves the name of the stepmother open.[29] An Alexandrian scholion to this poem provides some further precious information:

> *And from a stepmother's godless bolts*] For he (sc. Phrixos) was maltreated because of his stepmother having lusted after him, and he was plotted against, so he ran away. In his *Hymns*, Pindar says that this woman is Damodika, but Hippias says she is Gorgopis; Sophokles in his *Athamas* says (she is) Nephele; Pherekydes says (she is) Themisto. He also says that he (Phrixos) willingly gave himself over to be sacrificed when the crops failed of their own accord.[30]

Our first witness for Ino sharing in the guilt of Athamas is Herodotos, who may have added a passing note to his *Histories* towards the end of his life (ca. 427 BC):

> When Xerxes had come to Halus in Achaia, his guides, desiring to inform him of all they knew, told him the story which is related in that country concerning the worship of Laphystian Zeus, namely how Athamas, son of Aiolos, plotted Phrixos' death with Ino, and further, how the Achaians by an oracle's bidding compel Phrixos descendants to certain tasks.[31]

The ensuing literary tradition is very heterogeneous. Only some of its branches represent Ino in a negative light,[32] while others either focus on her role as victim or on Themisto as the antagonist of Phrixos' sons. Quite telling are the two Argonautic epics that survive: Apollonios Rhodios (3rd century BC) fades out completely the question of guilt for the fate of Phrixos, whereas Valerius Flaccus (1st century AD) avoids explicit blame, plays with subtle ambiguities and concludes with some kind of reconciliation between Ino on the one hand and Helle and Phrixos on the other.[33] We can thus be sure that, when Leukothea's cult began spreading through the Mediterranean world in the 8th and 7th centuries, the common versions of the legend presented her as an unstained maritime divinity. There is no reason to assume that conditions had changed, when Greek settlers (probably from Phokaia) introduced Leukothea to the Black Sea in the 6th century. The unique conflation of her cult with that of Phrixos was not induced by any events that form part of the preserved mythical narratives, but most likely reflects the physical take-over of her sanctuary

29 Pind. *Pyth.* 4.159–164, with Dräger 2001, 19 and West 2003, 157 for the date.
30 Pherekydes, *BNJ* 3 F 98 = *Scholia (BDEGQ) on Pindar*, 4.288a: ἔκ τε ματρυιᾶς ἀθέων βελέων] ἐκακώθη γὰρ διὰ τὴν μητρυιὰν ἐρασθεῖσαν αὐτοῦ καὶ ἐπεβουλεύθη ὥστε φυγεῖν. ταύτην δὲ ὁ μὲν Πίνδαρος ἐν Ὕμνοις Δαμοδίκαν· Ἱππίας δὲ Γοργῶπιν· Σοφοκλῆς ἐν Ἀθάμαντι Νεφέλην· Φερεκύδης Θεμιστώ. ὃς καί φησι τῶν καρπῶν φθειρομένων ἐκ ταὐτομάτου ἐθελούσιον δοῦναι ἑαυτὸν εἰς σφαγήν. Greek text and English translation from Morison 2011; cf. Hippias of Elis, *BNJ* 6 F 11 ed. by Węcowski 2009.
31 Hdt. 7.197.1: ἐς Ἆλον δὲ τῆς Ἀχαιίης ἀπικομένῳ Ξέρξῃ οἱ κατηγεμόνες τῆς ὁδοῦ βουλόμενοι τὸ πᾶν ἐξηγέεσθαι ἔλεγόν οἱ ἐπιχώριον λόγον, τὰ περὶ τὸ ἱρὸν τοῦ Λαφυστίου Διός, ὡς Ἀθάμας ὁ Αἰόλου ἐμηχανήσατο Φρίξῳ μόρον σὺν Ἰνοῖ βουλεύσας, μετέπειτα δὲ ὡς ἐκ θεοπροπίου Ἀχαιοί. Greek text and English translation (adapted) from Godley 1920.
32 Thus also the painting on a Tarentine red-figure vase of ca. 340: Staatliche Museen zu Berlin – Stiftung Preußischer Kulturbesitz 1984.41; see Bruneau 1994, 399 no. 1; Giuliani 1988, 6–10. For this and other (seeming) representations of Ino on vases, see Coşkun, in preparation a.
33 Apollonios' silence is conspicuous in 2.1140–1156. For ambiguity, see Val. Flacc. 1.277–292; 1.521f.; also 8.1–23; for a conciliatory tone, see 2.605–607; cf. Coşkun, in preparation a.

by either Milesian (2nd half of 6th century) or Sinopean colonists (6th or 4th century). We can only speculate about narrative aetiologies that would have been spun locally, but there is no good reason to believe that such nuances would have disconnected Leukothea's cult from the sea.[34]

2. Leukothea in Her Kolchian Context: Mother, River or Maritime Goddess?

The most serious obstacle of investigating the cult of Leukothea in Kolchis is that no specific monument dedicated to her has so far been identified. We thus lack the usual additional evidence that coins, inscriptions or plastic votive offerings might provide us with. In addition, the goddess barely seems to have played a role in the Black Sea region at large. Strabo's reference to her sanctuary is completely isolated. Given the density of our knowledge of the cultic landscape along the Euxine coast, the silence of our remaining sources is quite telling.[35]

As laid out in the introduction, however, the most eminent scholars have proposed either Vani or the Borjomi area for the location of the Leukotheion. If either of these were accepted, we would have to picture the sanctuary far inland, if not high up in the mountains, and thus disconnected from the sea. Accordingly, Leukothea's Kolchian emanation would have been less important as a maritime divinity, but probably one of the many syncretic manifestations of the Asian Mother Goddess.[36] The Black Sea region is very rich in evidence for the veneration of the Ephe-

---

34 For the Phokaians as the promoters of Leukothea in the Mediterranean, see Morel 2006, 380, 407. For the engagement of the Milesians and Sinopeans along the Black Sea coast, see, e.g., Strab. *Geogr.* 1.2.37 (46C), quoted below, n. 51; 1.2.39 (46C); 11.2.17f. (498f.C); 12.3.11 (546C); Pomp. Mela 1.19.104 on Kytissoros / Kytoros, discussed below, n. 50; cf. Ehrhardt 1988; Braund 1994, 8–39; 1998; 2005; Lordkipanidze 1996, 18–66; Tsetskhladze 1998; 2013; Burcu Erciyas 2007. Batumistsikhe may have hosted the earliest 'Greek colonies in Colchis ... somewhere between 610 and 570 BC'; see Tsetskhladze 2018b, 512–514.

35 Cf. Ehrhardt 1988 on Milesian colonies or *Ancient Sacral Monuments in the Black Sea* by Petropulos & Maslennikov (2010). The most recent monograph on *Goddesses in the Bosporan Kingdom* (Braund 2018) does not even have a lemma *Leukothea* in its index. Word searches for the same have also been in vain in Robu & Birzescu 2016 and Fornasier 2016 (who, in contrast, mentions Apollo 87 and Achilles 10 times). The survey of male deities by Saprykin 2010 also takes note of their female companions, but once more there is no reference to Leukothea. For Phasis, see Lordkipanidze 2000, 62–99 and Braund 2010. The latter emphasizes the key role of the Phasis River in its male and female emanations, calling 'both profoundly local to the landscape' (see below, with ns. 39–42), besides Artemis and Apollo. For the fortress of Apsaros (see below, n. 66), see Kakhidze & Mamuladze 2010; while they make no explicit reference to Leukothea, one might tentatively relate the topaz intaglio of a silver ring (2nd–3rd century AD), which shows a dolphin, although the authors (p. 459) connect it with Neptune.

36 Cf. the vague considerations of Lordkipanidze 1996, 265, claiming that 'unter dem Namen der Leukothea eine örtliche Gottheit zu verstehen ist, deren Attribute sehr jenen der griechischen Leukothea – der weißen Göttin, der Bewohnerin des Meeres, einer der eigenartigsten Gestalten der griechischen Mythologie – ähneln. Auch sonst sind die Griechen bei der Beschreibung religiöser Kulte fremder Völker oft auf diese Weise verfahren ...' But then he goes on to specify – arbitrarily, as I find, – that both Leukothea and the anonymous local divinity were in charge

sian Artemis, Artemis Parthenos, Artemis Hekate, Aphrodite Urania, or Kybele. We may add Egyptian Isis as another variation celebrated throughout the Hellenistic kingdoms and the Roman Empire, often as the consort of Serapis.[37] Her Near-Eastern counterparts were Babylonian Ishtar and Syro-Phoenician Astarte, whereas she went by the name *Anahid* or *Anahita* in Persian and Armenian societies, who had more immediate relevance for Kolchis. Strabo calls this eastern Mother *Anaïtis*, presenting her as the most popular divinity in Armenia, famous not least for ritual prostitution. Precisely this detailed knowledge of the geographer should caution us not to accept that the Kolchian Leukothea is a variation of *Anaïtis*.[38]

An intriguing alternative might be *Theos Phasiane*, a distinct Kolchian version of the Magna Mater, which Arrian introduces as follows:

> The Statue of the Goddess Phasiane is placed to the left of the entrance into the Phasis; which Deity we may reasonably conjecture, from her figure and appearance, to be Rhea, as she holds in her hands a cymbal, has lions under her throne, and is seated in the same manner as the statue by Pheidias in the temple of Kybele at Athens. / An anchor, said to be of the ship Argo, is shown here; but as it is of iron, it does not seem to be ancient; it differs indeed both in size and shape from those at present in use, but nevertheless appears to me to be of later date than the Argonautic period. They also show there some fragments of an ancient stone anchor, which are more likely than the other to be the remains of the anchor of the Argo. No other monument is now to be found there of the fabulous history of Jason.[39]

of fertility, agriculture and viticulture (266–268). Moreover, he suggests identifying her by reference to a statue base beside one of the city gates, which a graffito reveals as 'eindeutig' dedicated to Leukothea, although the inscription only implies a female figure, whether goddess or queen (*anassa*) (pp. 267f.). Lordkipanidze goes one step back when denying that Leukothea could have been the main goddess as a maritime divinity, before downplaying his theory as 'rein hypothetisch' (p. 268). This is all quite confusing. Cf. Blázquez 2005, 235 for Leukothea as a local goddess; Belfiore 2009, 170 n. 80: 'forse assimilate ad una dea di origine assira' and 173 n. 82: 'identificabile con una dea assira delle acque'. Against such speculations, see Nawotka 2005, 236. Bremmer 1999 presents Leukothea as a goddess of 'initiation and rites of reversal', focusing on practices known from the wider context of a few cult sites, irrespective of the literary evidence. For a more balanced approach, see Krauskopf 1981, quoted below, n. 46.

37 For all of them in the Bosporan Kingdom, see Braund 2018; also Maslennikov 2010, 211; 217 etc. as well as Molev & Moleva 2010. Add Lazarenko et al. 2010 on the 'Pontic Mother' in Dionysopolis; Rousyaeva 2010, 74–76 and Krapivina 2010 on mother goddesses (esp. Aphrodite) in Olbia; Moga 2012 on Artemis in Pontos, Licheli 2007, 1091–1093 on Hekate in Kolchis and ritual pits in Vani and Sairkhe. For Aphrodite and Demeter as well as Hekate and Hestia in Milesian colonies, see Ehrhardt 1988, 164–167 and 173–176. For Serapis and Isis in Kolchis, see Kakhidze & Mamuladze 2010, 456–458; for Pontos, see Saprykin 2010, 492–498.

38 Strab. *Geogr.* 11.14.16 (532f.C). *Pace* Chaumont in Boyce et al. 1989/2012, ch. iii: 'Regarding the Caucasian countries adjacent to Armenia, Strabo (*Geography* 11.2.17) states that there was a temple dedicated to Leucothea, obviously an analogue of the Iranian goddess (O. G. von Wesendonck, *Caucasica* I, 1924, p. 87) in the land of the Moschi in Colchis. The legendary and late-dated *Life of the Apostle St. Andrew* mentions a cult of Apollo and Artemis, that is, Mithra and Anāhitā, in the same region.' But for the children of Leto, we should rather think of a Milesian background, see Ehrhardt 1988, 127–161 on the Apolline triad; also above, n. 35.

39 Arr. *PPE* 9.1f.: Εἰσβαλλόντων δὲ εἰς τὸν Φᾶσιν ἐν ἀριστερᾷ ἵδρυται ἡ Φασιανὴ θεός. Εἴη δ᾽ ἂν ἀπό γε τοῦ σχήματος τεκμαιρομένῳ ἡ Ῥέα· καὶ γὰρ κύμβαλον μετὰ χεῖρας ἔχει καὶ λέοντας ὑπὸ τῷ θρόνῳ, καὶ κάθηται ὥσπερ ἐν τῷ Μητρῴῳ Ἀθήνησιν ἡ τοῦ Φειδίου. / Ἐνταῦθα καὶ ἡ ἄγκυρα

Arrian's description of Phasiane stands out in his otherwise mostly dry report.[40] It is a beautiful example of Kolchian syncretism. On the one hand, he explicitly refers to the Athenian Mother of the Gods under her names Kybele and Rhea, supporting this identification with her typical features (cymbal, lions and throne). On the other hand, the name and site unfailingly betray Phasiane's local roots. In addition, her gender also reveals her non-Greek origin, since river gods in the Hellenic world are normally represented as bull-horned father gods.[41] Of particular interest is the anchor and its motivation: there would actually be no need to reference the Argonauts here, since the protection of seafaring would be sufficiently plausible for the river goddess, especially in a place where the Phasis releases sailors onto the open sea or receives those coming in from the Euxine. The link between Phasiane and the anchor was old and strong, when Arrian wrote those lines, a conclusion which its antiquated shape and its twofold manifestation in iron and stone both endorse.[42]

In theory, the anchor might allow us to muse about a potential connection of this Phasiane with Leukothea, but not a single detail in the accounts of Arrian and Strabo matches. After all, it would be hazardous to identify the statue of a composite chthonic-fluvial goddess in the estuary of the Phasis (superficially linked to the Argonauts) with the sanctuary of a sea goddess (opaquely connected with Phrixos) in the Moschian Mountains.[43] The description of Phasiane thus sets the expectations for how Strabo or Arrian might have introduced a sanctuary of Leukothea as a local variation of the Mother Goddess or of Phasiane – had only she been such.

Considering the evidence we have, as lacunose as it may be, I see no reason to doubt that Kolchian Leukothea largely followed Mediterranean traditions. Strabo, at least, did not feel the need to alert his readers to any local peculiarity, except perhaps the unique circumstance that her sanctuary was regarded as founded by Phrixos.[44] We therefore have to consider her a sea goddess, a quality so genuine to Leukothea that some ancient authors turned the offspring of Kadmos into one of the

---

δείκνυται τῆς Ἀργοῦς. Καὶ ἡ μὲν σιδηρᾶ οὐκ ἔδοξέ μοι εἶναι παλαιά – καίτοι τὸ μέγεθος οὐ κατὰ τὰς νῦν ἀγκύρας ἐστίν, καὶ τὸ σχῆμα ἀμηγέπη ἐξηλλαγμένη –, ἀλλὰ νεωτέρα μοι ἐφάνη εἶναι τοῦ χρόνου. Λιθίνης δέ τινος ἄλλης θραύσματα ἐδείκνυτο παλαιά, ὡς ταῦτα μᾶλλον εἰκάσαι ἐκεῖνα εἶναι τὰ λείψανα τῆς ἀγκύρας τῆς Ἀργοῦς. Ἄλλο δὲ οὐδὲν ὑπόμνημα ἦν ἐνταῦθα τῶν μύθων τῶν ἀμφὶ τὸν Ἰάσονα. The text follows the ed. of Wirth 1967 (cf. Silberman 1995; Liddle 2003); also see Szwajcer (undated =Arr. *PPE* 11). The translation has been adapted from the one by Falconer 1805 (=Arr. *PPE* 7) and compared with that of Silberman and Liddle.

40 Silberman 1995, 29f. explains it with Arrian's particular interest in the Mother Goddess in general; cf. Belfiore 2009, 171f. n. 180. Tsetskhladze 1998, 11 emphasizes her function as city founder. Lordkipanidze 2000, 90–96, explores here role as mother goddess and (p. 96) focuses on her position at the city's gate. Cf. Licheli 2007, 1090; Braund 2010, 434f.

41 See Braund 2010, 433; also see Lordkipanidze 2000, 81–90 on the male emanation of Phasis.

42 For antiquarian comments, see Silbermann 1995, 30; Liddle 2003, 100; Belfiore 2009, 172f. n. 81, referencing Paus. 1.3.4 (for another mention of the statue) and Apollon. 1.955–960 (as the potential reason for Arrian's concern with the anchor's authenticity).

43 The awareness of Phasiane's nature is also implied in the custom that sailors entering the Phasis were expected to pour out their old water supplies, due to superstitious believes: Arr. *PPE* 8.5.

44 Also note how closely familiar Strabo was with various branches of the Argonautic myth; see Lordkipanidze 1996, 18–38. On the location of the Phrixeion, also see below, with ns. 50–57.

sea nymphs, although these were known as the daughters of Nereus (Nereids).[45] Besides, the article in Pauly-Wissowa compiles 29 cities, islands or territories for which her cult is attested. Although nearly a century old, the list is still impressive and meaningful for us, since the entirety of these locations endorse Leukothea's direct relevance for seafaring. Ingrid Krauskopf has presented a revised list in the hope of elucidating the Etruscan sanctuary of Pyrgi built around 500 BC for the mother goddess Uni (later identified with Eileithyia, Mater Matuta and Leukothea, instead of her more typical counterpart Juno). She explores possible non-maritime functions Leukothea may have fulfilled, such as that of *kurotrophos* (in her role as foster mother of Dionysos), guide for initiation rites, or helper in birth or death; despite observing a significant potential for local variation, she admits that the surest parts of our evidence relate her to the sea, with all known sanctuaries and a coherent literary tradition from Homer to Late Antiquity.[46] Hence, we should also expect the Kolchian branch of her cult to have had an immediate connection with the Euxine, if not to have been visible from the sea itself. Strong evidence to the contrary should be required to have us reject this assumption.

III. ARGUMENT 2:
THE TOPOGRAPHICAL IMPLICATIONS OF STRABO'S ACCOUNT

1. The Textual and Topographical Context of the Leukotheion

Let us now examine whether Strabo's topographical details support the conclusion that I have drawn from the mythical and cultic traditions. For a better understanding, I include some relevant context preceding the brief notes on the sanctuary itself, starting with the description of the Phasis. The sections that will be important in my subsequent discussion are printed in bold:

> Further, the greater part of the remainder of Kolchis **is on the sea**. Through it flows the **Phasis**, a large river having its sources in Armenia and receiving the waters of the **Glaukos** and the **Hippos**, which issue from the neighbouring mountains. It is navigated as far as Sarapana, a fortress capable of admitting the population even of a city. From here people go by land to the

---

45 A loose connection with the Nereids is expressed as early as the 5th century BC (Pind. *Olymp.* 2.29f.); she is counted among them by Philostratos the Elder, *Imagines* 2.16 (3rd century AD), and Nonnos, *Dionysiaka* 10.122. Cf. Eitrem 1925, 2300; Nercessian 1990, 659f.; *theoi.com*, with further references.

46 Eitrem 1925, 2293–2297. And Krauskopf 1981, e.g., 140: 'Leukothea wird also in der antiken Dichtung einhellig als Göttin der Schiffahrt und der Seeleute, als Retterin in Seenot, als freundliche, hilfreiche Meeresgottheit geschildert. Andere Funktionen sind, soweit ich sehe, nicht überliefert – nicht berücksichtigt wurden hier die Stellen, die sich auf Mater Matuta und nur indirekt durch sie auf Leukothea beziehen.' Also 148: 'Sicher erscheint mir aber, daß die Leukothea, die Griechen im 4. Jh. mit der Göttin von Pyrgi identifizierten, nicht eine der Mater Matuta völlig entsprechende Göttin war, sondern ganz wesentlich eine See– und Schiffahrtsgottheit, und daß dieser Aspekt nicht außer acht gelassen werden darf.' Braund 1994, 149 only concedes that 'the cult of Leucothea would most naturally be associated with water', but rivers and seas are sufficiently different, as are their divine personifications.

Kyros in four days by a wagon road. On the Phasis is situated a city bearing the same name, an emporion of the Kolchians, which is protected on one side by the river, on another by a lake, and on another by the sea. Thence people go to Amisos and Sinope by sea (a voyage of two or three days), because the shores are soft around the **outlets of the rivers**. The country is excellent both in respect to its produce – except its honey, which is generally bitter – and in respect to everything that pertains to ship-building; for it not only produces quantities of timber, but also brings it down **on rivers**. And the people make linen in quantities, and hemp, wax, and pitch. Their linen industry has been famed far and wide; for they used to export linen to outside places; and some writers, wishing to show forth a kinship between the Kolchians and the Egyptians, confirm their belief by this. **Above the aforesaid rivers in the Moschian Mountain lies the temple of Leukothea, founded by Phrixos, and the Oracle of Phrixos, where a ram is never sacrificed. It was once rich, but it was robbed in our time by Pharnakes, and a little later by Mithradates of Pergamon.** For when a country is devastated, "things divine are in sickly plight and wont not even to be respected", says Euripides.

... Now the Moschian Mountain, in which the sanctuary is located, is divided into three parts: one part is held by the Kolchians, another by the Iberians, and another by the Armenians. There is also a small city in Iberia, the City of Phrixos, the present Ideëssa, well fortified, on the confines of Kolchis.[47]

A first, superficial reading of the text seems to be suggestive of a site in the coastal area. This is not only a possible implication of the first sentence, which describes Kolchis as mainly coastal, but it is also compatible with the structure of the account: e.g., the treatment of the Phasis River ends with Phasis City, which is said to touch the sea; the route to Amisos and Sinope is by sea following the sandy shore; and the produce of the plain is transported to the coast on the rivers. Strabo's main perspective obviously stands in the tradition of the *periplus* literature.

---

47   Strab. *Geogr*. 11.2.17f. (498f.C). The translation has been adapted from Jones 1924. The Greek text follows Meineke 1877, slightly corrected according to Radt 2004: καὶ ἡ λοιπὴ δὲ Κολχὶς **ἐπὶ τῇ θαλάττῃ** ἡ πλείων ἐστί: διαρρεῖ δ᾽ αὐτὴν ὁ **Φᾶσις**, μέγας ποταμὸς ἐξ Ἀρμενίας τὰς ἀρχὰς ἔχων, δεχόμενος τόν τε **Γλαῦκον** καὶ τὸν **Ἵππον** ἐκ τῶν πλησίον ὀρῶν ἐκπίπτοντας: ἀναπλεῖται δὲ μέχρι Σαραπανῶν ἐρύματος δυναμένου δέξασθαι καὶ πόλεως συνοικισμόν, ὅθεν πεζεύουσιν ἐπὶ τὸν Κῦρον ἡμέραις τέτταρσι δι᾽ ἁμαξιτοῦ. ἐπίκειται δὲ τῷ Φάσιδι ὁμώνυμος πόλις, ἐμπόριον τῶν Κόλχων, τῇ μὲν προβεβλημένη τὸν ποταμὸν, τῇ δὲ λίμνην, τῇ δὲ τὴν θάλατταν. ἐντεῦθεν δὲ πλοῦς ἐπ᾽ Ἀμισοῦ καὶ Σινώπης τριῶν ἡμερῶν ἢ δύο (*) διὰ τὸ τοὺς αἰγιαλοὺς μαλακοὺς εἶναι κατὰ **τὰς τῶν ποταμῶν ἐκβολάς**. ἀγαθὴ δ᾽ ἐστὶν ἡ χώρα καὶ καρποῖς πλὴν τοῦ μέλιτος (πικρίζει γὰρ τὸ πλέον) καὶ τοῖς πρὸς ναυπηγίαν πᾶσιν: ὕλην τε γὰρ καὶ φύει καὶ **ποταμοῖς κατακομίζει**, λίνον τε ποιεῖ πολὺ καὶ κάνναβιν καὶ κηρὸν καὶ πίτταν. ἡ δὲ λινουργία καὶ τεθρύληται: καὶ γὰρ εἰς τοὺς ἔξω τόπους ἐξεκόμιζον, καί τινες βουλόμενοι συγγένειάν τινα τοῖς Κόλχοις πρὸς τοὺς Αἰγυπτίους ἐμφανίζειν ἀπὸ τούτων πιστοῦνται. **ὑπέρκειται δὲ τῶν λεχθέντων ποταμῶν ἐν τῇ Μοσχικῇ τὸ τῆς Λευκοθέας ἱερὸν Φρίξου ἵδρυμα, καὶ μαντεῖον ἐκείνου, ὅπου κριὸς οὐ θύεται, πλούσιόν ποτε ὑπάρξαν, συληθὲν δὲ ὑπὸ Φαρνάκου καθ᾽ ἡμᾶς καὶ μικρὸν ὕστερον ὑπὸ Μιθριδάτου τοῦ Περγαμηνοῦ.** κακωθείσης γὰρ χώρας "νοσεῖ τὰ τῶν θεῶν οὐδὲ τιμᾶσθαι θέλει", φησὶν Εὐριπίδης. / ... **ἡ δ᾽ οὖν Μοσχική, ἐν ᾗ τὸ ἱερόν, τριμερής ἐστι: τὸ μὲν γὰρ ἔχουσιν αὐτῆς Κόλχοι, τὸ δὲ Ἴβηρες, τὸ δὲ Ἀρμένιοι.** ἔστι δὲ καὶ πολίχνιον ἐν τῇ Ἰβηρίᾳ Φρίξου πόλις ἡ νῦν Ἰδήεσσα, εὐερκὲς χωρίον ἐν μεθορίοις τῆς Κολχίδος.   * Radt indicates a lacuna after δύο; Nicolai & Traina 2000, 32 and 90 read τεττάρων with *lacuna* (for which they reference Lassère 1975, criticising him for the indication of the *lacuna* albeit). Perhaps δυοῖν?

In order to reject Vani as the sanctuary's site, Braund has claimed: 'Strabo explicitly locates the temple of Leucothea at the common border of Colchis, Iberia,
and Armenia: he states that it was administered jointly by these three peoples.'[48]
Obviously, this statement draws on the last sentence quoted in bold letters. But there
is no mention of a joint administration of the Leukotheion, nor even of its location
close to the triple border. Strabo rather states that the sanctuary was to be found in
the Moschian Mountain, and that the latter extended over three distinct geopolitical
units. And since the context of the paragraph deals with Kolchis (11.2.13–19 [496–
499C]), we should assume that the Leukotheion was located somewhere in the
Kolchian part of the Lesser Caucasus. Only the City of Phrixos is explicitly located
in Iberian territory.[49] There may even have been another sanctuary of Phrixos near
the city of Phasis, which is mentioned by Pomponius Mela in the 1st century AD. It
is difficult to decide whether it escaped Strabo's (or his sources') attention, did not
yet exist at his time or was located too far in the hinterland.[50]

2. The Leukotheion and Phrixeion in Kolchis versus the City of Phrixos in Iberia

This said, Braund may be right to link the Leukotheion and the Phrixeion: at their
first mention, they are united not only through the authority of Phrixos, the founder
of the former and the oracular god of the latter, but also by their location in the
Moschian Mountains. We should therefore leave it open whether the sanctuary of
Leukothea was also the place where oracles of Phrixos could be obtained or whether
there was a distinct Phrixeion not too far from the Leukotheion. When Strabo
repeats his mention of the sanctuary (sc. of Leukothea) a few lines below, he also
adds another reference to Phrixos, but this time to the City of Phrixos among the
Iberians. He does not specify whether this hosted the Oracle of Phrixos mentioned
before. While it is plausible to assume that a city named after a hero had a sanctuary
dedicated to its mythical *ktistes*, I am inclined to distinguish it from the Oracle,
since this was mentioned only in a Kolchian context, if not as part of the
Leukotheion. In a very different context, however, Strabo mentions once more 'a

---

48 Braund 1994, 148f.; cf. 170.
49 For Strabo on Kolchis, see *Geogr.* 11.2.13–19 (496–499C). The City of Phrixos is not identified
    by Braund 1997/2000, *BA* 88, cf. *Directory* p. 1283, with reference to *Phrixou polis* and Lord-
    kipanidze 1996, 275.
50 Pomp. Mela 1.19.108. Lordkipanidze 2000, 98f. questions that Mela requires us to locate this
    Phrixeion within the city boundaries of Phasis and suggests identifying *Phrixi templum et lucus*
    with the Phrixeion that Strab. *Geogr.* 1.2.39 (46C) locates in the mountainous border zone be-
    tween Kolchis and Iberia. Mela is admittedly vague, but searching for it close to the Phasis
    River would still be most intuitive, since Mela continues with a description of the Caucasian
    mountains (1.19.109: *hinc orti montes* ...). This means that the temple and grove were in the
    Kolchian plain, most likely not far from the mouth of the Phasis River. The Phasis was regarded
    as Phrixos' destination by the early 5th century, see Pind. *Pyth.* 4.211f. of 462 BC.

Phrixeion on the boundary of Kolchis and Iberia', which must be referring to the City of Phrixos, again.[51]

But there is further evidence which may seem to support Braund's interpretation. In his *Annals*, Tacitus contrasts Pharasmanes, king of the Iberians and ally of the Albanians and Armenians, with the Parthian Prince Orodes, boasting of the military versatility of the Caucasian peoples as follows:

> And their (sc. the Parthians') strength only resided in their cavalry: Pharasmanes also had a vigorous infantry, since the Iberians and Albanians, as dwellers of forested mountains, had become accustomed to roughness and toil; they boasted that they hailed from Thessalians, at the time when Jason, after he had carried away Medeia and she had given birth to his children, returned to the then empty palace of Aiëtes and the abandoned (territory of the) Kolchians. They praise his name and the Oracle of Phrixos much, and no one would ever have sacrificed a ram, for there is the belief that Phrixos had ridden on it, whether this was an animal or a ship of distinction.[52]

While this additional testimony may explain the temptation to draw a close connection between the two cults mentioned by Strabo and to regard them as largely shared between the various Caucasian peoples, there is reason for pause. First, Tacitus' explanation can barely be accepted as authentic. It is obvious that he only adduces the exotic oracle of Phrixos to exemplify the alleged genealogical link, which was intended to give some clout to the otherwise less glamorous enemies of the Parthians. This is unlikely to reflect the content of the original diplomatic exchange, but rather constitutes historiographical flourish.[53] For its design, Tacitus drew on a literary source, perhaps Strabo himself (whether the passages we have

---

51 Strab. *Geogr.* 1.2.39 (45C): καὶ ἔστιν ὑπομνήματα τῆς ἀμφοῖν στρατείας τό τε Φρίξειον τὸ ἐν τοῖς μεθορίοις τῆς τε Κολχίδος καὶ τῆς Ἰβηρίας, καὶ τὰ Ἰασόνεια, ἃ πολλαχοῦ καὶ τῆς Ἀρμενίας καὶ τῆς Μηδίας καὶ τῶν πλησιοχώρων αὐταῖς τόπων δείκνυται. But also see Roller 2018, 38, who locates the Oracle of Phrixos somewhere on the upper Phasis and the City of Phrixos further inland; the same, however, locates this city 'presumably on the upper Phasis', though at the same time in the Moschian Mountains, which is inconsistent. Sens 2009, 166 identifies the Leukotheion and the Oracle of Phrixos.

52 Tac. *Ann.* 6.34.1f.: *... atque illis sola in equite vis: Pharasmanes et pedite valebat. nam Hiberi Albanique saltuosos locos incolentes duritiae patientiaeque magis insuevere; feruntque se Thessalis ortos, qua tempestate Iaso post avectam Medeam genitosque ex ea liberos inanem mox regiam Aeetae vacuosque Colchos repetivit. multaque de nomine eius et oraclum Phrixi celebrant; nec quisquam ariete sacrificaverit, credito vexisse Phrixum, sive id animal seu navis insigne fuit.* The Latin text has been drawn from *The Latin Library* (cf. Woodman 2017); the translation is mine (cf. Martin 2001). For the idea of Jason's (or Medeia's) return to Kolchis, also see Just. 42.2.12; 42.3.5; Koestermann 1965, 323; De Siena 2001; Woodman 2017, 236.

53 More optimistically, Radt 2008, 255 believes that the oracle 'existierte offenbar noch zu Tacitus' Zeit'; likewise, Keyßner 1941, 768 regards Tacitus as an accurate witness. Woodman 2017, 235–237 does not directly address the question, but discusses the grammatical implication of *qua tempestate ... repetivit* (indicative instead of subjunctive might imply that this is an authorial addition rather than part of the *oratio obliqua*) – but this does not change much, since *feruntque ... celebrant* clearly surmises the Iberians' point of view, though in the historiographical construction of Tacitus. Woodman also talks about the topical nature of mythical digressions and their entertaining character, without discussing whether recourse to them may still be seen as an authentic part of the diplomacy or only literary flourish.

cited in the *Geography* or a similar reference in his lost *Histories*). This is betrayed by the tralatitious mention of the avoidance of ram sacrifices, which here lacks the motivation that it still had in its former ethnographic context. Tacitus quite obviously conflates Strabo's information on the Oracle of Phrixos with that on the City of Phrixos, both of which the geographer has located in the Moschian Mountains, but only the latter also in Iberian territory.

The evidence for Phrixos among the Iberians thus comes down to their possession of Ideëssa, possibly a Kolchian foundation ascribed to Phrixos. We hear no more of the relevance of Phrixos for the Iberians after Strabo except for the playful reference in Tacitus' account. More widespread was the link between Medeia (and Jason) with the Medes (and Armenians), based on a quite transparent folk etymology, but taken seriously by Strabo and many other ancient writers regardless.[54] Be this as it may, even if my argument should not be followed and Tacitus' testimony be accepted as credible, I would emphasize that he only mentions the Oracle of Phrixos. It did not occur to him to also suggest the sea goddess Leukothea's popularity among the mountain-dwelling Iberians or Albanians.[55]

All of this said, the exclusion of Tacitus from our evidence does not yet speak against Braund's location of the Leukotheion (with or without the Oracle) around Akhaltsikhe or in Atsquri. But, on closer inspection, the Mtkvari valley cannot be an option. According to Strabo, the upper course of the Kyros belonged to Armenia, whereas its middle reaches flowed through Iberia, before the lower Kyros crossed (or flanked) Albania and, after uniting with the Araxes, empties into the Caspian Sea. As a result, no part of the Kyros touched Kolchis. Strabo mentions the fortress Sarapana / Shorapani as the easternmost of the noteworthy settlements of Kolchis, whence Iberia could be accessed. There is no evidence for Kolchis having extended much farther. On the contrary, Sarapana had been under Iberian control in the 3rd century BC, as is known from a Georgian chronicle.[56] Perhaps the City of Phrixos was also located somewhere in the west-end of the plain, where it would have changed hands between the Kolchians and Iberians, possibly even more than once.[57]

3. Above or beyond the Aforesaid Rivers?

Two questions remain to be asked in order to tease out some further information from Strabo's testimony: which are the 'mentioned rivers', and how exactly should

---

54   Strabo adds even further detail in *Geogr.* 11.13.10 (526C) and 11.14.14 (531C); also see 1.2.37 (46C), quoted above, n. 51.

55   Roller 2018, 641 conveys the contrary impression.

56   Strab. *Geogr.* 11.2.17 (498C) and 11.3.2, 4 (500C), cf. 11.1.5 (491C) on the upper and middle course of the Kyros; 11.4.2 (501C), cf. 11.1.5 (491C) on its lower course. And Strab. *Geogr.* 11.2.17 (498C) and 11.3.4 (500C) on Sarapana and Iberia. Cf. Braund 1994, 145 and Plontke-Lüning 2001, both with reference to *Kartlis Tskhoureba* 24 (1.34).

57   Radt 2008, 256 finds it confusing that an Iberian city (attested nowhere else) is mentioned in the context of Kolchis.

we understand 'above / beyond ... lies ...' (ὑπέρκειται δὲ τῶν λεχθέντων ποταμῶν)? As far as I see, scholars have taken for granted that those rivers were the Phasis and its tributaries, of which Strabo names the *Glaukos* and *Hippos*. We can leave the latter's exact identities open, as long as we agree that these must have been two of the major rivers springing from the northern Caucasus and running south.[58]

Was Lordkipanidze's choice of Vani meant to emphasize the height of the hill site?[59] The city unfolds into the outer slopes of the Lesser Caucasus, and at least its acropolis 'hangs over' the Kolchian plain. It would have provided a view on the nearby Phasis, but not on any of the latter's northern tributaries (the Sulori as a southern tributary is disqualified).[60] If Vani were the location we are looking for, then the literal meaning of the prefix ὑπέρ- would have been well preserved in the contrast of the two different altitude levels,[61] but the reference to the 'mentioned rivers' would be quite vague.

Even more problematic is the identification by Braund. Admittedly, Akhaltsi-khe and Atsquri somehow lie 'beyond' the Phasis and its tributaries, and with its

---

58 The identity of the Hippos with the Tskhenistsqali (Tskhenistskhali) is widely acknowledged due to the continuity of the onomastic theme 'Horse River'; see Kießling 1913, 1915: 'Pfer-defluß' / 'Čenistsquali'; Braund 1994, 300; Lordkipanidze 1996, 108f.; Braund & Sinclair 1997/2000, *BA* 87 and *Directory*, p. 1229; Radt 2008, 254. Prokop. *Goth.* 8.1.1.6 attests the *Hippis* (sic) as located in the Mocheresis, fordable by men and horses (a folk etymology?); Prokop. *Goth.* 4.14.6.45 specifies the Mocheresis as the best land of Lazika, explicitly including a part of the Rheon and the city Kotaïs / Kytaion / Kutaisi. Controversial is the Glaukos. Lord-kipanidze 1996, 109 and 253–255 suggests identifying it with the Tekhuri, Kvirila, Rheon or Sulori (rejecting the latter on p. 109 n. 174). Similarly, Belfiore 2009, 173 n. 82 (following Lordkipanidze 1972) equates the Glaukos with the Tekhuri, but strangely posits (as one of two possible solutions) a location north of the Tekhuri, Tskhenistsqali and Kyros, thought to be close to Samtredia, but this is on the northern bank of the Phasis, located between the conflu-ence of the Kvirila and the Tekhuri. Roller 2018, 641 decides for the Tekhuri, Radt 2008, 254 for the 'Ziva' (Tsivi?). Koestermann 1965, 323 remains uncertain in both cases ('an Nebenflü-ssen des Phasis'). I have argued elsewhere that the river names Glaukos and Hippos formed part of a flexible Argonautic landscape, defining the location of Aia, just as Lykos and Kyaneos. As such, we must reckon with changing identifications over time. For an older tradition, see Plin. *NH* 6.4.13: *maxime autem inclaruit Aea, XV (milia passuum) a mari, ubi Hippos et Cya-neos vasti amnes e diverso in eum (sc. Phasim) confluunt (... et alios accipit fluvios magnitudine numeroque mirabiles, inter quos Glaucum)*; cf. Ps.-Skylax, *Asia* 81 (on the home of Medeia, without the river names). They seem to be referring to an Aia close to Senaki, which would speak for the Tsivi and Tekhuri as Hippos and Kyaneos. The name Hippos probably moved east with the identification of Kytaion / Kutaisi as the royal city of Kolchis, which yielded the Tskhenistsqali as Hippos and Rheon / Rioni as Glaukos. See Coşkun 2019c and 2020a.

59 This is not explicit in Lordkipanidze 1996, 253, since his formulation 'oberhalb der erwähnten Flüsse' seems to be relating to a location on the middle course of the Rioni.

60 For Vani and its acropolis, see above, n. 15.

61 See LSJ (1996, p. 1865) *s.v.* ὑπέρκειμαι 'to lie above, ... to be placed or situated above or beyond' with some examples. For the notion of 'beyond', see, e.g., Polyb. 4.29.1 for the 'bar-barians dwelling beyond the Macedonians'. For the implication of a difference of altitude ('overhanging' or 'impending'), see, e.g., Polyb. 10.30.2. The 'eyebrow emerging over the eye' (Philostr. *Imag.* 2) aptly illustrates a certain immediacy of the contrast. Also cf. the usage of the verb in Strab. *Geogr.* 12.3.18 (548C), quoted below, n. 67.

altitude of over 1,000 m above sea level, it may, in theory, even claim to do justice to ὑπέρ- twice. This notwithstanding, ὑπέρκειται would have been a counter-intuitive word choice: the Borjomi area is blocked from the Kolchian plain by the northern ranges of the Lesser Caucasus, which reach altitudes of over 2,000 m. There was not even a theoretical sight onto the Phasis, but only potentially on the Kyros, whose bed, however, extended towards the Caspian Sea.

That Lordkipanidze's and Braund's readings imply a high degree of vagueness has an additional reason: their identification of τῶν λεχθέντων ποταμῶν with the Phasis and its tributaries.[62] Before Strabo introduces the location of the Leukotheion with these words, he has already ended his short digression on the course of the Phasis from its Armenian springs to the homonymous city on the coast. He then moves on to describe the southern plains of Kolchis, which those who sail from Phasis to Amisos and Sinope will encounter. In this context, Strabo mentions some anonymous rivers twice, first to explain the smoothness of the Euxine due to their calm and sandy estuaries (which probably contrasted with the outpouring of the Phasis),[63] and second in their capacity as waterways towards the sea. These rivers

---

62 Thus explicitly Lordkipanidze 1996, 253.

63 Many uncertainties relate to the estuaries of Kolchis: how strong were the currents, especially of the Phasis and the Akampsis? In how far did such currents or sand dunes impede navigation along the coast? Sedimentation is particularly strong at the estuaries along the eastern Euxine, as satellite images from *Google Maps* demonstrate; cf. Coşkun 2019a and 2020b. Radt 2008, 254 treats the literary tradition of the sandy shores and considers a *lacuna* in the transmitted text (with Lassère 1981). Dan 2016, 250–255 regards the Phasis mouth as largely stagnant, see Hippokr. *Aer.* 15 (*pace* Philostr. *Imag.* 8 on p. 245) and early modern travel accounts. Arrian's discussion of the sweetness of the sea water near the mouth of the Phasis (*PPE* 8) may also imply the strong force of the stream. The two themes (sweetness and current into the Euxine) are explicitly connected by Prokop. *Goth.* 2.30.4.25f. More difficult to explain is his description of the Akampsis estuary in *Goth.* 8.2.1.8f. (ed. Dewing 1914–1928): Ἄκαμψιν γὰρ αὐτὸν τὸ λοιπὸν καλοῦσιν οἱ ἐπιχώριοι, τούτου δὴ ἕνεκα, ὅτι δὴ κάμψαι αὐτὸν τῇ θαλάσσῃ ἀναμιχθέντα ἀμήχανά ἐστιν, ἐπεὶ ξὺν ῥύμῃ τοσαύτῃ καὶ ὀξύτητι τοῦ ῥοῦ τὰς ἐκβολὰς ποιεῖται, ταραχὴν τοῦ ῥοθίου πολλὴν ἐπίπροσθεν ἐργαζόμενος, ὥστε ὡς πορρωτάτω τῆς θαλάσσης ἰὼν ἄπορον ποιεῖται τὸν ταύτῃ διάπλουν· οἵ τε ναυτιλλόμενοι ἐνταῦθα τοῦ Πόντου, εἴτε Λαζικῆς εὐθὺ πλέοντες εἴτε καὶ ἐνθένδε ἀπάραντες, οὐκέτι ἑξῆς διαπλεῖν δύνανται. / κάμψαι γὰρ τοῦ ποταμοῦ τὸν ῥοῦν οὐδαμῇ ἔχουσιν, ἀλλὰ πορρωτάτω μὲν ἀναγόμενοι τοῦ ἐκείνῃ πελάγους, ἐπὶ μέσον δέ που τὸν Πόντον ἰόντες, οὕτω δὴ ἀπαλλάσσεσθαι τῆς τοῦ ποταμοῦ ἐκβολῆς δύνανται. τὰ μὲν οὖν ἀμφὶ ποταμὸν Βόαν τοιαῦτά ἐστι. 'The locals hereafter call it the Akampsis, and they so name it, obviously, because it is impossible to resist as it enters the sea, given that it discharges its stream with such force and swiftness, causing a great disturbance in the water before it, that it goes out for a very great distance into the sea and makes it impossible to coast along at that point. Those who are navigating in that part of the sea, whether sailing toward Lazike or putting out from there, are not able to hold a straight course in their voyage; / for they are quite unable to cut through the river's current, but they are forced to go out a great distance into the sea there, going somewhere near the middle of it, and only in this way can they escape the force of the river's discharge.' Translation by Dewing & Kaldellis 2014. Prokopios did not have a firm knowledge of the topography, a condition which also resulted in the confusion of the Boas / Akampsis with the Phasis in *Goth.* 2.29.3.19, 23–25; cf. Dewing & Kaldellis 2014, 138 n. 272 and p. 464 n. 740; Coşkun forthcoming b. The lack of familiarity largely results from the fact that most trade fleets avoided sailing along the coast, cutting through from Amisos or Trapezus

therefore include, from north to south, the Mogros / Supsa, Isis / Natanebi, Akinases
/ Kintrishi and possibly the Bathys / Qorolitsqali, to mention the most important
ones between the Phasis and the first navigable river in the south-eastern corner of
the Black Sea, the Akampsis / Tchorokhi / Çoruh Nehri.[64]

4. The South-Eastern Corner of Kolchis with Its 'Green Cape'

It is not entirely clear what formed the southern boundary of Kolchis, but perhaps
it is naïve to surmise that its territory remained stable for centuries, when Milesian,
Achaimenid, Athenian, Sinopean, Seleukid, Mithradatic, Polemonid and Roman
hegemony followed one upon the other. Every single change could have influenced
geopolitics and toponomy. The question is further complicated by the confused but
tenacious literary tradition that the *chora* of Trapezus extended as far as Kolchis.
Strabo is most explicit in this regard.[65] At any rate, it is somewhat surprising that
he does not mention the Akampsis as the first navigable river on the eastern coast
of the Euxine or the strategically important fortress Apsaros (Gonio) near its southern estuary. We do not hear of them either in his outline of Trapezus and the eastern-
Pontic tribes.[66] While uncertainty remains, Strabo confines Kolchis by the Moschi-

to Phasis or Dioskurias / Sebastopolis, see Coşkun, 2020a and 2020b. Prokopios' comment on
the Akampsis mouth is best explained by a Greek reinterpretation ('unbending', 'inflexible')
of a Caucasian name (see Lordkipanidze 2000, 12).

64   The suggested identifications follow Braund & Sinclair 1997/2000, *BA* 87R. *Qorolitsqali* is
rendered *Korilistskali* in *Google Maps* (2018). Our best but still incomplete ancient source is
Arr. *PPE* 7.4–8, which provides a detailed list from all the rivers merging into the Black Sea
between Trapezus and Phasis: 15 stades from the fort of Apsaros to the estuary of the Akampsis,
another 75 to the Bathys, further 90 to the Akinases, moreover 90 to the Isis, once more 90 to
the Mogros, which merges into the Euxine yet another 90 stades before the Phasis. A bit trou-
bled is Plin. *NH* 6.4.12: after mentioning the river *Absarrum* beside the namesake castle, he
lists *flumina Acampseon, Isis, Mogrus, Bathys*. He does not seem to notice that *Absarrum* and
Acampseon denote (at least in part) the same river. Ps.-Skylax, *Asia* 81 further mentions a Leis-
ton, which Braund 1994, 44; 88; 184f.; 349 equates with the Akampsis, but I suggest one of the
rivers to the north instead; see Coşkun forthcoming b.
65   Strab. *Geogr.* 11.2.14 (497C); 11.2.17, 18 (499C); 12.3.28, 29 (555C). Scholars have found
this to be unproblematic in principle (Radt 2008, 252; Roller 2018, 704), although Strabo's
description of Deiotaros' territory in *Geogr.* 12.3.13 (547C) has yielded major debates, beginning with Niese 1883, 579; cf. Lassère 1981, 161; Unger 1896, 249f. (μέχρι Κολχίδος καὶ
τ<ὴν> μικρ<ὰν> Ἀρμενία<ν>), followed, e.g., by Magie 1950, 1237f. and Radt 2008, 364; see
Coşkun, chapter X in this volume, for a new approach. The problem may also relate to an often-
overlooked cartographic inconsistency, which mislocates Trapezus further to the north-east;
see Podossinov 2012. I suggest elsewhere that the confusion roots in Trapezuntine (or Sinopean) ideological toponymy, which pictured Trapezus as located *within* the territory of Kolchis,
as Xen. *An.* 4.8.22 spells it out.
66   Strab. *Geogr.* 11.2.14–19 (497–499C) for Kolchis and 12.3 for Pontos, esp. 12.3.17f. (548C)
for Trapezus and 12.3.18–42 (548-563C) for the non-Greek tribes in the mountainous hinterland; §§ 21–27 are mythical reflections, starting with the Amazons. The digression repeatedly
addresses, if only in passing, Kolchis (§§ 18, 28 bis, 29) and the Moschian Mountains (18). The
vague idea of a shared boundary is further supported by the fact that Strabo, although listing

an Mountains twice in the sections quoted above in the introduction, and once more in his treatment of eastern Pontos. After listing the Tibarenoi, Chaldaioi, Sannoi and Appaitai in the hinterland of Pharnakeia and Trapezus, he goes on as follows:

> 'two mountains cross the country of these people, not only the Skydises, a very rugged mountain, which joins the Moschian Mountains above Kolchis (its heights are occupied by the Heptakometai), but also the Paryadres, which extends from the region of Sidene and Themiskyra to Lesser Armenia and forms the eastern side of Pontos.'[67]

Accordingly, the easternmost extension of the Pontic Mountains (Skydises) are adjacent to the southern boundary of Kolchis (Moschian Mountains). In addition, Strabo regards both Pontos and Kolchis as also contiguous with Armenia.[68] It therefore appears that the lower course of the Akampsis formed the most natural divide between Kolchis and Pontos, whereby the land enclosed by the Akampsis and Apsaros / Acharistsqali formed the northern tip of Armenia.[69] If this division is accepted, the north-western foothills of the Lesser Caucasus reach the sea just north of the mouth of the Akampsis. Its 'Green Cape' (Mtsvane Kontskhi) touches the Euxine slightly north of Batumi between the Bathys (Qorolitsqali) and Chakvistskali Rivers (the latter's ancient name is unknown).

Let us now return to our search for the Leukotheion. Following up the Kolchian coastline north of the Akampsis, the 'Green Cape' takes a prominent position among the outliers of the Lesser Caucasus. It rises steeply from the sea and almost immediately reaches heights of up to 220 m. This hill site some 9 km north of the city centre of Batumi has been occupied by the Batumi Botanical Garden since 1912.[70] It easily allows for views on the Chakvistskali and Kintrishi (Akinases) Rivers. A feasible alternative would be Tsikhisdziri some 8 km north, just past the plain estuary of the Chakvistskali. This is where the foothills of the Lesser Caucasus first touch the shoreline. It is thus the earliest significant elevation for those who sail southwards along the Kolchian Plain. Although the hills that rise immediately above the beach are much lower than the 'Green Cape' and the visibility of a

---

the various tribes (esp. § 18, see below), does not mention the minor Greek coastal settlements between Trapezus and Apsaros (or even Phasis): Hyssou Limen, Rhizaion, and Athenai, on which see Arr. *PPE* 3–7; Prokop. *Goth.* 8.2; Braund & Sinclair 1997/2000; Coşkun 2019a. On the history of the fortress of Apsaros, see Coşkun forthcoming b.

67 Strab. *Geogr.* 12.3.18 (548C): τῆς δὲ Τραπεζοῦντος ὑπέρκεινται καὶ τῆς Φαρνακίας Τιβαρανοί τε καὶ Χαλδαῖοι καὶ Σάννοι, οὓς πρότερον ἐκάλουν Μάκρωνας, καὶ ἡ μικρὰ Ἀρμενία: καὶ οἱ Ἀππαῖται δέ πως πλησιάζουσι τοῖς χωρίοις τούτοις οἱ πρότερον Κερκῖται. διήκει δὲ διὰ τούτων ὅ τε Σκυδίσης, ὄρος τραχύτατον συνάπτον τοῖς Μοσχικοῖς ὄρεσι τοῖς ὑπὲρ τῆς Κολχίδος, οὗ τὰ ἄκρα κατέχουσιν οἱ Ἑπτακωμῆται, καὶ ὁ Παρυάδρης ὁ μέχρι τῆς μικρᾶς Ἀρμενίας ἀπὸ τῶν κατὰ Σιδήνην καὶ Θεμίσκυραν τόπων διατείνων καὶ ποιῶν τὸ ἑωθινὸν τοῦ Πόντου πλευρόν. Translation adapted from Jones 1924.

68 Strab. *Geogr.* 12.3.1 (540f.C); cf. 12.3.13 (547C), where he mentions Trapezusia, Kolchis and Lesser Armenia, on which see above, n. 65.

69 This is at least compatible with Braund & Sinclair 1997/2000, *BA* 87R, although this map puts the names much farther to the west or east respectively.

70 See the website *Batumi Botanical Garden* for more information, though without mentioning the Leukotheion

sanctuary from the open sea would have been more limited, its heights still permit one to gaze as far as Poti, when the wheather is clear.[71]

Either position would thus be a very good match for Strabo's description ὑπέρκειται δὲ τῶν λεχθέντων ποταμῶν ἐν τῇ Μοσχικῇ, which we might translate as '(the sanctuary) lies *beyond and above* the aforesaid rivers, in the Moschian Mountain'. Especially the 'Green Cape' would have provided the sea goddess with the most impressive maritime view from within the Kolchian territory. Of course, my suggestion has to remain hypothetical as long as we do not have corroborating evidence from the ground. But even if such confirmation might never come forth, the two sites aptly illustrate what kind of location we should be looking for.

IV. ARGUMENT 3:
HISTORICAL IMPLICATIONS OF THE TOPOGRAPHY

The two previous sections have established the southern coastline of Kolchis as the most plausible location of the Leukotheion after reflecting on the implications of the mythological and cultic traditions as well as revisiting Strabo's topographical indications. In order to avoid circularity, I have held back an additional argument, namely that the speed of Pharnakes' campaign excludes the possibility of a detour of more than one or two days. The reconstruction of his itinerary is a matter of dispute, more so than that of Mithradates of Pergamon, which is less relevant for us. Having investigated their campaigns in more detail elsewhere,[72] I shall here confine myself to providing concise evidence for my specific claims that Pharnakes passed by the sanctuary twice, in the summers of 48 and 47 BC, that he did not have a week or more to spare in 48 (nor the intention to loot the goddess' treasury), and that he would not have invested more than a day in 47 (when he was in need of refilling his coffers, but even more desperate to join forces with his new recruits in the eastern parts of the Bosporan Kingdom). I shall proceed by addressing previous scholarship grouped according to three different chronological choices.

a) Some historians date Pharnakes' invasion of Kolchis in close proximity to his early conquests which precede his more famous Pontic campaign.[73] There are

71 Archaeological remains in the area go back to the early Iron Age. The near-common opinion also locates the Byzantine fortress of *Petra (Pia Iustiniana)* at Tsikhisdziri; see Inaishvili 1991; Braund 1994, 117 with n. 190; 276 n. 31; 290–295; Braund & Sinclair 1997/2000, *BA* 87 and *Directory*, p. 1237; Tsetskhladze 2013, 294 n. 5 with further bibliography. However, I am arguing elsewhere that Petra is better looked for on the southern bank of the Phasis estuary, where the accounts of Arrian and Prokopios are pointing to; see Coşkun forthcoming b.
72 Coşkun 2019a and forthcoming a.
73 For around 60 BC, see Veh & Brodersen 1987, 475 ns. 590f.; Stein-Kramer 1988, 60. Mid-50s BC: Sullivan 1990, 156; Ballesteros Pastor 2017, 297; 300f. But MacDonald 2005, 45f. questions the relevance of titulature. Saprykin 2002, 34: 55/50 BC. As a theoretical argument, one might add that Appian's account of the Pontic War does not mention Kolchis and even conveys the impression that Pharnakes had sailed straight from the Bosporos to Sinope (App. *Mith.*

indeed good arguments for disconnecting the siege of Phanagoreia from the attack of his Pontic homeland, and linking it with his subjection of Tanaïs and the Scythians on the Taman peninsula. The evidence is too vague to give a precise year for this, but Pharnakes' adoption of the title 'Great King of Kings' around 55 BC provides a probable *terminus ad quem* for his wars on the eastern coast of the Sea of Azov.[74] Kolchis, however, cannot have been affected, because it continued to be ruled by Aristarchos, the appointee of Pompey. The scarce numismatic evidence we have attests him to have been in power at least until 54/53.[75] Since the Kolchians supported Pompey at the Battle of Pharsalos in 48,[76] we can further exclude any other year before 48.

b) The majority's view is that the capture of Kolchis formed the beginning of Pharnakes' Pontic campaign.[77] He marched his field army from the Bosporos – perhaps from Gorgippia or Anapa – along the eastern coast of the Euxine. His infantry and cavalry were flanked by his fleet for easy supplies, also providing swift passage of the rivers through improvised ship bridges. This support was available at least until the mouth of the Iris / Yeşil Irmak in (Armenian or Kappadokian) Pontos was reached. Instead of crossing over into the territory of Amisos, by then a Roman provincial city which resisted the king, they turned south towards the Kappadokian kingdom of Ariobarzanes III. Negotiations with the Roman proconsul Cn. Domitius Calvinus induced Pharnakes to withdraw north to the confluence of the Iris and the Lykos / Kelkit Çayı, following the latter eastwards to Nikopolis. It is in this city's territory that he first confronted and defeated the Romans and their allies. The same battle also provides us with a first chronological marker for Pharnakes' activities in Asia Minor. Cassius Dio notes that Calvinus led away his legions quickly after the combat, 'before winter approached', which means before the first snow fell. This hints at a day in later December (or later October respectively, if adjusted to the

---

120.591). But a comparison with the more detailed narrative of *BAlex.* 34–78 and the brief plot of Cass. Dio 42.45.3 proves that Appian is misleadingly selective; see Coşkun forthcoming a.

74 See App. *Mith.* 108.505–511; 113.555; 114.560f.; Oros. 6.5.2; *Suda*, s.v. *Kastor*; cf. Ballesteros Pastor 1996, 278f. But Gajdukevič 1971, 324 dates the siege of Phanagoreia to ca. 50 BC. Olbrycht 2001, 437 speaks of 'um 48 v.Chr.' (despite his reference to Gajdukevič). Hoben 1969, 12–14 and 15f. is undecided as to how much before the Pontic campaign Phanagoreia was besieged; likewise, MacDonald 2005, 45. Previously, I also opted for the 'Anfangsphase des römischen Bürgerkrieges' (Coşkun 2014, 135). Von Bredow 2000, in turn, dates the conquest of Phanagoreia after Pharsalos, as the other attacks. Abramzon & Kuznetsov 2011, 70 leave the time open, but relate the destruction of the city to the revolt around 63 BC.

75 The latest explicit evidence for Aristarchos is a coin dated to year 12, which Lordkipanize 1996, 293 n. 487 dates to 52/51 BC. More likely, Pompey had appointed him in late in 65, so that 65/4 should be counted as his 1st and 54/3 as his 12th year, see Coşkun, chapter X in this volume. Braund 1994, 169 is undecided between those years, but favours ca. 52/1 BC. For more general information on Aristarchos, see Coşkun 2007/19; Biffi 2010, 54f. and 72.

76 See Cic. *Att.* 9.9.2=176 SB on 48 BC; cf. App. *BCiv.* 2.51.211; cf. Yoshimura 1961, 483. The Tolistobogian king Deiotaros was the protector of Anatolia, probably including its eastern extensions; see Coşkun, chapter X in this volume.

77 See above, n. 74.

Julian calendar).[78] The next noteworthy deed of the king was the siege he laid to Amisos. The turning point of his military endeavours was the Battle at Zela, where Caesar destroyed most of his army: this was on 2 August 47 (or 21 May, Julian)[79] and led to the king's speedy escape from Pontos.

Most scholars are seduced by Appian's vague account, which summarizes the main events of the war 'at the time when Pompey and Caesar were contending against each other'.[80] This timeframe would include any time after the news of the Final Decree of the Senate (7 January 49) had reached Pantikapaion, and most likely result in early spring 49 as the beginning of the march. By this account, there would have been plenty of time to 'visit' the Leukotheion at whichever location in Kolchis.[81] But as we shall see, this early departure is incompatible with a variety of other ancient sources.

c) I agree with those scholars who regard the Battle of Pharsalos as *terminus a quo* for Pharnakes' campaign. This was fought on 9 August 48 BC (7 June, Julian).[82] More accurately, we should speak of the arrival of the news that Pompey had suffered a crushing defeat in the Bosporan Kingdom. The latter is not only implied in the abovementioned coinage from Kolchis, but also in the pro-Caesarian *Bellum Alexandrinum*. According to this war account, Pharnakes' ambassadors were trying to assuage Caesar prior to the Battle of Zela by reminding him that 'Pharnakes had refused to provide Pompey troops against Caesar'.[83] Cassius Dio not only repeats the king's (unsuccessful) diplomacy, but also specifies the acquisition of Kolchis as the first step of Pharnakes' campaign:

> He also acquired Kolchis without any difficulty, and the entire (part of) Armenia which had belonged to Deiotaros, while the latter was absent, and he subjugated <part> of Kappadokia, and some cities of Pontos that had been assigned to the province of Bithynia.[84]

---

78 Cass. Dio 42.46.2f.: καὶ ὁ χειμὼν προσῄει. And see the next n. on the Julian calendar.
79 The Roman calendar had fallen behind the solar year by about two and a half months in 47 BC, see Groebe 1906, 814–817; cf. Judeich 1885; Gelzer 1960, 220–241; 267f.; Bennett 2004, 174; Coşkun forthcoming a.
80 App. *Mith.* 120.591: ᾧ χρόνῳ Πομπήιος καὶ Καῖσαρ ἐς ἀλλήλους ἦσαν. Also see 120.592: ἐπολέμησε δὲ καὶ αὐτῷ Καίσαρι καθελόντι Πομπήιον, ἐπανιόντι ἀπ' Αἰγύπτου. 'He fought with Caesar himself, when the latter had overthrown Pompey and returned from Egypt.'
81 See, e.g., Hoben 1969, 17f.; Stein-Kramer 1988, 60f.; Lordkipanize 1996, 292–295; Freber 1993, 81 n. 388. Goukowsky 2003, 253 n. 1109 speaks of a 'synchronism' as vaguely as Appian, leaving many questions open.
82 The date has been transmitted in the *Fasti Amiterni*, see CIL I², p. 244; cf. Gelzer 1960, 240 with n. 316; also see above, n. 79, for references. The battle is accepted as the *terminus a quo*, e.g., by Gelzer 1960, 235; von Bredow 2000.
83 *BAlex.* 69: *Maximeque commemorabant nulla Pharnacen auxilia contra Caesarem Pompeio dare voluisse, cum Deiotarus, qui dedisset, tamen ei satisfecisset.* Cf. 70: *Monuit autem ... legatos, ne ... nimis eo gloriarentur beneficio, quod auxilia Pompeio non misissent.* Also see Cass. Dio 42.47.3.
84 Cass. Dio 42.45.3: τήν τε Κολχίδα ἀκονιτὶ προσηγάγετο καὶ τὴν Ἀρμενίαν ἀπόντος τοῦ Δηιοτάρου πᾶσαν, τῆς τε Καππαδοκίας <μέρος> καὶ τῶν τοῦ Πόντου πόλεών τινας, αἳ τῷ τῆς Βιθυνίας νόμῳ προσετετάχατο, κατεστρέψατο. Greek text from Cary & Foster 1914 (cf.

The references to the ease of the conquest (ἀκονιτί)[85] and to Deiotaros' absence make it very clear that Kolchis and the Galatian part of the former Pontic Kingdom ('Armenia') were easy prey to the invader due to the Roman civil war and the engagement (or slaughter) of Pompey's allies at Pharsalos.

As a result, Pharnakes did not leave the Bosporos prior to the middle of August 48. This limits the time span leading up to the Battle of Nikopolis to little more than three months. Such a march looks quite ambitious, but Pharnakes was apparently well prepared: he facilitated the march of his land forces by supplying and assisting them from the sea, and was also fortunate enough not to encounter any noteworthy resistance before reaching the mouth of the Iris River. Instead of laying siege to Amisos and thus halting the speed of his attack, he decided to postpone the capture of this city (and of Sinope). He rather turned south, where the Pontic hinterland, Micro-Armenia and Kappadokia lay open to him. Altogether, his foot soldiers may have covered some 1,600 km in about 100 days, which is decent, but by no means spectacular.[86] Note that, still in the 5th century BC, Herodotos surmised 30 days for the march from the Maiotis (Sea of Azov) to the Phasis river (ca. 690 km), which equals an average of 23 km per day.[87] This would have come close to the speed of Pharnakes' men until reaching the Iris valley, whereas further progress through the Pontic mountains would have been slower.

At all events, Pharnakes was trying to occupy as much of his 'inherited' kingdom as possible before Caesar might return from Egypt. A detour to Vani, let alone the Borjomi area, – for whatever riches he might have hoped to find there – is highly implausible. While Vani was of course much quicker to reach, the acropolis on the Akhvlediani's Hill has been called a masterpiece of Hellenistic fortifications, so that its siege and destruction might have taken weeks or months.[88] And yet, I hesitate to concede that he looted the sanctuary on his way along the Kolchian coast. There is no reason to doubt that he was well resourced and further hopeful of expanding his realm. Why, then, should he have harmed a sanctuary that was going to be his anyway, when he was expecting to be received, if not welcomed, as the legitimate king?

Boissevain 1898, vol. 2, 63). The translation is mine and rejects the widespread view that Cass. Dio was speaking of 'Armenia in its entirety' (τὴν Ἀρμενίαν ... πᾶσαν). See Coşkun forthcoming a.

85   Without these arguments, the conquest of Kolchis is normally explained as the beginning of Pharnakes' Pontic campaign, e.g., by Magie 1950, 408f.; Stein-Kramer 1988, 61; von Bredow 2000. Saprykin 2002, 45f. differs, only explaining the speedy conquests in Asia Minor 48/7 BC with the Roman civil war.

86   He may have begun his campaign in Gorgippia or Anapa. *Google Maps* calculate the current land route from Anapa to Samsun (Amisos) as 1,240 km.

87   Hdt. 1.104.1. *Google Maps* calculate 682 km for the route from Temryuk to Poti.

88   See Tstskhladze 1998, 141: 'Es kann also als bewiesen gelten, dass die Siedlung von Vani im 3. Jh. v.Chr. über eine der ingenieurtechnisch vollkommensten Fortifikationsanlagen der hellenistischen Welt verfügte.' We have no reason to believe that the acropolis was no longer functional in the (mid-) 1st century. Also see Balandier 2005, although his account focuses on the 5th and 4th centuries.

d) Against this background, I would like to suggest that the most realistic context for the pillaging of the Leukotheion would be Pharnakes' flight from Asia Minor to the Bosporos. After his defeat by Caesar, he had lost all his stakes in Anatolia, and there was not even a realistic chance of keeping Kolchis, which had been ruled or at least controlled indirectly from Pontos over the last half-century.[89] In addition, his land army, together with most of his cavalry, were lost. As we learn from Appian, he escaped to Sinope together with 1,000 men on horseback. His situation was so desperate that he had all the horses killed before embarking on ships to sail home. The reason for this deed is unclear, but it seems that the land route appeared no longer safe to him, whereas the required number of ships and the according amount of supplies for the horses were not available in Sinope. We should further account for how exhausted the horses would have been after their flight; some of them might have died at all events. Whatever result his negotiations with Calvinus had yielded, there was no trust between the two men, for otherwise Pharnakes would at least have spared the horses.[90]

Until recently, the commonly-held view was that he sailed straight from Sinope to Theodosia on the European side of the Bosporos. There are, however, strong reasons to doubt this. He only had 1,000 men at his disposal when leaving Sinope, and these were badly equipped at that. He would scarcely have launched an immediate attack on the domains that the usurper Asandros was holding. Instead, Appian tells us that he recruited Scythian and Sarmatian reinforcements. He could not find them in the southern parts of the Crimea. Most likely, these were Aorsoi and Sirakoi, who inhabited the north of the Taman Peninsula, and they were certainly mobilized as soon as Pharnakes had heard of the revolt in Pantikapaion in spring 47.[91] Accordingly, we are safe to assume that he left Sinope sailing westward along the Pontic and Kolchian coast. This is a context in which looting sanctuaries would have made sense, if only they lay on his way and would not cause any major delay. He had no hopes of ever returning to Pontos or Kolchis, but if any of their resources could be made available to support his reconquest of the Bosporos, there was no reason for scruples.

The timeline for his return from Sinope to Pantikapaion is about three times as tight as it had been for his outbound way. After his defeat at Zela on 2 August 47 BC (21 May, Julian), he may have reached Sinope on 3 or 4 August, and set to sea one or two days later. Caesar was informed about his death in Nikomedia about the first week of September. Therefore, Pharnakes had at the utmost four weeks to sail to Anapa, Gorgippia or possibly Phanagoreia, unite with his fresh recruits, set over

---

89 *Pace* Braund 1994, 170, who emphasize its connection with the Bosporos, but see n. 76 above for a different view. Braund 1997, 169 dates the sack of the temple by Pharnakes to 47 BC. I assume that he was thinking of a time between the Battles of Pharsalos and Zela.

90 App. *Mith.* 120.590–596, with Coşkun forthcoming a.

91 App. *Mith.* 120.594 on Pharnakes' recruitment of Scythians and Sauromatians; Strab. *Geogr.* 11.5.8 (506C) on the Aorsoi and Sirakoi; cf. Plin. *NH* 4.80. See Coşkun forthcoming a for details. Previously, the recruitments have either been dated to before Pharnakes' Pontic Campaign or located in the European part of the kingdom: cf. Gajdukevič 1971 323f.; Stein-Kramer 1988, 60f.; Braund 1996, 1204; Mielczarek 1999, 73; 80; Olbrycht 2001, 436–438.

to the Crimea, receive the submission of Theodosia and Pantikapaion and engage in battle with Asandros.[92] Once again, this may appear very ambitious, but it was certainly doable, if we remember that Strabo considers the average sailing time from Phasis to Sinope to be between two or three days.[93] Pharnakes would have needed a bit longer, partly due to less favourable currents on his way north, partly because he had to find and potentially fight for supplies on his way. If this is accepted – and I do not see a plausible alternative that accounts for all the sources we have – then there was definitely no time to get to Vani, let alone Borjomi. But he would have regarded a wealthy sanctuary in an exposed coastal location, such as Batumi's Green Cape, as an 'invitation' to help himself.

V. EPILOGUE

I started my search for the Leukotheion in order to test my reconstruction of the campaign of Pharnakes II. The two most authoritative suggestions for its location, Vani and the Borjomi area, are incompatible with the speed with which the king passed by the Kolchian coast in 47 BC. Revisiting our only source, Strabo, and reflecting on the cultic as well as topographic implications, I have concluded that we should start looking for the sanctuary in the sea-facing foothills of the Lesser Caucasus north of Apsaros and south of Pichvnari. My suggestion of the most impressive elevation that touches the sea, the 'Green Cape' which now hosts the Batumi Botanical Garden can, of course, only be tentative. There are other suitable candidates in its environs, such as Tsikhisdziri, whence the Kolchian Plain begins.

I foresee that yet other suggestions for locating the Leukotheion without supporting it by hard evidence from the ground will meet with hesitation. I hope, however, that my arguments maintain their merit, even if my hypothetical identification with Batumi's 'Green Cape' or my alternative suggestion of Tsikhisdziri are not accepted. My study will have served its purpose well, if colleagues feel encouraged to reconsider some important facets of the ancient history of Georgia: the local character of a Leukotheion in Kolchis deserves to be appreciated in its broader mytho-historical context, including potential implications for the origin of its Greek settlers.[94] We should also look at the sites of Vani and Atsquri with a fresh view, being open to the possibilities that their history may be understood better without connections to Leukothea, Phrixos and Pharnakes.

---

92 App. *Mith.* 120.594f. Pharnakes obviously found only little resistance in Theodosia and Pantikapaion, because their subjection cannot be explained by a siege, for which he had neither the time nor the resources.

93 Strab. *Geogr.* 11.2.17 (498C), as quoted above. For these and other examples, cf. Dan 2016, 250.

94 See especially above, n. 34. Cf. Braund 1998, who emphasizes the scarcity of our historical sources and the need to examine more closely the mythical tradition, in order to understand better the conceptualization of Archaic Greek colonization in general (p. 287) as well as the re-invention of foundation myths throughout antiquity (p. 293). His Kolchian examples are, however, confined to the Argonauts (pp. 289, 295).

Dedication & Acknowledgments

I would like to dedicate this chapter to my friend and colleague Dr. Mackenzie Lewis (Toronto / Waterloo), who gave encouraging feedback on this piece at the Waterloo Study Day in August 2019. We were hoping to cooperate more closely on ancient colonialism, but he was suddenly taken away from his family and friends on 8 March 2020. His kindness and generosity will not be forgotten.
My thanks go to Luis Ballesteros Pastor (Sevilla), Eckart Olshausen (Stuttgart), Joanna Porucznik (Wrocław), Vera Sauer (Stuttgart) and Tassilo Schmitt (Bremen) for their constructive criticism and encouraging feedback on earlier drafts of this paper as well as to Mae Fernandez (Waterloo) and Augustine Dickinson (Toronto) for their editorial support. I am very grateful to Chen Stone (Waterloo) for producing the accompanying map (Map 4 at the end of the volume).

Bibliography – Ancient Sources

Belfiore, S. 2009: *Il Periplo del Ponto Eusino di Arriano e altri testi sul Mar Nero e il Bosforo. Spazio geografico, mito e dominio ai confine dell' Impero Romano*, Venice.

Cary, E. & Foster, H.B. 1914: *Cassius Dio*, London. (Loeb ed., drawn from the Perseus Collection)

Dewing, H.B. & Kaldellis, A. 2014: *Prokopios, The Wars of Justinian*. Translated by H.B. Dewing. Revised and Modernized, with an Introduction and Notes, by A. Kaldellis. Maps and Genealogies by I. Mladiov, Indianapolis.

Dewing, H.B. 1914–1928: *Procopius*, London. (Drawn from the Perseus Collection)

Falconer, W. 1805: *Arrian's Voyage around the Black Sea*, Oxford. (Drawn from Wikisource)

Godley, A.D. 1920: *Herodotus*. With an English Translation, Cambridge.

Goukowsky, P. 2003: *Appien, Histoire Romaine*, vol. 7: *Livre XII: La Guerre de Mithridate*, 2nd ed. Paris.

Hamilton, H.C. & Falconer, W. 1903/6: *Strabo*, 3 vols., 1st ed. by H. Bohn, 1854–1857; 2nd ed. London 1903–1906.

Jones, H.L. 1924: *The Geography of Strabo*, vol. 5, Cambridge, MA.

Koestermann, E. 1965: *Cornelius Tacitus, Annalen*, vol. 2: *Buch 4–6*. Erläutert und mit einer Einleitung versehen, Heidelberg.

Lassère, F. 1975: *Strabon, Géographie*, vol. 8: *Livre XI*, Paris.

Lassère, F. 1981: *Strabon, Géographie*, vol. 9: *Livre XII*, Paris.

Liddle, A. 2003: *Arrian. Periplus Ponti Euxini*, London.

Martin, R. 2001: *Tacitus, Annals V & VI*, edited with an introduction, translation and commentary, Warminster.

Meineke, A. 1877: *Strabonis Geographica*, vol. 2, Leipzig.

Radt, S. 2004/8: *Strabons Geographika*, 10 vols., Göttingen 2002–2011 (vol. 3, 2004; vol. 7, 2008).

Roller, D.W. 2014: *The Geography of Strabo*. Translated by D.W. Roller, Cambridge.

Roller, D.W. 2018: *A Historical and Topographical Guide to the Geography of Strabo*, Cambridge.

Silberman, A. 1995: *Arrien, Périple du Pont-Euxin*, Paris.

Szwajcer Marc (ed.) (online edition, undated): *Le Périple de la Mer Noire*. URL: http://remacle.org/bloodwolf/historiens/arrien/periplegr.htm. The edition follows the Greek text of Henry Chotard, *Le Périple de la Mer Noire par Arrien*, Paris 1860.

Veh, O. & Brodersen, K. 1987: *Appian von Alexandria, Römische Geschichte, Erster Teil: Die römische Reichsbildung*. Übersetzt von Otto Veh, durchgesehen, eingeleitet und erläutert von Kai Brodersen, Stuttgart.

Woodman, A.J. 2017: *The Annals of Tacitus, Books 5 and 6*, edited with a commentary, Cambridge.

Bibliography – Modern Scholarship

Abramzon, M. & Kuznetsov, V. 2011: 'The Phanagorian Revolt against Mithridates VI Eupator', in N. Povalahev & V. Kuznetsov (eds.), *Phanagoreia und seine historische Umwelt. Von den Anfängen der griechischen Kolonisation (8. Jh. v.Chr.) bis zum Chasarenreich (10. Jh. n.Chr.)*, Göttingen, 15–90.

Antoni, S. 2006/18: 'Ino [2]', *BNP Online*.

Balandier, C. 2005: 'Les défenses de la terrasse Nord de Vani (Géorgie). Analyse architecturale', in D. Kacharava, M. Faudot & É. Geny (eds.), *Pont-Euxine et Polis. Polis Hellenis and Polis Barbaron. Actes du Xᵉ Symposium de Vani – 23–26 septembre 2002*, Besançon, 245–264.

Ballesteros Pastor, L. 1996: *Mithrídates Eupátor, rey del Ponto*, Granada.

Ballesteros Pastor, L. 2008/19a: 'Pharnakes II, King of Pontos', *APR, s.v.*

Ballesteros Pastor, L. 2008/19b: 'Mithradates (VII) of Pergamon, King of the Trokmoi, King Designate of Kolchis and the Bosporos', *APR, s.v.*

Ballesteros Pastor, L. 2017: 'Pharnakes II and the Title "King of Kings"', *AEW* 16, 297–303.

Bennett, C. 2004: 'Two Notes on the Chronology of the Late Republic', *ZPE* 147, 169–174.

Biffi, N. 2010: *Scampoli die Mithridatika nella Geografia di Strabone*, Bari.

Blázquez, José María 2005: 'The Sanctuary-City of Vani and Its Parallels in the West', in D. Kacharava, M. Faudot & É. Geny (eds.), *Pont-Euxine et Polis. Polis Hellenis and Polis Barbaron. Actes du Xᵉ Symposium de Vani – 23–26 septembre 2002*, Besançon, 235–244.

Boffo, L. 1985: *I re ellenistici e i centri religiosi dell'Asia Minore*, Florence.

Boissevain, U.P. 1895/8: *Cassii Dionis Cocceiani Historiarum Romanarum quae supersunt*, 2 vols., Berlin.

Boyce, M., Chaumont, M.L. & Bier, C. 1989/2012: 'ANĀHĪD', *EncIr*, 1.9, 1989, 1003–1011; last update: 2011; upload: 2012. URL: http://www.iranicaonline.org/articles/anahid.

Braund, D. 1994: *Georgia in Antiquity: a History of Colchis and Transcaucasian Iberia, 550 BC–AD 562*, Oxford.

Braund, D. 1996/2000: 'Map 84 Maeotis' (1996), *BA Directory* 2000, 1201–1212.

Braund, D. 1997/2000: 'Map 88 Caucasia' (1997), *BA Directory* 2000, 1255–1267.

Braund, D. 1998: 'Writing and Re-Writing Colonial Origins', in G.R. Tsetskhladze (ed.), *The Greek Colonisation of the Black Sea Area. Historical Interpretation of Archaeology*, Stuttgart, 287–296.

Braund, D. 2010: 'The Religious Landscape of Phasis', in Petropoulos & Maslennikov 2010, 431–439.

Braund, D. 2018: *Greek Religion and Cults in the Black Sea Region: Goddesses in the Bosporan Kingdom from the Archaic Period to the Byzantine Era*, Cambridge.

Braund, D. & Sinclair, T. 1997/2000: 'Map 87 Pontus-Phasis' (1997), *BA Directory* 2000, 1226–1242.

Bremmer, J.N. 1999: 'Leucothea', *DNP* 7, 110.

Bruneau, P. 1994: 'Phrixos et Helle', *LIMC* 7.1, 398–404 (text) and 7.2, 332–338 (illustrations).

Burcu Erciyas, D. 2007: 'Cotyora, Kerasus and Trapezus: The Three Colonies of Sinope', in D.V. Grammenos & E.K. Petropoulos (eds.), *Ancient Greek Colonies in the Black Sea 2*, vol. 2, Oxford, 1195–1206.

Coşkun, A. 2007/19: 'Aristarchos, Dynast von Kolchis', *APR, s.v.*

Coşkun, A. 2014: 'Kastor von Phanagoreia, Präfekt des Mithradates und Freund der Römer', in N. Povalahev (ed.), *Phanagoreia und darüber hinaus ... – Festschrift für Vladimir Kuznetsov*, Göttingen, 131–138.

Coşkun, A. 2018a: 'Brogitaros and the Pessinus-Affair – Some Considerations on the Galatian Background of Cicero's Lampoon against Clodius in 56 BC (Harusp. resp. 27–29)', *Gephyra* 15, 117–131. URL: https://dergipark.org.tr/en/pub/gephyra/issue/31130.

Coşkun, A. 2018b: 'Mithradates, King of Kolchis', *APR, s.v.*

316 Altay Coşkun

Coşkun, A. 2019a: 'The Date of the Revolt of Asandros and the Relations between the Bosporan Kingdom and Rome under Caesar', in M. Nollé, P.M. Rothenhöfer, G. Schmied-Kowarzik, H. Schwarz & H.C. von Mosch (eds.), *Panegyrikoi Logoi. Festschrift für Johannes Nollé zum 65. Geburtstag*, Bonn, 125–146.

Coşkun, A. 2019b: 'Pontic Athens – An Athenian Emporion in Its Geo-Historical Context', *Gephyra* 18, 11–31. URL: https://dergipark.org.tr/tr/pub/gephyra/issue/49781.

Coşkun, A. 2019c: 'Phasian Confusion: Notes on Kolchian, Armenian and Pontic River Names in Myth, History and Geography', *Phasis* 21–22, 73–118 (with maps on pp. 111a, 111b). URL: http://phasis.tsu.ge/index.php/PJ/issue/view/569.

Coşkun, A. 2020a: '(Re-) Locating Greek & Roman Cities along the Northern Coast of Kolchis'. Part I: 'Identifying Dioskourias in the Recess of the Black Sea', *VDI* 80.2, 354–376.

Coşkun, A. 2020b: '(Re-) Locating Greek & Roman Cities along the Northern Coast of Kolchis'. Part II: 'Following Arrian's Periplous from Phasis to Sebastopolis', *VDI* 80.3, 654–674.

Coşkun, A. forthcoming a: 'The Course of Pharnakes' Pontic and Bosporan Campaigns in 48/47 BC'. *Phoenix* 73.1–2, 2019 (ca. Dec. 2020), 86–113.

Coşkun, A. forthcoming b: 'Akampsis, Boas, Apsarus, Petra, Sebastopolis: Rivers and forts on the southern littoral of Colchis', forthcoming in J. Boardman, A. Avram, J. Hargrave and A. Podossinov (eds.), *Connecting East and West. Studies Presented to Prof. Gocha R. Tsetskhladze*, Leuven 2021, chapter 15.

Coşkun, A. in preparation a: 'Stains on the Mytho-Biography of Ino-Leukothea: The Three Wives of Athamas (Nephele, Themisto, Ino) and the Plot against Phrixos Revisited'.

Dan, A. 2016: 'The Rivers Called Phasis', *AWE* 15, 245–277.

de Callataÿ, F. 1997: *L'histoire des guerres mithridatiques vue par les monnaies*, Louvain-la-Neuve.

De Siena, A.A. 2001: 'Medea e Medos, eponimi della Media', in G. Traina (ed.), *Studi sull'XI libro dei Geographika di Strabone*, Lecce, 85–94.

Dräger, P. 2001: *Die Argonautika des Apollonios Rhodios. Das zweite Zorn-Epos der griechischen Literatur*, Leipzig.

Dundua, G.F. & Lordkipanidze, G.A. 1979: 'Hellenistic Coins from the Site of Vani, in Colchis (Western Georgia)', *NC* 7.19 (139), 1–5.

Ehrhardt, N. 1988: *Milet und seine Kolonien*, 2nd ed. Frankfurt.

Eitrem, S. 1925: 'Leukothea', *RE* 12.2, 2293–2306.

Freber, P.-S.G. 1993: *Der hellenistische Osten und das Illyricum unter Caesar*, Stuttgart.

Gajdukevič, V.F. 1971: *Das Bosporanische Reich*, 2nd ed. Berlin.

Gantz, T. 1993: *Early Greek Myth. A Guide to Literary and Artistic Sources*, Baltimore.

Gamkrelidze, G.A. 1992: 'Hydroarchaeology in the Georgian Republic (the Colchian littoral)', *International Journal of Nautical Archaeology* 21, 101–109.

Gelzer, M. 1960: *Caesar. Der Politiker und Staatsmann*, Stuttgart 1921, 6th ed. Wiesbaden 1960.

Graves, R. 1952/71: *The White Goddess. A Historical Commentary of Poetic Myth*, 3rd ed. London 1952, repr. 1971.

Groebe, P. 1906/64: 'Der römische Kalender in den Jahren 65–43 v.Chr.', in W. Drumann (ed.), *Geschichte Roms in seinem Übergange von der republikanischen zur monarchischen Verfassung, oder: Pompeius, Caesar, Cicero und ihre Zeitgenossen nach Geschlechtern und mit genealogischen Tabellen*, vol. 3, 2nd ed. by P. Groebe, Berlin 1906, repr. Hildesheim 1964, 755–827.

Heinen, H. 1994: 'Mithradates von Pergamon und Caesars bosporanische Pläne. Zur Interpretation von Bellum Alexandrinum 78', in R. Günther & S. Rebenich (eds.), E fontibus haurire. *Beiträge zur römischen Geschichte und zu ihren Hilfswissenschaften*. FS H. Chantraine, Paderborn, 63–79.

Henkel, M. 2007: 'Medeas Gold. Neue Funde aus Georgien. Presse-Ankündigung der Ausstellung', Staatliche Museen zu Berlin, Generaldirektion, 27.2.2007. URL: https://www.smb.museum/ausstellungen/detail/medeas-gold.

Herrmann, A. 1933: 'Moschoi', *RE* 16.1, 351f.

Hoben, W. 1969: *Untersuchungen zur Stellung kleinasiatischer Dynasten in den Machtkämpfen der ausgehenden römischen Republik*, Diss. Mainz.

Inaishvili, N. 1991: *Petra-Tsikhisdziri?*, Tbilisi. (*non vidi*)

Judeich, W. 1885: *Caesar im Orient. Kritische Übersicht der Ereignisse vom 9. August 48 bis October 47*, Leipzig.

Kakhidze, E. & Mamuladze, S. 2010: 'Specimens Related to Cult from the Fort of Apsaros', in Petropoulos & Maslennikov 2010, 455–464.

Keyßner, K. 1941: 'Phrixos (1)', *RE* 19.1, 763–769.

Kießling, E. 1913: 'Hippos (6)', *RE* 8.2, 1915–1917.

Krapivina, V.V. 2010: 'Home Sanctuaries in the Northern Black Sea Littoral', in Petropoulos & Maslennikov 2010, 127–170.

Krauskopf, I. 1981: 'Leukothea nach den antiken Quellen', in *Akten des Kolloquiums zum Thema "Die Göttin von Pyrgi", Tübingen 16.–17.1.1979*, Florence, 137–148.

Lazarenko, I., Mircheva, E., Encheva, R. & Sharankov, N. 2010: 'The Temple of the Pontic Mother of Gods in Dionysopolis', in Petropoulos & Maslennikov 2010, 13–62.

Licheli, V. 2007: 'Hellenism and Ancient Georgia', in D.V. Grammenos & E.K. Petropoulos (eds.), *Ancient Greek Colonies in the Black Sea*, vol. 2, Oxford 2007, 1083–1142.

Lordkipanidze, O. (ed.) 1972: *Vani. Arqeologiuri gatkhrebi* [Vani. Archaeological Excavations], vol. 1, Tbilisi (in Georgian, with Russian and English summaries). (*non vidi*)

Lordkipanidze, O. 1991: 'Vani: An Ancient City of Colchis', *GRBS* 32, 151–195.

Lordkipanidze, O. 1996: *Das alte Georgien (Kolchis und Iberien) in Strabons Geographie. Neue Scholien*, deutsch von Nino Begiaschwili, Amsterdam.

Lordkipanidze, O. 2000: *Phasis. The River and the City*, Stuttgart.

MacDonald, D. 2005: *An Introduction to the History and Coinage of the Kingdom of the Bosporus. Including the Coinage of Panticapaeum (with "Apollonia" and "Myrmecium"), Phanagoria, Gorgippia, Sindicus Limen or the Sindoi, Nymphaeum, Theodosia, and the Kings of the Cimmerian Bosporus*, Lancaster.

Maslennikov, A.A. 2010: 'Ancient Rural Sanctuaries of the Crimean Azov Coast', in Petropoulos & Maslennikov 2010, 171–282.

Mielczarek, M. 1999: *The Army of the Bosporan Kingdom*, Łódź.

Moga, I. 2012: 'Strabo on the Persian Artemis and Mên in Pontus and Lydia', in G.R. Tsetskhladze (ed.), *The Black Sea, Paphlagonia, Pontus and Phrygia in Antiquity. Aspects of Archaeology and Ancient History*, Oxford, 191–195.

Molev, E.A. & Moleva, N.V. 2010: 'Sacral Complexes of Kytaia', in Petropoulos & Maslennikov 2010, 295–334.

Morel, J.-P. 2006: 'Phoenician Colonization', in G.R. Tsetskhladze (ed.), *Greek Colonisation. An Account of Greek Colonies and Other Settlements Overseas*, vol. 1, Leiden, 358–428.

Morison, W.S. 2011: 'Pherekydes', *BNJ* 3. (28 Dec. 2018)

Nawotka, K. 2005: Rev. of Otar Lordkipanidze, Phasis. The River and City in Colchis, *Revue belge de philologie et d'histoire*, 83.1, 234–236.

Nercessian, A. 1990: 'Ino', *LIMC* 5.1, 657–661 (text) and 5.2, 440f. (illustrations).

Nicolai, R. & Traina, G. 2000: *Strabone. Geografia. Il Caucaso e l'Asia Minore, libri XI–XII introduzione, traduzione, note e indici*, Milan.

Niese, B. 1883: 'Straboniana', *RhM* 38, 567–602.

Olbrycht, M.J. 2001: 'Die Aorser, die Oberen Aorser und die Siraker bei Strabon. Zur Geschichte und Eigenart der Völker im nordostpontischen und nordkaukasischen Raum im 2.–1. Jh. v.Chr.', *Klio* 82.2, 425–450.

Petropoulos, E.K. & Maslennikov, A.A. (eds.) 2010: *Ancient Sacral Monuments in the Black Sea*, Thessalonike.

Plontke-Lüning, A. 2001: 'Sarapanis', *DNP* 11, 52f.

Podossinov, A.V. 2012: 'Bithynia, Paphlagonia and Pontus on the Tabula Peutingeriana', in G.R.
  Tsetskhladze (ed.), *The Black Sea, Paphlagonia, Pontus and Phrygia in Antiquity. Aspects of
  Archaeology and Ancient History*, Oxford, 203–206.
Robu, A. & Bîrzescu, I. (eds.) 2016: *Mégarika: nouvelles recherches sur Mégare, les cités de la
  Propontide et du Pont-Euxin. Archéologie, épigraphie, histoire. Actes du colloque de Mangalia
  (8–12 juillet 2012)*, Paris.
Rousyaeva, A.S. 2010: 'Sanctuaries in the Context of the Cultural and Historical Development of
  Olbia Pontica', in Petropoulos & Maslennikov 2010, 63–92.
Saprykin, S.J. 2002: *Bosporskoe tsarstvo na rubezhe dvukh épokh* (The Kingdom of Bosporus on
  the Verge of Two Epochs), Moscow.
Saprykin, S.J. 2010: 'Male Deities and Their Cults on the South Black Sea Coast: Hellenistic and
  Roman Periods', in Petropoulos & Maslennikov 2010, 465–514.
Sens, Ulrich 2009: *Kulturkontakt an der östlichen Schwarzmeerküste*, Langenweißbach.
Stein-Kramer, M. 1988: *Die Klientelkönigreiche Kleinasiens in der Außenpolitik der späten
  Republik und des Augustus*, Berlin.
Sullivan, R.D. 1990: Near Eastern Royalty and Rome, 100–30 BC, Toronto.
Tsetskhladze, G.R. 1995: Rev. of Braund 1994, *CR* 45, 358–360.
Tsetskhladze, G.R. 1998: *Die Griechen in der Kolchis*, Amsterdam.
Tsetskhladze, G.R. 2013: 'The Greeks in Colchis Revisited', *Il Mar Nero* 8, 2010/1 (2013), 293–
  306.
Tsetskhladze, G.R. 2018: 'The Colchian Black Sea Coast: Recent Discoveries and Studies', in M.
  Manoledakis, G.R. Tsetskhladze & I. Xydopoulos (eds.), *Essays on the Archaeology and
  Ancient History of the Black Sea Littoral*, Leuven, 425–545.
Unger, G.F. 1896: 'Umfang und Anordnung der Geschichte des Poseidonios, V. Zeit der Reise an
  den Ocean', *Philologus* 55, 245–256.
von Bredow, I. 2000: Pharnakes (2), *DNP* 9, 752f.
Węcowski, M. 2009: 'Hippias of Elis', *BNJ* 6, 2009.
West, S. 2003: '"The Most Marvellous of All Seas": the Greek Encounter with the Euxine', *G&R*
  50, 151–167.
Yoshimura, T. 1961: 'Die Auxiliartruppen und die Provinzialklientel in der Römischen Republik',
  Historia 10, 1961, 473–495.

Open-Access Websites Cited

Batumi Botanical Garden 2016. URL: http://www.bbg.ge/en/home.
Google Maps. URL: https://www.google.com/maps.
Perseus Collection – Greek and Roman Materials. URL: http://www.perseus.tufts.edu/hopper/
  collection?collection=Perseus%3Acollection%3AGreco-Roman.
The Latin Library. URL: thelatinlibrary.com.
theoi.com. URL: http://www.theoi.com/Pontios/Leukothea.html.
Wikipedia: https://www.wikipedia.org.
Wikisource: https://en.wikisource.org.

D. Cultural Change in Late Antiquity

CHRISTIANITY AND URBAN CHANGES
IN LATE ROMAN SCYTHIA MINOR

Dan Ruscu

Abstract: The rise of Christianity in the Black Sea region resulted in a series of changes in the life of the cities situated on its shores. In previous scholarship, the process of urban 'Christianization' has been described through a number of developments: the emergence of churches, either in the peripheries or in more central locations of cities; the 'conquest' of the socially relevant areas of the polis by Christian monuments; the disappearance of the public buildings connected to the autonomous status of the city; the expansion of the patronage of the local bishop over civic life; and the transformation of the entire urban structure. Focusing on the main cities of the province of Scythia Minor from its creation under Diocletian and Constantine until its demise at the beginning of the 7th century, one ought to re-evaluate how much the rise of Christianity determined urban change in Late Antiquity and to what degree the development actually followed the lines listed above. I argue that factors other than Christianity caused more of the changes in the urban landscape of Scythia Minor. The Christian buildings erected in the 4th and 5th centuries rather seem to have filled voids, in a context where the Classical urban structures had already disappeared due to the destructions of the second half of the 3rd century and the reconstructions under Diocletian and Constantine.

Абстракт: Христианство и городские перемены в позднеримской Малой Скифии: Развитие христианства в Черноморском регионе привело к ряду изменений в жизни городов, расположенных на его берегах. В предыдущих исследованиях процесс городской «христианизации» описывался рядом событий: появлением церквей на перифериях или в центральных районах городов; «завоеванием» социально значимых районов полиса христианскими памятниками; исчезновением общественных зданий, связанных с автономным статусом города; расширением покровительства местного епископа общественной жизни; преобразование всей городской структуры. Настоящая глава посвящена основным городам провинции Малая Скифия от ее создания при Диоклетиане и Константине до их кончины в начале VII-го века. Цель главы состоит в том, чтобы установить, в какой степени городские изменения в поздней античности были обусловлены ростом христианства, а в какой степени развитие фактически следовало пунктам, перечисленным выше. Автор утверждает, что изменения в городском ландшафте Малой Скифии были вызваны в основном, помимо христианства, другими факторами. Христианские здания, возведенные в IV-м и V-м веках, скорее всего, заполнили пустые места, в то время как классические городские структуры уже исчезли из-за разрушений во второй половине III-го века и из-за реконструкций города при Диоклетиане и Константине.

I. INTRODUCTION

The Christianization of the Roman Empire resulted in a thorough transformation of society, which affected many areas of private and public life over time. The aim of the present paper is to explore the process of the Christianization of the urban space in the province of Scythia Minor.[1] The chronological frame of our investigation begins with the creation of the province under Diocletian (284–305) and Constantine (306–337) and is taken until its demise at the beginning of the 7th century. The most tangible change is certainly the emergence of churches especially in the course of the 4th and 5th centuries, either in the peripheries or in more central locations of the towns. Traditionally, these developments were regarded as a kind of a 'conquest' of the socially relevant areas of the polis by Christian monuments. However, I would like to revisit this perceived wisdom and ask more precisely if these first Christian monuments indeed transformed urban space in Scythia Minor. When can we truly talk of a 'Christianization' of urban structures? And which would be its indicators?

This kind of research faces several difficulties. To assess social changes caused by religion and reflected in the material culture involves a certain level of subjectivity and speculation.[2] For Scythia Minor (see Map 5 at the end of this volume), an additional challenge is the limited accessibility of archaeological data. In two of the main cities of the province, Tomis and Kallatis, the ancient structures are covered by the modern cities, which seriously constrains the possibility of archaeological research. Moreover, the results of previous excavations were not rarely published quite unsystematically. Apart from a few recent exceptions, Christian monuments were the focus of excavations and have been interpreted mostly separate from their context, as isolated monuments. With this present paper, I would like to present a more integrated approach and call for a shift of perspective on the late-antique 'Christian' cities. I shall concentrate on the three colonies on the Black Sea shore Tomis, Kallatis and Istros and further on Tropaeum Traiani. Our premise is that these settlements are representative for the general evolution of the late-Roman cityscape of Scythia Minor.

II. TOMIS

The cities of Scythia Minor were greatly affected by the events at the end of the 3rd and the beginning of the 4th centuries, from the Gothic raids to the reorganization of the Empire and the restructuring of the northern frontier under Diocletian

---

1   For the transformation and Christianization of the city in Late Antiquity, see Spieser 1986; *Actes XI Congr.*; Liebeschuetz 2001; Brands & Severin 2003. For the Late Roman province of Scythia Minor, see Zahariade 2006. For Christianity in Scythia Minor, see Oppermann 2010; Born 2012.
2   See Brenk 2003, 3–6; Bowden 2001.

and Constantine.[3] When the new province of Scythia Minor was established under the first Tetrarchy, Tomis became the residence of the provincial governor.[4] The new status of the city was marked, among other things, by the construction of a new wall, doubling the urban area and including the Hellenistic and Roman cemeteries.[5] The building of the new defence structures was conducted under the supervision of the *dux Scythiae*, a sign of imperial involvement and, implicitly, of the fact that the civic community was no longer the only actor in this kind of activity.[6] During the 4th century, construction work was also done in the harbour area, indicating prosperity and strong commercial engagements.[7]

The Christian community of Tomis was already led by a bishop in the 3rd century[8] and the city remained the only episcopal residence of Scythia Minor until the end of the 5th, when it became the metropolitan centre of a larger ecclesiastical organization.[9] Tomis also enjoyed a high prestige among the communities of Scythia Minor due to the significant number of its martyrs during the great persecutions.[10] Written accounts by Sozomenos and Zosimos mention church buildings in Tomis beginning with the second half of the 4th century, under Valens (364–378) and Theodosius I (379–395). But these cannot be identified archaeologically,[11] so that nothing can be said about their position or importance in the context of the late antique city.

The earliest archaeologically attested ecclesiastical building dates from the end of the 4th or the beginning of the 5th century, and is only partially preserved as a round apse and a crypt with painted decoration. It has been discovered in the courtyard of the 'M. Eminescu' gymnasium (Fig. 1, T1).[12] In the same area, fragments of an ambo from the 5th–6th centuries, made of Proconnesian marble, were discovered. They are supposed to have belonged to a late reconstruction phase of the same monument.[13] Archaeological excavations revealed other remains of a single-aisled basilica with an adjacent room. Its location is the area of the Palace Hotel, including the remains of a plastered floor and some marble panels, probably dating from the 5th century (Fig. 1, T2).[14]

---

3   Suceveanu & Barnea 1991, 178–208; Zahariade 2006, 61–114.
4   Suceveanu & Barnea 1991, 195f.; Zahariade 2006, 71f.
5   Rădulescu 1995/6, 84f.; Papuc & Lungu 1998, 207; Toma 2010, 60.
6   *IGLR* 3: the *dux* of Scythia, C. Aurelius Firminianus, supervised the reconstruction of the city gates; Suceveanu & Barnea, 1991, 195f.; Born 2012, 22–25. Nonetheless, the polis institutions were still functioning in this period, as attested epigraphically: *IGLR* 1, 4.
7   Poulter 2001, 115; Zahariade 2006, 76–78; Buzoianu & Bărbulescu 2012, 90–100; 196–206.
8   Popescu 1994, 201f.
9   Popescu 1994, 124–156; 201f.; 211–214.
10 Bratož 2004, 236–239.
11 Sozom. *Hist. eccl.* 6.21.3f. (vol. IV, p. 739f., ed. Hansen); Zos. 4.40.5 (p. 197, ed. Mendelssohn). See Born 2012, 47–49.
12 I. Barnea 1979, 132; Oppermann 2010, 61 with the previous literature; Born 2012, 55.
13 I. Barnea 1979, 132, Pl. 48; Barsanti 1989, 197; Oppermann 2010, 61; Born 2012, 55.
14 Lungu 1998, 456f.; Oppermann 2010, 60; Born 2012, 58.

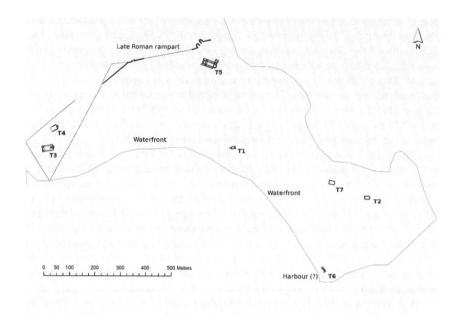

*Fig. 1: Churches of Late Roman Tomis. © Ioana Oltean 2019, with permission.
T1: 4ᵗʰ/5ᵗʰ-century church (now courtyard of M. Eminescu Gymnasium); T2: 5ᵗʰ-century church
(now Palace Hotel); T3: 'Great Basilica', 5ᵗʰ/6ᵗʰ cent.; T4: 'Small Basilica', 5ᵗʰ/6ᵗʰ cent.; T5:
church next to amphitheatre, 5ᵗʰ/6ᵗʰ cent.; T6: harbour basilica, 6ᵗʰ cent.; T7: 6ᵗʰ-century church
(now on str. 22 Decembrie 1989).*

At the end of the 5th century and the beginning of the 6th, during the reigns of Anastasius (491–518) and Justinian (527–565), the province of Scythia Minor enjoyed a period of prosperity, which is still reflected in the increased building activities.[15] For Tomis, this era meant not only the repairing and reconstructing of the city walls,[16] but also the erection of several new buildings by the Christian community.

The first to be mentioned are two churches constructed in the western area of the city, next to the city wall.[17] The so-called 'Great Basilica', dated to the time

15 I. Barnea 1960; Vulpe & Barnea 1968, 411; Suceveanu & Barnea 1991, 171.
16 *IGLR* 7, 8, 9; Vulpe & Barnea 1968, 423f.; Cheluță-Georgescu 1977; Rădulescu 1995/96, 86f.; Cliante 2006; Born 2012, 53.
17 The existence of two different alignments of the city wall in the western area of Tomis, as argued by Cheluță-Georgescu 1977, 258f. and Rădulescu 1995/6, 85f., is still to be confirmed. Toma 2010, 59 convincingly sustains a single alignment, prior to both churches.

from the second half of the 5th to the beginning of the 6th century,[18] was a three-aisled church, with a narthex, an atrium,[19] and a crypt under the altar area (Fig. 1, T3).[20] Whereas imported marble was used for the construction of the church,[21] the crypt was built using spolia, apparently from dismantled public buildings.[22] It is worth mentioning that the church was erected over a paved space, enclosing a well that dates from the 3rd–4th centuries. These structures most probably belonged to the first reconstruction phase of the city under the Tetrarchy.[23] A second church, called the 'Small Basilica', which most probably dates from the same period, was built to the north of the 'Great Basilica', next to the city wall (Fig. 1, T4). It was a single-aisled church with a crypt, built at least partially with spolia.[24] Both churches were erected after the construction of the city wall, and their location was thus adapted to the extant structure of this area of the city. This is indicated mainly by the 20° deviation of the 'Small Basilica' from the East-West direction, in comparison to the 6° of the 'Great Basilica',[25] a deviation most certainly related to the alignment of the city wall in the area.

The function of the two churches has been interpreted in many ways, from the 'Great Basilica' as a cathedral, to the 'Small' one as a cemeterial church, to both as part of an episcopal complex.[26] Their close location may very well indicate the fact that they belonged to a common ensemble. However, neither a baptistery nor any other elements indicating an episcopal residence have been discovered in the area, which rather seems to have housed craftsmen's workshops.[27] Therefore, the hypothesis of an *episcopium* cannot be supported.[28] This observation notwithstanding, the presence of crypts in both churches probably indicates that they served as memorial monuments. It is quite possible that they were even connected with the cemetery to the north of the corresponding sector of the city wall.[29]

---

18 Rădulescu 1965, 61; I. Barnea 1979, 128; Sâmpetru 1994, 91; Oppermann 2010, 58; Born 2012, 50.
19 Rădulescu 1965, 28–32; I. Barnea 1979, 128; Born 2012, 49.
20 Rădulescu 1965, 32–45.
21 I. Barnea 1979, 130; Barsanti 1989, 147; 163; 201f.; 208; Born 2012, 50.
22 Rădulescu 1965, 39f.; 60–62; Oppermann 2010, 58; Born 2012, 50.
23 Rădulescu 1965, 53–60; Papuc 2000–2001, 442 connects the well with a new water supply system constructed during the extensive changes in Tomis at the end of the 3rd or the beginning of the 4th centuries; Oppermann 2010, 59; Born 2012, 49f.
24 For the church: Rădulescu 1965, 23–27; Oppermann 2010, 58f.; Born 2012, 51. For the *spolia*, see Rădulescu 1965, 25; *ISM* II 90.
25 Rădulescu 1965, 23; 28.
26 For the 'Great Basilica': Rădulescu 1965, 77–82; Rădulescu 1995/6, 87f.; I. Barnea 1991, 272; Born 2012, 51. For the 'Small Basilica', see Cheluță-Georgescu 1977, 259 (cemeterial church); *contra*: Rădulescu 1995/96, 87f.; Oppermann 2010, 59.
27 Several kilns were unearthed around the two Christian buildings. Probably, there were workshops next to the city walls; see Rădulescu 1965, 6–23; Oppermann 2010, 59; Born 2012, 53 with several analogies.
28 For the structure of episcopal residences, see Caillet 2010; Real 2003.
29 Born 2012, 51f. For the cemetery, see Cheluță-Georgescu 1977; Rădulescu 1995/6, 88, n. 19.

The largest Christian building of Tomis was erected near the former amphitheatre of the city around the same time, perhaps on the verge from the 5th to the
6th centuries (Fig. 1, T5).[30] The entire area was organized as residential after the
building of the late antique perimeter wall. The church occupied quite a prominent
position in front of the main city gate.[31] Architectonic elements from the then
dismantled amphitheatre were used for the church's construction.[32] No doubt, we
here witness the replacement of the public spectacles by Christian ceremonies. In
this context, the church became a new point of social interaction for the citizens of
Tomis instead of the amphitheatre.[33] A commemorative function can be associated with the new basilica, related to the Christian martyrs attested for Tomis, many
of which could have met their death in the same amphitheatre.[34] Another factor to
be reckoned with is the influence of the bishop in the city at the time when the
basilica was constructed. He must have had considerable authority in order to use
this formerly public area as well as the materials from a public building for the
new construction.[35] Being the largest church discovered in Tomis, it is supposed
to be the city's cathedral.[36]

Also in the 6th century, a basilica was erected in the harbour area, superimpos-
ing a building from the period of the reconstruction of the harbour complex from
the 4th century (Fig. 1, T6).[37] And a crypt, found on the former Karl Marx street
(now '22 Decembrie 1989'), situated over a building with a heating installation,
and dating from the middle of the 6th century, belonged to another ecclesiastical
construction erected in this time of prosperity for the province (Fig. 1, T7).[38]

It is difficult to trace the penetration of Christianity into the public space of
Tomis, especially since the stratigraphy is not always known, and the archaeological excavations are permanently inhibited by the presence of the modern city.
Consequently, the changes that occurred in the urban structure of Tomis after the
reconstruction under Diocletian and Constantine cannot be fathomed. Since there
are no indications regarding the location of the public buildings of the city, we
cannot asses how significant the position of the two earliest churches (the build-

---

30 Rădulescu 1991, 29–34; Oppermann 2010, 59; Born 2012, 59–61.
31 For the gate, see Papuc & Lungu 1998; Toma 2010, 61.
32 Rădulescu 1991, 35.
33 A parallel may be the evolution of the theatres in the Eastern Roman Empire: they were still
   in use in the 4th–6th centuries, from Greece to the Middle East, but the new Christian leaders
   of the cities gradually replaced this kind of public performances with their own religious
   feasts and processions; see Jacobs 2014, 195–202. For similar situations regarding amphithea-
   tres in Moesia II and Macedonia, see Born 2012, 60.
34 Oppermann 2010, 59; Born 2012, 60f. For analogies, see Vaes 1984/6, 320f. For the martyrs
   of Tomis, see Bratož 2004, 125–128; 236–239.
35 Born 2012, 61. For the local authority of bishops concerning buildings under Justinian, see
   Brenk 2003a, 90.
36 Suceveanu & Barnea 1991, 197; Oppermann 2010, 59.
37 Oppermann 2010, 60 with the previous literature; Born 2012, 58.
38 Rădulescu & Lungu 1989, 2573–2576; Oppermann 2010, 59f.; Born 2012, 56f.; Achim &
   Cliante 2017, 73–78 with the previous literature.

ings identified at the 'Mihai Eminescu' gymnasium and near the Palace Hotel, T1 and T2) had in the context of the late antique city. Neither do we have stratigraphic information on what was underneath the two churches. The only aspect that can be evaluated is the fact that they were placed in an area formerly inhabited in the Roman Imperial era.[39] Most probably, some space had therefore to be made available for their erection. This indicates that the Christian community of Tomis, enjoying the authority of the single bishop of the province, gained an increased influence in the city towards the end of the 4th century, undoubtedly in connection with the dominant position attained by Christianity under the reign of Theodosius.[40] However, it is impossible to tell whether these earliest-attested Christian buildings occupied privileged positions and exerted a dominant influence on the reorganization of the civic space.

The changes that might have occurred in Tomis in the 5th and 6th centuries cannot be established either, but it is evident that the Christian community of the city arrived at a certain level of prestige and social status. The churches of this period superimposed or replaced older buildings, some of them public ones, like the amphitheatre. In most cases, the previous buildings had already ceased to function by the time the churches were erected, so that materials from them could be reused for the Christian monuments. The use of imported marble for the second phase of the church at the 'Mihai Eminescu' gymnasium and for the 'Great Basilica' indicates not only the prosperity of the Tomitan community – which was to be expected, given the status of the city – but also its willingness to invest in this kind of building.[41] This is a sign both of the importance of the churches for the city and of the influence of the local Church or its leadership.

The looks of Tomis under Anastasius and Justinian was, without any doubt, shaped by Christianity. The churches became landmarks everyone could discern when entering the city: The 'Great' and 'Small Basilica' complex as well as the church at the amphitheatre are situated in quite visible locations, near the city gates. They were thus connected to some of the main roads entering the city.[42] The churches of this period are scattered over the entire urban area, indicating that the Christians had taken it into possession. The number of churches functioning in this period and their positioning throughout the city area likewise indicate the dominant role Christianity was playing then. We know of several examples attesting Christian processions between the different churches of a city. They thus became cornerstones of the sacred cityscape, which had become fully Christian by then.[43] The initially peripheral locations of the churches need not suggest a marginal status of the Christian community in Tomis, but rather the dissolution of the older structure and the emergence of a new one, defined by Christianity.

---

39   For the different perimeters of Tomis in antiquity, see Toma 2010, 60f.
40   Jacobs 2014, 195; 204.
41   See Bowden 2001, 61; 65f.
42   Toma 2010, 61–63; 68–71.
43   Baldovin 1987, 229–234; Severin 2003, 257; Andrade 2010; Jacobs 2014, 206f.; 214.

III. KALLATIS

If the archaeological data for Tomis are not very rich, they are downright sparse for Kallatis. There is a single known church, next to the reconstructed city wall,[44] superimposed over some older structures from the Hellenistic period through to the 4th century.[45] The building, having a basilical plan, with analogies in Syria, was erected in the 5th century. Its form was affected by the limited space between the city rampart to the north and the street to the south.[46] The church was thus erected in an already established urban structure, without modifying it in any respect. Two capitals of Proconnesian marble were found in its atrium, dating from the 6th century, most probably destined for an intended but never fulfilled reconstruction or redecoration.[47]

IV. ISTROS

The city of Istros was destroyed in the second half of the 3rd century during the massive raids of the Goths in the Black Sea region.[48] It was first restored under the Emperor Probus,[49] suffered a new devastation in 295, again at the hands of Goths and Carpi, before its renewed reconstruction under the first Tetrarchy.[50] At some point afterwards, the city area was drastically reduced and the public space was reorganized consequently.[51] A so-called 'official district', a group of buildings consisting of several granaries and a centre of military command, was established next to the new city wall, in a rather peripheral position in its new context (Fig. 2, OD).[52] In the southern part of the city, probably the crafts and business area, workshops and storage rooms were built (Fig. 2, ED).[53] The street network was repaired, but apparently not changed. In the 6th century, the walls underwent new refurbishment. A new public area was then organized in front of the main gate, enclosing one of the granaries of the 'official district' – which at this time

---

44 Theodorescu 1963, 275.
45 Theodorescu 1963, 276; 280f.
46 Theodorescu 1963, 285–287; I. Barnea 1979, 134–136; Duval 1980, 322–326; Oppermann 2010, 65f.; Born 2012, 104f.
47 Theodorescu 1963, 271; 278; 280f.; Duval 1980, 324; 326; Barsanti 1990, 167f.; 206; Oppermann 2010, 66.
48 SHA *Max. Balb.* 16.3 (vol. 2, 70, ed. Hohl); Suceveanu & Barnea 1991, 34.
49 Suceveanu 2007, 89; Born 2010, 77.
50 On the topography of late antique Istros, see Suceveanu 1982; Suceveanu 2007, 86–107; Born 2012, 77. For the identification of the buildings in the 'official district', see Dintchev 2005.
51 For the changes that occurred in the urbanism of Istros in Late Antiquity, see Achim 2012.
52 One building was considered a basilica of the Severan age by Suceveanu 2007, 91f., but its recent re-interpretation as a granary is more convincing, see Dintchev 2005, 280–282. Cf. Born 2012, 89.
53 Suceveanu 2007, 90–95.

had been replaced by private dwellings –[54] and a new Christian church (now the 'Florescu Basilica'), as its main landmarks.[55]

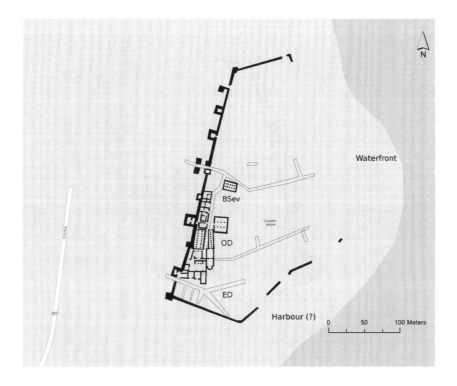

Fig. 2: Istros in the 4ᵗʰ Century. © Ioana Oltean 2019, with permission.
BSev: 'Severan Basilica'; OD: 'official district'; ED: 'economic district'.

The first recognizably Christian building is attested at Istros at the end of the 4ᵗʰ century: it is the so-called basilica C, a structure discovered beneath the later cathedral of the city (Fig. 3, BEp).[56] The church was situated in a central position of the reorganized late-Roman city, which, as mentioned above, does not correspond with the political and administrative centre albeit. Two distinct older structures, which had functioned until the end of the 4ᵗʰ century, yielded to the Christian basilica.[57] A slight deviation between the nave and the apse,[58] as well as the continu-

54 Suceveanu 2007, 99; Born 2012, 89.
55 Suceveanu 2007, 98–100.
56 Bounegru 1993, 195f.; Băjenaru 2003/5; Suceveanu 2007, 17f.; 96f.; Oppermann 2010, 70; Achim 2012, 131–133; Born 2012, 94f.
57 Băjenaru 2003/5, 160f.

ity of the street network in the area indicate that the church was integrated into an already extant urban structure, to which it adapted.[59] Destroyed around the middle of 5th century during the Hunnic raids, the church was restored and continued to function until the 6th century.[60]

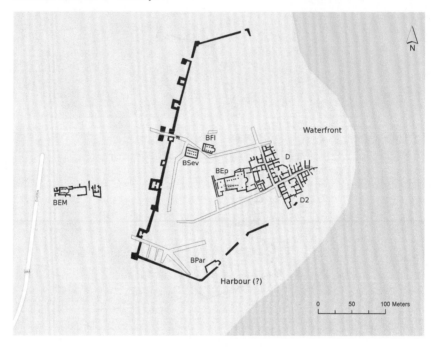

Fig. 3: Istros in the 6th Century. © Ioana Oltean 2019, with permission.
BEM: Basilica extra muros, late 5th cent.; BSev: 'Severan Basilica'; BFl: 'Florescu Basilica', 6th cent.; BEp: Episcopal Basilica, 6th cent.; BPar: 'Pârvan Basilica', 5th/6th cent.; D-D2: 'domus' area.

As in the case of Tomis, most of the Christian buildings of Istros date from the times of Anastasius to Justinian. While this period was one of political stability for Scythia Minor, it saw a particular evolvement of the ecclesiastical organization. For a long time, the province had been under the supervision of a single bishop, who resided in Tomis. In this later period, the church province was raised to metropolitan status, with Tomis as the residence of the archbishop. Istros became one of the several suffragan bishoprics.[61] The new status of the city was represented in many ways, most prominently by the constructions of new church

58   Băjenaru 2003/5, 161; Oppermann 2010, 70; Born 2012, 94.
59   Achim 2012, 156.
60   Băjenaru 2003/5, 163f.; Oppermann 2010, 70.
61   Popescu 1994, 124–156; 201–202; 211–214.

buildings. The first to be mentioned is the Basilica *extra muros*, built most probably under Anastasius (Fig. 3, BEM).[62] It has been interpreted as a cemeterial
church, but the semicircular tiered structure in the liturgical apse (*synthronon*) and
the annexes to the north as well as to the south of the nave rather seem to indicate
a regular community church.[63] Spolia were used in the construction of this building.[64] It is situated over a monument from the Severan age, destroyed at the
beginning of the 4th century,[65] when the area outside the city walls became a cemetery.[66] A reconstruction after the middle of the 6th century (perhaps after the attack by the Cutrigur Huns in 559)[67] shows that the building was then still in use.[68]

In the 6th century, an open square was established in front of the main gate,
flanked to the south by a granary of the 4th century and to the east by the newly
erected 'Florescu Basilica' (Fig. 3, BFl).[69] Its plan follows the street network of
the area.[70] Together with the new square and the granary, they seem to represent a
unity, connected, perhaps, with the administrative tasks of the Church in the
city.[71] The entire structure was placed in a pre-eminent, visually dominant location.[72] The area taken by the new basilica had formerly been occupied by ancient
structures, erected since the beginning of the 2nd century. Some of them probably
belonged to an older religious complex (a temple of Apollo Ietros?).[73] This or
other parts of the previous structures survived until the beginning of the 5th century.[74] Apparently, between the demise of these buildings and the erection of the
basilica some time lapsed.[75] We can therefore safely assume that the previous
building was not destroyed by or for the construction of the church, but that the
spolia employed in the construction of the new church were drawn from ruins.[76]

---

62 Rusu-Bolindeţ & Bădescu 2003/5, 107–109; 111; Oppermann 2010, 77; Achim 2012, 145–
    154; Born 2012, 79–81.
63 Cemeterial church: Popescu 1994, 312f., who argues that it could also function as a commu-
    nity church; *contra*: Duval 1980, 330. Oppermann 2010, 78 and Born 2012, 81–84 think that
    it most probably was a community church belonging to a settlement of *foederati* outside the
    city walls. Rusu-Bolindeţ & Bădescu 2003/5, 112 and Achim 2015, 294f. regard the funerary
    character as certain only for the second phase of the cemetery.
64 Born 2012, 80.
65 Nubar 1971; Rusu-Bolindeţ & Bădescu 2003/5, 116–118; Born 2012, 81.
66 Pippidi 1959: 294f.; Nubar 1971, 204–209; 210–214; Rusu-Bolindeţ & Bădescu 2003/5, 118;
    Achim 2012, 152f.; Achim 2015, 294f.
67 Agath. *Hist.* 5.11 (p. 301, 6–11 ed. Niebuhr).
68 Oppermann 2010, 77–79; Achim 2012, 153.
69 Duval 1980, 326; Popescu 1994, 314–316; Achim 2003/5, 180–194; Oppermann 2010, 76;
    Achim 2012, 141–145; Born 2012, 84–90; Achim 2014. See also above, n. 52.
70 Achim 2005, 96; Born 2012, 87. The southern side of the church superimposes a previous
    street: Achim 2014, 267.
71 For the connection between churches and granaries, see Barkoczi & Salamon 1984, 176–178.
72 Achim 2012, 154; 2014, 266–267; 276.
73 Achim 2003–2005, 181–186; Born 2012, 87f.; Achim 2014, 269–271.
74 Achim 2014, 271f.
75 Achim 2005, 94; 2014, 271f.
76 Achim 2012, 142; Born 2012, 85; Achim 2014, 276.

Another ecclesiastic monument, the so-called 'Pârvan Basilica' (Fig. 3, BPar), was built next to the southern rampart, most probably in connection with the reconstruction of the city fortifications either under Anastasius or under Justinian.[77] Spolia were used for the erection of the church,[78] which leans on the city wall and was adapted to its alignment.[79] Researchers have linked the building to the craft and business quarter in the south-eastern part of Istros.[80] Recent excavations discovered an older church, built at the beginning of the 5th century, under the 'Pârvan Basilica'.[81]

By far the most imposing Christian monument of Istros is the Episcopal Basilica (Fig. 3, BEp).[82] The replacement of the ancient C-basilica with a much larger church reflects the abovementioned change at the end of the 5th or the beginning of the 6th century, when Istros became a suffragan bishopric of Tomis. The cathedral was erected in a place of visual impact,[83] with imported marble.[84] The monument is, indeed, a manifestation of the social importance gained by the Christian community and its willingness to demonstrate its status. Although the entrance was positioned in accordance with the already existing street,[85] and despite the reuse of the space previously occupied by the basilica C, the differences in location and in dimensions between the two monuments suggest some changes in the urban design. The honorary, votive, and funerary inscriptions used as spolia show that the church was constructed after the dissolution of the classical urban structure.[86] Built in the first half of the 6th century,[87] it was seriously damaged in 559 by the Cutrigur attack on the Lower Danube.[88] Towards the end of the 6th or the beginning of the 7th century, it seems to have suffered its final destruction.[89]

The first visible sign of a Christian presence in the urban context of Istros is therefore basilica C, erected at the end of the 4th century, after the reorganization

---

77 Suceveanu 2007, 101; Born 2012, 91. For the reconstruction of the wall, see *IGLR* 112f.; I. Barnea 1960, 365f.; Suceveanu 2007, 99.
78 Florescu 1954, 158; Oppermann 2010, 76f.
79 Achim 2012, 129f.
80 Suceveanu & Barnea 1991, 270; Suceveanu 2007, 99f.; Born 2012, 92.
81 Angelescu & Bottez 2009; Oppermann 2010, 76f. with previous literature; Achim 2012, 128–133 with previous literature; Born 2012, 90–92.
82 Suceveanu 2007; Oppermann 2010, 71–74; Achim 2012, 134–141; Born 2012, 93–99.
83 Achim 2012, 154.
84 Milošević 2007, 195–202; Bounegru & Iaţcu 2007, 64; Oppermann 2010, 73; Achim 2012, 155; Born 2012, 99.
85 Suceveanu 2006, 18f.; Born 2012, 94.
86 For the *spolia*, see Suceveanu 1998, 109, 114; Suceveanu 2007, 145–151; Oppermann 2010, 73; Born 2012, 96. For the non-epigraphic *spolia*, see Milošević 2006, 55.
87 Suceveanu 2007, 29: beginning of the 6th century, most probably under Anastasius. Băjenaru 2003/5, 164: the basilica C still functioned in the first decades of the 6th century, therefore the cathedral was more likely built under Justinian.
88 Suceveanu 2007, 38; Oppermann 2010, 72; Achim 2012, 141; Born 2012, 95.
89 Băjenaru & Bâltâc 2000/1, 486f.: certainly after 592/3; Suceveanu 2007, 38: perhaps 592/3, with another destruction before 602; Achim 2012, 141.

of the urban area. It is difficult to establish the importance of its position in the new frame of the city, since the physical centre was then no longer identical with the political. Probably towards the end of the 5th century, a second Christian monument appeared in the city – the first phase of the 'Pârvan Basilica', situated in a peripheral location, next to the southern rampart. Even if much of the structure of late-antique Istros remains unknown, there is no indication that the Christian community influenced the urbanistic evolution of this period in an active and deliberate way.

Beginning with the reigns of Anastasius and Justinian, the new Episcopal Basilica certainly became the most prominent building of the city. Together with the newly erected Florescu and Pârvan *basilicae*, it made the Christian presence in Istros definitely more visible, as the churches became landmarks of the new city. The 'Florescu Basilica' and the older *horreum*, together with the open space between them, thus formed an ensemble of public relevance.[90] In the same period, some elegant houses were built to the east of the cathedral – the so-called 'Domus' area (Fig. 3, D).[91] One of these structures (Fig. 3, D2), with a peristyle and a longitudinal room with an apse, has been identified as the residence of the local bishop.[92] The whole complex of four buildings no doubt belonged to members of the local elite. As such, it reveals the importance this area had gained with the erection of the cathedral. From then on, Christian leadership clearly had a strong influence on the fate of the city.

Although the use of imported marble for the churches in Tomis could be expected in the face of the city's status and wealth, its employment in Istros constituted an extraordinary effort. This not only shows a period of prosperity,[93] but also the increased influence of the church and the high dedication of the decision-makers to invest in splendid Christian architecture.[94]

V. TROPAEUM TRAIANI

The last city to be discussed here is Tropaeum Traiani, a Roman city in the interior of Scythia Minor. Parts of its former structure from the Imperial era were preserved into Late Antiquity: the main axial streets remained the lines of reference

90 Born 2012, 89: The structures next to the Western rampart of the city, in the so-called 'administrative area', were transformed in private houses in the same period.
91 Sâmpetru 1994, 55–70; Suceveanu 2007, 102f.
92 Sâmpetru 1994, 62; Popescu 1994, 322–324; Bounegru & Lungu 2003–2005, 175–177; Suceveanu 2007, 103; 125; Oppermann 2010, 75; Born 2012, 102. The room with an apse was identified by Duval 1980, 328 as a *stibadium / triclinium*, judging from the presence of a *hypocaustum* and of a large quantity of *amphorae*.
93 Achim 2012, 155.
94 Bowden 2001, 67.

for all later building activities,[95] when the city was reorganized under Constantine and Licinius (308–324).[96]

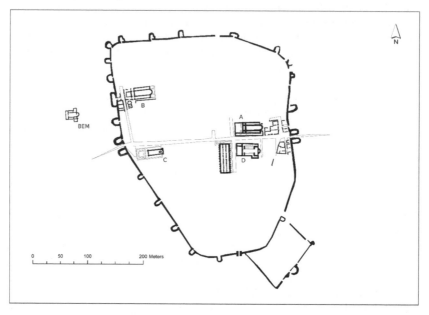

Fig. 4: Tropaeum Traiani and Its Churches. © Ioana Oltean 2019, with permission.
BEM: Basilica extra muros, 5ᵗʰ/6ᵗʰ cen.; B: 'Marble Basilica', 4ᵗʰ cent., reconstructed in 6ᵗʰ cent.;
C: 'Cistern Basilica', 4ᵗʰ cent.; A: 'Simple Basilica', 6ᵗʰ cent.; D: 'Transept Basilica', 6ᵗʰ cent.

Not long after this restructuring, a Christian church (basilica B, the 'Marble Basilica', Fig. 4, B)[97] was erected in the vicinity of the western alignment of the city wall, north of the *via principalis*, on the foundation of a previous building, already abandoned at the end of the 3ʳᵈ century and systematically dismantled at the beginning of the 4ᵗʰ.[98] A baptistery was built next to the church, extending over a colonnaded street from the 2ⁿᵈ–3ʳᵈ centuries,[99] indicating a change in the urban plan. At the beginning of the 6ᵗʰ century, the church received a new revetment of imported marble.[100] Around the same time (or perhaps a little earlier) the whole

95   Suceveanu & Barnea 1991, 271.
96   *IGLR* 170; Bogdan Cătăniciu 1979; Bogdan Cătăniciu 1998, 141f.; Born 2012, 112.
97   Bogdan Cătăniciu 1995, 578; Bogdan Cătăniciu 2006, 241.
98   Duval 1980, 335f.; Bogdan Cătăniciu 1995; Bogdan Cătăniciu 2006; Oppermann 2010, 99f.; Born 2012, 120–123.
99   Bogdan Cătăniciu 1995, 580; Bogdan Cătăniciu 2006, 246; Born 2012, 121f.
100   Mărgineanu-Cârstoiu & Barnea 1979, 139; I. Barnea 1979, 216f.; Bogdan Cătăniciu 1995, 583; Bogdan Cătăniciu 2006, 244.

complex was probably connected with the grain depot in the nearby tower of the city wall.[101] This transformation seems to correspond with the foundation of the new bishopric at Tropaeum Traiani in agreement with the aforementioned reorganization of the church structure in Scythia Minor.[102] In this context, the 'Marble Basilica', together with the baptistery and the grain depot nearby, became the episcopal residence of the city.[103]

In the second half of the 4th century, a cistern near the western rampart of the city, south of the *via principalis,* was filled in and a church was built over it, for which *spolia* were employed.[104] This 'Cistern Basilica' (also basilica C, Fig. 4, C) had a small crypt, indicating the cult of a martyr, although built in a later phase.[105] In the 6th century, burials occurred around the church,[106] which reveals that the boundaries between the 'city of the dead' and the 'city of the living' had disappeared under increasing Christian influence.

In the same period, a basilica with a crypt (*basilica coemeterialis*) was built in the northern cemetery of the city over a Roman funerary structure.[107] The latter, a *hypogeum* constructed at the beginning of the 4th century, was already out of use, when the church was erected.[108] The first phase of this cemeterial basilica had a single nave and was dated – without strong arguments – in the 4th century. Later, towards the end of the 5th or in the first part of the 6th century, the church was reconstructed as a three-aisled basilica.[109]

As with the Greek cities on the Pontic shore, several changes occurred in Tropaeum Traiani in the period of Anastasius and Justinian. Among other changes, two new Christian basilicae were erected in the city, and another one *extra muros*. The second church outside the city walls was built in front of the western gate, also in a cemetery (*basilica extra muros* – Fig. 4, BEM). In the first phase, it was a simple basilica with a single nave, erected most probably at the end of the 5th century.[110] At the beginning of the 6th century, it was transformed into a structure with three aisles and a transept.[111] The church belonged to a non-Roman settle-

---

101 Bogdan Cătăniciu 1995/6, 206f. dates the connection of the tower to the church already in the 4th century; Born 2012, 122 argues for a later date sometime at the end of the 5th century. For a later connection of the tower with the basilica and baptistery, also note the enclosure around the church, erected aparently under Anastasius or Justinian; see Born 2012, 122f.

102 Born 2012, 122.

103 Bogdan Cătăniciu 1995/6, 206f. with the observations of Born 2012, 121f.

104 I. Barnea 1977; Mărgineanu-Cârstoiu 1977; Oppermann 2010, 99 with previous literature; Born 2012, 117–120.

105 I. Barnea 1977, 221f.; 233; I. Barnea 1981, 499; Oppermann 2010, 99; Born 2012, 118.

106 I. Barnea 1977, 226f.; 233; Born 2012, 119f.; Achim 2015, 300; 305, Table 2.

107 Sâmpetru 1994, 83; Oppermann 2010, 105 with previous literature; Born 2012, 115.

108 *Cronica* 1997, Adamclisi; *Cronica* 1998, Adamclisi, cetate; Oppermann 2010, 105; Born 2012, 115.

109 *Cronica* 1997, Adamclisi; Oppermann 2010, 105; Born 2012, 115.

110 Lungu 1998, 451–452 proposes the 4th century; see also Oppermann 2010, 105, note 1020. *Cronica* 1997, Adamclisi; Achim 2016, 56 argues for the end of the 5th century.

111 *Cronica* 1997, Adamclisi; Lungu 1998, 451f.; Oppermann 2010, 105; Born 2012, 116.

ment discovered in the area, inhabited most probably by *foederati* (Goths or Huns).[112]

The new churches erected inside the city walls of Tropaeum Traiani have a far more prominent location than those of the 4th century. The 'Simple Basilica' (also basilica A, Fig. 4, A),[113] which has a crypt, is situated north of the *via principalis*, at the crossroads with the still functioning *cardo maximus* of the Imperial period. It was integrated into the already existing urban structure, since the western side of the atrium follows the 4th-century *cardo*.[114] The church superimposes the remains of a previous structure, which functioned between the 2nd and the 4th centuries.[115] Between the demise of this building and the construction of the church, therefore, some time must have passed, which means that the Christian community took possession of the place only after it had been in disuse for a certain period.

The 'Transept Basilica' (basilica D, Fig. 4, D), the largest of the city, was built in the same period. It is located north of the same *via principalis*, directly opposite the 'Simple Basilica'.[116] The church was a three-aisled and had a small crypt erected over an older structure with a hypocaust installation. This may have been public baths, which were destroyed on the verge from the 4th to the 5th centuries.[117] The architectonic material discovered indicates the use of Proconnesian marble for the basilica.[118]

It is important to mention that the streets of Tropaeum were also repaired at the beginning of the 6th century. For their colonnades, however, local marble and spolia were used.[119] The distinctive employment of imported and local marble respectively shows the importance of those Christian monuments and the priorities of the citizens of Tropaeum. The Christian character of the city in this period was also made explicit by the use of building blocks marked with crosses for the repair of the city wall.[120]

VI. CONCLUSION

All the presented cases demonstrate that the changes in the city landscape of Scythia Minor were, until the 6th century, influenced largely by other factors than Christianity: the rebuilding of the fortifications, the diminution or increase of the urban areas, the dissolution of local administrative organisation. Most of the earli-

---

112 Born 2012, 116.
113 I. Barnea 1978; I. Barnea 1979, 154–157; Duval 1980, 333; Oppermann 2010, 101–103.
114 I. Barnea 1975, 62; 1978, 183.
115 I. Barnea 1978, 184f.; Bogdan Cătăniciu 1995/6, 207; 1998, 144; Oppermann 2010, 102.
116 I. Barnea 1979, 158; 166; Duval 1980, 333; Olariu 2009; Oppermann 2010, 103f. with previous literature; Born 2012, 125–128.
117 *Cronica* 2001, Adamclisi, Sector basilica D; Olariu 2009, 165; Born 2012, 127.
118 Barsanti 1989, 128, n. 143.
119 A. Barnea 1970, 689f.; 1969, 595; 1977, 349f.; Born 2012, 123.
120 Oppermann 2010, 98; Papuc 1979, 185.

est Christian buildings, no doubt due to the progress of Christianity under Theodosius I, appeared towards the end of the 4th century, in an already structured urban context.[121] They filled empty spaces in the frame of the late antique cities, without determining structural changes in the urban environment.

The great period of Christian architecture in Scythia Minor occurred at the end of the 5th and the beginning of the 6th century. It was a period of prosperity for the province, under the reigns of Anastasius and Justinian, when its ecclesiastical status was also changed by the establishment of several new bishoprics. For the Eastern Roman Empire, the reign of Justinian marked the definitive demise of paganism and a new emphasis on Christianity.[122] The churches erected in Scythia during this period are situated in visually prominent positions in the structure of their respective cities – in front of the main gates, as at Tomis or Istros, or at main crossroads as at Tropaeum. They are spread through the entire area of the respective cities and their number increased notably. This phenomenon marks the dissolution of the ancient urban ideal, of a city area centred around civil and religious public buildings representing the entire citizen body. In fact, we may speak of a fragmentation of urban space, as several of these churches became the new centres of specific areas of the city.[123] With the dissolution of the classical structures, parts of the former buildings were used as building materials for the new Christian monuments. The employment of imported marble for these constructions indicates their priority in their respective communities. There is reason to believe that the Christian buildings represented the new centres of the respective cities in this period, regardless of their physical position in the original urban grid. By the mid-6th century, the Christian character of the cities of Scythia Minor was immediately apparent to all their visitors.

Acknowledgments

This study was supported by a grant of the Romanian National Authority for Scientific Research, CNCS-UEFISCDI, project number PN-III-P4-ID-PCE-2016-0279. The copyright for the maps belongs to Dr. Ioana Oltean (University of Exeter), to whom I would like to express my gratitude for sharing them with me.

Abbreviations

Actes XI Congr.	*Actes du XIème Congrès International d'Archéologie Chrétienne*, Lyon – Vienne – Grenoble – Geneve – Aoste (21–28 septembre 1996), Rome 1989.
Cronica	Ministerul Culturii, Comisia Naţională de Arheologie, Direcţia Generală a Patrimoniului Cultural Naţional, *Cronica cercetărilor arheologice* (Ministry

---

121 For the general process, see Jacobs 2014, 204, n. 73.
122 Stein 1949, 368–375; Evans 2001, 65–78.
123 Jäggi & Meier 1997, 194–198.

of Culture, National Commission for Archaeology, General Directorate for National Cultural Patrimony, *Chronicle of Archaeological Research*). URL: http://cronica.cimec.ro.

IGLR   E. Popescu (ed.), *Inscripţiile greceşti şi latine din secolele IV–XIII descoperite în România* (The Greek and Latin Inscriptions of the 4ᵗʰ–13ᵗʰ Centuries, Discovered in Romania), Bucharest 1976.

ISM II   I. Stoian (ed.), *Inscripţiile din Scythia Minor greceşti şi latine* (The Inscriptions of Scythia Minor, Greek and Latin), vol. II, Bucharest 1987.

SCIV   *Studii şi Cercetări de Istorie Veche* (changed to *SCIVA* in 1984).
SCIVA   *Studii şi Cercetări de Istorie Veche şi Arheologie.*

Bibliography – Ancient Sources

Agath. *Hist.*   *Agathiae Myrinaei historiarum libri quinque*, ed. B.G. Niebuhr, Bonn 1828.
SHA *Max. Balb.*   *Vita Maximi et Balbini*, in Scriptores Historiae Augustae, ed. H. Hohl, vol. 2, Stuttgart 1997, 57–72.
Sozom. *Hist. eccl.* Sozomenos, *Historia ecclesiastica / Kirchengeschichte*, 4 vols., ed. G.C. Hansen, Turnhout 2004.
Zos.   *Zosimi comitis et exadvocati fisci historia nova*, ed. L. Medelssohn, Leipzig 1887.

Bibliography – Modern Scholarship

Achim, I. & Cliante, L. 2017: 'Anciennes recherches, nouvelles considerations sur la crypte repérée dans le périmètre de la rue Karl Marx à Constanţa', *Materiale şi Cercetări Arheologice* (serie nouă) 13, 73–88.
Achim, I. 2003/5: 'Histria. Basilica 'Florescu'. Noi cercetări (2002–2004)' (Histria. The Florescu Basilica. New Research, 2002–2004), *SCIVA* 54–56, 179–194.
Achim, I. 2005: 'Études d'archéologie chrétienne en Scythie Mineure: la basilique à crypte d'Histria', in *Mélanges Jean-Pierre Sodini. Travaux et mémoires* 15, Paris, 85–97.
Achim, I. 2012: 'Paysage urbain trado-antique à Histria: les églises paléochrétiennes entre le cadre architectural et la liturgie', *Dacia* N.S. 56, 125–167.
Achim, I. 2014: 'La basilique à crypte d'Istros: Dix campagnes de fouilles (2002–2013)', *Materiale şi Cercetări Arheologice* (serie nouă) 10, 265–287.
Achim, I. 2015: 'Churches and Graves of the Early Byzantine Period in Scythia Minor and Moesia Secunda. The Development of a Christian Topography at the Periphery of the Roman Empire', in J.R. Brandt, M. Prusac & H. Roland (eds.), *Death and Changing Rituals. Function and Meaning in Ancient Funerary Practices*, Oxford, 287–342.
Achim, I. 2016: 'Réflexions sur les monuments de culte chrétien dans l'urbanisme. Scénographie urbaine en Scythie entre adaptation et transformation durant l'antiquité tardive', in I. Topalilov & B. Georgiev (eds.), *Transition from Late Paganism to Early Christianity in the Architecture and Art in the Balkans*, Šumen, 47–71.
Andrade, N. 2010: 'The Processions of John Chrysostom and the Contested Spaces of Constantinople', *Journal of Early Christian Studies* 18.2, 161–189.
Angelescu, M. & Bottez, V. 2009: 'Histria. The Basilica 'Pârvan' Sector. (I). The Sector Archaeological Topography (2001–2007)', *Pontica* 42, 192–212.
Baldovin, J. F. 1987: *The Urban Character of Christian Worship. The Origins, Development, and Meaning of Stational Liturgy*, Rome.

Bárkoczi, L. & Salamon, Á. 1984: 'Tendenzen der strukturellen und organisatorischen Änderungen pannonischer Siedlungen im 5. Jahrhundert', *Alba Regia* 21, 147–187.

Barnea, A. 1969: 'Trei altare inedite de la Tropaeum Traiani' (Three Unpublished Altars from Tropaeum Traiani), *SCIV* 20, 595–609.

Barnea, A. 1970: 'Piese arhitecturale inedite din cetatea Tropaeum Traiani' (Unpublished Architectural Fragments from the City of Tropaeum Traiani), *SCIV* 21, 683–695.

Barnea, A. 1977: 'Descoperiri epigrafice noi in cetatea Tropaeum Traiani' (New Epigraphic Discoveries in the City of Tropaeum Traiani), *Pontica* 10, 349–354.

Barnea, I. 1960: 'Contributions to Dobrudja History under Anastasius I', *Dacia* N.S. 4, 363–374.

Barnea, I. 1975: 'Atriile basilicilor paleocreştine de la Tropaeum Traiani' (The Atria of the Early Christian Basilicae of Tropaeum Traiani), *SCIVA* 26, 1, 57–67.

Barnea, I. 1977: 'La basilique citerne de Tropaeum Traiani à la lumiere des dernières fouilles archéologiques', *Dacia* N.S. 21, 221–233.

Barnea, I. 1978: 'Bazilica 'simplă' (A) de la Tropaeum Traiani' (The 'Simple Basilica' (A) of Tropaeum Traiani), *Pontica* 11, 181–187.

Barnea, I. 1979: *Christian Art in Romania*, Bucharest.

Barnea, I. 1981: 'Le cripte delle basiliche paleocristiane della Scizia Minore', *RESEE* 19.3, 489–505.

Barnea, I. 1991: 'Consideraţii privind cele mai vechi monumente creştine de la Tomis' (Considerations about the Earliest Christian Monuments of Tomis), *Pontica* 24, 269–275.

Barnea, A. & Barnea, I. (eds.) 1979: *Tropaeum Traiani*, vol. 1: *Cetatea*, Bucharest.

Barsanti, C. 1989: 'L'esportazione di marmi dal Proconneso nelle regioni pontiche durante il IV–VI secolo', *Rivista dell'Istituto Nazionale d'Archeologia e Storia dell'Arte* 12, 91–220.

Băjenaru, C. & Bâltâc, A. 2000/1: 'Depozitul de candele din sticlă descoperit la bazilica episcopală de la Histria' (The Lamp Depot Discovered by the Episcopal Basilica of Histria), *Pontica* 33–34, 469–513.

Băjenaru, C. 2003/5: 'Histria, "Bazilica C". Rezultate preliminare' (Histria. The 'C Basilica'. Preliminary Results), *SCIVA* 54–56, 149–165.

Bogdan Cătăniciu, I. 1979: 'Incinta. Structură, datare, istoric' (The Precincts. Structure, Dating, History), in Barnea & Barnea 1979, 47–63.

Bogdan Cătăniciu, I. 1995: 'Note sur l'évolution architecturale de la basilique de marbre (B) de Tropaeum Traiani', in *Akten des XII. Internationalen Kongresses für Christliche Archäologie*, Bonn 22.–28. September 1991, Teil 1, Münster, 578–585.

Bogdan Cătăniciu, I. 1995/6: 'Semnificaţia ultimelor schimbări în urbanismul de la Tropaeum Traiani' (The Significance of the Last Changes in the Urbanism of Tropaeum Traiani), *Pontica* 28–29, 201–214.

Bogdan Cătăniciu, I. 1998: 'Les dernières modifications de l'urbanisme à Tropaeum Traiani', in *Acta XIII Congressus Internationalis Archaeologiae Christianae*, Split – Poreč (25.9–1.10. 1994), vol. 3, Vatican, 141–148.

Bogdan Cătăniciu, I. 2006: 'The Marble Basilica (B) in Tropaeum Traiani', *Dacia* N.S. 50, 235–254.

Born, R. 2012: *Die Christianisierung der Provinz Scythia Minor. Ein Beitrag zum spätantiken Urbanismus auf dem Balkan*, Wiesbaden.

Bounegru, O. & Iaţcu, I. 2007: 'Éléments de décoration intérieure', in Suceveanu 2007, 75–72.

Bounegru, O. & Lungu, V. 2003–2005: 'Histria. Cercetări recente în cartierul Domus' (Histria. Recent Research in the Domus Neighbourhood), *SCIVA* 54–56, 167–178.

Bounegru, O. 1993: 'Contributions stratigraphiques concernant la chronologie des édifices de Scythia Minor. La basilique chrétienne du IVᵉ siècle d'Histria', in *La politique édilitaire dans les provinces de l'Empire romain, IIᵉᵐᵉ – IVᵉᵐᵉ siècles après J.C. Actes du 1ᵉʳ Colloque Roumano-Suisse*, Deva 1991, Cluj-Napoca, 195–200.

Bowden, W. 2001: 'A New Urban Élite? Church Builders and Church Building in Late-Antique Epirus', in L. Lavan (ed.), *Recent Research in Late-Antique Urbanism*, Portsmouth, RI, 57–68.

Brands, G. & Severin, H.-G. (eds.) 2003: *Die spätantike Stadt und ihre Christianisierung*. Symposion vom 14. bis 16. Februar 2000 in Halle / Saale, Wiesbaden.

Bratož, R. 2004: 'Die diokletianische Christenverfolgung in den Donau- und Balkanprovinzen', in A. Demandt, A. Goltz & H. Schlange-Schöningen (eds.), *Diokletian und die Tetrarchie. Aspekte einer Zeitenwende*, Berlin, 115–140; 209–251.

Brenk, B. 2003: *Die Christianisierung der spätrömischen Welt*, Wiesbaden.

Brenk, B. 2003a: 'Zur Christianisierung der spätrömischen Stadt im östlichen Mittelmeerraum', in Brands & Severin 2003, 85–95.

Buzoianu, L. & Bărbulescu, M. 2012: *Tomis. Comentariu istoric şi arheologic / Tomis. Historical and Archaeological Commentary*, Constanţa.

Caillet, J.-P. 2010: 'Remarques sur la problématique des 'palais épiscopaux' à la din de l'antiquité', in N. Duval & V. Popović (eds.), *Caričin Grad III, L'acropole et ses monuments (cathédrale, baptistère et bâtiments annexes)*, Rome, 508–523.

Canarache, V. & Aricescu, A. & Barbu, V. & Rădulescu, A. 1963: *Tezaurul de sculpturi de la Tomis* (The Sculpture Hoard of Tomis), Bucharest.

Cheluţă-Georgescu, N. 1977: 'Contribuţii la topografia Tomisului în sec. VI e. n.' (Contributions to the Topography of Tomis in the 6th century C. E.), *Pontica* 10, 253–260.

Cliante, L. 2006: 'Un nou segment al incintei tomitane tîrzii' (A New Segment of the Late Tomitan Precincts Wall), *Pontica* 39, 249–258.

Dintchev, V. 2005: 'Kasnoantichni obschestveni skladove ot Thracia i Dacia' (Late Antique Public Granaries from Thracia and Dacia), in *Stephanos Archaeologicos in honorem Professoris Ludmili Getov, Studia Archaeologica Universitatis Serdicensis*, Suppl. IV, Sofia, 277–295.

Duval, N. 1980: 'L'archéologie chrétienne en Roumanie à propos de deux livres récents de I. Barnea', *Revue Archéologique*, N.S. 2, 313–340.

Evans, J.A.S. 2001: *The Age of Justinian. The Circumstances of Imperial Power*, London.

Florescu, G. 1954: 'Sectoarele I–VI' (Sectors I–VI), in *Histria I. Monografie arheologică* (Histria I. Archaeological Monograph), 99–162.

Jacobs, I. 2014: 'A Time for Prayer and a Time for Pleasure. Christianity's Struggle with the Secular World', in D. Engels & P. Van Nuffelen, *Religion and Competition in Antiquity*, Brussels, 192–219.

Jäggi, C. & Meier, H.-R. 1997: 'Zum Kirchenboom am Ende der Spätantike', in R.L. Collela & M.J. Gill & L.A. Jenkens & P. Lamers (eds.), *Pratum Romanum. Richard Krautheimer zum 100. Geburtstag*, Wiesbaden, 181–198.

Liebeschuetz, J.H.W.G. 2001: *Decline and Fall of the Roman City*, Oxford.

Lungu, V. 1998: 'L'evoluzione tipologica delle basiliche della Scythia Minor', in *Acta XIII Congressus Internationalis Archaeologiae Christianae*, Split – Poreč (25.9.–1.10.1994), vol. 3, Vatican, 451–462.

Mărgineanu-Cârstoiu, M. & Barnea, A. 1979: 'Piese de arhitectură din cetatea Tropaeum Traiani' (Architectonic Fragments from the City of Tropaeum Traiani), in Barnea & Barnea 1979, 129–176.

Mărgineanu-Cârstoiu, M. 1977: Problèmes d'architecture concernant la citerne romaine et la basilique chrétienne de Tropaeum Traiani, *Dacia* N.S. 21, 236–250.

Milošević, G. 2007: 'Pièces d'architecture', in Suceveanu 2007, 194–202.

Nastasi, I. 2014: 'Elemente de infrastructură tomitană: observaţii preliminare asupra reţelei stradale în perioada romană şi în cea bizantină timpurie' (Infrastructure Elements of Tomis: Preliminary Observations on the Street Network in the Roman and Early Byzantine Periods), *Pontica* 47, 187–202.

Nubar, H. 1971: 'Contribuţii la topografia cetăţii Histria în epoca romano-bizantină' (Contributions to the Topography of the City of Histria in the Roman-Byzantine Period), *SCIV* 22.2, 199–215.

Olariu, C. 2009: 'Tropaeum Traiani: The Basilica D Sector (With Transept) and the Surrounding Area (Archaeological Researches, 2000–2006)', in M. Rakocija (ed.), *Niš & Byzantium. Symposium 7*, Niš, 163–177.

Oppermann, M. 2010: *Das frühe Christentum an der Westküste des Schwarzen Meeres und im anschliessenden Binnenland. Historische und archäologische Zeugnisse*, Langenweißbach.

Papuc, Gh. & Lungu, L. 1998: 'Poarta mare a cetăţii Tomis' (The Great Gate of the City of Tomis), *Pontica* 31, 201–208.

Papuc, Gh. 1979: 'Sectorul de sud-vest al zidului de incintă' (The South-Western Sector of the Ramparts), in Barnea & Barnea 1979, 64–78.

Papuc, Gh. 1979a: 'Tropaeum Traiani 1978. Sectorul "poarta de vest"' (Tropaeum Traiani 1978. The 'Western Gate' Sector), *Materiale şi Cercetări Arheologice* 13, 183–187.

Papuc, Gh. 2000–2001, 'Tomis – aprovizionarea cu apă potabilă în epoca romană şi romană tîrzie' (Tomis – the Drinking Water supply in the Roman and Late Roman Periods), *Pontica* 33–34, 425–449.

Pippidi, D.M. 1959: 'Raport asupra activităţii şantierului Histria în campania 1956' (Report on the Activity of the Archaeological Excavation Histria in the Campaign of 1956), *Materiale şi Cercetări Arheologice* 5, 283–328.

Pippidi, D.M. 1998, 'Sfârşitul păgânismului în Sciţia Mică' (The End of Paganism in Scythia Minor), in idem, *Studii de istorie a religiilor antice* (Studies of History of the Ancient Religions), Bucharest, 339–370.

Popescu, E. 1994: *Christianitas Daco-Romana. Florilegium studiorum*, Bucharest.

Poulter, A. 1983–1984: 'Roman Towns and the Problem of Late Roman Urbanism: The Case of the Lower Danube', *Hephaistos* 5–6, 109–132.

Poulter, A. 1998: 'L'avenir du passé. Recherches sur la transition entre la période romaine et le monde protobyzantin dans la région du Bas-Danube', *Antiquité Tardive* 6, 329–343.

Poulter, A. 2001: 'The Use and Abuse of Urbanism in the Danubian Provinces During the Later Roman Empire', in J. Rich (ed.), *The City in Late Antiquity*, London, 99–135.

Rădulescu, A. & Lungu, V. 1989: 'Le christianisme en Scythie Mineure à la lumière des dernières découvertes archéologiques', in *Actes du XIᵉ Congrès International d'Archéologie Chrétienne*, vol. 3, Rome, 2561–2615.

Rădulescu, A. 1965: *Monumente romano-bizantine din sectorul de vest al cetăţii Tomis* (Romano-Byzantine Monuments from the Western Sector of the City of Tomis), Constanţa.

Rădulescu, A. 1991: 'Recherches archéologiques récentes dans le périmètre de la cite de Tomis', in: E. Popescu & O. Iliescu & T. Teoteoi (eds.), *Études byzantines et post-byzantines*, vol. 2, Bucharest 23–45.

Rădulescu, A. 1995/6: 'Zidul de apărare al Tomisului, de epocă tîrzie, în reconstituirea sa actuală' (The Defense Rampart of Tomis in Late Antiquity in its Present Reconstruction), *Pontica* 28–29, 83–93.

Real, U. 2003: 'Die Bischofsresidenz in der spätantiken Stadt', in Brands & Severin 2003, 219–237.

Rusu-Bolindeţ, V. & Bădescu, A. 2003/5: 'Histria. Sectorul Basilica extra muros' (Histria. The Extra Muros Basilica Sector), *SCIVA* 54–56, 103–130.

Sâmpetru, M. 1994: *Oraşe şi cetăţi romane târzii la Dunărea de Jos* (Late Roman Cities and Forts on the Lower Danube), Bucharest.

Severin, H.-G. 2003: 'Aspekte der Positionierung der Kirchen in oströmischen Städten', in Brands & Severin 2003, 249–258.

Spieser, J.-M. 1986: 'La christianisation de la ville dans l'Antiquité tardive', *Ktèma* 11, 49–55.

Stein, E. 1949: *Histoire du Bas-Empire*, vol. 2, Paris.

Suceveanu, A. & Barnea, A. 1991: *La Dobrudja romaine*, Bucharest.

Suceveanu, A. 1982: 'Les thermes et l'urbanisme d'Histria. Considérations historiques', in idem (ed.), *Histria VI. Les thermes romains*, Bucharest & Paris, 75–92.

Suceveanu, A. 1998: 'Două inscripții inedite de la Histria' (Two Unpublished Inscriptions from Histria), *Pontica* 31, 109–117.

Suceveanu, A. 2007: Histria XIII. La basilique épiscopale, Bucharest.

Theodorescu, D. 1963: 'L'édifice romano-byzantin de Callatis', *Dacia* N.S. 7, 257–300.

Toma, N. 2010: 'Tomis – Kustendje – Constanța. Topografia antică tomitană în hărți și însemnări de călătorie din epoca modernă' (Tomis – Kustendje – Constanța. The Ancient Tomitan Topography in Maps and Travel Notices from the Modern Era), *Caiete ARA* 1, 53–71.

Vaes, J. 1984/6: 'Christliche Wiederverwendung antiker Bauten: Ein Forschungsbericht', *AncSoc* 15–17, 305–443.

Vulpe, R. & Barnea, I. 1968: *Din istoria Dobrogei*. Vol. 2: *Romanii la Dunărea de jos* (From the History of Dobruja. Vol. 2: The Romans at the Lower Danube), Bucharest.

Zahariade, M. 2006: *Scythia Minor. A History of a Later Roman Province (284–681)*, Amsterdam.

AGRICULTURAL DECISION MAKING ON THE SOUTH COAST OF THE BLACK SEA IN CLASSICAL ANTIQUITY

Hugh Elton

Abstract: The choices made by farmers in Paphlagonia and Pontos about which crops to plant during classical antiquity depended on many factors including administrative systems, landscapes, markets, and attitudes to risk and profit. The various literary sources for ancient agriculture have both strengths and limitations, prompting an analysis of the wide range of crops available, change over time between antiquity and the early modern era, variations in choices between different parts of northern Anatolia and the challenges of getting these goods to market while competing with other producers. Although some recent research has focused on the impact of climate changes or plagues in antiquity, these factors were less significant than changes in political structures in determining which crops to plant. For this part of northern Anatolia, the establishment of the Mithradatic kingdom, its replacement by Roman direct rule, the foundation of Constantinople and the Arab wars of the 7th century all had greater impact on agriculture than any variations in the climate of the region.

Абстракт: Сельское хозяйство и изменение климата в позднем римском Понте: В этой главе автор обсуждает решения, принятые древними крестьянами в Пафлагонии и Понте, касающиеся того, какие растения выращивать. Эти решения зависели от многих факторов, включая административные системы, ландшафты, рынки и отношения к риску и прибыли. Автор обсуждает сильные и слабые стороны различных литературных источников, посвященных древнему сельскому хозяйству, а также широкий спектр доступных сельскохозяйственных культур, изменения, которые произошли между древностью и ранней современной эпохой, изменения в выборе между различными регионами северной Анатолии, а также проблемы, связанные с транспортированием товаров на рынок и конкуренцию с другими производителями. Хотя некоторые недавние исследования были сосредоточены на влиянии изменений климата в древности, в этой статье автор утверждает, что при определении того, какие культуры сажать, такие факторы как изменение климата или эпидемии были менее значимыми, чем изменения в политических структурах. Для этой части северной Анатолии такие события, как создание царства Митридата, его замена римским непосредственным правлением, основание Константинополя и арабские войны VII-го века оказали большее влияние на сельское хозяйство, чем любые изменения в климате региона.

INTRODUCTION

The decisions that farmers make about which crops to plant are determined by many factors.[1] With respect to Anatolia much scholarly interest has focussed on the impact of changes in climate and the transformation of the Roman Empire in

---

1   Van der Veen 2010; Stone 2005.

the 7th century AD on farmers' decision making. This paper looks at these processes as they relate to the south coast of the Black Sea, i.e. ancient Paphlagonia and Pontos in classical antiquity.

This research involves combining differing types of evidence, with textual and archaeological sources relating to agriculture needing to be interpreted in the light of other data, particularly pollen cores, regarding climate change. When examining the relationships between these varying types of evidence and between climate events, agricultural change, natural disasters like plagues, and political events, it is good to avoid confusing correlation with causation.[2] Thus the 7th century AD was a transformative period which saw both the end of the classical city in Anatolia and a major change in Roman farming patterns, but what is less clear is whether these changes were the result of military activity (in the form of the Persian and Arab Wars) or the result of changing climate or culture. A second issue is that generalisations regarding Anatolia or the Eastern Mediterranean do not always apply to all regions or even to all of a region. The majority of recent work on Anatolian climate has focussed on Kappadokia and Pisidia, both of which are very different ecological zones from the Black Sea coast (see Fig. 1 for a map).

II. THE PONTIC AND PAPHLAGONIAN ECONOMY
IN A GEOHISTORICAL PERSPECTIVE

Before the earliest Greek settlements in the 7th century BC, the Black Sea coast and its immediate interior were inhabited by indigenous tribal communities. This unurbanized region appears like Kappadokia, in that the cities typical for contemporary Phrygia and Lydia were absent. It was not until the establishment of Greek colonies that cities began to appear in the region. The first cities were coastal colonies, initially at Sinope and then at Trapezus. There is no sign of Persian control of these areas as Xenophon and the Ten Thousand passed through at the end of the 5th century BC. Xenophon's account emphasizes the problems that the Ten Thousand had in getting enough food from these cities to feed themselves.[3]

Details of 4th- and 3rd-century-BC Pontos are scarce, and this did not change much under Mithradatic rule, which gradually included Paphlagonia and Pontos in the course of the 3rd and 2nd centuries.[4] Throughout the Hellenistic period there is little evidence for these poorly urbanized regions. According to Stephen Mitchell, 'It is possible to list and describe the entire epigraphic harvest of pre-Roman Galatia, Cappadocia, and Pontus, excluding the south coastal cities of the Black Sea, in a single paragraph'.[5] Much of the kingdom's territory was directly administered by the monarchy. There were several types of economic exploitation, of which timber production, fishing, and agriculture were the most significant.

---

2   Grey 2019; more generally, van Bavel et al. 2019.
3   Doonan 2018; Xen. *An.* 5.1.6; 5.2.1f.; 6.1.1; 6.5.7.
4   Erciyas 2006, 9–16; see also Payen, chapter VII in this volume.
5   Mitchell 1993, vol. 1, 86.

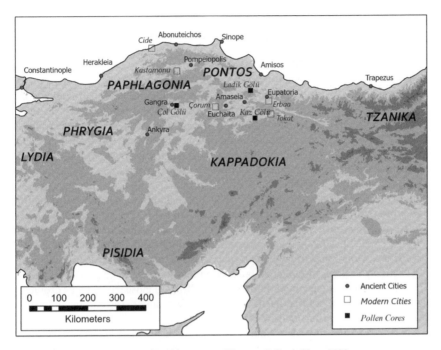

Fig. 1: Map of Paphlagonia and Pontos. © Hugh Elton, 2020.

The imposition of Roman administration by Pompey resulted in the creation of a number of small cities (albeit with large rural territories) in the 1st century BC. Lower levels of population in comparison with western Asia Minor are suggested by the fact that Roman settlements on the sites of the modern cities of Kastamonu and Çorum did not have civic status. In the 2nd century AD the satirist Lucian was dismissive of Abonuteichos (Ionopolis) in Paphlagonia, with one of his characters describing its inhabitants as 'thickwits and fools', while both Gangra and Euchaita were used as places of exile for bishops in the 5th and 6th centuries.[6] Prokopios wrote of Tzanika during the mid-6th century that, in the mountains east of Pontos,

> It is not possible either to irrigate the land or to harvest wheat; one cannot find meadow-land in that region. Indeed even the trees which grow in Tzanika bear no fruit and are entirely unproductive, for seasons do not regularly follow one another, and the earth is not visited at one period by a cold wet season, while at another the sun's heat quickens it, but the land is held in the grip of an endless winter and buried under everlasting snows.[7]

---

6 Marek 1993; 2003. Also see Coşkun, chapter XI in this volume, on the cities in Pontus-Bithynia.
7 Lukian, *Alex.* 9; Prokop. *Aed.* 3.6.5.

These anecdotes suggest low levels of population and perceptions of low levels of culture throughout the Roman period, i.e. it was a region very different from the west coast.[8]

Although our lack of detailed knowledge limits us to generalizations about agricultural practices and social structures, it is still possible to ask questions regarding the changing agricultural opportunities open to the inhabitants of Paphlagonia and Pontos. Economic activity was a city-based process before the 4th century. Then, under the Pontic monarchy, royal demand drove the generation of surpluses in kind, but under Roman administration the need to pay taxes to the Roman state encouraged a market economy and created a number of new markets.[9] Not all Romans were the same, of course, and the phrase 'Roman farmers' covers a great variety of practices. However, a useful structure is a division into two groups, those concerned mostly for their own subsistence and those concerned to maximize profit from their estates, i.e. peasants and landlords. Peasants tend to be risk averse and thus diversify their production, whereas landlords are more able to profit by specialisation and can afford to take more risks. Diversification usually results in a wider range of crops than specialisation, but lower productivity.[10]

Farmers seeking to monetize production had to move their goods to markets that could accommodate them. Pontos was a difficult landscape and thus problems of transportation meant that much of the region's produce was consumed locally. Along the Black Sea is a narrow plain where the largest ancient cities were Herakleia (Ereğli) Sinope (modern Sinop), Amisos (Samsun), and Trapezus (Trabzon). South of this plain lay three ranges of the Küre Mountains, the Ilgaz Mountains, and the Köroğlu Mountains in the west, the Pontic Alps in the east. All of these mountain ranges ran east-west with some peaks at over 2500 m. South of these mountains lay the rolling hills of the Anatolian plateau, with significant cities including Pompeiopolis (Taşköprü), Gangra (Çankırı), and Amaseia (Amasya). Moving between the coast and the interior could be difficult since these mountains are very steep and in many cases wagons could not have been used. Although there were rivers running from the interior to the coast, the Halys (Kızılırmak) in the west of the region and the Iris (Yeşilırmak) in the east, neither appears to have been well-suited to navigation. This is not to say that agricultural goods were not moved across the mountains, only that doing so was slow and costly, so that they may have been less competitive when sold, though these costs were of less of a concern in the royal economy.[11]

Because of the east-west orientation of the Pontic Alps, land communication in these directions was better. A major Roman route ran from Ankyra to Amaseia and then on to Kappadokia. Though politically and militarily important, it was of little significance for long-distance commerce.[12] The mountains thus separated two different agricultural zones, the interior and the coast. They also formed their

8   Mitchell 1993, vol. 1, 91–94; Koder 2017.
9   Hopkins 1980; Temin 2001; Bowman & Wilson 2009; 2014.
10   Grey 2011, 60–63.
11   Decker 2009, 247–251; cf. Bekker-Nielsen 2021.
12   Haldon, Elton & Newhard 2018, 72–99, 236–241.

own economic zone, dominated by timber, honey, and animal herding with limited production of field crops. Although the mountains and the interior could not easily export agricultural surpluses, the coastal zone could export goods by sea.

Some indications of the varying levels of economic connectivity between coastal and interior Pontos and the wider Mediterranean world can be seen in the distribution of ceramics. Imports from outside the Black Sea were usually available along the Pontic coast, but were not found as often inland, a reflection of the low levels of urbanism as well as the difficulty of access, though this picture may be modified by further fieldwork. In the Classical and early Hellenistic period many of these imports were Athenian vessels, perhaps return cargoes linked to grain exports to Athens. In the environs of the later Roman city Pompeiopolis, around 150 km inland, there were only a few local imitations of Rhodian amphora forms from the 3rd century BC, and no Rhodian examples. At coastal Cide and at Sinope, for example, there were numerous amphorae from the Aegean and from north Africa, but at Pompeiopolis and also at Euchaita (Avkat) in Pontos, around 160 km from the coast, there were few imported amphorae, despite the large quantities produced at Sinope. However, in the late Roman period Avkat had some miniature LR3 amphorae, holding about a litre and so probably used for luxuries rather than staples, similar to the few imported *unguentaria* and Mediterranean amphorae at Pompeiopolis.[13]

The pattern of Pontic exports is also informative about economic connectivity. In the 5th and 4th centuries BC, Athens was the largest market for goods produced in the Black Sea, especially for grain that the state imported from the northern Black Sea region.[14] However, since Sinope lay about 1000 km from Athens, Pontic merchants here had to compete with merchants based in the Aegean who had a much shorter distance to market. The same factors worked in reverse, and in the Hellenistic period Sinope was responsible for many of the trade goods that were sent to the Caucasus and the north coast of the Black Sea.[15] Lying between the Aegean and the Black Sea, Byzantion was in good position to serve as a redistribution centre. Polybios wrote in the 2nd century BC:

For those commodities which are the first necessaries of existence, cattle and slaves, are confessedly supplied by the districts round the Pontos in greater profusion, and of better quality, than by any others: and for luxuries, they supply us with honey, wax, and salt-fish in great abundance; while they take our superfluous stock of olive oil and every kind of wine. As to corn there is a mutual interchange, they supply or take it as it happens to be convenient.[16]

In the late-Hellenistic period and early-Roman Empire, Ephesos had replaced Athens as the closest large market for Pontos and Paphlagonia, where Pontic merchants faced similar challenges as their classical predecessors.[17] Then, from the 4th

13   Düring & Glatz 2015, 260–293; Haldon, Elton & Newhard 2018, 203; Opaiț 2018, 701–713; Domżalski 2011, 168; Xen. *An.* 6.4.6.
14   Braund 2007; Moreno 2013.
15   Rempel & Doonan 2019; Davis et al. 2018.
16   Polyb. 4.38.4f.
17   Lund 2007.

century AD, the foundation of Constantinople provided a large market which changed commercial patterns in the Black Sea. Constantinople lay about 550 km west of Sinope, similar to the distance to Athens or Ephesos. Although much of the grain to feed Constantinople was imported directly from Egypt by a state-subsidized fleet, making it difficult for Pontic grain producers to compete, other goods had a similar journey to market as those imported from the Aegean. Like its predecessor Byzantion, Constantinople continued to act as a redistribution centre.

Thus, a 5th-century-AD shipwreck near Sinop which was carrying a mixed cargo of Gazan, Ephesian, and North African amphorae, had probably sailed from Constantinople. Both Late Roman (LR) 1 amphorae produced in Kilikia and Cyprus and Sinopean amphorae were found at military bases on the Danube, probably transhipped in Constantinople, while Sinopean amphorae were also found at Zeugma on the Euphrates in the 7th century.[18] Constantinople's role as a market and a transhipment centre lasted until the crisis of the 7th century AD. Then the city shrank rapidly, in part because the loss of Roman control of Egypt also meant that the subsidized shipments of the 4th to 6th centuries AD no longer took place. Thus, a changing political situation in the Mediterranean may have made it easier for Pontic merchants to compete in the Black Sea. At the same time, the creation of the thematic system with the army of the Armeniakon now based in northern Anatolia gave these cities new opportunities to supply Roman troops directly, transporting goods inland.[19] Imperial geopolitics and distribution systems thus had the greatest impact on farmers' decision-making processes. Other factors, it will be suggested, are far less important.

III. CROP CHOICES IN PAPHLAGONIA AND PONTOS

1. Crop Choices in (Early) Modern Times

With this in mind, what can we say about the crops that farmers chose to grow? The southern coast of the Black Sea is temperate and well-watered, with rain typically falling at both Sinop and Amasya in every month of the year. Thus, crop failure because of insufficient rain is not a major concern, very different from the eastern Mediterranean. Ancient descriptions of what crops were grown come from various periods. Xenophon described the coast at Kalpe near Herakleia (modern Ereğli) at the start of the 4th century BC and after stressing the excellence of its timber, added that 'the land produces barley, wheat, beans of all sorts, millet and sesame, a sufficient quantity of figs, an abundance of grapes which yield a good sweet wine, and in fact everything except olives'.[20] Four hundred years later, when Strabo (who came from Amaseia) described Pontos in the Augustan period, he also mentioned its deciduous timber and then wrote that Sinope 'is all planted

---

18   Opaiț et al. 2019; Swan 2010; Reynolds 2013, vol. 2, 105.
19   Haldon 2018; Teall 1959.
20   Xen. *An.* 6.4.4–6.

with olive-trees which are grown a little above the sea'. The textual evidence for the ancient production of olives at Sinop is supported by archaeological finds of olive presses and some traces of ancient olive pollen.[21]

Ancient writers were less enthusiastic about the interior. An 11th-century bishop, John Mauropous, described Euchaita, saying that 'It abounds, however, in the production of grain, although this also is achieved with much toil, but with regard to wine and oil and next to such products, the land is unfortunate owing to its utter poverty and want'. Despite this gloom, a recent archaeological survey here found numerous wine presses, but no olive presses.[22] Strabo wrote similarly about the territory south of Amisos, noting it 'is generally bare and produces wheat.' However, the interior also contained a few micro-climatic zones where conditions were more favourable. One of these was the Phanaroia, a fertile valley 200–300 m above sea level, containing the rivers Iris and Lykos and the ancient city of Eupatoria (later Magnopolis), an area that Strabo noted 'is planted with olive trees, abounds in wine, and has all the other virtues'. An epigram by the 6th-century poet Marianos Scholastikos also mentioned vineyards and olive trees.[23]

Item	Sinop	Erbaa	Ereğli
Figs	182	99	
Apples	1478	1556	819
Pears	602	146	229
Quince	29	115	78
Apricot		31	
Cherries	171	319	488
Peaches	5	364	29
Plums	348	357	
Cranberry	43	72	
Kiwi	8		24
Strawberries	2	338	495
Mulberry	120	64	33
Muşmula	33	74	
Pomegranate	2	52	
Persimmon	13	46	
Chestnut	95		500
Hazelnut	95	2607	20926
Walnut	298	911	350
Olives	33	5	
Grapes		13314	

Table 1: 2019 Fruit Production in Metric Tons in Sinop, Ereğli, and Erbaa İlçeleri)[24]

21 Strab. *Geogr.* 12.3.12 (546C); Emery-Barbier 2010; Doonan 2002; 2015.
22 John Mauropous, *Letters* 64.55–62; Haldon, Elton & Newhard 2018, 101.
23 Strab. *Geogr.* 12.3.30 (556C), 38 (560C), cf. 2.1.15 (73C); Marianos Scholastikos in *Anthologia Graeca* 9.668; Erciyas 2006, 45f.; Mitchell 2015, 289f.; Aktaş 2018.
24 Source: https://biruni.tuik.gov.tr/medas/.

Although these ancient descriptions of timber on the coast and vast fields of cereals in the interior still hold true today, simplistic visions of the premodern world being the same as an unchanging ancient ecology need to be challenged. Both modern and early-modern agricultural statistics suggest the varieties of crops that can be grown. Xenophon's description of the crops at Kalpe sounds comprehensive, at least until compared to modern reports on productivity in this region at Ereğli. Moreover, regional variations can be significant, and the choices open to farmers here were different from those at Sinop and Erbaa (the modern centre of the Phanaroia), as may be exemplified with the harvest data of 2019 (Table 1).

In addition to change over space, change over time impacted decisions about which crops to plant. When the French geographer Vital Cuinet produced estimates of the value of typical production of agricultural goods in Turkey in 1890, he noted production of tobacco at Sinop and at Samsun and then added that it used

Item	Estimated value in oke	% of total	Item	Estimated value in oke	% of total
Wheat	40,000,000	48.341	'yellow seed'	100,000	0.121
Barley	10,000,000	12.085	Tragacanth	32,000	0.039
Rye	1,150,000	1.390	Salep	5,000	0.006
Oats	100,000	0.121	Tobacco	1,000,000	1.209
Maize	10,010,410	12.098	Anise	1,000	0.001
Millet	6,340,000	7.662	Opium	2,500	0.003
Rice	90,000	0.109	Poppy seed	100,000	0.121
Green beans	15,500	0.019	Hemp	100,000	0.121
Haricot beans	2,000,000	2.417	Wax	3,500	0.004
Lentils	1,000,000	1.209	Honey	700	0.001
Chickpeas	2,000,000	2.417	Mohair	41,500	0.050
Okra	215,000	0.260	Mahaleb cherry	40,000	0.048
Raisins	1,000,000	1.209	Almonds	900,000	1.088
Apples and Pears	2,000,000	2.417	Dry Raisins	500,000	0.604
Various fruits	3,000,000	3.626	Wine	180,000	0.218
			Pekmez	819,000	0.990
			Total	**82,746,110**	**100.00**

Table 2: Cuinet's estimated annual production figures for the sanjak of Tokat, vol. 4.715f.
(1 oke = 1.28 kg).

to be higher, before changes in taxation (Table 2). Cuinet did not provide estimates for production at Erbaa, but he did note a preference for hemp and opium. He also mentioned that the region did not grow many vines despite being well-suited for viticulture. Cuinet's work is a reminder not only of how many different crops might be grown, but also of other factors drving crop selection.[25] His mention of tobacco also draws attention to the new crops not available to ancient farmers, including tobacco, cotton, potatoes, maize (corn), and tomatoes.[26]

2. Crop Choices in Antiquity Based on Pollen-Coring

The majority of ancient evidence about crop choices is very anecdotal, with the lists of crops provided by Xenophon, Strabo, or Mauropous saying little about change over time. A differing perspective is provided by analysing layers of pollen grains recovered from lake beds, which gives information about the whole range of vegetation in a region over long periods of time. However, pollen-coring can only occur where environmental conditions are suitable, generally reflects only wind-pollenated plants, and can only rarely be focused annually. The typical Mediterranean crop mixture of cereals, vines and olives, known as the Beyşehir Occupation Phase (BOP), shows up well in pollen cores.[27] However, there are few pollen cores with usable data from antiquity for Pontos. A pollen core from Ladik Gölü on the north slopes of the Pontic Alps at about 800 m shows BOP agriculture beginning in the early Iron Age, c. 3200 before present (BP). Vine and olive pollen stops in late antiquity, but there is continuity of cereal cultivation after this. A pollen core from Kaz Gölü near Zela at an altitude of around 500 m showed BOP agriculture, including some olive pollen, starting in the Hellenistic period (2220 BP ± 90) and continuing until late antiquity. The dating of the Kaz and Ladik cores is not very precise, so that it is probably best to say only that at these sites the BOP ended between the 6th and 8th century.

And finally, the Çöl Gölü core, from 12 km east of Çankırı at about 1000 m, contains data only from the past 2000 years, with an estimated start date of 100 BC / AD 200. This shows no evidence of olive or vine pollen, but high levels of cereal production for a period of about two centuries under the Roman Empire. Like the comments of Strabo and Mauropous regarding the interior, this is similar to modern cultivation patterns.[28] But as with all these cores, there are considerable uncertainties regarding dating.[29] Overall, the Pontic cores suggest that BOP agriculture in Pontos started much earlier than the major expansion of cities under the Romans in the 1st century BC, while on many sites it ends around the time of the Islamic raids and the political transformation of the 7th century. The pollen data is useful for understanding the style of agriculture at various periods, but is less useful for understanding choices made by farmers over short periods of time.

---

25 Cuinet 1890/5, vol. 4, 429, 431f.; 567; 569 on Sinop and vol. 1, 727–729 on Erbaa.
26 Watson 1983; Sallares 1991.
27 Eastwood, Roberts & Lamb 1998; Izdebski 2013; Roberts 2018.
28 Ladik and Kaz: Bottema, Woldring & Aytuğ 1993/4; Çöl: Roberts, Eastwood & Carolan 2009; for summaries and interpretations of this data, see Izdebski 2013, 194–198 and Haldon, Elton & Newhard 2018, 39f.
29 Izdebski 2012, 48–53; Roberts 2018, 58–61.

3. The Uncertain Impact of Plagues

Short-term events like plagues at various eras (the 2nd-century AD Antonine plague, the 3rd-century Cyprianic plague, or the 6th-century Justinianic plague) are sometimes adduced as significant factors in terms of agricultural change. Methodologically, any assessment that plague had an impact on agriculture needs to be accompanied by consideration of other factors producing change.[30] There is, moreover, currently little consensus among scholars regarding the impact of the various plagues.[31]

4. Climate Change as a Factor

This finally brings us to climate change, which has been suggested as being critical to the collapse of the Roman Empire in the 7th century, the fall of the classical city, and the end of the BOP. As elsewhere, evidence for ancient climate and for changes in ancient climate in northern Anatolia depend on proxy indicators. Many of these proxies come from outside this region, though new data is continually arriving.[32] In very general terms, there was little change in the Mediterranean climate between classical antiquity and 1900 so that the same sorts of crop remained viable. Over this long period, the majority of variations in European climate fell within a small range of ±2 degrees C or ± 100 mm rainfall.[33] For most farmers, these sorts of long-term climate changes were less important than variations in temperature or rainfall caused by weather. However, they were significant at the margins of crop ranges, e.g., the upper elevation limit for vines and olives might be affected, or wheat production might have been more or less possible in arid regions. Even the Roman Climatic Optimum of the early Roman Empire appears relatively unimportant at this scale. Both the literary and archaeological data for the micro-climates at Sinope and the Phanaroia show olive production throughout classical antiquity, suggesting that it was not the Roman Climatic Optimum that enabled olives to be grown where they do not occur now. Historical climate change should be thought of in terms of modifying the edges of production zones, not of forcing wholescale changes in production systems.

---

30 E.g., Marzano 2013 is excellent on the impact of barrels replacing amphorae in the archaeological record, but leans heavily on the Antonine plague as an explanatory mechanic, whereas Mitchell 2015 focuses mainly on culture and imperial impact.

31 See, on the one hand, Harper 2018; on the other hand, Haldon et al. 2018; Mordechai & Eisenberg 2019; and in between, Meier 2020.

32 Izdebski et al. 2016; Finné et al. 2011.

33 Büntgen et al. 2011, Fig. 4; current efforts to minimize global warming focus on keeping temperatures no more than 1.5 to 2 degrees C above this pre-industrial range.

IV. CONCLUSIONS

Farmers in all periods respond to many factors in deciding which crops to plant. As elsewhere in the ancient world, on the Black Sea coast of Anatolia, these factors included the availability of land, labour and capital, markets, and technology, and taxation systems, though many of these factors are hard to see in our sources and harder to quantify. Changes in climate over decades and centuries were much harder for ancient farmers to see and to respond to, though they did produce quantifiable data for modern researchers. Human factors such as political change and military threats are often easily visible in the evidence that we do have, though their impact is hard to quantify. Thus in this region there were major changes in opportunities for local farmers with the change of authority from the Mithradatic kingdom to the Roman Empire, the foundation of Constantinople, and the Arab wars, all of which were probably more profound than changes imposed by climatic variation. This paper suggests that these factors were the most significant for agricultural decision-making in classical antiquity.

Acknowledgements

My thanks to Altay Coşkun for an invitation to speak at the Black Sea Workshop in the University of Waterloo in November 2018, to the audience for their perceptive questions, and the comments of the anonymous reviewers.

Bibliography

Aktaş, E. 2018: *Roma İmparatorluk Döneminde Orta ve Doğu Karadeniz Bölgesi Yerleşimleri*, unpublished PhD, Adnan Menderes University, Aydın, Turkey.

Bekker-Nielsen, T. 2021: 'Navigable Rivers in Northern Anatolia', forthcoming in A. Dan, H-J. Gehrke & A. Podossinov (eds.), *The Black Sea in Late Antiquity: Spaces, Peoples, Transfers between Changes and Continuity*, Stuttgart.

Bottema, S., Woldring, H. & Aytuğ, B. 1993/4: 'Late Quaternary Vegetation of Northern Turkey', *Palaeohistoria* 35/6, 13–72.

Bowman, A. & Wilson, A. (eds.) 2009: *Quantifying the Roman Economy*, Oxford.

Bowman, A. & Wilson, A. (eds.) 2014: *The Roman Agricultural Economy*, Oxford.

Braund, D. 2007: 'Black Sea Grain for Athens? From Herodotus to Demosthenes', in V. Gabrielsen & J. Lund (eds.), *The Black Sea in Antiquity: Regional and Interregional Economic Exchanges*, Aarhus, 39–68.

Büntgen, U., Tegel, W., Nicolussi, K., McCormick, Frank, D., Trouet, V., Kaplan, J.O., Herzig, F., Heussner, K.U., Wanner, H., Luterbacher, J. & Esper, J. 2011: '2500 Years of European Climate Variability and Human Susceptibility', *Science* 331, 578–582.

Cuinet, V. 1890/5: *La Turquie d'Asie*, 4 vols., Paris.

Davis, D., Brennan, M.L., Opaiṭ, A. & Beatrice, J.S. 2018: 'The Ereğli E Shipwreck, Turkey: An Early Hellenistic Merchant Ship in the Black Sea', *International Journal of Nautical Archaeology* 47.1, 57–80.

Decker, M. 2009: *Tilling the Hateful Earth*, Oxford.

Domżalski, K. 2011: 'Late Roman Pottery from Pompeiopolis', in L. Summerer (ed.), *Pompeiopo-lis I*, Langenweißbach, 163–177.

Doonan, O. 2002: 'Production in a Pontic Landscape: The Hinterland of Greek and Roman Si-nope', in M. Faudot, A. Fraysse & É. Geny (eds.), *Pont-Euxine et commerce. La genèse de la 'route de la soie'*, Paris, 185–198.

Doonan, O. 2015: 'Settlement and Economic Intensification in the Late Roman / Early Byzantine Hinterland of Sinop', in K. Winther-Jacobsen & L. Summerer (eds.), *Landscape Dynamics and Settlement Patterns in Northern Anatolia during the Roman and Byzantine Period*, Stuttgart, 43–59.

Doonan, O. 2018: 'Xenophon in a Black Sea Landscape: Settlement Models for the Iron Age on the Sinop Promontory (Turkey)', *European Journal of Archaeology* 2018, 1–20.

Düring, B.S. & Glatz, C. (eds.) 2015: *Kinetic Landscapes: The Cide Archaeological Project: Sur-veying the Turkish Western Black Sea Region*, Berlin.

Eastwood, W.J., Roberts, N. & Lamb, H.F. 1998: 'Palaeoecological and Archaeological Evidence for Human Occupance in Southwest Turkey: The Beyşehir Occupation Phase', *Anatolian Studies* 48, 69–86.

Emery-Barbier, A. 2010: 'Vegetation actuelle et passée de la région de Sinope: Apports des ana-lyses palynologiques et anthracologiques du site de Demirci à la reconstitution de la couver-ture végétale au Ier millénaire AD', in D. Kassab Tezgör (ed.), *Les fouilles et le matériel amphorique de l'atelier de Demirci près de Sinope* (= *Varia Anatolica* 22), 27–40.

Erciyas, D.B. 2006: *Wealth, Aristocracy, and Royal Propaganda under the Hellenistic Kingdom of Mithradatids in the Central Black Sea Region in Turkey*, Leiden.

Finné, M., Holmgren, K., Sundqvist, H.S., Weiberg, E. & Lindblom, M. 2011: 'Climate in the Eastern Mediterranean, and Adjacent Regions, during the Past 6000 Years. A Review', *Jour-nal of Archaeological Science* 38, 3153–3173.

Grey, C. 2011: *Constructing Communities in the Late Roman Countryside*, Cambridge.

Grey, C. 2019: 'Climate Change and Agrarian Change between the Fourth and Sixth Centuries: Questions of Scale, Coincidence, and Causality', in J.W. Drijvers & N. Lenski (eds.), *The Fifth Century: Age of Transformation*, Bari, 35–48.

Haldon, J.F. 2018: 'Some Thoughts on Climate Change, Local Environment, and Grain Production in Byzantine Northern Anatolia', in A. Izdebski & M. Mulryan (eds.), *Environment and Soci-ety in the First Millennium AD*, Leiden, 18–24.

Haldon, J., Elton, H., Huebner, S.R., Izdebski, A., Mordechai, L. & Newfield, T.P. 2018: 'Plagues, Climate Change and the End of an Empire. A Response to Kyle Harper's *The Fate of Rome*', *History Compass*. URL: https://onlinelibrary.wiley.com/doi/10.1111/hic3.12508.

Haldon, J., Elton, H. & Newhard, J. (eds.) 2018: *Archaeology and Urban Settlement in Late Ro-man and Byzantine Anatolia*, Cambridge.

Harper, K. 2018: *The Fate of Rome*, Princeton.

Hopkins, K. 1980: 'Taxes and Trade in the Roman Empire (200 BC–AD 400)', *JRS* 70, 101–125.

Izdebski, A. 2012: 'The Changing Landscapes of Byzantine Northern Anatolia', *Archaeologia Bulgarica* 16.1, 47–66.

Izdebski, A. 2013: *A Rural Economy in Transition: Asia Minor from Late Antiquity into the Early Middle Ages*, Warsaw.

Izdebski, A., Pickett, J., Roberts, N. & Waliszewski, T. 2016: 'The Environmental, Archaeological and Historical Evidence for Regional Climatic Changes and Their Societal Impacts in the Eastern Mediterranean in Late Antiquity', *Quaternary Studies Review* 136, 189–208.

Koder, J. 2017: 'Historical Geography', in P. Niewöhner (ed.), *The Archaeology of Byzantine Anatolia*, Oxford, 9–27.

Lund, J. 2007: 'The Circulation of Ceramic Fine Wares and Transport Amphorae from the Black Sea Region in the Mediterranean, c. 400 BC–AD 200', in V. Gabrielsen & J. Lund (eds.), *The Black Sea in Antiquity*, Aarhus, 183–194.

Marek, C. 1993: *Stadt, Ära und Territorium in Pontus-Bithynia und Nord-Galatia*, Tübingen.

Marek, C. 2003: *Pontus et Bithynia. Die römischen Provinzen im Norden Kleinasiens*, Mainz.

Marzano, A. 2013: 'Capital Investment and Agriculture: Multi-Press Facilities from Gaul, the Iberian Peninsula, and the Black Sea Region', in A. Bowman & A. Wilson (eds.), *The Roman Agricultural Economy: Organisation, Investment, and Production*, Oxford, 107–141.

Meier, M., 2020: 'The 'Justinianic Plague': An "Inconsequential Pandemic"? A Reply', *Medizinhistorisches Journal*.55.2, 172–199.

Mitchell, S. 1993: *Anatolia: Land, Men, and Gods in Asia Minor*, vol. 1: *The Celts and the Impact of Roman Rule*, Oxford.

Mitchell, S. 2015: 'Food, Culture and Environment in Ancient Asia Minor', in J. Wilkins and R. Nadeau (eds.), *A Companion to Food in the Ancient Worlds*, London, 285–295.

Mordechai, L. & Eisenberg, M. 2019: 'Rejecting Catastrophe: The Case of the Justinianic Plague', *Past and Present* 244, 3–50.

Moreno, A. 2013: *Feeding the Democracy: the Athenian Grain Supply in the Fifth and Fourth Centuries BC*, Oxford.

Opaiţ, A. 2018: 'Local and Imported Wine at Pompeiopolis, Paphlagonia', *Rei Cretariae Romanae Fautorum Acta* 45, 701–713.

Opaiţ, A., Davis, D., Brennan, M.L. & Kofahl, M. 2019: 'The Sinop F Shipwreck in the Black Sea: An International Cargo from Late Antiquity', in H. Kaba, G. Kan Şahin, B.M. Akarsu & O. Bozoğlan (eds.), *International Symposium on Sinope and Black Sea Archaeology: 'Ancient Sinope and the Black Sea' Proceedings Book*, Sinop, 77–89.

Rempel, J.E. & Doonan, O. 2019: 'Rural Hinterlands in the Black Sea during the Fourth Century BCE: Expansion, Intensification and New Connections', *AS* 69, 1–25.

Reynolds, P. 2013: 'Transport Amphorae of the First to Seventh Centuries', in W. Aylward (ed.), *Excavations at Zeugma, Conducted by Oxford Archaeology*, vol. 2, Los Altos, 93–161.

Roberts, C.N. 2018: 'Revisiting the Beyşehir Occupation Phase: Land-Cover Change and the Rural Economy in the Eastern Mediterranean during the First Millennium AD', in A. Izdebski & M. Mulryan (eds.), *Environment and Society in the First Millennium AD*, Leiden, 53–68.

Roberts, C.N., Eastwood, W.J. & Carolan, J. 2009: 'Palaeolimnological Investigations in Paphlagonia', in R. Matthews & C. Glatz (eds.), *At Empires' Edge: Project Paphlagonia: Regional Survey in North-Central Turkey*, London, 64–73.

Sallares, R. 1991: *The Ecology of the Ancient Greek World*, London.

Stone, D. 2005: *Decision-Making in Medieval Agriculture*, Oxford.

Swan, V.G. 2010: 'Dichin (Bulgaria): The Destruction Deposits and the Dating of Black Sea Amphorae in the Fifth and Sixth Centuries A.D.', *Varia Anatolica* 21, 107–118.

Teall, J. 1959: 'The Grain Supply of the Byzantine Empire, 330–1025', *Dumbarton Oaks Papers* 13, 87–139.

Temin, P. 2001: 'A Market Economy in the Early Roman Empire', *JRS* 91, 169–181.

van Bavel B.J.P., Curtis, D.R., Hannaford, M.M., Roosen, J. & Soens, T. 2019: 'Climate and Society in Long-Term Perspective: Opportunities and Pitfalls in the Use of Historical Datasets', *WIREs Climate Change*; e611. URL: https://doi.org/10.1002/wcc.611.

Van der Veen, M. 2010: 'Agricultural Innovation: Invention and Adoption or Change and Adaptation?', *World Archaeology* 42.1, 1–12.

Watson, A.M. 1983: *Agricultural Innovation in the Early Islamic World*, Cambridge.

ABOUT THE AUTHORS

Luis **Ballesteros Pastor** is Associate Professor (Profesor Titular) at the University of Seville. He joined the Department of Ancient History in 1994 after gaining his PhD at the same University in the same year. His research interests are mainly focused on the figure of Mithridates Eupator, the history of the kingdoms of Pontos and Kappadokia, and in general Anatolia and the Black Sea in Hellenistic times. He has also dealt with the Late Roman Republic, Quintus Curtius Rufus and above all Justin's *Epitome* of Pompeius Trogus. He is the author of the monographs *Mitrídates Eupátor, rey del Ponto* (1996), and *Pompeyo Trogo, Justino y Mitrídates. Comentario al Epítome de las Historias Filípicas (37.1.6–38.8.1)* (2013).

Altay **Coşkun** is Professor of Classical Studies at the University of Waterloo, Ontario. He joined the department in 2009, after gaining his PhD (1999) and Habilitation (2007) at Trier University and holding research positions at Oxford, Trier & Exeter. His interests range from Ancient Anatolia over the Greek poleis, Hellenistic kingdoms, Roman diplomacy and citizenship to the Late Roman state. His latest books are *Fremd und rechtlos? Zugehörigkeitsrechte Fremder von der Antike bis zur Gegenwart* (with Lutz Raphael, 2014), *Interconnectivity in the Mediterranean and Pontic World* (co-edited with Victor Cojocaru and Madalina Dana, 2014), *Seleukid Royal Women* (co-edited with Alex McAuley, 2016) and *Rome and the Seleukid East* (co-edited with David Engels, 2019). He is currently preparing edited volumes on ancient Galatia and the Bosporan kingdom, besides writing monographs on the *Resilience of the Seleukid Dynasty* (2nd century BC), the *First Book of Maccabees* and *An Analysis of the Book of Daniel* (with Rabbi Ben Scolnic).

Madalina **Dana** is Professor of Greek History at the University of Lyon 3, France. She was previously Associate Professor of Hellenistic History at the University Paris 1 Sorbonne (2011–2019). She gained her PhD at EHESS, Paris (2008) and *Habilitation à diriger des recherches* at the University of Paris Nanterre (2018). Her research focuses on cultural history with special interest in epigraphic practices, mobility and networks, and local history. Her publications include *Mobilité et culture dans le Pont-Euxin* (2011), *La cité interconnectée dans le monde gréco-romain* (co-edited with I. Savalli, 2019) and *Another Way of Being Greek* (co-edited with M. Costanzi, 2020).

Hugh **Elton** is a Professor in the Ancient Greek and Roman Studies program at Trent University, Ontario. He came to Trent in 2006, after earning a DPhil from Oxford in 1990, teaching at Rice University, Trinity College, and Florida International University in the US, and serving as Director of the British Institute at Ankara in Turkey. His research interests include field survey archaeology, the late Roman army, and the administration and politics of the late Roman Empire in the east. Recent books include *The Roman Empire in Late Antiquity*, (2018), *Archaeology and Urban Settlement in Late Roman and Byzantine Anatolia* (co-written with John Haldon and Jim Newhard, 2018), and *Asia Minor in the Long Sixth Century* (co-edited with Ine Jacobs, 2019).

Philip A. **Harland** gained his BA at the University of Waterloo and his MA and PhD at the University of Toronto. He is now Full Professor in the Departments of Humanities and History at York University. His research focuses on ancient Christianity, Judaism and ethnicity. His main publications include *Greco-Roman Associations: Texts, Translations, and Commentary. II. North Coast of the Black Sea, Asia Minor*, Berlin 2014. *Associations, Synagogues, and Congregations: Claiming a Place in Ancient Mediterranean Society*. 2nd ed., Kitchener, ON 2013 (1st ed. 2003). *Dynamics of Identity in the World of the Early Christians: Associations, Judeans, and Cultural Minorities*, London 2009. And, co-written with Richard Last, *Group Survival in the Ancient Mediterranean: Rethinking Material Conditions in the Landscape of Jews and Christians* (London 2020).

Valentina **Mordvintseva** is Associate Professor at the National Research University 'Higher School of Economics' (since 2019) and Senior Researcher at the Institute of World History of the Russian Academy of Sciences (since 2015), both in Moscow. She earned her PhD at the Institute of Material Culture of the Russian Academy of Sciences (St. Petersburg 1996). Her several publications include the books *Sarmatische Phaleren*, Rahden, Westf. 2001; *Polychrome Animal Style*, Simferopol 2003; *Toreutik und Schmuck im Nördlichen Schwarzmeergebiet: 2. Jh. v.Chr. – 2. Jh. n.Chr.*, vol. 3, Simferopol 2007 (co-edited with Mikhail Treister); the proceedings of the Humboldt-Kolleg *Archäologie und Sprachwissenschaft. Austausch von Wissen in der Geschichte der Menschheit*, Kiev 2014 (co-edited with Heinrich Härke & Tetiana Shevchenko); *Crimean Scythia in a System of Cultural Relations between East and West*, Simferopol 2017 (co-edited with Askold Ivantchik).

Marta **Oller Guzmán** is Associate Professor in Greek Philology at the Autonomous University of Barcelona (Spain), where she obtained her PhD in 2004. Her research focuses on heroic cults, colonisation and condition of foreigners in the ancient Greek world. Her publications include *Tierra, territorio y población en la Grecia antigua: aspectos institucionales y míticos*, Mering 2017 (co-edited with Jordi Pàmias & Carlos Varias) and *Contacto de poblaciones y extranjería en el Mundo Griego antiguo. Estudio de fuentes*, Bellaterra 2013 (co-edited with Rosa-Araceli Santiago Álvarez).

Germain **Payen** is a postdoctoral fellow at the University of Cologne, Germany. He started his fellowship in 2019, after gaining his PhD (2016) at the University of Sorbonne, France and at the University Laval, Quebec, and holding a first post-doctoral position at the University of Waterloo, Ontario (2017–2018). His research interests range from Hellenistic kingdoms and Anatolian geopolitics and

Roman expansion in the eastern Mediterranean to the Black Sea and the Bosporan kingdom in the Roman Imperial period. He has published various articles on the dynastic politics in Asia Minor and the Bosporan kingdom. His most important publication to date is the book on *Les suites géopolitiques du traité d'Apamée en Anatolie* (Quebec 2020).

Alexandr V. **Podossinov** is Professor and Head of the Department of Classical Languages at the Moscow Lomonosov State University (since 2008) and a Leading Research Fellow of the Institute of World History of the Russian Academy of Sciences (since 1972). He holds a PhD in Classical Philology (Moscow State University, 1980) and a Habilitation (Institute of World History, 1997). His international visiting fellowships took him to Munich, Trier, Erfurt, Oxford and Melbourne. His research concentrates on the Northern Black Sea area, with an interest in the sources of its ancient history, geography and cartography. His latest books include *The Periphery of the Classical World in Ancient Geography and Cartography*, Leuven 2014. *Scythia in the Historical and Geographical Tradition of Antiquity and the Middle Ages*, Moscow 2016 (in Russian, co-edited with Tatjana Jackson & Irina Konovalova); *Rhipaean Mountains in Ancient and Medieval Geography*, Moscow 2019 (in Russian, co-edited with Tatjana Jackson & Irina Konovalova).

Joanna **Porucznik** gained her PhD within an international Polish-British programme at the Universities of Liverpool & Wrocław (2015). As a postdoctoral fellow at Wrocław, she researched 'The Greek City in the Hellenistic and Roman Age and the Territorial Powers' (2015–2019). In 2019, she was appointed Assistant Professor in the Department of History at the University of Opole, Poland. She has published various articles, including 'The Cult of Chersonasos in Tauric Chersonesos. Numismatic and Epigraphic Evidence Revisited', *ACSS*, 23, 2017, 63–89; 'Heuresibios Son of Syriskos and the Question of Tyranny in Olbia Pontike (5th–4th c. BC)', *BSA* 2018, 1–16. Her thesis *Cultural Identity within the Northern Black Sea Region in Antiquity: (De)constructing Past Identities* is currently in preparation for publication as a volume of *Colloquia Antiqua*.

Dan **Ruscu** is Assistant Professor of Church History at the Babeş-Bolyai University in Cluj, Romania. He holds a PhD in Ancient History from the same university. His research focuses on Christianity in Late Antiquity. His recent publications include 'The Ecclesiastical Network of the Regions on the Western and Northern Shores of the Black Sea in Late Antiquity', in G.R. Tsetskhladze, A. Avram & J. Hargrave (eds.), *The Danubian Lands between the Black, Aegean and Adriatic Seas (7th Century BC – 10th Century AD)*, Oxford 2015, 189–195; 'Scythians and Places of Exile: The Black Sea in Early Christian Literature', in V. Cojocaru & A. Rubel (eds.), *Mobility in Research on the Black Sea Region*, Cluj-Napoca 2016, 367–383; 'Cultural Identities and Personal Relationships in Structuring the Ecclesiastical Networks around the Black Sea in the 4th–6th Centuries', in *Costellazioni geo-ecclesiali da Costantino a Giustiniano: Dalle chiese 'principali' alle chiese patriarcali*, Rome 2017, 481–490.

Gaius **Stern** taught History at San Jose State University and Classics at UC Berkeley. He took a PhD in Ancient History from UC Berkeley after writing a thesis on the politics of the Augustan Age in 13 BC, using the *Ara Pacis Augustae* and coinage as contemporary evidence to indicate the 'official version' of the power dynamic in Rome. Much of his research looks at the *Ara Pacis* as it should be understood, as a Roman monument that does not proclaim monarchy to an audience of Romans of 13 BC, unaware of who would later rule Rome. Other topics of research include Prisoners of War, Classical Greece, and archaic Rome to the Punic Wars. He has organized a series of Punic War panels at CAMWS and other conferences in the past few years.

INDEX

1. NAMES

Roman individuals are listed under their most common name, mostly their cognomen.

2. LITERARY SOURCES

3. DOCUMENTARY SOURCES

Map 1   377

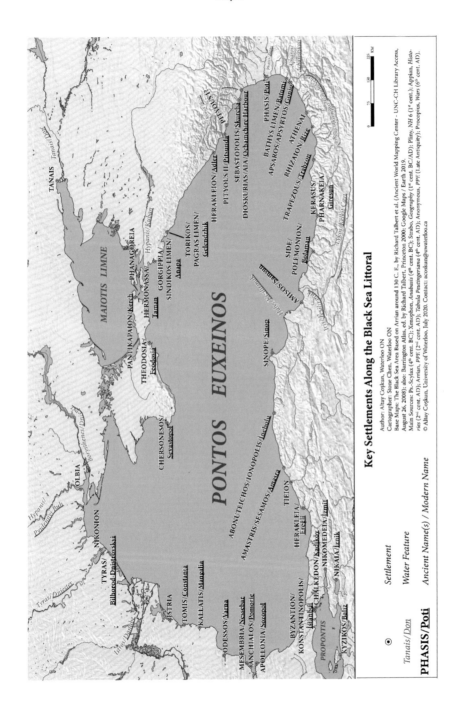

Key Settlements Along the Black Sea Littoral

Author: Altay Coşkun, Waterloo ON
Cartographer: Stone Chen, Waterloo ON
Base Maps: The Black Sea Area Based on Arrian around 130 C. E., by Richard Talbert et al. (Ancient World Mapping Center – UNC-CH Library Access,
August 26, 2008); also: Barrington Atlas, ed. by Richard Talbert, Princeton 2000; Google Maps / Earth 2019.
Main Sources: Ps.-Scylax (4th cent. BC); Xenophon, Anabasis (4th cent. BC); Strabo, Geography (1st cent. BC/AD); Pliny, NH 6 (1st cent.); Appian, Histo-
ries (2nd cent. AD); Arrian, PPE (2nd cent. AD); Tabula Peutingeriana (4th cent. AD); Anonymous, PPE (Late Antiquity); Procopius, Wars (6th cent. AD).
© Altay Coşkun, University of Waterloo, July 2020. Contact: acoskun@uwaterloo.ca

⊙   Settlement

Tantais / Don   Water Feature

PHASIS/Poti   Ancient Name(s) / Modern Name

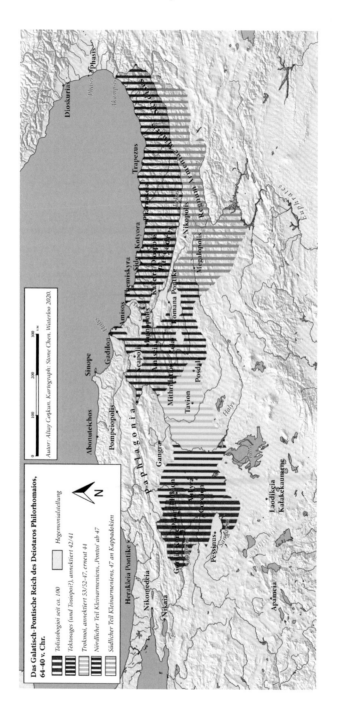

Map 3   379

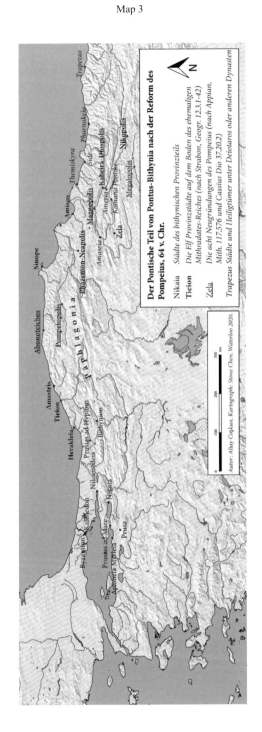

Der Pontische Teil von Pontus-Bithynia nach der Reform des Pompeius, 64 v. Chr.

Nikaia   *Städte des bithynischen Provinzteils*

Tieion   *Die Elf Provinzstädte auf dem Boden des ehemaligen Mithradates-Reiches (nach Strabon, Geogr. 12.3.1-42)*

Zela   *Die acht Neugründungen des Pompeius (nach Appian, Mith. 117.576 und Cassius Dio 37.20.2)*

Trapezus Städte und Heiligtümer unter Deiotaros oder anderen Dynasten

Autor: Altay Coşkun. Kartograph: Stone Chen, Waterloo 2020.

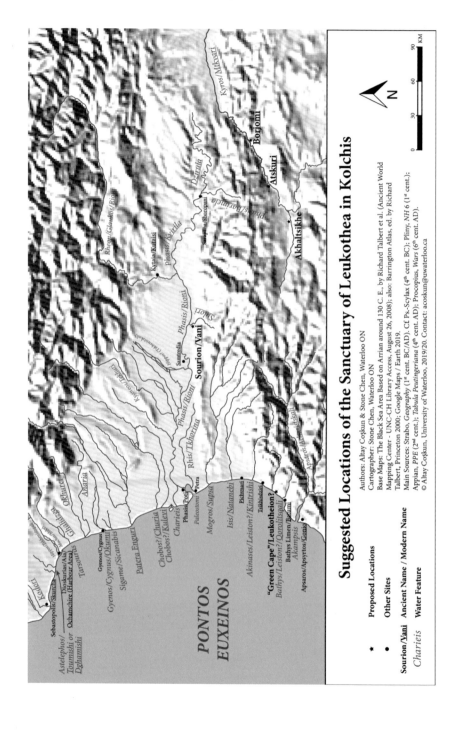

Suggested Locations of the Sanctuary of Leukothea in Kolchis

Authors: Altay Coşkun & Stone Chen, Waterloo ON
Cartographer: Stone Chen, Waterloo ON
Base Maps: The Black Sea Area Based on Arrian around 130 C. E., by Richard Talbert et al. (Ancient World Mapping Center - UNC-CH Library Access, August 26, 2008); also: Barrington Atlas, ed. by Richard Talbert, Princeton 2000; Google Maps / Earth 2019.
Main Sources: Strabo, *Geography* (1st cent. BC/AD). Cf Ps.-Scylax (4th cent. BC); Pliny, *NH* 6 (1st cent.); Appian, *PPE* (2nd cent.); *Tabula Peutingeriana* (4th cent. AD); Procopius, *Wars* (6th cent. AD).
© Altay Coşkun, University of Waterloo, 2019/20. Contact: acoskun@uwaterloo.ca

Proposed Locations
Other Sites
Sourion/Yani Ancient Name / Modern Name
Charieis **Water Feature**

Map 5 381